Kirchhoff Equations

Kirchhoff Equations: A Variational Approach is primarily focused on recent results concerning existence, multiplicity, and the asymptotic behavior of solutions to some stationary Kirchhoff problems, involving fractional integro-differential elliptic operators, and presenting difficulties relating to an intrinsic lack of compactness, which are elucidated upon within the text. These operators appear in a quite natural way in many different applications, such as continuum mechanics, phase transition phenomena, population dynamics, and game theory, as they are the typical outcome of stochastically stabilization of Lévy processes.

This book will be of interest to postgraduates in applied mathematics, with a particular emphasis for those working in differential and partial differential equations. It will also find an audience among researchers interested in the qualitative, quantitative, and asymptotic analysis of various types of solutions to the Kirchhoff equation.

Features

- Each chapter concludes with a detailed glossary and set of open problems.
- Rigorous proofs and illustrative examples.
- Broad-spectrum appeal to both applied mathematicians and those in other qualitative disciplines such as engineering.

Vicenţiu D. Rădulescu is a Professor at the AGH University of Krakow, Brno University of Technology and Professorial Fellow at the Simion Stoilow Institute of Mathematics of the Romanian Academy. He was also a Distinguished Visiting Scientist at the University of Ljubljana (2008), Distinguished Adjunct Professor at the King Abdulaziz University in Jeddah (2014–2021), and Highly Cited Researcher (2014, 2019–2021). He is a member of the Accademia Peloritana dei Pericolanti (since 2014), Accademia delle Scienze dell'Umbria (since 2017), Senior Research Fellow of the City University of Hong Kong (2015), and Senior Research Fellow of the Central South University (2024-2026). He has editorial positions at the De Gruyter Series in Nonlinear Analysis and Applications, Journal of Geometric Analysis, Mathematical Methods in the Applied Sciences, Asymptotic Analysis, Complex Variables and Elliptic Equations, and Rendiconti del Circolo Matematico di Palermo. Vicenţiu D. Rădulescu is also Editor-in-Chief of Bulletin of Mathematical Sciences, Opuscula Mathematica, and Boundary Value Problems.

Kirchhoff Equations

A Variational Approach

Vicenţiu D. Rădulescu

CRC Press
Taylor & Francis Group
Boca Raton London New York

CRC Press is an imprint of the
Taylor & Francis Group, an **informa** business
A CHAPMAN & HALL BOOK

Designed cover image: Shutterstock

First edition published 2026
by CRC Press
2385 NW Executive Center Drive, Suite 320, Boca Raton, FL 33431

and by CRC Press
4 Park Square, Milton Park, Abingdon, Oxon, OX14 4RN

CRC Press is an imprint of Taylor & Francis Group, LLC

ISBN: 978-1-041-35186-3 (hbk)
ISBN: 978-1-041-35319-5 (pbk)
ISBN: 978-1-003-79715-9 (ebk)

DOI: 10.1201/9781003797159

Typeset in CMR10 font
by KnowledgeWorks Global Ltd.

Publisher's note: This book has been prepared from camera-ready copy provided by the authors.

To my parents Dumitru and Ana,
with emotion and gratitude

To Teodora and Bianca,
with love

Contents

II Normalized Solutions 59

III Concentration Phenomena 137

Foreword

My interest in the study of Kirchhoff-type problems started in 2018, after my joint papers [115, 116, 117] with Mingqi Xiang and Binlin Zhang. A particular role in my interest to the analysis of Kirchhoff systems is played by my joint work [129] with Patrizia Pucci as guest editors of the special issue "Progress in nonlinear Kirchhoff problems" [*Nonlinear Anal.*, 186 (2019), 258 pp.]. I thank all of them for developing my interest to study this kind of nonlinear problems.

I am also grateful for our fruitful collaboration in this research field to Xianhua Tang, Sitong Chen, Dongdong Qin, Jian Zhang, Wen Zhang, Vincenzo Ambrosio, Chao Ji, Sihua Liang, Marius Ghergu, Giovany Figueiredo, Quanqing Li, Jianjun Zhang, Lixi Wen, Shuai Yuan, Xuexiu Zhong, Li Cai, Lintao Liu, Giovanni Molica Bisci, and Zhisu Liu. I would like to thank Peng Jin for the careful check of the galley proofs.

In the last years, I was fortunate to work with Jerzy Stochel, who always gave me his expert advice. I am grateful to Jerzy for his intellectual generosity and for the collaborative spirit he always brings.

Some of the results included in this book were presented at some international conferences or in scientific seminars at various institutions, including the AGH University of Kraków, Brno University of Technology, University of Perugia, Central South University, Tsinghua University, Accademia Peloritana dei Pericolanti, Accademia delle Scienze dell'Umbria, Université de Lille, Beijing Normal University, East China Normal University, East China University of Science and Technology, Minzu University of China, Guangzhou University, South China Normal University, etc.

This research is supported by the AGH University of Kraków under grant no. 16.16.420.054, funded by the Polish Ministry of Science and Higher Education.

Vicenţiu D. Rădulescu, February 2026
AGH University of Kraków
Brno University of Technology
"Simion Stoilow" Institute of Mathematics of the Romanian Academy
`http://math.ucv.ro/~radulescu`

Preface

This book is devoted to the mathematical analysis of the stationary Kirchhoff equation. This class of nonlinear problems was introduced by the German physicist and mathematician Gustav Kirchhoff [*Mechanik*, Teubner, Leipzig, 1883] in relationship with the study of the motion of a rigid body in an ideal fluid. Subsequently, the Kirchhoff equation appeared as a basic tool for the understanding and the rigorous analysis of basic processes arising in other applied sciences, including electrical engineering or chemical engineering.

Mathematics has always benefited from its involvement with developing sciences. Each successive interaction revitalizes and enhances the field. That is why the subject of this book assumes the correspondingly increased dialogue between mathematics and interdisciplinary disciplines that study models described by the modern mathematical machinery. The framework we develop and study in this volume is inspired by concrete and relevant applications from various disciplines.

This book is a research monograph, and it is essentially based on original results. This volume is focused on recent results concerning existence, multiplicity, and the asymptotic behavior of solutions to some stationary Kirchhoff problems, involving fractional integro-differential elliptic operators, and presenting also difficulties due to intrinsic lacks of compactness, which arise from different reasons. These operators appear in a quite natural way in many different applications, such as continuum mechanics, phase transition phenomena, population dynamics, and game theory, as they are the typical outcome of stochastic stabilization of Lévy processes.

The subject of this monograph is about nonlocal stationary equations of Kirchhoff-type. A good question is:

Why nonlocal problems are important for mathematics?

Indeed, during the last centuries, science was governed by local equations, where interactions happen only between immediate neighbors. However, reality is more complicated: memories remain in materials, and relationships span time and location. This intricate connection is captured by nonlocal equations, which radically alter our perception of reality. The content of this book lies at the interplay between pure mathematics and applied fields such as electrical engineering, chemistry, or fluids (either Newtonian or non-Newtonian). The main purpose is to build a bridge between mathematics and other disciplines, due to the fact that problems and models arising in many applications are using increasingly sophisticated mathematical techniques. This monograph intends to emphasize that Kirchhoff's equations illustrate that the most powerful mathematical structures often emerge from the simplest conservation principles, proving that even the humble electrical circuit contains universes of abstract beauty. This is achieved by applying refined mathematical methods lying at the interplay between the calculus of variations, pure and applied nonlinear analysis, topology, and mathematical physics.

The content of this volume is divided into three Parts, which contain nine Chapters. In such a way, the main themes described in this volume are concerned with the following topics: least-energy solutions, normalized solutions, and concentration properties for several classes of Kirchhoff-type equations.

Part I is devoted to the mathematical analysis of least-energy solutions for Kirchhoff problems with critical growth. The study of this type of solutions is at the heart of the modern calculus of variations and nonlinear analysis. In simple terms, least-energy solutions (also called ground state solutions) are the nontrivial solutions that are absolute minimizers of the associated energy functional among all possible nontrivial solutions. The physical meaning of least-energy solutions is that they represent the ground state, which is the most stable, lowest-energy configuration of the system. In short, least-energy solutions are the most fundamental and stable critical points of the associated Kirchhoff energy. Finding and characterizing these solutions is the first major step in understanding the variational structure of the Kirchhoff system.

Part I contains two chapters, and the analysis carried out here covers the study of least-energy solutions for the following classes of Kirchhoff equations:

(a) *Critical Kirchhoff problems with logarithmic reaction.* Here, two fundamental mathematical concepts compete: nonlocal differential operator (Kirchhoff), and critical logarithmic growth nonlinearity. Problems with logarithmic reaction have their origins in quantum mechanics (logarithmic Schrödinger equations model Bose-Einstein condensates with unusual particle interactions), information theory, and in the study of patterns with nonlinearities that saturate rather than blow-up.

(b) *Planar Kirchhoff equations with critical exponential growth and trapping potential.* This combination is challenging because it creates a mathematically optimal competition:

   (i) *Kirchhoff*: nonlocality in the source term;

   (ii) *critical exponential growth*: maximal growth in two dimensions;

   (iii) *trapping potential*: confinement and compactness.

   In summary, planar Kirchhoff systems with critical exponential growth (Trudinger-Moser type) and trapping potential offer deep insights into how nonlocality, criticality, and confinement interact in low-dimensional nonlinear systems. Physical patterns of this type include Bose-Einstein condensates in 2D traps, Kirchhoff-type dispersion from nonlocal nonlinearities (thermal, molecular), and exponential response in photoreactive materials.

Part II contains three chapters, and it is devoted to the qualitative analysis of normalized solutions (also called solutions with prescribed mass). There are several reasons to study solutions of this type. Firstly, for their physical necessity, since prescribing the mass is often required to define a specific physical system, for instance, because the gravitational interaction strength depends directly on the mass. Secondly, normalized stationary states define naturally the eigenvalue problem, which is a nonlinear eigenvalue problem constrained by mass normalization. Thirdly, problems dealing with solutions having a prescribed mass yield a key prediction. Indeed, solving this kind of problem for different masses reveals the mass-radius relationship, a critical signature used to test the model against astrophysical observations (e.g., for fuzzy dark matter).

We are mainly concerned with the study of normalized solutions for the following classes of Kirchhoff equations:

(a) *Nonautonomous Kirchhoff problems.* The analysis developed in this book covers both the subcritical and supercritical abstract settings. Nonautonomous Kirchhoff equations are special because they represent a hierarchical structure of nonlocalities and nonlinearities:

(i) *Mesoscopic nonlocality*: Kirchhoff nonlinearity in the source term;
(ii) *Local-nonlocal interplay in the reaction*: the nonlinear reaction can have such double behavior;
(iii) *Macroscopic nonlocality*: spatial dependence through the potential describing the reaction.

In summary, the nonautonomous case moves Kirchhoff equations from model problems to realistic physical patterns where spatial inhomogeneity is the rule, not the exception.

(b) *Autonomous Kirchhoff equations with Sobolev critical exponent.* A feature of this study developed in the three-dimensional case is that it unifies the sub- and super-critical regimes. This case is fundamental because the autonomous Kirchhoff equation with Sobolev critical exponent sits at the intersection of three deep mathematical theories:

(i) *Calculus of Variations*: critical growth, lack of compactness;
(ii) *Geometric Analysis*: best constants, extremals, bubbling;
(iii) *Nonlocal PDEs*: Kirchhoff-type nonlocality.

The basic abstract tools in this study are the following: truncation method and fiber mapping (to analyze the energy levels), concentration-compactness principle (for handling the lack of compactness), and genus theory (to establish the multiplicity of solutions).

(c) *Kirchhoff equations with double-behavior reaction.* A feature of this study is that it addresses the existence of normalized solutions for Kirchhoff equations in the following fundamental cases:

(i) if the reaction is L^2-subcritical (mass-like at small amplitudes) at the origin and L^2-critical at infinity;
(ii) if the reaction is L^2-critical at the origin and L^2-supercritical (energy-dominant at large amplitudes) at infinity.

The case of a double-behavior reaction is important because this specific interplay between the growth of the reaction near the origin and at infinity relative to the L^2-critical exponent is a central technical feature, which creates a competition between two different variational geometries. A major effort consists in handling the lack of compactness due to the Sobolev supercritical growth at infinity (which is still subcritical in the Sobolev sense) through the interplay with the mass constraint and the Kirchhoff term.

Part III contains four chapters, and it is concerned with concentration properties of solutions to several classes of Kirchhoff systems. This analysis is important because it goes to one of the main analytical challenges for the Kirchhoff equation. Indeed, concentration properties of solutions are crucial for the Kirchhoff equation because they determine whether the energy functional associated with the equation satisfies the Palais-Smale compactness condition. This condition is the cornerstone of applying powerful variational methods to establish the existence of weak solutions.

The analysis carried out in this book covers the study of concentration solutions for the following classes of Kirchhoff equations:

(a) *Fractional p-Kirchhoff equations.* The features of this analysis developed in the present volume are distinct from other related studies on the Kirchhoff equation due to its focus on potential well concentration rather than critical exponent bubbling. Some reasons for which the concentration analysis is important in this case are listed below.

(i) *Problem complexity*: the study of this class of Kirchhoff equations tackles a triply nonlocal problem (fractional + Kirchhoff + potential well) on the whole space.
(ii) *Methodological innovation*: the analysis provides a blueprint for asymptotic analysis via the limit problem strategy and uniform estimates.
(iii) *Theoretical impact*: the study highlights new theoretical ideas based on potential-induced concentration, distinct from critical exponent bubbling.

(b) *Magnetic Kirchhoff equations with critical growth.* Moving from Kirchhoff to magnetic Kirchhoff equations shifts the focus from classical elasticity to quantum mechanics, introduces the deep physical principle of gauge invariance, and creates a much more challenging and interesting set of mathematical problems centered on new types of functional spaces and concentration phenomena. Introducing the magnetic field in the study of Kirchhoff equations creates a host of new analytical difficulties, such as:

(i) *Loss of compactness*: the standard embedding property is no longer compact in the magnetic case, and new techniques like the magnetic concentration-compactness principle are needed to overcome this failure.
(ii) *Concentration properties*: the magnetic field can influence the precise location and shape of the concentration of solutions.
(iii) *Interaction of nonlocalities*: the problem features a fascinating interplay between the Kirchhoff nonlocality and the magnetic nonlocality.

(c) *Fractional Kirchhoff equations with discontinuous reaction.* This study is important since it shifts the problem from the smooth world of standard critical point theory to the more complex terrain of nonsmooth analysis, specifically for applications where the response is not continuous or the forcing term is not a function but a multivalued map. The key feature of this analysis is that it combines three major analytical complexities (nonlocality, nondifferentiability, and potential-driven localization) into a single challenging problem. This triple interaction implies the following challenge layers:

(i) The operator is nonlocal in two ways, making compactness hard to obtain.
(ii) The discontinuous reaction implies that the associated energy is not differentiable (only locally Lipschitz), so standard critical point theory fails.
(iii) In order to establish the concentration of solutions and to prove localization, one must show decay outside a prescribed set and rule out energy leaking to infinity.

(d) *Mass critical fractional Kirchhoff equations.* These problems represent a delicate and significant frontier in nonlinear analysis, where three fundamental constraints interact: fractional diffusion, nonlocal Kirchhoff effects, and mass conservation. This combination makes these equations important for modeling quantum phenomena and mathematically challenging due to the complete lack of compactness. This analysis represents a state-of-the-art contribution for several reasons:

(i) *It solves a natural open problem*: the combination of "mass constraint + Sobolev criticality + Kirchhoff nonlocality" had been a major open challenge. This study likely provides a near-complete picture for the ground state case.

(ii) *Foundation of dynamics*: the detailed properties of constraint minimizers are essential for studying the orbital stability of standing waves and the blow-up dynamics to the associated time-dependent fractional Kirchhoff equation.

(iii) *Main results*: the analysis establishes sharp thresholds, characterizes the precise concentrating behavior of minimizers at criticality, and breaks new ground by proving a uniqueness result in this highly nonlocal and nonlinear context.

This volume is addressed to wide categories of readers in mathematics, and it is designed for two basic audiences:

1. The book is intended as a post-graduate course for students in applied mathematics. They will be interested in some of the recent advances in this field, with a particular emphasis for those working in differential and partial differential equations.
2. The monograph is addressed to senior and junior researchers interested in the qualitative, quantitative and asymptotic analysis of various types of solutions to the Kirchhoff equation, which is a basic tool in the understanding of basic phenomena arising in the applied sciences, such as mathematical physics, engineering, physical science, computer science, numerical methods, and even mathematical biology and economics.

I would like to acknowledge the AGH University of Kraków for the exceptional resources and infrastructure that supported this project. Access to the AGH University Library was essential to my research. Furthermore, the stimulating discussions with colleagues within the faculty continuously enriched my thinking. I remain also grateful to my editor Callum Fraser for his professional support during the preparation of this volume.

Vicenţiu D. Rădulescu
Kraków, February 2026

Author

Vicenţiu D. Rădulescu received his Ph.D. at the Université Paris Sorbonne in 1995 under the supervision of Haim Brezis. For his PhD thesis "Analyse de quelques problèmes liés à l'équation de Ginzburg-Landau", he was awarded the highest distinction "très honorable avec félicitations". In 2003, he defended his Habilitation "Analyse de quelques problèmes aux limites elliptiques non linéaires" at the Université Paris Sorbonne. In 2023, he obtained the Habilitation degree (with distinction) at the AGH University of Kraków with the Work "Local and nonlocal problems in nonlinear analysis".

Since 2018, Vicenţiu D. Rădulescu has been employed as Professor at the AGH University of Kraków. Vicenţiu D. Rădulescu is also Professor at the University of Craiova and at the Brno University of Technology, as well as Professorial Fellow at the Institute of Mathematics "Simion Stoilow" of the Romanian Academy. He was awarded the Prize of the Romanian Academy (1999) and the Prize of the Prime Minister of Poland (2024).

The author published more than 500 research papers in nonlinear analysis and several books, including *Variational and Nonvariational Methods in Nonlinear Analysis and Boundary Value Problems* (Kluwer, 2003), *Singular Elliptic Problems: Bifurcation and Asymptotic Analysis* (Oxford University Press, 2008), *Problems in Real Analysis: Advanced Calculus on the Real Axis* (Springer, 2009), *Variational Principles in Mathematical Physics, Geometry and Economics: Qualitative Analysis of Nonlinear Equations and Unilateral Problems* (Cambridge University Press, 2010), *Nonlinear PDEs: Mathematical Models in Biology, Chemistry and Population Genetics* (Springer, 2012), *Variational Methods for Nonlocal Fractional Problems* (Cambridge University Press, 2015) *Partial Differential Equations with Variable Exponents: Variational Methods and Qualitative Analysis* (CRC Press, 2015), *Equilibrium Problems and Applications* (Academic Press, 2018), and *Nonlinear Analysis–Theory and Methods* (Springer Monographs in Mathematics, Springer, 2019).

Vicenţiu D. Rădulescu was a Distinguished Visiting Scientist at the University of Ljubljana (2008), Distinguished Adjunct Professor at the King Abdulaziz University in Jeddah (2014-2021), and *Highly Cited Researcher* (2014, 2019–2021). He is a member of the *Accademia Peloritana dei Pericolanti* (since 2014), *Accademia delle Scienze dell'Umbria* (since 2017), Senior Research Fellow of the City University of Hong Kong (2015), and Senior Research Fellow of the Central South University (2024 and 2025). He has editorial positions at the *De Gruyter Series in Nonlinear Analysis and Applications*, *Journal of Geometric Analysis*, *Mathematical Methods in the Applied Sciences*, *Asymptotic Analysis*, *Complex Variables and Elliptic Equations*, and *Rendiconti del Circolo Matematico di Palermo.* Vicenţiu D. Rădulescu is also Editor-in-Chief of *Bulletin of Mathematical Sciences*, *Opuscula Mathematica*, and *Boundary Value Problems.*

Kraków, February 2026

List of Tables

Part I

Least-energy Solutions

1

Critical Kirchhoff problems with logarithmic reaction

This chapter deals with the existence of least-energy sign-changing solutions for a class of fractional Kirchhoff problems with logarithmic and critical nonlinearities. *Why least-energy solutions?* Because the study of this type of solutions is at the heart of the modern calculus of variations and nonlinear analysis. In simple terms, least-energy solutions (also called ground state solutions) are the nontrivial solutions that are absolute minimizers of the associated energy functional among all possible nontrivial solutions. The physical meaning of least-energy solutions is that they represent the ground state, which is the most stable, lowest-energy configuration of the system. In short, least-energy solutions are the most fundamental and stable critical points of the associated Kirchhoff energy. Finding and characterizing these solutions is the first major step in understanding the variational structure of Kirchhoff-type systems.

In this chapter, we consider the following Kirchhoff equation

$$\begin{cases} \left(a + b[u]_{s,p}^p\right)(-\Delta)_p^s u = \lambda |u|^{q-2} u \ln |u|^2 + |u|^{p_s^*-2} u & \text{in } \Omega, \\ u = 0 & \text{in } \mathbb{R}^N \setminus \Omega, \end{cases}$$

where $s \in (0,1)$, $p > 1$, $\Omega \subset \mathbb{R}^N$ $(N \geq 3)$ is a bounded domain with Lipschitz boundary, $N > sp$, λ is a positive parameter, $p_s^* = Np/(N - ps)$ is the critical fractional Sobolev exponent, and

$$[u]_{s,p}^p = \int_\Omega \int_\Omega \frac{|u(x) - u(y)|^p}{|x - y|^{N+ps}} dxdy.$$

The presence of the term $a + b[u]_{s,p}^p$ makes this equation *nonlocal*, in the sense that the Laplacian's coefficient depends on the global energy of the unknown u. At the same time, the energy functional associated to this problem does not satisfy the Palais-Smale condition in the standard way, due to the combined effect of critical growth and the logarithm.

In this chapter, two fundamental mathematical concepts compete:

1. nonlocal differential operator (Kirchhoff);
2. critical logarithmic growth nonlinearity.

Problems with logarithmic reaction have their origins in quantum mechanics (logarithmic Schrödinger equations model Bose-Einstein condensates with unusual particle interactions), information theory, and in the study of patterns with nonlinearities that saturate rather than blow-up.

The main result in this chapter establishes the existence of at least-energy *sign-changing* (also called *nodal*) solution u_b. This is done by combining constraint variational methods, topological degree theory, and quantitative deformation arguments. Moreover, *energy*

DOI: 10.1201/9781003797159-1

doubling is established, that is, the energy of the sign-changing solution u_b is strictly greater than twice the ground state energy for all $\lambda > 0$. Finally, we consider the nonlocal coefficient b as a parameter and study the convergence property of the least-energy sign-changing solution as $b \to 0$.

The analysis carried out in this chapter extends the mountain pass/Nehari manifold methodology to a highly non-standard setting by handling the triple difficulty of nonlocality, critical exponent, and logarithmic nonlinearity. In such a way, we provide a complete variational characterization (least-energy property) of the nodal solution.

1.1 Statement of the problem

In this chapter, we are interested in the existence, energy estimates, and the convergence property of the least-energy sign-changing solution for the following fractional Kirchhoff problems with logarithmic and critical nonlinearity:

$$\begin{cases} \left(a + b[u]_{s,p}^p\right)(-\Delta)_p^s u = \lambda |u|^{q-2} u \ln |u|^2 + |u|^{p_s^*-2} u & \text{in } \Omega, \\ u = 0 & \text{in } \mathbb{R}^N \setminus \Omega, \end{cases} \tag{1.1}$$

where $s \in (0,1)$, $p > 1$, $\Omega \subset \mathbb{R}^N$ $(N \geq 3)$ is a bounded domain with Lipschitz boundary, $N > sp$, λ is a positive parameter, $p_s^* = Np/(N - ps)$ is the critical fractional Sobolev exponent, and

$$[u]_{s,p}^p = \int_\Omega \int_\Omega \frac{|u(x) - u(y)|^p}{|x - y|^{N+ps}} dxdy.$$

We denote by $(-\Delta)_p^s$ the fractional p Laplace operator which, up to a normalization constant, is defined as

$$(-\Delta)_p^s \varphi(x) = 2 \lim_{\varepsilon \to 0^+} \int_{\mathbb{R}^N \setminus B_\varepsilon(x)} \frac{|\varphi(x) - \varphi(y)|^{p-2}(\varphi(x) - \varphi(y))}{|x - y|^{N+ps}} dy, \quad x \in \mathbb{R}^N,$$

for all $\varphi \in C_0^\infty(\mathbb{R}^N)$. Henceforward, $B_\varepsilon(x)$ denotes the open ball of $\mathbb{R}^N$ centered at $x \in \mathbb{R}^N$ and radius $\varepsilon > 0$.

Let us recall some preliminary results on the fractional Sobolev space $W_0^{s,p}(\Omega)$ with respect to the norm $\|u\| = [u]_{s,p}$. We refer to Appendix A for the definition and some basic properties of fractional Sobolev spaces.

By Sobolev embeddings, $W_0^{s,p}(\Omega)$ is compactly embedded into the Lebesgue space $L^r(\Omega)$ endowed the norm $|u|_r = \left(\int_\Omega |u|^r\, dx\right)^{\frac{1}{r}}$ for all $p < r < p_s^*$. Denote by S_r the best Sobolev constant for this embedding, that is,

$$S_r |u|_r \leq \|u\|, \quad \forall u \in W_0^{s,p}(\Omega). \tag{1.2}$$

In particular, if S is the best constant for the embedding $W_0^{s,p}(\Omega) \hookrightarrow L^{p_s^*}(\Omega)$, then it is defined by

$$S = \inf_{u \in W_0^{s,p}(\Omega) \setminus \{0\}} \frac{\int_\Omega \int_\Omega \frac{|u(x)-u(y)|^p}{|x-y|^{N+ps}} dxdy}{\left(\int_\Omega |u|^{p_s^*}\, dx\right)^{\frac{p}{p_s^*}}}. \tag{1.3}$$

Definition 1.1 *We say that $u \in W_0^{s,p}(\Omega)$ is a (weak) solution of problem (1.1) if*

$$\left(a+b[u]_{s,p}^{p}\right) L(u,v) = \lambda \int_\Omega |u|^{q-2}uv \ln|u|^2 dx + \int_\Omega |u|^{p_s^*-2}uv\, dx, \tag{1.4}$$

where

$$L(u,v) := \int_\Omega\int_\Omega \frac{|u(x)-u(y)|^{p-2}(u(x)-u(y))}{|x-y|^{N+sp}}(v(x)-v(y))dxdy \tag{1.5}$$

for any $v \in W_0^{s,p}(\Omega)$.

The energy functional $I_b^\lambda : W_0^{s,p}(\Omega) \to \mathbb{R}$ corresponding to problem (1.1) is defined by

$$I_b^\lambda(u) = \frac{a}{p}[u]_{s,p}^p + \frac{b}{2p}[u]_{s,p}^{2p} + \frac{2\lambda}{q^2}\int_\Omega |u|^q dx - \frac{\lambda}{q}\int_\Omega |u|^q \ln|u|^2 dx - \frac{1}{p_s^*}\int_\Omega |u|^{p_s^*} dx. \tag{1.6}$$

Then I_b^λ is a functional of class $C^1(W_0^{s,p}(\Omega),\mathbb{R})$ and the critical points of I_b^λ are solutions of problem (1.1). Furthermore, if we write

$$u^+(x) = \max\{u(x),0\} \quad \text{and} \quad u^-(x) = \min\{u(x),0\}$$

for $u \in W_0^{s,p}(\Omega)$, then every solution $u \in W_0^{s,p}(\Omega)$ of problem (1.1) with the property that $u^\pm \neq 0$ is a sign-changing solution of problem (1.1).

We remark that if $u^\pm \not\equiv 0$, then

$$\begin{aligned} I_b^\lambda(u) &= I_b^\lambda(u^+) + I_b^\lambda(u^-) + \tfrac{b}{p}\|u^+\|^p\|u^-\|^p, \\ \langle (I_b^\lambda)'(u), u^+\rangle &= \langle (I_b^\lambda)'(u^+), u^+\rangle + b\|u^+\|^p\|u^-\|^p, \\ \langle (I_b^\lambda)'(u), u^-\rangle &= \langle (I_b^\lambda)'(u^-), u^-\rangle + b\|u^+\|^p\|u^-\|^p. \end{aligned} \tag{1.7}$$

Our goal in this chapter is to study the least-energy sign-changing solutions of problem (1.1). Consider the following standard semilinear elliptic equation

$$-\Delta u + V(x)u = f(x,u) \text{ in } \Omega, \tag{1.8}$$

where Ω is an open subset of $\mathbb{R}^N$. A standard method to establish the existence of sign-changing solutions relies on the following decompositions:

$$J(u) = J(u^+) + J(u^-), \tag{1.9}$$

$$\langle J'(u), u^+\rangle = \langle J'(u^+), u^+\rangle, \langle J'(u), u^-\rangle = \langle J'(u^-), u^-\rangle, \tag{1.10}$$

where J is the energy functional of (1.8) given by

$$J(u) = \frac{1}{2}\int_\Omega (|\nabla u|^2 + V(x)u^2)dx - \int_\Omega F(x,u)dx.$$

In our case, if $b > 0$, the associated energy functional I_b^λ does not possess decompositions as (1.9) and (1.10). In fact, a straightforward computation yields that

$$I_b^\lambda(u) > I_b^\lambda(u^+) + I_b^\lambda(u^-), \tag{1.11}$$

$$\langle (I_b^\lambda)'(u), u^+\rangle > \langle (I_b^\lambda)'(u^+), u^+\rangle \quad \text{and} \quad \langle (I_b^\lambda)'(u), u^-\rangle > \langle (I_b^\lambda)'(u^-), u^-\rangle$$

for $u^\pm \neq 0$. Therefore, the method to obtain sign-changing solutions for the *local* problem (1.8) does not seem applicable to the *nonlocal* equation (1.1). That is why we follow a different approach in this chapter, and we define the constrained set

$$\mathcal{M}_b^\lambda = \{u \in W_0^{s,p}(\Omega) \colon u^\pm \neq 0 \text{ and } \langle (I_b^\lambda)'(u), u^+\rangle = \langle (I_b^\lambda)'(u), u^-\rangle = 0\}. \tag{1.12}$$

The main idea in what follows is to study a constrained minimization problem of I_b^λ on the manifold $\mathcal{M}_b^\lambda$. A key tool in the arguments developed in this chapter is the Poincaré-Miranda existence theorem, which is a generalization of the intermediate value property to higher dimensions. This theorem is named after Henri Poincaré, who stated it in 1883 (with a hint of the proof), and by Carlo Miranda [118], who proved in 1940 that it is equivalent to Brouwer's fixed point theorem. We also obtain in this chapter that the minimizer of the constrained problem is a sign-changing solution of the Kirchhoff equation. This will be achieved via the quantitative deformation lemma and related degree theory arguments.

1.1.1 Existence, energy doubling, and convergence

The first result in this chapter establishes the existence of least-energy sign-changing solutions in the case of large perturbations of the logarithmic reaction.

Theorem 1.1 *There exists $\lambda^* > 0$ such that for all $\lambda \geq \lambda^*$, problem* (1.1) *has a least-energy nodal solution $u_b \in \mathcal{M}_b^\lambda$ with precisely two nodal domains such that $I_b^\lambda(u_b) = \inf_{u \in \mathcal{M}_b^\lambda} I_b^\lambda(u)$.*

Another goal of this chapter is to establish the *energy doubling* property (cf. Weth [152]), that is, the energy of any sign-changing solution of the *semilinear elliptic equation* (1.1) is strictly greater than twice the ground state energy. In the *local* setting described by problem (1.8), the conclusion is trivial. Indeed, if we denote the Nehari manifold associated to the elliptic equation (1.8) by

$$\mathcal{N} = \{u \in W_0^{s,p}(\Omega) \setminus \{0\} : \ \langle J'(u), u\rangle = 0\}$$

and define

$$c = \inf_{u \in \mathcal{N}} J(u) \tag{1.13}$$

then it is easy to verify that $u^\pm \in \mathcal{N}$ for any sign-changing solution $u \in W_0^{s,p}(\Omega)$ of problem (1.8). We deduce that

$$J(u) = J(u^+) + J(u^-) \geq 2c. \tag{1.14}$$

We point out that the minimizer of (1.13) is a ground state solution of problem (1.8), and $c > 0$ is the least-energy of all weak solutions of problem (1.8). Therefore, by (1.14), it follows that the energy of any sign-changing solution of problem (1.8) is larger than twice of the least energy.

We are interested in establishing whether the energy doubling property (1.14) is still true for the *nonlocal* equation (1.1). In this case, even if we know that the energy I_b^λ associated with an arbitrary solution u satisfies (1.11), it is not clear that the energy estimate (1.14) is still valid. Indeed, for any sign-changing solution u of problem (1.1), it follows from (1.7) that $u^\pm \notin \{v \in W_0^{s,p}(\Omega) \setminus \{0\} : \ \langle J'(v), v\rangle = 0\}$.

The next result shows that the energy doubling property remains valid in the case of large perturbations of the logarithmic nonlinearity in the reaction term of the Kirchhoff problem (1.1).

Theorem 1.2 *Set* $c^* := \inf_{u\in\mathcal{N}_b^\lambda} I_b^\lambda(u) > 0$. *Then there exists* $\lambda^{**} > 0$ *such that for all* $\lambda \geq \lambda^{**}$, c^* *is achieved and* $I_b^\lambda(u) > 2c^*$, *where*

$$\mathcal{N}_b^\lambda = \left\{u \in W_0^{s,p}(\Omega) \setminus \{0\} : \ \langle (I_b^\lambda)'(u), u\rangle = 0\right\}$$

and u *is the least-energy sign-changing solution obtained in Theorem 1.1. In particular,* $c^* > 0$ *is achieved either by a positive or a negative function.*

It is obvious that the energy of the sign-changing solution u_b obtained in Theorem 1.1 depends on b. In the following, we give a convergence property of u_b as $b \to 0$, which reflects some relationship between the nonlocal fractional case $b > 0$ and the limit fractional setting $b = 0$ for problem (1.1).

Theorem 1.3 *For any sequence of real numbers* $\{b_n\}$ *converging to zero, there exists a subsequence, still denoted by* $\{b_n\}$, *such that* $\{u_n\}$ *converges to* u_0 *strongly in* $W_0^{s,p}(\Omega)$ *as* $n \to \infty$, *where* u_0 *is a least-energy sign-changing solution of the following problem*

$$\begin{cases} a(-\Delta)_p^s u = \lambda |u|^{q-2} u \ln |u|^2 + |u|^{p_s^*-2} u & \text{in } \Omega, \\ u = 0 & \text{in } \mathbb{R}^N \setminus \Omega. \end{cases} \tag{1.15}$$

Section 1.2 is concerned with some constrained minimum and maximum problems, as well as with a related convergence property. Next, in Section 1.3, we give detailed proofs of the main qualitative and asymptotic properties established in this chapter.

1.2 Constrained minimum and maximum problems

Fix $u \in W_0^{s,p}(\Omega)$ with $u^\pm \neq 0$. Define the function $\psi_u : [0,\infty) \times [0,\infty) \to \mathbb{R}$ and the mapping $T_u : [0,\infty) \times [0,\infty) \to \mathbb{R}^2$ by

$$\psi_u(\alpha, \beta) = I_b^\lambda(\alpha u^+ + \beta u^-) \tag{1.16}$$

and

$$T_u(\alpha,\beta) = \left(\langle (I_b^\lambda)'(\alpha u^+ + \beta u^-), \alpha u^+\rangle, \langle (I_b^\lambda)'(\alpha u^+ + \beta u^-), \beta u^-\rangle\right). \tag{1.17}$$

Lemma 1.1 *For any* $u \in W_0^{s,p}(\Omega)$ *with* $u^\pm \neq 0$, *there is a unique maximum point pair* (α_u, β_u) *of the function* ψ_u *such that* $\alpha_u u^+ + \beta_u u^- \in \mathcal{M}_b^\lambda$.

Proof. The proof will be divided into three steps.

Step 1. For any $u \in W_0^{s,p}(\Omega)$ with $u^\pm \neq 0$, we establish the existence of α_u and β_u.

By our assumptions, we have

$$\lim_{t\to 0}\frac{|t|^{q-1}\ln|t|^2}{|t|^{p-1}}=0 \quad \text{and} \quad \lim_{t\to\infty}\frac{|t|^{q-1}\ln|t|^2}{|t|^{r-1}}=0 \tag{1.18}$$

for all $r\in(q,p_s^*)$. Then for any $\varepsilon>0$, there exists $C_\varepsilon>0$ such that

$$|t|^{q-1}\ln|t|^2\le\varepsilon|t|^{p-1}+C_\varepsilon|t|^{r-1}. \tag{1.19}$$

Since $4\le 2p<q<p_s^*$, it follows from (1.19) and the Sobolev embedding theorem that

$$\begin{aligned}\langle (I_b^\lambda)'(\alpha u^++\beta u^-),\alpha u^+\rangle &= a\alpha^p\|u^+\|^p+b\alpha^{2p}\|u^+\|^{2p}+b\alpha^p\beta^p\|u^+\|^p\|u^-\|^p\\ &\quad-\lambda\int_\Omega|\alpha u^+|^q\ln|\alpha u^+|^2dx-\alpha^{p_s^*}\int_\Omega|u^+|^{p_s^*}dx\\ &\ge a\alpha^p\|u^+\|^p+b\alpha^{2p}\|u^+\|^{2p}+b\alpha^p\beta^p\|u^+\|^p\|u^-\|^p\\ &\quad-\lambda\alpha^p\varepsilon\int_\Omega|u^+|^pdx-\lambda C_\varepsilon\alpha^r\int_\Omega|u^+|^rdx-\alpha^{p_s^*}\int_\Omega|u|^{p_s^*}dx.\end{aligned}$$

We conclude that

$$\begin{aligned}&\langle (I_b^\lambda)'(\alpha u^++\beta u^-),\alpha u^+\rangle\\ &\ge a\alpha^p\|u^+\|^p+b\alpha^{2p}\|u^+\|^{2p}-\lambda\alpha^p\varepsilon C_1\|u^+\|^p-\lambda C_\varepsilon\alpha^rC_2\|u^+\|^r-C_3\alpha^{p_s^*}\|u^+\|^{p_s^*}\\ &=(a-\lambda\varepsilon C_1)\,\alpha^p\|u^+\|^p+b\alpha^{2p}\|u^+\|^{2p}-\lambda C_\varepsilon\alpha^rC_2\|u^+\|^r-C_3\alpha^{p_s^*}\|u^+\|^{p_s^*}.\end{aligned}$$

Choose $\varepsilon>0$ such that $a-\lambda\varepsilon C_1>0$. Since $p_s^*,r>2p$, we have that $\langle (I_b^\lambda)'(\alpha u^++\beta u^-),\alpha u^+\rangle>0$ for α small enough and all $\beta\ge 0$.

Similarly, we obtain that $\langle (I_b^\lambda)'(\alpha u^++\beta u^-),\beta u^-\rangle>0$ for β small enough and all $\alpha\ge 0$. Thus, there exists $\delta_1>0$ such that for all $\alpha,\beta\ge 0$

$$\langle (I_b^\lambda)'(\delta_1u^++\beta u^-),\delta_1u^+\rangle>0,\quad \langle (I_b^\lambda)'(\alpha u^++\delta_1u^-),\delta_1u^-\rangle>0. \tag{1.20}$$

Choose $\alpha=\delta_2^*>\delta_1$ large enough. If $\beta\in[\delta_1,\delta_2^*]$, it follows that

$$\begin{aligned}\langle (I_b^\lambda)'(\delta_2^*u^++\beta u^-),\delta_2^*u^+\rangle &\le a(\delta_2^*)^p\|u^+\|^p+b(\delta_2^*)^{2p}\|u^+\|^{2p}+b(\delta_2^*)^{2p}\|u^+\|^p\|u^-\|^p\\ &\quad-(\delta_2^*)^{p_s^*}\int_\Omega|u^+|^{p_s^*}dx\le 0.\end{aligned}$$

Similarly, we have that

$$\begin{aligned}\langle (I_b^\lambda)'(\alpha u^++\delta_2^*u^-),\delta_2^*u^-\rangle &\le (\delta_2^*)^p\|u^-\|^p+b(\delta_2^*)^{2p}\|u^+\|^{2p}+b(\delta_2^*)^{2p}\|u^+\|^p\|u^-\|^p\\ &\quad-(\delta_2^*)^{p_s^*}\int_\Omega|u^-|^{p_s^*}dx\le 0.\end{aligned}$$

We conclude that if $\delta_2>\delta_2^*$ is large enough, then for all $\alpha,\beta\in[\delta_1,\delta_2]$

$$\langle (I_b^\lambda)'(\delta_2^*u^++\beta u^-),\delta_2^*u^+\rangle<0 \quad\text{and}\quad \langle (I_b^\lambda)'(\alpha u^++\delta_2^*u^-),\delta_2^*u^-\rangle<0. \tag{1.21}$$

Combining relations (1.20) and (1.21) with the Poincaré-Miranda theorem, there exists $(\alpha_u,\beta_u)\in(0,+\infty)\times(0,+\infty)$ such that $T_u(\alpha,\beta)=(0,0)$, hence $\alpha u^++\beta u^-\in\mathcal{M}_b^\lambda$.

We notice that the Poincaré-Miranda existence theorem was announced by Henri Poincaré without proof in 1883 in the following version.

Let $f_1, \ldots, f_n$ be n continuous functions of n variables $x_1, \ldots, x_n$: the variable x_i is subjected to vary between the limits $-a_i$ and $+a_i$. Let us suppose that for $x_i = a_i$, f_i is constantly positive, and that for $x_i = -a_i$, f_i is constantly negative. Then there exists a system of values of x where f's vanish.

Step 2. We prove the uniqueness of the pair (α_u, β_u). We distinguish the following cases.
(a) *Case* $u \in \mathcal{M}_b^\lambda$.
If $u \in \mathcal{M}_b^\lambda$, then

$$\|u^+\|^p + b\|u^+\|^{2p} + b\|u^+\|^p\|u^-\|^p = \lambda \int_\Omega |u^+|^q \ln |u^+|^2 dx + \int_\Omega |u^+|^{p_s^*} dx \tag{1.22}$$

and

$$\|u^-\|^2 + b\|u^-\|^{2p} + b\|u^+\|^p\|u^-\|^p = \lambda \int_\Omega |u^-|^q \ln |u^-|^2 dx + \int_\Omega |u^-|^{p_s^*} dx. \tag{1.23}$$

We show that $(\alpha_u, \beta_u) = (1,1)$ is the unique pair such that $\alpha_u u^+ + \beta_u u^- \in \mathcal{M}_b^\lambda$.
Let $(\alpha_0, \beta_0) \in [0,\infty) \times [0,\infty)$ such that $\alpha_0 u^+ + \beta_0 u^- \in \mathcal{M}_b^\lambda$ with $0 < \alpha_0 \le \beta_0$. Hence

$$\begin{aligned} \alpha_0^p\|u^+\|^p + b\alpha_0^{2p}\|u^+\|^{2p} + b\alpha_0^p\beta_0^p\|u^+\|^p\|u^-\|^p &= \lambda \int_\Omega |\alpha_0 u^+|^q \ln |\alpha_0 u^+|^2 dx \\ &+ \alpha_0^{p_s^*} \int_\Omega |u^+|^{p_s^*} dx \end{aligned} \tag{1.24}$$

and

$$\begin{aligned} \beta_0^p\|u^-\|^p + b\beta_0^{2p}\|u^-\|^{2p} + b\alpha_0^p\beta_0^p\|u^+\|^p\|u^-\|^p &= \lambda \int_\Omega |\beta_0 u^-|^q \ln |\beta_0 u^-|^2 dx \\ &+ \beta_0^{p_s^*} \int_\Omega |u^-|^{p_s^*} dx. \end{aligned} \tag{1.25}$$

Since $0 < \alpha_0 \le \beta_0$ and using (1.25), we have

$$\begin{aligned} \frac{\|u^-\|^p}{\beta_0^p} + b\|u^-\|^{2p} + b\|u^+\|^p\|u^-\|^p &\ge \lambda \int_\Omega \frac{|\beta_0 u^-|^q \ln |\beta_0 u^-|^2}{\beta_0^{2p}} dx \\ &+ \beta_0^{p_s^*-2p} \int_\Omega |u^-|^{p_s^*} dx. \end{aligned} \tag{1.26}$$

If $\beta_0 > 1$, by (1.23) and (1.26), one has that

$$\begin{aligned} 0 > \left(\frac{1}{\beta_0^p} - 1\right) \|u^-\|^p &\ge \lambda \int_\Omega \left[\frac{|\beta_0 u^-|^q \ln |\beta_0 u^-|^2}{\beta_0^{2p}} - |u^-|^q \ln |u^-|^2 \right] dx \\ + (\beta_0^{p_s^*-2p} - 1) \int_\Omega |u^-|^{p_s^*} dx &\ge (\beta_0^{p_s^*-2p} - 1) \int_\Omega |u^-|^{p_s^*} dx > 0. \end{aligned}$$

This is a contradiction. Therefore, we conclude that $0 < \alpha_0 \le \beta_0 \le 1$.

Similarly, by (1.24) and $0 < \alpha_0 \le \beta_0$, we have that

$$\left(\frac{1}{\alpha_0^p} - 1\right) \|u^+\|^p \le \lambda \int_\Omega \left[\frac{|\alpha_0 u^+|^q \ln |\alpha_0 u^+|^2}{\alpha_0^{2p}} - |u^+|^q \ln |u^+|^2 \right] dx + (\beta_0^{p_s^*-2p} - 1) \int_\Omega |u^+|^{p_s^*} dx.$$

This fact implies that $\alpha_0 \geq 1$. Consequently, $\alpha_0 = \beta_0 = 1$.

(b) *Case* $u \notin \mathcal{M}_b^\lambda$.

Suppose that there exist (α_1, β_1) and (α_2, β_2) such that

$$\omega_1 = \alpha_1 u^+ + \beta_1 u^- \in \mathcal{M}_b^\lambda \quad \text{and} \quad \omega_2 = \alpha_2 u^+ + \beta_2 u^- \in \mathcal{M}_b^\lambda.$$

Hence

$$\omega_2 = \left(\frac{\alpha_2}{\alpha_1}\right)\alpha_1 u^+ + \left(\frac{\beta_2}{\beta_1}\right)\beta_1 u^- = \left(\frac{\alpha_2}{\alpha_1}\right)\omega_1^+ + \left(\frac{\beta_2}{\beta_1}\right)\omega_1^- \in \mathcal{M}_b^\lambda.$$

By $\omega_1 \in \mathcal{M}_b^\lambda$, we deduce that

$$\frac{\alpha_2}{\alpha_1} = \frac{\beta_2}{\beta_1} = 1.$$

Hence, $\alpha_1 = \alpha_2$, $\beta_1 = \beta_2$.

Step 3. We prove that (α_u, β_u) is the unique maximum point of ψ_u on $[0, \infty) \times [0, \infty)$.

We first observe that

$$2\tau^q - q\tau^q \ln|\tau|^2 \leq 2 \quad \text{for all } \tau \in (0, \infty). \tag{1.27}$$

Let $\Omega^+ = \{x \in \Omega : u(x) > 0\}$ and $\Omega^- = \{x \in \Omega : u(x) < 0\}$. For all $u \in W_0^{s,p}(\Omega)$ with $u^\pm \neq 0$, we have

$$\begin{aligned}\int_\Omega |\alpha u^+ + \beta u^-|^q \ln|\alpha u^+ + \beta u^-|^2 dx &= \int_{\Omega^+} |\alpha u^+ + \beta u^-|^q \ln|\alpha u^+ + \beta u^-|^2 dx \\ &\quad + \int_{\Omega^-} |\alpha u^+ + \beta u^-|^q \ln|\alpha u^+ + \beta u^-|^2 dx \\ &= \int_{\Omega^+} |\alpha u^+|^q \ln|\alpha u^+|^2 dx + \int_{\Omega^-} |\beta u^-|^q \ln|\beta u^-|^2 dx \\ &= \int_\Omega \left[|\alpha u^+|^q \ln|\alpha u^+|^2 + |\beta u^-|^q \ln|\beta u^-|^2\right] dx. \end{aligned} \tag{1.28}$$

Combining (1.27) and (1.28), we have

$$\begin{aligned}\psi_u(\alpha,\beta) &= I_b^\lambda(\alpha u^+ + \beta u^-) = \frac{1}{p}\|\alpha u^+ + \beta u^-\|^p + \frac{b}{2p}\|\alpha u^+ + \beta u^-\|^{2p} + \frac{2\lambda}{q^2}\int_\Omega |\alpha u^+ + \beta u^-|^q dx \\ &\quad - \frac{\lambda}{q}\int_\Omega |\alpha u^+ + \beta u^-|^q \ln|\alpha u^+ + \beta u^-|^2 dx - \frac{1}{p_s^*}\int_\Omega |\alpha u^+ + \beta u^-|^{p_s^*} dx \\ &= \frac{\alpha^p}{p}\|u^+\|^p + \frac{\beta^p}{p}\|u^-\|^p + \frac{b\alpha^{2p}}{2p}\|u^+\|^{2p} + \frac{b\beta^{2p}}{2p}\|u^-\|^{2p} + \frac{b\alpha^p\beta^p}{p}\|u^+\|^p\|u^-\|^p \\ &\quad + \frac{\lambda}{q^2}\int_\Omega \left(2|\alpha u^+|^q - q|\alpha u^+|^q \ln|\alpha u^+|^2\right) dx + \frac{\lambda}{q^2}\int_\Omega \left(2|\beta u^-|^q - q|\beta u^-|^q \ln|\beta u^-|^2\right) dx \\ &\quad - \frac{\alpha^{p_s^*}}{p_s^*}\int_\Omega |u^+|^{p_s^*} dx - \frac{\beta^{p_s^*}}{p_s^*}\int_\Omega |u^-|^{p_s^*} dx.\end{aligned}$$

We deduce that

$$\begin{aligned}\psi_u(\alpha,\beta) &\leq \frac{\alpha^p}{p}\|u^+\|^p + \frac{\beta^p}{p}\|u^-\|^p + \frac{b\alpha^{2p}}{2p}\|u^+\|^{2p} + \frac{b\beta^{2p}}{2p}\|u^-\|^{2p} + \frac{b\alpha^p\beta^p}{p}\|u^+\|^p\|u^-\|^p \\ &\quad + \frac{4}{q^2}\lambda|\Omega| - \frac{\alpha^{p_s^*}}{p_s^*}\int_\Omega |u^+|^{p_s^*} dx - \frac{\beta^{p_s^*}}{p_s^*}\int_\Omega |u^-|^{p_s^*} dx,\end{aligned}$$

which implies that $\lim_{|(\alpha,\beta)|\to\infty}\psi(\alpha,\beta)=-\infty$, since $p_s^*>2p$. Hence, (α_u,β_u) is the unique critical point of ψ_u in $[0,\infty)\times[0,\infty)$. So, it is sufficient to check that a maximum point cannot be achieved on the boundary of $[0,\infty)\times[0,\infty)$. By contradiction, we suppose that $(0,\beta_0)$ is a maximum point of ψ_u with $\beta_0\geq 0$. It follows that

$$\begin{aligned}\psi_u(\alpha,\beta_0) &= \frac{\alpha^p}{p}\|u^+\|^p+\frac{b\alpha^{2p}}{2p}\|u^+\|^{2p}+\frac{2\alpha^q\lambda}{q^2}\int_\Omega |u^+|^q dx-\frac{\alpha^q\lambda}{q}\int_\Omega |u^+|^q\ln|\alpha u^+|^2dx\\ &-\frac{\alpha^{p_s^*}}{p_s^*}\int_\Omega |u^+|^{p_s^*}dx+\frac{\beta_0^p}{p}\|u^-\|^p+\frac{b\beta_0^{2p}}{2p}\|u^-\|^{2p}+\frac{2\beta^q\lambda}{q^2}\int_\Omega |u^-|^q dx\\ &-\frac{\beta^q\lambda}{q}\int_\Omega |u^-|^q\ln|\beta u^-|^2dx-\frac{\beta^{p_s^*}}{p_s^*}\int_\Omega |u^-|^{p_s^*}dx+\frac{b\alpha^p\beta_0^p}{p}\|u^+\|^p\|u^-\|^p.\end{aligned}$$

We conclude that for all $\alpha>0$ is small enough.

$$\begin{aligned}(\psi_u)'_\alpha(\alpha,\beta_0) &= \alpha^{p-1}\|u^+\|^p+b\alpha^{2p-1}\|u^+\|^{2p}+b\alpha^{p-1}\beta_0^p\|u^+\|^p\|u^-\|^p\\ &+\frac{2\alpha^{q-1}\lambda}{q}\int_\Omega |u^+|^q dx-\alpha^{q-1}\int_\Omega |u^+|^q\ln|\alpha u^+|^2dx-\alpha^{p_s^*-1}\int_\Omega |u^+|^{p_s^*}dx\\ &>0.\end{aligned}$$

It follows that ψ_u is an increasing function with respect to α if α is small enough. This yields the contradiction. Similarly, ψ_u cannot achieve its global maximum on $(\alpha,0)$. □

Lemma 1.2 *Let $u\in W_0^{s,p}(\Omega)$ with $u^\pm\neq 0$ such that $\langle (I_b^\lambda)'(u),u^\pm\rangle\leq 0$. Then, the unique maximum point of ψ_u on $[0,\infty)\times[0,\infty)$ satisfies $0<\alpha_u,\beta_u\leq 1$.*

Proof. Without loss of generality, we assume that $\alpha_u\geq\beta_u>0$. On the one hand, by $\alpha_u u^++\beta_u u^-\in\mathcal{M}_b^\lambda$, we have

$$\begin{aligned}a\alpha_u^p\|u^+\|^p+b\alpha_u^{2p}\|u^+\|^{2p}+b\alpha_u^{2p}\|u^+\|^p\|u^-\|^p &= \lambda\alpha_u^q\int_\Omega |u^+|^q\ln|\alpha_u u^+|^2dx\\ &-\alpha_u^{p_s^*}\int_\Omega |u^+|^{p_s^*}dx.\end{aligned}\tag{1.29}$$

On the other hand, by $\langle (I_b^\lambda)'(u),u^+\rangle\leq 0$, we have

$$a\|u^+\|^p+b\|u^+\|^{2p}+b\|u^+\|^p\|u^-\|^p\leq\lambda\int_\Omega |u^+|^q\ln|u^+|^2dx+\int_\Omega |u^+|^{p_s^*}dx.\tag{1.30}$$

So, according to (1.29) and (1.30), we have

$$\begin{aligned}\left(\frac{1}{\alpha_u^p}-1\right)a\|u^+\|^p &\geq\lambda\int_\Omega\left[\alpha_u^{q-2p}|u^+|^q\ln|\alpha_u u^+|^2-|u^+|^q\ln|u^+|^2\right]dx\\ &+(\alpha_u^{p_s^*-2p}-1)\int_\Omega |u^+|^{p_s^*}dx.\end{aligned}\tag{1.31}$$

If $\alpha_u\geq 1$, we obtain

$$\int_\Omega\left[\alpha_u^{q-2p}|u^+|^q\ln|\alpha_u u^+|^2-|u^+|^q\ln|u^+|^2\right]dx\geq 0.$$

This fact, together with (1.31), yields

$$0>\left(\frac{1}{\alpha_u^p}-1\right)a\|u^+\|^p\geq(\alpha_u^{p_s^*-2p}-1)\int_\Omega |u^+|^{p_s^*}dx>0.$$

This contradiction shows that $\alpha_u\leq 1$. Thus, we obtain that $0<\alpha_u,\beta_u\leq 1$. □

Lemma 1.3 *Let* $c_b^\lambda = \inf_{u\in\mathcal{M}_b^\lambda} I_b^\lambda(u)$, *then* $\lim_{\lambda\to\infty} c_b^\lambda = 0$.

Proof. For any $u \in \mathcal{M}_b^\lambda$, we have

$$a\|u^\pm\|^p + b\|u^\pm\|^{2p} + b\|u^+\|^p\|u^-\|^p = \lambda\int_\Omega |u^\pm|^q \ln|u^\pm|^2 dx + \int_\Omega |u^\pm|^{p_s^*} dx.$$

Then, by (1.19) and the Sobolev theorem, we have

$$a\|u^\pm\|^p \le \lambda\int_\Omega |u^\pm|^q \ln|u^\pm|^2 dx + \int_\Omega |u^\pm|^{p_s^*} dx \le \lambda\varepsilon C_1\|u^\pm\|^p + \lambda C_\varepsilon C_2\|u^\pm\|^r + C_3\|u^\pm\|^{p_s^*}.$$

Thus, we get

$$(1-\lambda\varepsilon C_1)a\|u^\pm\|^p \le \lambda C_\varepsilon C_2\|u^\pm\|^r + C_3\|u^\pm\|^{p_s^*}.$$

Since $r, p_s^* > p$, choosing ε small enough, there exists $\rho > 0$ such that

$$\|u^\pm\| \ge \rho \quad \text{for all} \quad u \in \mathcal{M}_b^\lambda. \tag{1.32}$$

On the other hand, for any $u \in \mathcal{M}_b^\lambda$, we have $\langle (I_b^\lambda)'(u), u\rangle = 0$. Then, we have

$$\begin{aligned} I_b^\lambda(u) &= I_b^\lambda(u) - \frac{1}{q}\langle (I_b^\lambda)'(u), u\rangle \\ &= \left(\frac{1}{p}-\frac{1}{q}\right)a\|u\|^p + \left(\frac{1}{2p}-\frac{1}{q}\right)b\|u\|^{2p} + \left(\frac{1}{q}-\frac{1}{p_s^*}\right)\int_\Omega |u|^{p_s^*}dx + \frac{2\lambda}{q^2}\int_\Omega |u|^q dx \\ &\ge \left(\frac{1}{p}-\frac{1}{q}\right)a\|u\|^p. \end{aligned}$$

Summing up the above remarks, we obtain that $I_b^\lambda(u) > 0$ for all $u \subset \mathcal{M}_b^\lambda$. Therefore, I_b^λ is bounded from below on $\mathcal{M}_b^\lambda$, hence $c_b^\lambda = \inf_{u\in\mathcal{M}_b^\lambda} I_b^\lambda(u)$ is well defined.

Let $u \in W_0^{s,p}(\Omega)$ with $u^\pm \neq 0$ be fixed. By Lemma 1.1, for each $\lambda > 0$, there exist $\alpha_\lambda, \beta_\lambda > 0$ such that $\alpha_\lambda u^+ + \beta_\lambda u^- \in \mathcal{M}_b^\lambda$. By using Lemma 1.1 again, we obtain

$$\begin{aligned} 0 \le c_b^\lambda &= \inf_{u\in\mathcal{M}_b^\lambda} I_b^\lambda(u) \le I_b^\lambda(\alpha_\lambda u^+ + \beta_\lambda u^-) \\ &\le \frac{1}{p}a\|\alpha_\lambda u^+ + \beta_\lambda u^-\|^p + \frac{b}{2p}\|\alpha_\lambda u^+ + \beta_\lambda u^-\|^{2p} \\ &\le a\alpha_\lambda^p\|u^+\|^p + a\beta_\lambda^p\|u^-\|^p + pb\alpha_\lambda^{2p}\|u^+\|^{2p} + pb\beta_\lambda^{2p}\|u^-\|^{2p}. \end{aligned}$$

In virtue of this relation, to conclude the proof, it remains to show that $\alpha_\lambda \to 0$ and $\beta_\lambda \to 0$ as $\lambda \to \infty$. Let

$$\mathcal{T}_u = \{(\alpha_\lambda, \beta_\lambda) \in [0,\infty)\times[0,\infty) : T_u(\alpha_\lambda, \beta_\lambda) = (0,0), \lambda > 0\},$$

where T_u is defined in (1.17). By (1.18), we have

$$\begin{aligned} &\alpha_\lambda^{p_s^*}\int_\Omega |u^+|^{p_s^*}dx + \beta_\lambda^{p_s^*}\int_\Omega |u^-|^{p_s^*}dx \\ &\le \alpha_\lambda^{p_s^*}\int_\Omega |u^+|^{p_s^*}dx + \beta_\lambda^{p_s^*}\int_\Omega |u^-|^{p_s^*}dx + \lambda\alpha_\lambda^q\int_\Omega |u^+|^q\ln|u^+|^2dx + \lambda\beta_\lambda^q\int_\Omega |u^+|^q\ln|u^+|^2dx \\ &= \|\alpha_\lambda u^+ + \beta_\lambda u^-\|^p + b\|\alpha_\lambda u^+ + \beta_\lambda u^-\|^{2p} \\ &\le p\alpha_\lambda^p\|u^+\|^p + p\beta_\lambda^p\|u^-\|^p + 2pb\alpha_\lambda^{2p}\|u^+\|^{2p} + 2pb\beta_\lambda^{2p}\|u^-\|^{2p}. \end{aligned}$$

Hence, $\mathcal{T}_u$ is bounded since $2p < p_s^*$.

Let $\{\lambda_n\} \subset (0, \infty)$ be such that $\lambda_n \to \infty$ as $n \to \infty$. Then, there exist α_0 and β_0 such that $(\alpha_{\lambda_n}, \beta_{\lambda_n}) \to (\alpha_0, \beta_0)$ as $n \to \infty$.

We now claim that $\alpha_0 = \beta_0 = 0$. Suppose, by contradiction, that $\alpha_0 > 0$ or $\beta_0 > 0$. By $\alpha_{\lambda_n} u^+ + \beta_{\lambda_n} u^- \in \mathcal{M}_b^{\lambda_n}$, for any $n \in \mathbb{N}$, we have

$$\begin{aligned}&\|\alpha_{\lambda_n} u^+ + \beta_{\lambda_n} u^-\|^p + b\|\alpha_{\lambda_n} u^+ + \beta_{\lambda_n} u^-\|^{2p} = \\ &\lambda_n \int_\Omega |\alpha_{\lambda_n} u^+ + \beta_{\lambda_n} u^-|^q \ln|\alpha_{\lambda_n} u^+ + \beta_{\lambda_n} u^-|^2 dx + \int_\Omega |\alpha_{\lambda_n} u^+ + \beta_{\lambda_n} u^-|^{p_s^*} dx. \qquad (1.33)\end{aligned}$$

Thanks to $\alpha_{\lambda_n} u^+ \to \alpha_0 u^+$ and $\beta_{\lambda_n} u^- \to \beta_0 u^+$ in $W_0^{s,p}(\Omega)$, relation (1.19) and the Lebesgue dominated convergence theorem, we obtain

$$\int_\Omega |\alpha_{\lambda_n} u^+ + \beta_{\lambda_n} u^-|^q \ln|\alpha_{\lambda_n} u^+ + \beta_{\lambda_n} u^-|^2 dx \to \int_\Omega |\alpha_0 u^+ + \beta_0 u^-|^q \ln|\alpha_0 u^+ + \beta_0 u^-|^2 dx > 0$$

as $n \to \infty$. Since $\lambda_n \to \infty$ as $n \to \infty$ and $\{\alpha_{\lambda_n} u^+ + \beta_{\lambda_n} u^-\}$ is bounded in $W_0^{s,p}(\Omega)$, we obtain a contradiction with equality (1.33). Hence, $\alpha_0 = \beta_0 = 0$. We conclude that $\lim_{\lambda\to\infty} c_b^\lambda = 0$. □

Lemma 1.4 *There exists $\lambda^* > 0$ such that for all $\lambda \geq \lambda^*$, the infimum c_b^λ is achieved.*

Proof. By the definition of c_b^λ, there exists a sequence $\{u_n\} \subset \mathcal{M}_b^\lambda$ such that

$$\lim_{n\to\infty} I_b^\lambda(u_n) = c_b^\lambda.$$

Obviously, $\{u_n\}$ is bounded in $W_0^{s,p}(\Omega)$. Thus, up to a subsequence, still denoted by $\{u_n\}$, there exists $u \in W_0^{s,p}(\Omega)$ such that $u_n \rightharpoonup u$. Since the embedding $W_0^{s,p}(\Omega) \hookrightarrow L^t(\Omega)$ is compact for all $t \in (p, p_s^*)$, we have

$$u_n \to u \quad \text{in} \quad L^t(\Omega), \qquad u_n \to u \quad \text{a.e.} \quad x \in \Omega.$$

Hence

$$\begin{aligned} u_n^\pm &\rightharpoonup u^\pm \quad \text{in } W_0^{s,p}(\Omega), \\ u_n^\pm &\to u^\pm \quad \text{in } L^t(\Omega), \\ u_n^\pm &\to u^\pm \quad \text{a.e. } x \in \Omega. \end{aligned}$$

By Lemma 1.1, we have for all $\alpha, \beta \geq 0$

$$I_b^\lambda(\alpha u_n^+ + \beta u_n^-) \leq I_b^\lambda(u_n).$$

On the one hand, the Vitali convergence theorem yields that

$$\lim_{n\to\infty} \int_\Omega |u_n|^q \ln|u_n|^2 dx = \int_\Omega |u|^q \ln|u|^2 dx. \qquad (1.34)$$

On the other hand, since $u_n \to u$ in $L^q(\Omega)$, we have

$$\lim_{n\to\infty} \int_\Omega |u_n|^q dx = \int_\Omega |u|^q dx. \qquad (1.35)$$

Combining relations (1.34), (1.35) with the Brezis-Lieb lemma and the weak semicontinuity of the norm, we have

$$\begin{aligned}\liminf_{n\to\infty} I_b^\lambda(\alpha u_n^+ + \beta u_n^-) \geq{}& \frac{\alpha^p}{p}\lim_{n\to\infty}(\|u_n^+ - u^+\|^p + \|u^+\|^p) + \frac{\beta^p}{p}\lim_{n\to\infty}(\|u_n^- - u^-\|^p + \|u^-\|^p)\\ &+ \frac{b\alpha^{2p}}{2p}\left[\lim_{n\to\infty}(\|u_n^+ - u^+\|^p + \|u^+\|^p)\right]^2\\ &+ \frac{b\beta^{2p}}{2p}\left[\lim_{n\to\infty}(\|u_n^- - u^-\|^p + \|u^-\|^p)\right]^2\\ &- \frac{\alpha^{p_s^*}}{p_s^*}\lim_{n\to\infty}\left[\int_\Omega |u_n^+ - u^+|^{p_s^*}dx + \int_\Omega |u^+|^{p_s^*}dx\right]\\ &- \frac{\beta^{p_s^*}}{p_s^*}\lim_{n\to\infty}\left[\int_\Omega |u_n^- - u^-|^{p_s^*}dx + \int_\Omega |u^-|^{p_s^*}dx\right]\\ &+ \frac{2\lambda}{q^2}\int_\Omega |u|^q dx - \frac{\lambda}{q}\int_\Omega |u|^q \ln|u|^2 dx + \frac{b\alpha^2\beta^2}{2}\liminf_{n\to\infty}(\|u_n^+\|^2\|u_n^-\|^2).\end{aligned}$$

Therefore,

$$\begin{aligned}\liminf_{n\to\infty} I_b^\lambda(\alpha u_n^+ + \beta u_n^-) \geq{}& I_b^\lambda(\alpha u^+ + \beta u^-) + \frac{\alpha^p}{p}\lim_{n\to\infty}\|u_n^+ - u^+\|^p + \frac{\beta^p}{p}\lim_{n\to\infty}\|u_n^- - u^-\|^p\\ &+ \frac{b\alpha^{2p}}{p}\lim_{n\to\infty}\|u_n^+ - u^+\|^p\|u^+\|^p + \frac{b\beta^{2p}}{p}\lim_{n\to\infty}\|u_n^- - u^-\|^p\|u^-\|^p\\ &+ \frac{b\alpha^{2p}}{2p}(\lim_{n\to\infty}\|u_n^+ - u^+\|^p)^2 + \frac{bt^{2p}}{2p}(\lim_{n\to\infty}\|u_n^- - u^-\|^p)^2\\ &- \frac{\alpha^{p_s^*}}{p_s^*}\int_\Omega |u_n^+ - u^+|^{p_s^*}dx - \frac{\beta^{p_s^*}}{p_s^*}\int_\Omega |u_n^- - u^-|^{p_s^*}dx.\end{aligned}$$

We conclude that

$$\begin{aligned}\liminf_{n\to\infty} I_b^\lambda(\alpha u_n^+ + \beta u_n^-) \geq{}& I_b^\lambda(\alpha u^+ + \beta u^-) + \frac{\alpha^p}{p}A_1 + \frac{b\alpha^{2p}}{p}A_1\|u^+\|^p + \frac{b\alpha^{2p}}{2p}A_1^p - \frac{\alpha^{p_s^*}}{p_s^*}B_1\\ &+ \frac{\beta^p}{p}A_2 + \frac{b\beta^{2p}}{p}A_2\|u^-\|^p + \frac{b\beta^{2p}}{2p}A_2^p - \frac{\beta^{p_s^*}}{p_s^*}B_2,\end{aligned}$$

where

$$\begin{aligned} A_1 &= \lim_{n\to\infty}\|u_n^+ - u^+\|^p, \quad A_2 = \lim_{n\to\infty}\|u_n^- - u^-\|^p,\\ B_1 &= \lim_{n\to\infty}|u_n^+ - u^+|_{p_s^*}^{p_s^*}, \quad B_2 = \lim_{n\to\infty}|u_n^- - u^-|_{p_s^*}^{p_s^*}.\end{aligned}$$

It follows that for all $\alpha \geq 0$ and $\beta \geq 0$

$$\begin{aligned} I_b^\lambda(\alpha u^+ + \beta u^-) &+ \frac{\alpha^p}{p}A_1 + \frac{b\alpha^{2p}}{p}A_1\|u^+\|^p + \frac{b\alpha^{2p}}{2p}A_1^p - \frac{\alpha^{p_s^*}}{p_s^*}B_1\\ &+ \frac{\beta^p}{p}A_2 + \frac{b\beta^{2p}}{p}A_2\|u^-\|^p + \frac{b\beta^{2p}}{2p}A_2^p - \frac{\beta^{p_s^*}}{p_s^*}B_2 \leq c_b^\lambda. \end{aligned} \tag{1.36}$$

Claim: $u^\pm \neq 0$.

Since the case $u^- \neq 0$ is analogous, we only prove that $u^+ \neq 0$. By contradiction, we suppose $u^+ = 0$. Hence, let $\beta = 0$ in (1.32) and we have that for all $\alpha \geq 0$

$$\frac{\alpha^p}{p}A_1 + \frac{b\alpha^{2p}}{2p}A_1^2 - \frac{\alpha^{p_s^*}}{p_s^*}B_1 \leq c_b^\lambda. \tag{1.37}$$

Case 1: $B_1 = 0$.

If $A_1 = 0$, that is, $u_n^+ \to u^+$ in $W_0^{s,p}(\Omega)$, then by relation (1.37), we obtain $\|u^+\| > 0$, which contradicts our supposition. If $A_1 > 0$, by relation (1.37), we get

$$\frac{\alpha^p}{p}A_1 + \frac{b\alpha^{2p}}{2p}A_1^2 \leq c_b^\lambda$$

for all $\alpha \geq 0$, which is absurd by Lemma 1.3. Anyway, we obtain a contradiction.

Case 2: $B_1 > 0$.

By Lemma 1.3, there exists $\lambda^* > 0$ such that

$$c_b^\lambda < \frac{s}{N}S^{N/ps} \quad \text{for all } \lambda \geq \lambda^*, \tag{1.38}$$

where $S > 0$ is given by (1.3).

Since $B_1 > 0$, we obtain that $A_1 > 0$. Hence, in view of relation (1.37), we get

$$\frac{s}{N}S^{N/ps} \leq \max_{\alpha \geq 0}\left\{\frac{\alpha^p}{p}A_1 - \frac{\alpha^{p_s^*}}{p_s^*}B_1\right\} \leq \max_{\alpha \geq 0}\left\{\frac{\alpha^p}{p}A_1 + \frac{b\alpha^{2p}}{2p}A_1^2 - \frac{\alpha^{p_s^*}}{p_s^*}B_1\right\} \leq c_b^\lambda,$$

which is a contradiction. We conclude that $u^\pm \neq 0$.

Claim: $B_1 = B_2 = 0$.

Since the case $B_2 = 0$ is analogous, we only prove that $B_1 = 0$. By contradiction, we suppose that $B_1 > 0$.

Case 1: $B_2 > 0$.

According to $B_1, B_2 > 0$ and Sobolev embeddings, we obtain that $A_1, A_2 > 0$. Let

$$\phi(\alpha) = \frac{\alpha^p}{p}A_1 + \frac{b\alpha^{2p}}{2p}A_1^2 - \frac{\alpha^{p_s^*}}{p_s^*}B_1 \quad \text{for all } \alpha \geq 0.$$

Then $\phi(\alpha) > 0$ for $\alpha > 0$ small enough and $\phi(\alpha) < 0$ for $\alpha < 0$ large enough. Hence, by continuity, there exists $\hat{\alpha} > 0$ such that

$$\frac{\hat{\alpha}^p}{p}A_1 + \frac{b\hat{\alpha}^{2p}}{2p}A_1^2 - \frac{\hat{\alpha}^{p_s^*}}{p_s^*}B_1 = \max_{\alpha \geq 0}\left\{\frac{\alpha^p}{p}A_1 + \frac{b\alpha^{2p}}{2p}A_1^2 - \frac{\alpha^{p_s^*}}{p_s^*}B_1\right\}.$$

Similarly, there exists $\hat{\beta} > 0$ such that

$$\frac{\hat{\beta}^p}{p}A_2 + \frac{b\hat{\beta}^{2p}}{2p}A_2^2 - \frac{\hat{\beta}^{p_s^*}}{p_s^*}B_2 = \max_{\alpha \geq 0}\left\{\frac{\alpha^p}{p}B_1 + \frac{b\alpha^{2p}}{2p}B_1^2 - \frac{\alpha^{p_s^*}}{p_s^*}B_1\right\}.$$

Since $[0, \hat{\alpha}] \times [0, \hat{\beta}]$ is compact and ϕ is continuous, there exists $(\alpha_u, \beta_u) \in [0, \hat{\alpha}] \times [0, \hat{\beta}]$ such that

$$\phi(\alpha_u, \beta_u) = \max_{(\alpha,\beta) \in [0,\hat{\alpha}] \times [0,\hat{\beta}]} \phi(\alpha, \beta).$$

Now, we prove that $(\alpha_u, \beta_u) \in (0, \hat{\alpha}) \times (0, \hat{\beta})$.

Note that if β is small enough, we have that for all $\alpha \in [0, \hat{\alpha}]$

$$\phi(\alpha, 0) = I_b^\lambda(\alpha u^+) < I_b^\lambda(\alpha u^+) + I_b^\lambda(\beta u^-) \leq I_b^\lambda(\alpha u^+ + \beta u^-) = \phi(\alpha, \beta).$$

Hence, there exists $\beta_0 \in [0, \hat{\beta}]$ such that

$$\phi(\alpha, 0) \leq \phi(\alpha, \beta_0) \quad \text{for all } \alpha \in [0, \hat{\alpha}].$$

This means that any point $(\alpha, 0)$ with $0 \leq \alpha \leq \hat{\alpha}$ is not a maximizer of ϕ. Hence, $(\alpha_u, \beta_u) \notin [0, \hat{\alpha}] \times \{0\}$. Similarly, we obtain that $(\alpha_u, \beta_u) \notin \{0\} \times [0, \hat{\alpha}]$.

On the other hand, by straightforward computation,

$$\frac{\alpha^p}{p} A_1 + \frac{b\alpha^{2p}}{p} A_1 \|u^+\|^p + \frac{b\alpha^{2p}}{2p} A_1^2 - \frac{\alpha^{p_s^*}}{p_s^*} B_1 > 0 \tag{1.39}$$

and

$$\frac{\beta^p}{p} A_2 + \frac{b\beta^{2p}}{p} A_2 \|u^-\|^p + \frac{b\beta^{2p}}{2p} A_2^2 - \frac{\beta^{p_s^*}}{p_s^*} B_2 > 0 \tag{1.40}$$

for $\alpha \in (0, \hat{\alpha}]$, $\beta \in (0, \hat{\beta}]$.

Then, we have that

$$\begin{aligned}\frac{s}{N} S^{N/ps} \leq & \frac{\hat{\alpha}^p}{p} A_1 + \frac{b\hat{\alpha}^{2p}}{2p} A_1^2 - \frac{\hat{\alpha}^{p_s^*}}{p_s^*} B_1 + \frac{b\hat{\alpha}^{2p}}{p} A_1 \|u^+\|^p \\ & + \frac{\beta^p}{p} A_2 + \frac{b\beta^{2p}}{p} A_2 \|u^-\|^p + \frac{b\beta^{2p}}{2p} A_2^2 - \frac{\beta^{p_s^*}}{p_s^*} B_2\end{aligned}$$

and

$$\begin{aligned}\frac{s}{N} S^{N/ps} \leq & \frac{\tilde{\beta}^p}{p} A_2 + \frac{b\tilde{\beta}^{2p}}{2p} A_2^p - \frac{\tilde{\beta}^{p_s^*}}{p_s^*} B_2 + \frac{b\tilde{\beta}^{2p}}{p} A_2 \|u^-\|^p \\ & + \frac{\alpha^p}{p} A_1 + \frac{b\alpha^{2p}}{p} A_1 \|u^+\|^p + \frac{b\alpha^{2p}}{2p} A_1^2 - \frac{\alpha^{p_s^*}}{p_s^*} B_1,\end{aligned}$$

for all $\alpha \in [0, \hat{\alpha}]$ and all $\beta \in [0, \hat{\beta}]$.

Therefore, according to (1.36), we conclude that for all $\alpha \in [0, \hat{\alpha}]$ and all $\beta \in [0, \hat{\beta}]$

$$\psi(\alpha, \hat{\beta}) \leq 0, \quad \psi(\hat{\alpha}, \beta) \leq 0.$$

Thus, $(\alpha_u, \beta_u) \notin \{\hat{\alpha}\} \times [0, \hat{\beta}]$ and $(\alpha_u, \beta_u) \notin [0, \hat{\alpha}] \times \{\hat{\beta}\}$.

Finally, we get that $(\alpha_u, \beta_u) \in (0, \hat{\alpha}) \times (0, \hat{\beta})$. We conclude that (α_u, β_u) is a critical point of ψ_u.

Hence, $\alpha_u u^+ + \beta_u u^- \in \mathcal{M}_b^\lambda$. From (1.32), (1.35), and (1.36), we have that

$$\begin{aligned} c_b^\lambda \geq & I_b^\lambda(\alpha_u u^+ + \beta_u u^-) + \frac{\alpha_u^p}{p} A_1 + \frac{b\alpha_u^{2p}}{p} A_1 \|u^+\|^p + \frac{b\alpha_u^{2p}}{2p} A_1^2 - \frac{\alpha^{p_s^*}}{p_s^*} B_1 \\ & + \frac{\beta_u^p}{p} A_2 + \frac{b\beta_u^{2p}}{p} A_2 \|u^-\|^p + \frac{b\beta_u^{2p}}{2p} A_2^p - \frac{\beta^{p_s^*}}{p_s^*} B_2 \\ & > I_b^\lambda(\alpha_u u^+ + \beta_u u^-) \geq c_b^\lambda,\end{aligned}$$

which is a contradiction.

Case 2: $B_2 = 0$.

In this case, we can maximize on $[0, \hat{\alpha}] \times [0, \infty)$. Indeed, it is possible to show that there exists $\beta_0 \in [0, \infty)$ such that

$$I_b^\lambda(\alpha_u u^+ + \beta_u u^-) \leq 0 \quad \text{for all } (\alpha, \beta) \in [0, \hat{\alpha}] \times [\beta_0, \infty).$$

Hence, there exists $(\alpha_u, \beta_u) \in [0, \hat{\alpha}] \times [0, \infty)$ such that

$$\phi(\alpha_u, \beta_u) = \max_{(\alpha,\beta)\in[0,\hat{\alpha}]\times[0,\infty)} \phi(\alpha, \beta).$$

In the following, we prove that $(\alpha_u, \beta_u) \in (0, \hat{\alpha}) \times (0, \infty)$.

Notice that $\phi(\alpha, 0) < \phi(\alpha, \beta)$ for $\alpha \in [0, \hat{\alpha}]$ and β small enough, so $(\alpha_u, \beta_u) \notin [0, \hat{\alpha}] \times \{0\}$.

Meanwhile, $\phi(0, \beta) < \phi(\alpha, \beta)$ for $\beta \in [0, \infty)$ and α small enough, then we have $(\alpha_u, \beta_u) \notin \{0\} \times [0, \infty)$.

On the other hand, we have the following estimate

$$\frac{s}{N} S^{N/ps} \leq \frac{\hat{\alpha}^p}{p} A_1 + \frac{b\hat{\alpha}^{2p}}{2p} A_1^2 - \frac{\alpha^{p_s^*}}{p_s^*} B_1 + \frac{b\hat{\alpha}^{2p}}{p} A_2 \|u^+\|^p + \frac{\beta^p}{p} A_2 + \frac{b\beta^{2p}}{p} A_2 \|u^-\|^p + \frac{b\beta^{2p}}{2p} A_2^2$$

for all $\beta \in [0, \infty)$.

Hence, we have that $\phi(\hat{\alpha}, \beta) \leq 0$ for all $\beta \in [0, \infty)$. Thus, $(\alpha_u, \beta_u) \notin \{\hat{\alpha}\} \times [0, \infty)$. Thus, $(\alpha_u, \beta_u) \in (0, \hat{\alpha}) \times (0, \infty)$. We obtain that (α_u, β_u) is an interior maximum point of ϕ, hence $\alpha_u u^+ + \beta_u u^- \in \mathcal{M}_b^\lambda$.

Using relation (1.39), we obtain

$$\begin{aligned} c_b^\lambda &\geq I_b^\lambda(\alpha_u u^+ + \beta_u u^-) + \frac{\alpha_u^p}{p} A_1 + \frac{b\alpha_u^{2p}}{p} A_1 \|u^+\|^p + \frac{b\alpha_u^{2p}}{2p} A_1^2 - \frac{\alpha^{p_s^*}}{p_s^*} B_1 \\ &+ \frac{\beta_u^p}{p} A_2 + \frac{b\beta_u^{2p}}{p} A_2 \|u^-\|^p + \frac{b\beta_u^{2p}}{2p} A_2^2 > I_b^\lambda(\alpha_u u^+ + \beta_u u^-) \geq c_b^\lambda, \end{aligned}$$

which is a contradiction.

From the above arguments, we conclude that $B_1 = B_2 = 0$.

Final step: c_b^λ *is achieved.*

Since $u^\pm \neq 0$, by Lemma 1.1, there exist $\alpha_u, \beta_u > 0$ such that

$$\bar{u} := \alpha_u u^+ + \beta_u u^- \in \mathcal{M}_b^\lambda.$$

Furthermore, we observe that

$$\langle (I_b^\lambda)'(u), u^\pm \rangle \leq 0.$$

By Lemma 1.2, we obtain $0 < \alpha_u, \beta_u \leq 1$.

Since $u_n \in \mathcal{M}_b^\lambda$, according to Lemma 1.3, we get

$$I_b^\lambda(\alpha_u u_n^+ + \beta_u u_n^-) \leq I_b^\lambda(u_n^+ + u_n^-) = I_b^\lambda(u_n).$$

Thanks to $B_1 = B_2 = 0$ and by the lower semicontinuity of the norm in $W_0^{s,p}(\Omega)$, we obtain

$$\begin{aligned} c_b^\lambda &\leq I_b^\lambda(\bar{u}) - \frac{1}{q} \langle (I_b^\lambda)'(\bar{u}), \bar{u} \rangle \\ &\leq \left(\frac{1}{p} - \frac{1}{q}\right) a\|\bar{u}\|^p + \left(\frac{1}{2p} - \frac{1}{q}\right) b\|\bar{u}\|^{2p} + \frac{2\lambda}{q^2} \int_\Omega |\bar{u}|^q dx + \left(\frac{1}{q} - \frac{1}{p_s^*}\right) \int_\Omega |\bar{u}|^{p_s^*} dx \\ &= \left(\frac{1}{p} - \frac{1}{q}\right) a(\|\alpha_u u^+\|^p + \|\beta_u u^-\|^p) + \left(\frac{1}{2p} - \frac{1}{q}\right) b(\|\alpha_u u^+\|^p + \|\beta_u u^-\|^p)^2 \\ &+ \frac{2\lambda}{q^2} \left[\int_\Omega |\alpha_u u^+|^q dx + \int_\Omega |\beta_u u^-|^q dx\right] + \left(\frac{1}{q} - \frac{1}{p_s^*}\right) \left[\int_\Omega |\alpha_u u^+|^{p_s^*} dx + \int_\Omega |\beta_u u^-|^{p_s^*} dx\right]. \end{aligned}$$

It follows that

$$c_b^\lambda \le \left(\frac{1}{p}-\frac{1}{q}\right)a\|u\|^p + \left(\frac{1}{2p}-\frac{1}{q}\right)b\|u\|^{2p} + \frac{2\lambda}{q^2}\int_\Omega |u|^q dx + \left(\frac{1}{q}-\frac{1}{p_s^*}\right)\int_\Omega |u|^{p_s^*}dx$$
$$\le \liminf_{n\to\infty}\left[I_b^\lambda(u_n) - \frac{1}{q}\langle (I_b^\lambda)'(u_n), u_n\rangle\right] \le c_b^\lambda.$$

We conclude that $\alpha_u = \beta_u = 1$, and c_b^λ is achieved by $u_b := u^+ + u^- \in \mathcal{M}_b^\lambda$. □

1.3 Least-energy nodal solution: qualitative properties

In this section, we prove our main results. First, we prove Theorem 1.1. In fact, thanks to Lemma 1.4, it remains to establish that the minimizer u_b for c_b^λ is indeed a sign-changing solution of problem (1.1).

1.3.1 Proof of Theorem 1.1: existence

Since $u_b \in \mathcal{M}_b^\lambda$, we have $\langle (I_b^\lambda)'(u_b), u_b^+\rangle = \langle (I_b^\lambda)'(u_b), u_b^-\rangle = 0$. By Lemma 1.4, for all $(\alpha,\beta) \in (\mathbb{R}_+ \times \mathbb{R}_+)\backslash(1,1)$, we have

$$I_b^\lambda(\alpha u_b^+ + \beta u_b^-) < I_b^\lambda(u_b^+ + u_b^-) = c_b^\lambda. \tag{1.41}$$

Arguing by contradiction, we assume that $(I_b^\lambda)'(u_b) \neq 0$. Then there exist $\delta > 0$ and $\iota > 0$ such that

$$\|(I_b^\lambda)'(v)\| \ge \iota \quad \text{for all } \|v - u_b\| \ge 3\delta.$$

Choose $\tau \in (0, \min\{1/2, \frac{\delta}{\sqrt{2}\|u_b\|}\})$. Let

$$D := (1-\tau, 1+\tau) \times (1-\tau, 1+\tau)$$

and define

$$g(\alpha,\beta) = \alpha u_b^+ + \beta u_b^- \quad \text{for all } (\alpha,\beta) \in D.$$

In view of (1.41), we have

$$\bar{c}_\lambda := \max_{\partial\Omega} I_b^\lambda \circ g < c_{b,\lambda}. \tag{1.42}$$

Let $\varepsilon := \min\{(c_b^\lambda - \bar{c}_\lambda)/3, \iota\delta/8\}$ and $S_\delta := B(u_b, \delta)$, according to Lemma 2.3 in [153], there exists a deformation $\eta \in C([0,1] \times D, D)$ such that

(a) $\eta(1,v) = v$ if $v \notin (I_b^\lambda)^{-1}([c_b^\lambda - 2\varepsilon, c_b^\lambda + 2\varepsilon] \cap S_{2\delta})$,

(b) $\eta(1, (I_b^\lambda)^{c_b^\lambda+\varepsilon} \cap S_\delta) \subset (I_b^\lambda)^{c_{b,\lambda}-\varepsilon}$,

(c) $I_b^\lambda(\eta(1,v)) \le I_b^\lambda)(v)$ for all $v \in W_0^{s,p}(\Omega)$.

First, from (b) and Lemma 1.2, we obtain

$$\max_{(\alpha,\beta)\in\bar{D}} I_b^\lambda(\eta(1, g(\alpha,\beta))) < c_b^\lambda. \tag{1.43}$$

Next, we prove that $\eta(1, g(D)) \cap \mathcal{M}_b^\lambda \neq \emptyset$, which contradicts the definition of c_b^λ.

Let $\gamma(\alpha,\beta) := \eta(1, g(\alpha,\beta))$. Define the mappings

$$\begin{aligned}\Psi_0(\alpha,\beta) : &= (\langle (I_b^\lambda)'(g(\alpha,\beta)), u_b^+\rangle, \langle (I_b^\lambda)'(g(\alpha,\beta)), u_b^-\rangle)\\ &= (\langle (I_b^\lambda)'(\alpha u_b^+ + \beta u_b^-), u_b^+\rangle, \langle (I_b^\lambda)'(\alpha u_b^+ + \beta u_b^-), u_b^-\rangle)\\ &=: (\varphi_u^1(\alpha,\beta), \varphi_u^2(\alpha,\beta))\end{aligned}$$

and

$$\Psi_1(\alpha,\beta) := \left(\frac{1}{\alpha}\langle (I_b^\lambda)'(\gamma(\alpha,\beta)), (\gamma(\alpha,\beta))^+\rangle, \frac{1}{\beta}\langle (I_b^\lambda)'(\gamma(\alpha,\beta)), (\gamma(\alpha,\beta))^-\rangle\right).$$

Since $u_b \in \mathcal{M}_b^\lambda$, we have

$$\begin{aligned}\frac{\varphi_u^1(\alpha,\beta)}{\partial\alpha}\Big|_{(1,1)} &= a(p-1)\|u_b^+\|^p + b(2p-1)\|u_b^+\|^{2p} + b(p-1)\|u_b^+\|^p\|u_b^-\|^p\\ &\quad - \lambda(q-1)\int_\Omega |u_b^+|^q \ln|u_b^+|^2 dx - 2\int_\Omega |u_b^+|^q dx - (p_s^*-1)\int_\Omega |u_b^+|^{p_s^*} dx\\ &= bp\|u_b^+\|^{2p} - \lambda(q-p)\int_\Omega |u_b^+|^q \ln|u_b^+|^2 dx - 2\int_\Omega |u_b^+|^q dx - (p_s^*-p)\int_\Omega |u_b^+|^{p_s^*} dx\end{aligned}$$

and

$$\frac{\varphi_u^1(\alpha,\beta)}{\partial\beta}\Big|_{(1,1)} = pb\|u_b^+\|^p\|u_b^-\|^p.$$

Similarly, we have

$$\frac{\varphi_u^2(\alpha,\beta)}{\partial\beta}\Big|_{(1,1)} = bp\|u_b^-\|^{2p} - \lambda(q-p)\int_\Omega |u_b^-|^q \ln|u_b^-|^2 dx - 2\int_\Omega |u_b^-|^q dx - (p_s^*-p)\int_\Omega |u_b^-|^{p_s^*} dx$$

and

$$\frac{\varphi_u^2(\alpha,\beta)}{\partial\alpha}\Big|_{(1,1)} = pb\|u_b^+\|^p\|u_b^-\|^p.$$

Let

$$M = \begin{bmatrix} \frac{\varphi_u^1(\alpha,\beta)}{\partial\alpha}\big|_{(1,1)} & \frac{\varphi_u^2(\alpha,\beta)}{\partial\alpha}\big|_{(1,1)} \\ \frac{\varphi_u^1(\alpha,\beta)}{\partial\beta}\big|_{(1,1)} & \frac{\varphi_u^2(\alpha,\beta)}{\partial\beta}\big|_{(1,1)} \end{bmatrix}.$$

It follows that

$$\det M = \frac{\varphi_u^1(\alpha,\beta)}{\partial\alpha}\Big|_{(1,1)} \times \frac{\varphi_u^2(\alpha,\beta)}{\partial\beta}\Big|_{(1,1)} - \frac{\varphi_u^1(\alpha,\beta)}{\partial\beta}\Big|_{(1,1)} \times \frac{\varphi_u^2(\alpha,\beta)}{\partial\alpha}\Big|_{(1,1)} \neq 0.$$

Since $\Psi_0(\alpha,\beta)$ is a C^1-function and $(1,1)$ is the unique isolated zero point of Ψ_0, by using the degree theory, we deduce that $\deg(\Psi_0, D, 0) = 1$.

Hence, combining (1.43) with (a), we obtain

$$g(\alpha,\beta) = \gamma(\alpha,\beta) \quad \text{on } \partial D.$$

Consequently, we obtain $\deg(\Psi_1, D, 0) = 1$. Therefore, $\Psi_1(\alpha_0,\beta_0) = 0$ for some $(\alpha_0,\beta_0) \in D$ so that

$$\eta(1, g(\alpha_0,\beta_0)) = \gamma(\alpha_0,\beta_0) \in \mathcal{M}_b^\lambda,$$

which contradicts relation (1.43).

From the above discussions, we deduce that u_b is a nodal solution of problem (1.1).

Finally, we prove that u has exactly two nodal domains. To this end, we assume by contradiction that

$$u_b = u_1 + u_2 + u_3,$$

where

$$u_i \neq 0,\ u_1 \geq 0,\ u_2 \leq 0,\ \Omega_1 \cap \Omega_2 = \emptyset,\ u_1|_{\Omega\setminus\Omega_1\cup\Omega_2} = u_2|_{\Omega\setminus\Omega_1\cup\Omega_2} = u_3|_{\Omega_1\cap\Omega_2} = 0,$$

$$\Omega_1 := \{x \in \Omega : u_1(x) > 0\} \quad \text{and} \quad \Omega_2 := \{x \in \Omega : u_2(x) < 0\} \quad \text{for } i \neq j,\ i,j = 1,2,3$$

are two connected open subsets of Ω, and

$$\langle (I_b^\lambda)'(u), u_i \rangle = 0 \quad \text{for } i = 1,2,3.$$

Setting $v := u_1 + u_2$, we see that $v^+ = u_1$ and $v^- = u_2$, that is, $v^\pm \neq 0$. Then, there exists a unique pair (α_v, β_v) of positive numbers such that

$$\alpha_v u_1 + \beta_v u_2 \in \mathcal{M}_b^\lambda.$$

Hence

$$I_b^\lambda(\alpha_v u_1 + \beta_v u_2) \geq c_b^\lambda.$$

Moreover, using the fact that $\langle (I_b^\lambda)'(u), u_i \rangle = 0$, we obtain

$$\langle (I_b^\lambda)'(v), v^\pm \rangle = -b\|v^\pm\|^p \|u_3\|^p < 0.$$

From Lemma 1.1 (ii), it follows that

$$(\alpha_v, \beta_v) \in (0,1] \times (0,1].$$

On the other hand, we have

$$\begin{aligned}
0 = \frac{1}{2p}\langle (I_b^\lambda)'(u), u_3 \rangle &= \frac{1}{2p} a\|u_3\|^p + \frac{b}{2p}\|u_1\|^p\|u_3\|^p + \frac{b}{2p}\|u_2\|^p\|u_3\|^p + \frac{b}{2p}\|u_3\|^{2p} \\
&\quad - \frac{1}{2p}\int_\Omega |u_3|^{p_s^*} dx - \frac{\lambda}{2p}\int_\Omega |u_3^+|^q \ln |u_3^+|^2 dx \\
&< I_b^\lambda(u_3) + \frac{b}{2p}\|u_1\|^2\|u_3\|^p + \frac{b}{2p}\|u_2\|^2\|u_3\|^p.
\end{aligned}$$

Hence, by (1.30), we obtain

$$\begin{aligned}
c_b^\lambda &\leq I_b^\lambda(\alpha_v u_1 + \beta_v u_2) = I_b^\lambda(\alpha_v u_1 + \beta_v u_2) - \frac{1}{2p}\langle (I_b^\lambda)'(\alpha_v u_1 + \beta_v u_2), (\alpha_v u_1 + \beta_v u_2)\rangle \\
&= \frac{1}{2p}(\|\alpha_v u_1\|^p + \|\beta_v u_2\|^p) + \frac{\lambda}{q^2}\left[\int_\Omega |\alpha_v u_1|^q dx + \int_\Omega |\beta_v u_2|^q dx\right] \\
&\quad + \left(\frac{1}{2p} - \frac{1}{q}\right)\lambda\left[\int_\Omega [|\alpha_v u_1|^q \ln |\alpha_v u_1|^2 dx + \int_\Omega |\beta_v u_2|^q \ln |\beta_v u_2|^2 dx\right] \\
&\quad + \left(\frac{1}{2p} - \frac{1}{p_s^*}\right)\int_\Omega \alpha_v^{p_s^*}|u_1|^{p_s^*} dx + \left(\frac{1}{2p} - \frac{1}{p_s^*}\right)\int_\Omega \beta_v^{p_s^*}|u_2|^{p_s^*} dx.
\end{aligned}$$

It follows that

$$
\begin{aligned}
c_b^\lambda &\leq \frac{1}{2p}(\|u_1\|^p + \|u_2\|^p) + \frac{\lambda}{q^2}\left[\int_\Omega |u_1|^q dx + \int_\Omega |u_2|^q dx\right] \\
&+ \left(\frac{1}{2p} - \frac{1}{q}\right)\lambda\left[\int_\Omega |u_1|^q \ln |u_1|^2 dx + \int_\Omega |u_2|^q \ln |u_2|^2 dx\right] \\
&+ \left(\frac{1}{2p} - \frac{1}{p_s^*}\right)\int_\Omega |u_1|^{p_s^*} dx + \left(\frac{1}{2p} - \frac{1}{p_s^*}\right)\int_\Omega |u_2|^{p_s^*} dx \\
&= I_b^\lambda(u_1 + u_2) - \frac{1}{2p}\langle (I_b^\lambda)'(u_1 + u_2), (u_1 + u_2)\rangle \\
&= I_b^\lambda(u_1 + u_2) + \frac{1}{2p}\langle (I_b^\lambda)'(u), u_3\rangle + \frac{b}{2p}\|u_1\|^p\|u_3\|^p + \frac{b}{2p}\|u_2\|^p\|u_3\|^p.
\end{aligned}
$$

This relation implies that

$$
\begin{aligned}
c_b^\lambda &< I_b^\lambda(u_1) + I_b^\lambda(u_2) + I_b^\lambda(u_3) + \frac{b}{2p}(\|u_2\|^p + \|u_3\|^p)\|u_1\|^p \\
&+ \frac{b}{2p}(\|u_1\|^p + \|u_3\|^p)\|u_2\|^p + \frac{b}{2p}(\|u_1\|^p + \|u_2\|^p)\|u_3\|^p = I_b^\lambda(u) = c_b^\lambda.
\end{aligned}
$$

This contradiction implies that $u_3 = 0$ and u_b has exactly two nodal domains. □

By Theorem 1.1, we obtain a least-energy sign-changing solution u_b of problem (1.1). Next, we prove that the energy of u_b is strictly greater than twice the ground state energy.

1.3.2 Proof of Theorem 1.2: energy doubling

Similar to the proof of Lemma 1.3, there exists $\lambda_1^* > 0$ such that for all $\lambda \geq \lambda_1^*$, and for each $b > 0$, there exists $v_b \in \mathcal{N}_b^\lambda$ satisfying $I_b^\lambda(v_b) = c^* > 0$. By Nehari manifold theory, the critical points of the functional I_b^λ on $\mathcal{N}_b^\lambda$ are critical points of I_b^λ in $W_0^{s,p}(\Omega)$. Thus, $(I_b^\lambda)'(v_b) = 0$, hence v_b is a ground state solution of problem (1.1).

According to Theorem 1.1, we know that the problem (1.1) has a least-energy sign-changing solution u_b, which changes sign only once when $\lambda \geq \lambda^*$.

Let

$$\lambda^{**} = \max\{\lambda^*, \lambda_1^*\}.$$

Suppose that $u_b = u_b^+ + u_b^-$. As in the proof of Lemma 1.1, there exist $\alpha_{u_b^+} > 0$ and $\beta_{u_b^-} > 0$ such that

$$\alpha_{u_b^+} u_b^+ \in \mathcal{N}_b^\lambda, \quad \beta_{u_b^-} u_b^- \in \mathcal{N}_b^\lambda.$$

Furthermore, Lemma 1.2 implies that $\alpha_{u_b^+}, \beta_{u_b^-} \in (0,1)$.

Therefore, in view of Lemma 1.1, we have that

$$2c^* \leq I_b^\lambda(\alpha_{u_b^+} u_b^+) + I_b^\lambda(\beta_{u_b^-} u_b^-) \leq I_b^\lambda(\alpha_{u_b^+} u_b^+ + \beta_{u_b^-} u_b^-) < I_b^\lambda(u_b^+ + u_b^-) = c_b^\lambda.$$

We conclude that $c^* > 0$ cannot be achieved by a sign-changing function. □

1.3.3 Proof of Theorem 1.3: convergence to the limit case

In what follows, we regard $b > 0$ as a parameter in problem (1.1).

We proceed through several steps by analyzing the convergence property of u_b as $b \to 0$, where u_b is the least-energy sign-changing solution obtained in Theorem 1.1.

Step 1. For any sequence $\{b_n\}$ as $b_n \searrow 0$, $\{u_{b_n}\}$ is bounded in $W_0^{s,p}(\Omega)$.

Choose a nonzero function $\omega \in C_0^\infty(\Omega)$ with $\omega^\pm \neq 0$. As in the proof of Lemma 1.2, for any $\lambda \in [0,1]$, there exists a pair positive numbers (λ_1, λ_2) independent of λ, such that

$$\langle (I_b^\lambda)'(\lambda_1\omega^+ + \lambda_2\omega^-), \lambda_1\omega^+\rangle < 0, \quad \langle (I_b^\lambda)'(\lambda_1\omega^+ + \lambda_2\omega^-), \lambda_2\omega^-\rangle < 0.$$

Then by virtue of Lemma 1.1, we obtain that for any $b \in [0,1]$, there exists a unique pair $(\alpha_\omega(b), \beta_\omega(b)) \in (0,1] \times (0,1]$ such that

$$\overline{\omega} := \alpha_\omega(b)\lambda_1\omega^+ + \beta_\omega(b)\lambda_2\omega^- \in \mathcal{M}_b^\lambda. \tag{1.44}$$

Thus, for any $\lambda \in [0,1]$, we have

$$\begin{aligned} I_b^\lambda(u_b) &= I_b^\lambda(\overline{\omega}) = I_b^\lambda(\overline{\omega}) - \frac{1}{2p}\langle (I_b^\lambda)'(\overline{\omega}), \overline{\omega}\rangle \\ &= \frac{1}{2p}\|\overline{\omega}\|^p + \left(\frac{1}{2p} - \frac{1}{p_s^*}\right)\int_\Omega |\overline{\omega}|^{p_s^*}dx + \frac{\lambda}{q^2}\int_\Omega |\overline{\omega}|^q dx + \left(\frac{1}{2p} - \frac{1}{q}\right)\lambda\int_\Omega |\overline{\omega}|^q \ln|\overline{\omega}|^2 dx. \end{aligned}$$

It follows that

$$\begin{aligned} I_b^\lambda(u_b) \leq & \frac{1}{2p}\|\overline{\omega}\|^p + \left(\frac{1}{2p} - \frac{1}{p_s^*}\right)\int_\Omega |\overline{\omega}|^{p_s^*}dx + \frac{\lambda}{q^2}\int_\Omega |\overline{\omega}|^q dx \\ & + \left(\frac{1}{2p} - \frac{1}{q}\right)\lambda\int_\Omega \left(C_1|\overline{\omega}|^p + C_2|\overline{\omega}|^r\right)dx. \end{aligned}$$

We conclude that

$$\begin{aligned} I_b^\lambda(u_b) \leq & \frac{1}{2p}\|\lambda_1\omega^+\|^p + \frac{1}{2p}\|\lambda_2\omega^-\|^p + \left(\frac{1}{2p} - \frac{1}{p_s^*}\right)\int_\Omega |\lambda_1\omega^+|^{p_s^*}dx + \left(\frac{1}{2p} - \frac{1}{p_s^*}\right)\int_\Omega |\lambda_2\omega^-|^{p_s^*}dx \\ & + \frac{\lambda}{q^2}\int_\Omega |\lambda_1\omega^+|^q dx + \frac{\lambda}{q^2}\int_\Omega |\lambda_2\omega^-|^q dx + \left(\frac{1}{2p} - \frac{1}{q}\right)\lambda\int_\Omega \left(C_1\lambda_1^2|\omega^+|^2 + C_2\lambda_2^r|\omega^+|^r\right)dx \\ & + \left(\frac{1}{2p} - \frac{1}{q}\right)\lambda\int_\Omega \left(C_1\lambda_1^2|\omega^-|^2 + C_2\lambda_2^r|\omega^-|^r\right)dx =: C^*, \end{aligned}$$

where $C^* > 0$ is a constant independent of λ. So, as $n \to \infty$, we obtain

$$C^* + 1 \geq I_{b_n}^\lambda(u_{b_n}) = I_{b_n}^\lambda(u_{b_n}) - \frac{1}{2p}\langle (I_{b_n}^\lambda)'(u_{b_n}), u_{b_n}\rangle \geq \frac{1}{2p}\|u_{b_n}\|^2,$$

which implies that $\{u_{b_n}\}$ is bounded in $W_0^{s,p}(\Omega)$.

Step 2. Problem (1.15) has a nodal solution u_0.

Since $\{u_{b_n}\}$ is bounded in $W_0^{s,p}(\Omega)$, according to Step 1, going if necessary to a subsequence, there exists $u_0 \in W_0^{s,p}(\Omega)$ such that

$$\begin{aligned} & u_{b_n} \rightharpoonup u_0 \ \text{ in } W_0^{s,p}(\Omega), \\ & u_{b_n} \to u_0 \ \text{ in } L^t(\Omega) \text{ for } t \in [p, p_s^*), \\ & u_{b_n} \to u_0 \ \text{ a.e. } x \in \Omega. \end{aligned} \tag{1.45}$$

Since u_{b_n} is a weak solution of problem (1.1) for $b = b_n$, we have for all $v \in C_0^\infty(\Omega)$

$$\left(a + b_n[u_{b_n}]_{s,p}^p\right) L(u_{b_n}, v) = \lambda\int_\Omega |u_{b_n}|^{q-2}u_{b_n}v\ln|u_{b_n}|^2 dx + \int_\Omega |u_{b_n}|^{p_s^*-2}u_{b_n}v\,dx, \tag{1.46}$$

where $L(u,v)$ is defined by relation (1.5).

From (1.45), (1.46) and Step 1, we find that

$$\left(a+b_n[u_0]_{s,p}^p\right) L(u_0,v)=\lambda\int_\Omega |u|^{q-2}u_0 v\ln|u_0|^2dx+\int_\Omega |u_0|^{p_s^*-2}u_0 v\,dx \tag{1.47}$$

for all $v\in C_0^\infty(\Omega)$, which in turn implies that u_0 is a weak solution of problem (1.15). By a similar argument as in the proof of Lemma 1.3, we conclude that $u_0^\pm\neq 0$. Therefore, we complete the proof of Step 2.

Step 3. Problem (1.15) possesses a least-energy sign-changing solution v_0, and there exists a unique pair $(\alpha_{b_n},\beta_{b_n})\in[0,\infty)\times[0,\infty)$ such that $\alpha_{b_n}v_0^+ +\beta_{b_n}v_0^-\in\mathcal{M}_{b_n}^\lambda$. Moreover, $(\alpha_{b_n},\beta_{b_n})\to(1,1)$ as $n\to\infty$.

By a similar argument as in the proof of Theorem 1.1, we deduce that problem (1.15) has a least-energy sign-changing solution v_0, where $I_0^\lambda(v_0)=c_0$ and $(I_0^\lambda)'(v_0)=0$. Then, by Lemma 1.1, we obtain the existence and uniqueness of the pair $(\alpha_{b_n},\beta_{b_n})$ such that $\alpha_{b_n}v_0^+ +\beta_{b_n}v_0^-\in\mathcal{M}_{b_n}^\lambda$. Moreover, we have $\alpha_{b_n}>0$ and $\beta_{b_n}>0$. Then the claim will follow once we can prove that $(\alpha_{b_n},\beta_{b_n})\to(1,1)$ as $n\to\infty$. In fact, since $\alpha_{b_n}v_0^+ +\beta_{b_n}v_0^-\in\mathcal{M}_{b_n}^\lambda$, we have that

$$\begin{aligned}\alpha_{b_n}^p a\|v_0^+\|^p+b_n\alpha_{b_n}^{2p}\|v_0^+\|^{2p}+b_n\alpha_{b_n}^p\beta_{b_n}^p\|v_0^+\|^p\|v_0^-\|^p&=\lambda\int_\Omega|\alpha_{b_n}v_0^+|^q\ln|\alpha_{b_n}v_0^+|^2dx\\&\quad+\int_\Omega|\alpha_{b_n}v_0^+|^{p_s^*}dx\end{aligned}\tag{1.48}$$

and

$$\begin{aligned}\beta_{b_n}^p a\|v_0^-\|^p+b_n\beta_{b_n}^{2p}\|v_0^-\|^{2p}+b_n\alpha_{b_n}^p\beta_{b_n}^p\|v_0^+\|^p\|v_0^-\|^p&=\lambda\int_\Omega|\beta_{b_n}v_0^+|^q\ln|\beta_{b_n}v_0^+|^2dx\\&\quad+\int_\Omega|\beta_{b_n}v_0^-|^{p_s^*}dx.\end{aligned}\tag{1.49}$$

From the convergence of b_n as $n\to\infty$, we deduce that the sequences $\{\alpha_{b_n}\}$ and $\{\beta_{b_n}\}$ are bounded. Up to a subsequence, suppose that $\alpha_{b_n}\to\alpha_0$ and $\beta_{b_n}\to\beta_0$. Then it follows from (1.48) and (1.49) that

$$\alpha_0^p a\|v_0^+\|^p=\lambda\int_\Omega|\alpha_0v_0^+|^q\ln|\alpha_0v_0^+|^2dx+\int_\Omega|\alpha_0v_0^+|^{p_s^*}dx \tag{1.50}$$

and

$$\beta_0^p a\|v_0^-\|^p=\lambda\int_\Omega|\beta_0v_0^-|^q\ln|\beta_0v_0^-|^2dx+\int_\Omega|\beta_0v_0^-|^{p_s^*}dx. \tag{1.51}$$

Since v_0 is a sign-changing solution of problem (1.15), we obtain

$$a\|v_0^\pm\|^2=\lambda\int_\Omega|v_0^\pm|^q\ln|v_0^\pm|^2dx+\int_\Omega|v_0^\pm|^{p_s^*}dx. \tag{1.52}$$

Hence, in view of (1.50)–(1.52), we obtain that $(\alpha_0,\beta_0)=(1,1)$, and the Step 3 follows.

Now, we can now give the proof of Theorem 1.3. We assert that u_0 obtained in Step 2 is a least-energy solution of problem (1.15). In fact, by virtue of Step 3 and Lemma 1.1, we find that

$$\begin{aligned}I_0^\lambda(v_0)\le I_0^\lambda(u_0)=\lim_{n\to\infty}I_{b_n}^\lambda(u_{b_n})&\le\lim_{n\to\infty}I_{b_n}^\lambda(\alpha_{b_n}v_0^+ +\beta_{b_n}v_0^-)\\&=I_0^\lambda(v_0^+ +v_0^-)=I_0^\lambda(v_0).\end{aligned}$$

The proof of Theorem 1.3 is now complete. □

1.4 Historical comments

One of the classical topics in the qualitative analysis of PDEs is the study of existence and multiplicity properties of solutions for both the Kirchhoff problems and the fractional Kirchhoff problems under various hypotheses on the nonlinearity. There is a vast recent literature concerning the existence and multiplicity of solutions for the following Dirichlet problem of Kirchhoff-type

$$\begin{cases} -\left(a+b\int_{\Omega}|\nabla u|^2dx\right)\Delta u = f(x,u) \text{ in } \Omega, \\ u|_{\partial\Omega}=0. \end{cases} \tag{1.53}$$

Problem (1.53) is a generalization of a model introduced by Kirchhoff [88]. More precisely, Kirchhoff proposed a model given by the equation

$$\rho\frac{\partial^2 u}{\partial t^2}-\left(\frac{\rho_0}{h}+\frac{E}{2L}\int_0^L\left|\frac{\partial u}{\partial x}\right|^2dx\right)\frac{\partial^2 u}{\partial x^2}=0, \tag{1.54}$$

where ρ, ρ_0, h, E, L are constants. This nonlocal model extends the classical D'Alembert's wave equation, by considering the effects of changes in the length of the strings during the vibrations.

Since Lions [99] introduced an abstract framework to Kirchhoff-type equations, the solvability of these nonlocal problems has been well studied in the general dimension by various authors. We refer to D'Ancona and Shibata [39] and D'Ancona and Spagnolo [38] for the global solvability of various classes of Kirchhoff-type problems. We also refer to Carrier [21] who used a more rigorous method to model transverse vibration via the coupled governing equation of planar vibration in order to recover the nonlinear integro partial-differential equation, in which a more general Kirchhoff function was considered. In addition, the nonlocal Kirchhoff problems of parabolic type can model several biological systems, such as population density, see for example Ghergu and Rădulescu [62].

Problem (1.53) is a nonlocal problem because the term $b\int_\Omega|\nabla u|^2dx\Delta u$ appears in the left-hand side of the equation, which results that (1.53) is not a pointwise identity. Moreover, the energy functional associated to (1.53) has different properties with respect to the local case corresponding to $b=0$, hence several mathematical difficulties are brought naturally out in the study of the nonlocal problems ($b\neq 0$) by means of variational methods. We refer to Fiscella and Valdinoci [57], who proposed a steady-state Kirchhoff model involving the fractional Laplacian by taking into account the nonlocal aspect of the tension arising from nonlocal measurements of the fractional length of the string, see [57, Appendix A] for more details. For a detailed analysis of nonlocal fractional problems, we refer to the monograph by Molica Bisci, Rădulescu and Servadei [119].

However, there are a few papers dealing with the existence and multiplicity of solutions for fractional problems involving logarithmic nonlinearity. In [40], d'Avenia *et al.* considered the following fractional logarithmic Schorödinger equation

$$(-\Delta)^s u+\omega u = u\log|u|^2 \text{ in } \mathbb{R}^N,$$

where $\omega > 0$. By employing the fractional logarithmic Sobolev inequality, in [40] it is obtained the existence of infinitely many solutions. Moreover, the regularity of solutions was also discussed in [40]. In [149], Truong studied the following problem with logarithmic nonlinearity

$$(-\Delta)^s_p u+V(x)|u|^{p-2}u=\lambda a(x)|u|^{p-2}u\ln|u| \text{ in } \mathbb{R}^N,$$

where a is a sign-changing function. Under some assumptions on V, a and λ, Truong [149] obtained two nontrivial solutions by using the Nehari manifold approach. Next, Xiang, Hu and Yang [154] considered the following Kirchhoff equation with combined nonlinearity of logarithmic and power type

$$\begin{cases} M([u]_{s,p}^p)(-\Delta)_p^s u = h(x)|u|^{\theta p-2}u\ln|u| + \lambda|u|^{q-2}u & \text{in } \Omega, \\ u = 0 & \text{in } \mathbb{R}^N \setminus \Omega, \end{cases}$$

where $s \in (0,1)$, $1 < p < N/s$, Ω is a bounded domain in $\mathbb{R}^N$ with Lipschitz boundary, $M([u]_{s,p}^p) = [u]_{s,p}^{(\theta-1)p}$ with $\theta \geq 1$, and h is a sign-changing function. When λ is sufficiently small, Xiang, Hu and Yang [154] obtained two nonnegative local least-energy solutions by using the Nehari manifold approach.

The existence of sign-changing solutions to nonlinear elliptic PDEs with power-type nonlinearities has been studied extensively both for the p-Laplacian operator and for the fractional p-Laplacian operator. Consider the nonlocal problem

$$\begin{cases} (-\Delta)_p^s u = f(x,u) \text{ in } \Omega, \\ u = 0 \text{ in } \mathbb{R}^N \setminus \Omega, \end{cases} \tag{1.55}$$

For $p = 2$, Chang and Wang [24] studied problem (1.55), where the fractional Laplacian operator is defined through spectral decomposition, in order to obtain the existence of sign-changing solutions.

The method of harmonic extension was introduced by Caffarelli and Silvestre [16] and it consists in transforming the nonlocal equation in Ω to a local problem in the half cylinder $\Omega \times (0,\infty)$, by using an equivalent definition of the fraction Laplacian operator, cf. Brändle *et al.* [13].

In [53], Figueiredo and Nascimento considered the following Kirchhoff equation

$$\begin{cases} -M\left(\displaystyle\int_\Omega |\nabla u|^2 dx\right)\Delta u = g(u) \text{ in } \Omega, \\ u|_{\partial\Omega} = 0, \end{cases} \tag{1.56}$$

where Ω is a bounded domain in $\mathbb{R}^3$, M is a general C^1 class function, and g is a superlinear C^1 class function with subcritical growth. By using the minimization argument and a quantitative deformation lemma, the existence of a sign-changing solution for this Kirchhoff equation was obtained. In unbounded domains, Figueiredo and Santos Júnior [52] studied a class of nonlocal Schrödinger-Kirchhoff problems involving only continuous functions. Using a minimization argument and a quantitative deformation lemma, they obtained a least-energy sign-changing solution to Schrödinger-Kirchhof problems.

Combining constraint variational methods and the quantitative deformation lemma, Shuai [136] proved that problem (1.53) has one least-energy sign-changing solution u_b and the energy of u_b strictly larger than the ground state energy. Moreover, the author investigated the asymptotic behavior of u_b as the parameter $b \searrow 0$. Later, under some more weak assumptions on g (especially, Nehari-type monotonicity condition been removed), with the aid of some new analytical skills and Non-Nehari manifold method, Tang and Cheng [146] improved and generalized some results obtained in [136].

In [45], Deng, Peng, and Shuai studied the following Kirchhoff problem:

$$-\left(a + b\int_{\mathbb{R}^3} |\nabla u|^2 dx\right)\Delta u = f(x,u) \text{ in } \mathbb{R}^3. \tag{1.57}$$

The authors obtained the existence of radial sign-changing solutions with prescribed numbers of nodal domains for Kirchhoff problem (1.57) in $H_r^1(\mathbb{R}^3)$, the subspace of radial functions of $H^1(\mathbb{R}^3)$ by using a Nehari manifold and gluing solution pieces together, when

$V(x) = V(|x|)$, $f(x, u) = f(|x|, u)$ and satisfies some conditions. Precisely, they proved the existence of a sign-changing solution, which changes signs exactly k times for any $k \in \mathbb{N}$. Moreover, they investigated the energy property and the asymptotic behavior of the sign-changing solution.

By using a combination of the invariant set method and the Ljusternik-Schnirelmann type minimax method, Sun et al. [142] obtained infinitely many sign-changing solutions for the Kirchhoff problem (1.57) when $f(x, u) = f(u)$ and f is odd in u. It is worth noticing that, in [142], the nonlinear term may not be 4-superlinear at infinity; in particular, it includes the power-type nonlinearity $|u|^{p-2}u$ with $p \in (2, 4]$. In [150], the authors obtained the existence of least-energy sign-changing solutions of Kirchhoff-type equation with critical growth by using the constraint variational method and the quantitative deformation lemma.

1.5 Final remarks, perspectives, and open problems

The analysis developed in this chapter establishes the existence of least-energy sign-changing solutions for a fractional Kirchhoff problem with logarithmic and critical nonlinearities. Using constraint variational methods, topological degree theory, and quantitative deformation arguments, the main results prove three basic properties:

(i) existence of at least one least-energy sign-changing solution u_b;

(ii) the energy of u_b is strictly larger than twice the ground state energy;

(iii) convergence properties of the solution as the Kirchhoff parameter b goes to zero.

The content of this chapter is based on the abstract and analytic setting developed in [119] and the contributions obtained in [96]. Related properties concerning the existence of least-energy solutions for two-dimensional Choquard equations with exponential growth can be found in [87].

The "energy doubling" phenomenon described in Theorem 1.2 establishes that the energy of the nodal solution is strictly greater than twice the ground state energy for large values of the positive parameter. A related property was established by Liu, Ouyang and Zhang [107] in the framework of standing waves for the 2D nonlocal gauged Schrödinger equation with Chern-Simons term. A related property in a different nonlocal setting was previously observed by Shuai [136].

Open Problem 1. The existence of λ^* in Theorem 1.1 is based on the convergence property found in Lemma 1.3. Is it possible to find the optimal value for λ^*?

Open Problem 2. In the statement of Theorem 1.1, does problem (1.1) have nodal solutions if $0 < \lambda < \lambda^*$?

Open Problem 3. Does the doubling energy property established in Theorem 1.2 remain true if $0 < \lambda < \lambda^{**}$?

The following problem is concerned with the asymptotic behavior as $p \to \infty$.

Open Problem 4. What happens when the exponent p becomes large? This question relates to p-Laplacian problems in the limit and possible connections to frozen or limiting profiles.

The current analysis works on bounded domains with Lipschitz boundary. The following research direction proposes to extend this analysis to other cases.

Open Problem 5. Develop a similar analysis to the whole space $\mathbb{R}^N$ $(N \geq 3)$. This presents challenges due to loss of compactness. Study the critical exponential growth in the two-dimensional case.

What about the case of operators with a variable structure? This corresponds to the $p(x)$-Laplacian logarithmic reaction and this study could open new perspectives in the understanding of Kirchhoff equations with logarithmic nonlinearity.

Open Problem 6. Extend the analysis developed in this chapter to more complex operators like the $p(x)$-Laplacian, which allows the growth rate to vary with the point in the domain.

Studying the supercritical cases of Kirchhoff equations is important because it is where the most interesting and physically relevant mathematical behaviors occur. This abstract setting describes the frontier where solutions can exhibit fundamentally different properties. This is the content of the following research direction.

Open Problem 7. Extend the results to the supercritical case, that is, study the Kirchhoff equation

$$\begin{cases} \left(a + b[u]_{s,p}^p\right)(-\Delta)_p^s u = \lambda |u|^{q-2} u \ln |u|^2 + |u|^{r-2} u & \text{in } \Omega, \\ u = 0 & \text{in } \mathbb{R}^N \setminus \Omega, \end{cases}$$

where $r > p_s^*$.

A very interesting problem is concerned with the study of the following logarithmic Kirchhoff equation:

$$\begin{cases} \left(a + b[u]_{s,p}^p\right)(-\Delta)_p^s u = \lambda |u|^{p_s^*-1} \ln |u|^2 + \mu |u|^q & \text{in } \Omega, \\ u = 0 & \text{in } \mathbb{R}^N \setminus \Omega. \end{cases} \tag{1.58}$$

A feature of this problem is that the logarithm weakens the critical nonlinearity near the origin and infinity, but it introduces a lack of monotonicity and sign changes, making variational methods delicate.

Open Problem 8. Study the existence of least-energy solutions for problem (1.58) with respect to the values of λ and μ. Consider also the singular case where $\mu < 0$.

1.6 Glossary

Least-energy Solution: The least-energy solution is a solution (of a partial differential equation or an optimization problem) that minimizes the associated energy functional.

Logarithmic Nonlinearity: Logarithmic nonlinearity refers to differential or partial differential equations where the nonlinear term involves the logarithm of the unknown function. Equations of this type arise in applied sciences and have distinctive mathematical properties due to the slow growth and singular behavior of the logarithm near zero or infinity.

Kirchhoff Nonlocality: It defines a class of equations where the governing physics at each point is tied to the global state of the system, creating a deeply interconnected and complex problem to solve. Namely, instead of a local property like the value of the unknown at a point, the equation's behavior is influenced by the overall energy or amplitude of the solution across the whole space.

Nodal Solution: Nodal solution, in the context of partial differential equations (especially elliptic PDEs), refers to a nontrivial solution that changes sign. Thus, the solution has a nonempty nodal set (the set where it vanishes), dividing the domain into at least two subdomains where the solution is positive and negative respectively. These are also called sign-changing solutions.

Nehari Manifold: This is a fundamental and powerful tool in modern variational analysis for finding nontrivial solutions to nonlinear PDEs. In simple words, the Nehari manifold is a constraint that cleverly restricts the search for critical points of an energy functional to a set where the geometry becomes much simpler to analyze.

Energy Doubling: This is a phenomenon which occurs in the analysis of solutions for some classes of local and nonlocal equations. It establishes that the energy of the nodal solution is strictly greater than twice the ground state energy.

2

Planar Kirchhoff equations with critical exponential growth

This chapter deals with the existence of nontrivial solutions and least-energy solutions for a class of planar Kirchhoff equations with critical exponential growth and trapping potential. This is achieved by combining some related variational, topological, and analytic techniques. A key feature of the analysis developed in this chapter is the absence of any monotonicity conditions on the reaction term. Another contribution in this chapter concerns the mountain pass characterization of the least-energy solution by constructing a fine path.

We study the following Kirchhoff equation

$$\begin{cases} -\left(a+b\int_{\mathbb{R}^2}|\nabla u|^2\mathrm{d}x\right)\Delta u+V(x)u=f(u) & \text{in } \mathbb{R}^2, \\ u\in H^1(\mathbb{R}^2), \end{cases}$$

where a, b are positive constants, $V \in \mathcal{C}(\mathbb{R}^2,(0,\infty))$ is a trapping potential, and f has critical exponential growth of Trudinger-Moser type. Indeed, in two dimensions, the maximal growth for $f(u)$ is governed by the Trudinger-Moser inequality (a behavior like $e^{\alpha u^2}$), which also leads to non-compactness. The common restriction $\liminf_{t\to+\infty}\frac{tf(t)}{e^{\alpha_0 t^2}}>0$ has been removed in this analysis, where α_0 is the critical value in the Trudinger-Moser inequality. This is a major relaxation of standard assumptions, significantly broadening the class of nonlinearities $f(u)$ the theory can handle.

This chapter integrates three mathematically challenging features into a single, physically relevant model, extending prior work. These three competing mathematical concepts are the following:

1. nonlocal coefficient $\left(\int_{\mathbb{R}^2}|\nabla u|^2\mathrm{d}x\right)\Delta u$;
2. critical exponential growth in $\mathbb{R}^2$;
3. a trapping potential $V(x)$ (it prevents the system from escaping to infinity).

The primary mathematical challenge in this analysis is to prove the existence of solutions despite several combined sources of non-compactness. In this setting, it is difficult to restore the compactness of minimizing sequences on the prescribed-mass manifold, which is especially hard when the potential $V(x)$ is *non-autonomous*, that is, V is not a constant. The positivity of the potential ensures coercivity, while its trapping nature creates an energy well. This breaks translation invariance, causing loss of compactness in the Sobolev embedding. The analysis developed in this chapter covers both autonomous and non-autonomous cases. Under minimal assumptions, this analysis establishes a clear structural link between the least-energy solution and the mountain pass level.

DOI: 10.1201/9781003797159-2

2.1 Statement of the problem

This chapter deals with the study of the following Kirchhoff equation

$$\begin{cases} -\left(a+b\int_{\mathbb{R}^2}|\nabla u|^2\mathrm{d}x\right)\Delta u+V(x)u=f(u) & \text{in } \mathbb{R}^2; \\ u\in H^1(\mathbb{R}^2), \end{cases} \tag{$\mathcal{K}$}$$

where a, b are positive real numbers constants.

A feature of the analysis developed in this chapter is that the potential $V \in \mathcal{C}(\mathbb{R}^2,(0,\infty))$ is assumed to be a Rabinowitz-type trapping potential, namely it satisfies

(V1) $V_0 \le V(x) \le \liminf_{|y|\to\infty} V(y) = V_\infty$ for all $x \in \mathbb{R}^2$.

This trapping condition was introduced by Rabinowitz [133].

We assume that the reaction $f \in \mathcal{C}(\mathbb{R},\mathbb{R})$ satisfies the following conditions:

(F0) $f(t) = 0$ for all $t \le 0$;

(F1) there exists $\alpha_0 > 0$ such that

$$\lim_{|t|\to\infty}\frac{|f(t)|}{e^{\alpha t^2}} = \begin{cases} 0, & \text{for all } \alpha > \alpha_0, \\ +\infty, & \text{for all } \alpha < \alpha_0; \end{cases} \tag{2.1}$$

(F2) $f(t) = o(t)$ as $t \to 0$ and $F(t) := \int_0^t f(s)\mathrm{d}s \ge 0$.

Definition 2.1 *We say that the function $f(t)$ has critical exponential growth at $t = \pm\infty$ if condition* (F1) *is fulfilled.*

It was shown by Trudinger [148] and Moser [120] that this kind of nonlinearity has the maximal growth that can be treated variationally in $H^1(\mathbb{R}^2)$, which is motivated by the following Trudinger-Moser inequality.

Lemma 2.1 *The following properties hold true.*

(i) If $\alpha > 0$ and $u \in H^1(\mathbb{R}^2)$, then

$$\int_{\mathbb{R}^2}\left(e^{\alpha u^2}-1\right)\mathrm{d}x < \infty.$$

(ii) If $u \in H^1(\mathbb{R}^2), \|\nabla u\|_2^2 \le 1, \|u\|_2 \le M < \infty$, and $\alpha < 4\pi$, then there exists a constant $\mathcal{C}(M,\alpha)$, which depends only on M and α, such that

$$\int_{\mathbb{R}^2}\left(e^{\alpha u^2}-1\right)\mathrm{d}x \le \mathcal{C}(M,\alpha). \tag{2.2}$$

From (F1) and (F2), it follows that

$$\lim_{t\to 0}\frac{F(t)}{t^2} = 0 \tag{2.3}$$

and

$$\lim_{t\to+\infty}\frac{t^2F(t)}{e^{\alpha t^2}}=\begin{cases}0, & \text{for all } \alpha>\alpha_0,\\ +\infty, & \text{for all } \alpha<\alpha_0.\end{cases} \tag{2.4}$$

Then for any $\varepsilon>0, \alpha>\alpha_0$ and $q>0$, there exists $C=C(\varepsilon,\alpha,q)>0$ such that

$$F(t)\le \varepsilon t^2+C|t|^q e^{\alpha t^2}, \quad \forall\, t\in\mathbb{R}. \tag{2.5}$$

Using relation (2.5), it follows that the associated energy functional $\Phi : H^1(\mathbb{R}^2)\to\mathbb{R}$ defined by

$$\Phi(u)=\frac{1}{2}\int_{\mathbb{R}^2}\left[a|\nabla u|^2+V(x)u^2\right]\mathrm{d}x+\frac{b}{4}\left(\int_{\mathbb{R}^2}|\nabla u|^2\mathrm{d}x\right)^2-\int_{\mathbb{R}^2}F(u)\mathrm{d}x \tag{2.6}$$

is of class $\mathcal{C}^1(H^1(\mathbb{R}^2),\mathbb{R})$ and

$$\begin{aligned}\langle\Phi'(u),v\rangle=&\int_{\mathbb{R}^2}\left[a\nabla u\cdot\nabla v+V(x)uv\right]\mathrm{d}x+b\int_{\mathbb{R}^2}|\nabla u|^2\mathrm{d}x\int_{\mathbb{R}^2}\nabla u\cdot\nabla v\mathrm{d}x\\ &-\int_{\mathbb{R}^2}f(u)v\mathrm{d}x, \quad \forall\, u,v\in H^1(\mathbb{R}^2).\end{aligned} \tag{2.7}$$

Hence, the solutions of problem $(\mathcal{K})$ are the critical points of the functional Φ.

2.1.1 Autonomous and non-autonomous cases

Let us first consider the *autonomous* case.

We study the following Kirchhoff equation

$$\begin{cases}-\left(a+b\displaystyle\int_{\mathbb{R}^2}|\nabla u|^2\mathrm{d}x\right)\Delta u+V_\infty u=f(u) & \text{in } \mathbb{R}^2;\\ u\in H^1(\mathbb{R}^2),\end{cases} \tag*{$(\mathcal{K})_\infty$}$$

where f satisfies (F0), (F1), (F2) and the following condition:

(F3) $\lim_{t\to+\infty}\frac{F(t)}{t^2}=+\infty$ and $f(t)t\ge 2F(t)$ for all $t\ge 0$.

Using a suitable minimization method, the first purpose is to establish both the existence of a positive least-energy solution for $(\mathcal{K})_\infty$ and to give its mountain pass characterization.

Define the energy functional $\Phi^\infty : H^1(\mathbb{R}^2)\to\mathbb{R}$ by

$$\Phi^\infty(u):=\frac{1}{2}\int_{\mathbb{R}^2}\left(a|\nabla u|^2+V_\infty u^2\right)\mathrm{d}x+\frac{b}{4}\left(\int_{\mathbb{R}^2}|\nabla u|^2\mathrm{d}x\right)^2-\int_{\mathbb{R}^2}F(u)\mathrm{d}x, \tag{2.8}$$

and let c^∞ be the mountain pass level of Φ^∞, that is,

$$c^\infty=\inf_{\gamma\in\Gamma^\infty}\max_{t\in[0,1]}\Phi^\infty(\gamma(t)), \tag{2.9}$$

where

$$\Gamma^\infty:=\left\{\gamma\in\mathcal{C}([0,1],H^1(\mathbb{R}^2)) : \gamma(0)=0, \Phi^\infty(\gamma(1))<0\right\}. \tag{2.10}$$

Definition 2.2 *A solution u of problem $(\mathcal{K})_\infty$ is called a least-energy solution if $\Phi^\infty(u)=m^\infty$ with*

$$m^\infty:=\inf\left\{\Phi^\infty(u) \ :\ u\in H^1(\mathbb{R}^2)\setminus\{0\}, (\Phi^\infty)'(u)=0\right\}. \tag{2.11}$$

In the autonomous case, the next result establishes the existence of a least-energy solution, provided that the positive potential is smaller than the Trudinger-Moser ratio.

Theorem 2.1 *Assume that f satisfies hypotheses* (F0)–(F3). *Then there exists $V^* \in (0, +\infty]$ such that for any $V_\infty \in (0, V^*)$, equation $(\mathcal{K})_\infty$ has a positive least-energy solution. Moreover, V^* is equal to the Trudinger-Moser ratio:*

$$C^*_{TM}(F) := \sup\left\{\frac{2}{\|u\|_2^2}\int_{\mathbb{R}^2} F(u)\mathrm{d}x \ : \ u \in H^1(\mathbb{R}^2)\setminus\{0\}, \|\nabla u\|_2^2 \leq \frac{4\pi}{\alpha_0}\right\}.$$

In particular, $V^* = +\infty$ is equivalent to $\lim_{t\to+\infty}\frac{t^2F(t)}{e^{\alpha_0 t^2}} = +\infty$.

The following property establishes a relationship between the any least-energy solution of problem $(\mathcal{K})_\infty$ and the level of the maximum of the associated energy at the optimal path.

Theorem 2.2 *Under the assumptions of Theorem* 2.1, *the least-energy level m^∞ is equal to the mountain pass level c^∞. Moreover, for any least-energy solution w of $(\mathcal{K})_\infty$, there exists a path $\tilde{\gamma} \in \Gamma^\infty$ such that $w \in \tilde{\gamma}([0,1])$ and*

$$\max_{t\in[0,1]} \Phi^\infty(\tilde{\gamma}(t)) = \Phi^\infty(w).$$

Next, we study the *non-autonomous* case and we are concerned with the existence of ground state solutions for the critical exponential growth Kirchhoff equation $(\mathcal{K})$ with the trapping potential V satisfying condition (V1). Some effective methods, treating the dimension $N \geq 3$, do not carry over our case $N = 2$ due to the simultaneous appearance of the nonlocal term and the nonlinear term with critical exponential growth.

Let us now introduce the following assumptions:

(F3′) $f(t)t \geq 2F(t)$ for all $t \in \mathbb{R}$, and

$$\frac{f(t)}{t} \geq V_0 \Rightarrow f(t)t - 2F(t) > 0;$$

(F4) there exist $M_0 > 0$ and $\beta_0 > 0$ such that $F(\beta_0) > 0$ and

$$F(t) \leq M_0|f(t)|, \quad \forall\ |t| \geq \beta_0.$$

Definition 2.3 *We say that a solution u of problem $(\mathcal{K})$ is a ground state solution (of Nehari type) if $\Phi(u) = c_{\mathcal{N}}$ with*

$$c_{\mathcal{N}} := \inf_{u\in\mathcal{N}} \Phi(u) \tag{2.12}$$

and $\mathcal{N}$ is the Nehari manifold defined by

$$\mathcal{N} := \{u \in H^1(\mathbb{R}^2)\setminus\{0\}: \ \langle\Phi'(u), u\rangle = 0\}. \tag{2.13}$$

The following result points out the importance of the level V_∞ of the trapping potential at infinity. This level acts as a fundamental energy threshold that governs ground state solutions, but also whether compactness can be recovered in the variational formulation.

Theorem 2.3 *Assume that V satisfies* (V1) *with $V_\infty \in (0, V^*)$, and f satisfies conditions* (F1), (F2), (F3′), (F4) *and* (F5). *Then problem $(\mathcal{K})$ has a ground state solution, where V^* is given by Theorem* 2.1.

Corollary 2.1 *Assume that $V \in \mathcal{C}(\mathbb{R}^2, [V_0, V_\infty])$ is 1-periodic in each of x_1, x_2 with $V_\infty \in (0, V^*)$, and f satisfies* (F1), (F2), (F3′), (F4) *and* (F5). *Then $(\mathcal{K})$ has a ground state solution, where V^* is given by Theorem* 2.1.

In the final part of this chapter, we consider the existence of nontrivial solutions to problem $(\mathcal{K})$ in the *radial* case, that is, if the trapping potential V satisfies

(V2) $V(x) = V(|x|)$ and $V(x) \geq V_0$ for all $x \in \mathbb{R}^2$.

In this, it is nontrivial to show that the weak limit of Cerami sequences is a weak solution because of the fact

$$u_n \rightharpoonup u \text{ in } H^1(\mathbb{R}^2) \not\Rightarrow \|\nabla u_n\|_2^2 \int_{\mathbb{R}^2} \nabla u_n \cdot \nabla \varphi \mathrm{d}x \to \|\nabla u\|_2^2 \int_{\mathbb{R}^2} \nabla u \cdot \nabla \varphi \mathrm{d}x, \quad \forall\, \varphi \in \mathcal{C}_0^\infty(\mathbb{R}^2).$$

To address this issue, a classical way is to restrict the energy functional in the subspace of radially symmetric functions $H_r^1(\mathbb{R}^2)$ belonging to $H^1(\mathbb{R}^2)$ since the limit

$$\int_{\mathbb{R}^2} [f(u_n) - f(u)](u_n - u)\mathrm{d}x \to 0 \tag{2.14}$$

can be easily deduced from the compactness of the embedding $H_r^1(\mathbb{R}^2) \hookrightarrow L^q(\mathbb{R}^2)$ for $q > 2$ if f is superlinear at zero and has polynomial growth. However, when f is of critical exponential growth, it is still unknown whether the limit (2.14) holds or not since the embedding of $H_r^1(\mathbb{R}^2)$ into the Orlicz space associated with the function $\varphi(s) = \exp(4\pi s^2) - 1$ is not compact. Thus, a deeper analysis is required for $(\mathcal{K})$ with the radial potential V in this direction, which is the focus in the last part of the present chapter. For this purpose, we introduce the following condition:

(F6) $f(t)t - 4F(t) + V_0 t^2 \geq 0$ for all $t \in \mathbb{R}$,

which is weaker than the Ambrosetti-Rabinowitz condition (F6′) and can be derived from the monotonicity condition (F5).

Theorem 2.4 *Assume that V satisfies* (V2) *with $V_\infty \in [V_0, V^*)$, and f satisfies conditions* (F1), (F2), (F3′), (F4) *and* (F6). *Then problem $(\mathcal{K})$ has a nontrivial radial solution, where V^* is given by Theorem* 2.1.

2.1.2 Sketch of the main proofs

For the proofs of Theorems 2.1 and 2.2, motivated by the Pohozaev identity for problem $(\mathcal{K})_\infty$, we introduce the auxiliary functional $J^\infty : H^1(\mathbb{R}^2) \to \mathbb{R}$ defined by

$$J^\infty(u) = V_\infty \|u\|_2^2 - 2\int_{\mathbb{R}^2} F(u)\mathrm{d}x, \tag{2.15}$$

the set

$$\mathcal{P}_\infty := \left\{u \in H^1(\mathbb{R}^2) \setminus \{0\} : \ J^\infty(u) = 0\right\} \tag{2.16}$$

and the constrained minimization problem

$$A^\infty := \inf_{u\in\mathcal{P}_\infty} \left(\frac{a}{2}\|\nabla u\|_2^2 + \frac{b}{4}\|\nabla u\|_2^4\right) = \inf_{u\in\mathcal{P}_\infty} \Phi^\infty(u). \tag{2.17}$$

Based on a sufficient and necessary condition for compactness of general nonlinear functionals, a key idea is to prove that A^∞ can be attained if V_∞ is less than the Trudinger-Moser ratio $C^*_{\mathrm{TM}}(F)$. At the same time, under a suitable change of scale combined with an analytical method, it is established that the minimizer is a least-energy solution of problem $(\mathcal{K})_\infty$. Different from the dimension $N \geq 3$, the minimum A^∞ has no saddle point structure with respect to the fiber $\{u(\cdot/t) : t > 0\} \subset H^1(\mathbb{R}^2)$, $u \in H^1(\mathbb{R}^2)$ since the Pohozaev functional $J^\infty(u)$ does not have a $\|\nabla u\|_2$-component. Thus, in the two-dimensional case, it is more complicated to establish the relationship between the minimum A^∞, the least-energy m^∞, and the mountain pass level c^∞. To address this issue, inspired by the idea of Jeanjean and Tanaka [80], we construct a new path belonging to Γ^∞ and derive the mountain pass characterization of the least-energy solution for $(\mathcal{K})_\infty$, which is the highlight of the proof of Theorem 2.2.

The proofs of Theorems 2.3 and 2.4 are based on the mountain pass theorem. A standard procedure in this case is to prove the boundedness of Cerami sequences, and verify that the weak limit of Cerami sequences is non-trivial and is also a weak solution.

Compared with previous works dealing with the Kirchhoff equation $(\mathcal{K})$ with trapping or radial potential in higher dimensions ($N \geq 3$), some new obstacles arising in the proofs are the following:

1. the lack of the monotonicity condition (F5$'$) and the Ambrosetti-Rabinowitz type condition (F6$'$) prevent us from using usual methods to prove the boundedness of Cerami sequences;
2. it is more difficult to rule out the concentration phenomena and the vanishing phenomena of Cerami sequences;
3. it does not work that the BL-splitting property for the energy functional along Cerami sequences caused by the appearance of the nonlinear term with critical growth, which is a powerful tool to restore the compactness of Cerami sequences.

In what follows, we make use of the following notations:

- $H^1(\mathbb{R}^2)$ denotes the usual Sobolev space equipped with the inner product and norm

$$(u, v) = \int_{\mathbb{R}^2} (\nabla u \cdot \nabla v + uv)\mathrm{d}x, \quad \|u\| = (u, u)^{1/2}, \quad \forall\, u, v \in H^1(\mathbb{R}^2);$$

- $H^1_r(\mathbb{R}^2)$ denotes the space of spherically symmetric functions belonging to $H^1(\mathbb{R}^2)$:

$$H^1_r(\mathbb{R}^2) := \{u \in H^1(\mathbb{R}^2) : \ u(x) = u(|x|) \text{ a.e. in } \mathbb{R}^2\};$$

- $L^s(\mathbb{R}^2)$ $(1 \le s < \infty)$ is the Lebesgue space with the norm $\|u\|_s = \left(\int_{\mathbb{R}^2} |u|^s \mathrm{d}x\right)^{1/s}$;
- For any $x \in \Omega$ and $r > 0$, $B_r(x) := \{y \in \Omega : |y - x| < r\}$ and $B_r = B_r(0)$;
- $C_1, C_2, \ldots$ denote positive constants possibly different in different places.

2.2 Least-energy solutions in the autonomous case

In this section, we consider the existence of least-energy solutions for problem $(\mathcal{K})_\infty$, and we establish its mountain pass characterization, which completes the proofs of Theorems 2.1 and 2.3.

2.2.1 Auxiliary properties of the energy functionals

First, by a truncation argument, we establish the Pohozaev-type identity for $(\mathcal{K})_\infty$ when f has critical exponential growth.

Lemma 2.2 *Assume that f satisfies hypotheses* (F1)–(F3). *Let $u \in H^1(\mathbb{R}^2)$ be a weak solution of $(\mathcal{K})_\infty$, then we have the following Pohozaev-type identity*

$$J^\infty(u) = V_\infty \|u\|_2^2 - 2 \int_{\mathbb{R}^2} F(u)\mathrm{d}x = 0.$$

Proof. Let $\psi \in \mathcal{C}^\infty([0, +\infty), [0, 1])$ such that $\psi(r) = 1$ for $r \in [0, 1]$ and $\psi(r) = 0$ for $r \in [2, +\infty)$. Define $\psi_n(x) := \psi(|x|^2/n^2)$ on $\mathbb{R}^2$ for $n \in \mathbb{N}$. Then there exists $C_1 > 0$ such that

$$0 \le \psi_n(x) \le C_1, \quad |x||\nabla \psi_n(x)| \le C_1 \quad \forall\, x \in \mathbb{R}^2. \tag{2.18}$$

By a standard regularity argument, we can show that $u \in H^2_{\mathrm{loc}}(\mathbb{R}^2)$.

Let $\bar{a} = (a + b\|\nabla u\|_2^2)$. It follows from $(\mathcal{K})_\infty$ that for every $n \in \mathbb{N}$,

$$0 = [-\bar{a}\Delta u + V_\infty u - f(u)]\, \psi_n (x \cdot \nabla u). \tag{2.19}$$

We also obtain, for every $n \in \mathbb{N}$,

$$-\psi_n f(u)(x \cdot \nabla u) = -\mathrm{div}(x\psi_n F(u)) + 2\psi_n F(u) + F(u)(x \cdot \nabla \psi_n), \tag{2.20}$$

$$\begin{aligned} -\psi_n \Delta u(x \cdot \nabla u) = &-\mathrm{div}\left\{\left[\nabla u(x \cdot \nabla u) - x\frac{|\nabla u|^2}{2}\right]\psi_n\right\} \\ &- \frac{|\nabla u|^2}{2}(x \cdot \nabla \psi_n) + (x \cdot \nabla u)(\nabla \psi_n \cdot \nabla u) \end{aligned} \tag{2.21}$$

and

$$\psi_n u(x \cdot \nabla u) = \frac{1}{2}\mathrm{div}\left(u^2 \psi_n x\right) - u^2 \psi_n - \frac{1}{2}u^2(x \cdot \nabla \psi_n) - \frac{1}{2}u^2 \psi_n. \tag{2.22}$$

Hence, for every $n \in \mathbb{N}$, it follows from relations (2.19), (2.20), (2.21), (2.22) and the divergence theorem that

$$\int_{\partial B_{2n}} \left\{ \frac{\bar{\alpha}|x \cdot \nabla u|^2}{2n} - \bar{\alpha} n |\nabla u|^2 - n V_\infty u^2 + 2nF(u) \right\} \psi_n \mathrm{d}\sigma$$
$$= - \int_{B_{2n}} \left[V_\infty u^2 - 2F(u) \right] \psi_n \mathrm{d}x - \frac{1}{2} \int_{B_{2n}} \left\{ \bar{\alpha}|\nabla u|^2 + V_\infty u^2 - 2F(u) \right\} (x \cdot \nabla \psi_n) \mathrm{d}x$$
$$+ \bar{\alpha} \int_{B_{2n}} (x \cdot \nabla u)(\nabla \psi_n \cdot \nabla u) \mathrm{d}x. \tag{2.23}$$

Using this relation together with the fact that $\psi_n|_{\partial B_{2n}} = 0$, we obtain

$$\int_{B_{2n}} \left[V_\infty u^2 - 2F(u) \right] \psi_n \mathrm{d}x = -\frac{1}{2} \int_{B_{2n}} \left\{ \bar{\alpha}|\nabla u|^2 + V_\infty u^2 - 2F(u) \right\} (x \cdot \nabla \psi_n) \mathrm{d}x$$
$$+ \bar{\alpha} \int_{B_{2n}} (x \cdot \nabla u)(\nabla \psi_n \cdot \nabla u) \mathrm{d}x$$
$$= -\frac{1}{2} \int_{B_{\sqrt{2}n} \setminus B_n} \left[\bar{\alpha}|\nabla u|^2 + V_\infty u^2 - 2F(u) \right] (x \cdot \nabla \psi_n) \mathrm{d}x$$
$$+ \bar{\alpha} \int_{B_{\sqrt{2}n} \setminus B_n} (x \cdot \nabla u)(\nabla \psi_n \cdot \nabla u) \mathrm{d}x. \tag{2.24}$$

From (2.18) and (2.24), we have

$$\left| \int_{\mathbb{R}^2} \left[V_\infty u^2 - 2F(u) \right] \mathrm{d}x \right| = \left| \lim_{n \to \infty} \int_{B_{2n}} \left[V_\infty u^2 - 2F(u) \right] \psi_n \mathrm{d}x \right|$$
$$\leq \frac{1}{2} \lim_{n \to \infty} \int_{B_{\sqrt{2}n} \setminus B_n} \left[3\bar{\alpha}|\nabla u|^2 + V_\infty u^2 + 2F(u) \right] |x||\nabla \psi_n| \mathrm{d}x$$
$$\leq \frac{C_1}{2} \lim_{n \to \infty} \int_{B_{\sqrt{2}n} \setminus B_n} \left[3\bar{\alpha}|\nabla u|^2 + V_\infty u^2 + 2F(u) \right] \mathrm{d}x = 0.$$

The proof is complete. □

We now solve the constrained minimization problem A^∞, given by (2.17).

Lemma 2.3 *Assume that f satisfies conditions* (F1)–(F3). *Then there exists a minimizing sequence $\{u_n\} \subset \mathcal{P}_\infty$ satisfying $\|u_n\|_2 = 1$ for A^∞. In particular,*

$$A^\infty = \inf_{u \in \mathcal{P}_\infty} \left(\frac{a}{2} \|\nabla u\|_2^2 + \frac{b}{4} \|\nabla u\|_2^4 \right) = \inf_{u \in \widetilde{\mathcal{P}}_\infty} \left(\frac{a}{2} \|\nabla u\|_2^2 + \frac{b}{4} \|\nabla u\|_2^4 \right), \tag{2.25}$$

where

$$\widetilde{\mathcal{P}}_\infty = \mathcal{P}_\infty \cap \{ u \in H^1(\mathbb{R}^2) : \|u\|_2 = 1 \} \tag{2.26}$$

and $\mathcal{P}_\infty$ is given by (2.16).

Proof. We first prove that $\mathcal{P}_\infty \neq \emptyset$. Let $u \in H^1(\mathbb{R}^2) \setminus \{0\}$ be fixed and define a function $\zeta(t) := J^\infty(tu)$ on $(0, \infty)$. Using (F1)–(F3), it is easy to check that $\zeta(t) > 0$ for small $t > 0$ and $\zeta(t) < 0$ for large $t > 0$. Then there exists $t_u > 0$ such that $\zeta(t_u) = J^\infty(t_u u) = 0$, and so $\mathcal{P}_\infty \neq \emptyset$. Thus, we can assume that there exists a minimizing sequence $\{u_n\} \subset \mathcal{P}_\infty$ satisfying

$$\frac{a}{2} \|\nabla u_n\|_2^2 + \frac{b}{4} \|\nabla u_n\|_2^4 \to A^\infty.$$

Let $\tilde{u}_n = u_n(\|u_n\|_2^{1/2} x)$, hence $\tilde{u}_n \in \mathcal{P}_\infty$, $\|\tilde{u}_n\|_2 = 1$ and $\|\nabla \tilde{u}_n\|_2 = \|\nabla u_n\|_2$. This shows that $\tilde{u}_n \in \widetilde{\mathcal{P}}_\infty$. From this and the fact that $\widetilde{\mathcal{P}}_\infty \subset \mathcal{P}_\infty$, relation (2.25) follows directly. The proof is complete. □

Lemma 2.4 *Assume that f satisfies hypotheses* (F1)–(F3).

(i) *If $u \in H^1(\mathbb{R}^2)$ is a critical point of Φ^∞ on the set $\mathcal{P}_\infty$, then it is a nontrivial solution of $(\mathcal{K})_\infty$ under a suitable change of scale.*

(ii) *If the infimum A^∞ is attained, then $A^\infty = m^\infty$.*

Proof. (i) Let $u \in H^1(\mathbb{R}^2)$ be a critical point of Φ^∞ on the set $\mathcal{P}_\infty$. Then there is a Lagrange multiplier $\lambda \in \mathbb{R}$ such that

$$-(a + b\|\nabla u\|_2^2)\Delta u + V_\infty u - f(u) = 2\lambda[V_\infty u - f(u)], \tag{2.27}$$

namely,

$$-(a + b\|\nabla u\|_2^2)\Delta u = (2\lambda - 1)[V_\infty u - f(u)]. \tag{2.28}$$

Since $u \neq 0$, we deduce from (2.28) that

$$2\lambda - 1 \neq 0 \quad \text{and} \quad V_\infty u - f(u) \neq 0. \tag{2.29}$$

For any $T > 0$, by (F1)–(F3), there exist $0 < t_1 < t_2 < T$ such that

$$2F(t) - f(t)t \leq 0, \quad \forall\, t \in [0, T], \quad \text{and} \quad 2F(t) - f(t)t < 0, \quad \forall\, t \in [t_1, t_2]. \tag{2.30}$$

Hence, it follows from (2.30) and the definition of $\mathcal{P}_\infty$ that

$$\int_{\mathbb{R}^2} [V_\infty u - f(u)]u\mathrm{d}x = \int_{\mathbb{R}^2} [2F(u) - f(u)u]\mathrm{d}x < 0. \tag{2.31}$$

This implies that there exists $w \in \mathcal{C}_0^\infty(\mathbb{R}^2)$ such that

$$\langle (J^\infty)'(u), w\rangle = \int_{\mathbb{R}^2} [V_\infty u - f(u)]w\mathrm{d}x < 0. \tag{2.32}$$

By multiplying relation (2.28) by w and integrating, we have

$$(a + b\|\nabla u\|_2^2) \int_{\mathbb{R}^2} \nabla u \cdot \nabla w\mathrm{d}x = (2\lambda - 1) \int_{\mathbb{R}^2} [V_\infty u - f(u)]w\mathrm{d}x. \tag{2.33}$$

Recalling that $J^\infty(u) = 0$, it follows from (2.32) that for small enough $\varepsilon > 0$,

$$J^\infty(u + \varepsilon w) < J^\infty(u) = 0. \tag{2.34}$$

Let

$$A(v) := \frac{a}{2}\|\nabla v\|_2^2 + \frac{b}{4}\|\nabla v\|_2^4, \quad \forall\, v \in H^1(\mathbb{R}^2).$$

Noticing that $A(u) = A^\infty$, by (2.33) and (2.34), we have

$$\begin{aligned} A(u + \varepsilon w) &= \frac{a}{2}\|\nabla u\|_2^2 + \frac{b}{4}\|\nabla u\|_2^4 + \varepsilon(a + b\|\nabla u\|_2^2) \int_{\mathbb{R}^2} \nabla u \cdot \nabla w\mathrm{d}x + \frac{a\varepsilon^2}{2}\|\nabla w\|_2^2 + \frac{b\varepsilon^4}{2}\|\nabla w\|_2^4 \\ &\quad + 2\varepsilon^2\|\nabla u\|_2^2\|\nabla w\|_2^2 + 4\varepsilon^2\left(\int_{\mathbb{R}^2} \nabla u \cdot \nabla w\mathrm{d}x\right)^2 + 4\varepsilon^3\|\nabla w\|_2^2 \int_{\mathbb{R}^2} \nabla u \cdot \nabla w\mathrm{d}x \\ &= A^\infty + \varepsilon(2\lambda - 1) \int_{\mathbb{R}^2} [V_\infty u - f(u)]w\mathrm{d}x + O(\varepsilon^2). \end{aligned} \tag{2.35}$$

We claim that $2\lambda - 1 < 0$. Otherwise, there exists $\varepsilon_0 > 0$ small enough such that

$$J^\infty(u + \varepsilon_0 w) < 0 \quad \text{and} \quad A(u + \varepsilon_0 w) < A^\infty, \tag{2.36}$$

due to (2.34) and (2.35).

Let $u_0 = u + \varepsilon_0 w$. Then $J^\infty(u_0) < 0$ and $J^\infty(su_0) > 0$ for $s > 0$ small enough as a consequence of (F2). Therefore, there exists $s_0 \in (0,1)$ such that $J^\infty(s_0 u_0) = 0$. Moreover, by (2.36), we have

$$A(s_0 u_0) = \frac{as_0^2}{2}\|\nabla u_0\|_2^2 + \frac{bs_0^4}{4}\|\nabla u_0\|_2^4 < s_0^2 A(u_0) < A^\infty. \tag{2.37}$$

This shows that $s_0 u_0 \in \mathcal{P}_\infty$ and $\Phi^\infty(s_0 u_0) < A^\infty$, which contradicts the definition of A^∞. Hence, we have $2\lambda - 1 < 0$, as claimed. Thus,

$$\tilde{u}(x) := u\left(\frac{x}{(1-2\lambda)^{1/2}}\right) \text{ for a.e. } x \in \mathbb{R}^2 \tag{2.38}$$

is a nontrivial solution of $(\mathcal{K})_\infty$.

(ii) Suppose that the infimum A^∞ is attained by $u \in H^1(\mathbb{R}^2)$. From (2.25), we see that $u \in H^1(\mathbb{R}^2)$ is a critical point of Φ^∞ on the set $\mathcal{P}_\infty$. Then (i) of this lemma shows that $\tilde{u} \in H^1(\mathbb{R}^2)$ defined by (2.38) is a nontrivial solution of $(\mathcal{K})_\infty$, and so $(\Phi^\infty)'(\tilde{u}) = 0$ and $A^\infty = \Phi^\infty(\tilde{u}) \geq m^\infty$.

To conclude the proof, it remains to show that $A^\infty \leq m^\infty$. Note that Lemma 2.2 shows that if $(\Phi^\infty)'(v) = 0$ for $v \in H^1(\mathbb{R}^2)$, then v satisfies the Pohozaev-type identity $J^\infty(v) = 0$, namely,

$$\left\{u \in H^1(\mathbb{R}^2) \setminus \{0\} : (\Phi^\infty)'(u) = 0\right\} \subset \mathcal{P}_\infty.$$

This implies that $A^\infty \leq m^\infty$. The proof is now complete. □

Before studying the attainability of A^∞, we recall necessary and sufficient conditions for the boundedness and the compactness of general nonlinear functionals in $H^1(\mathbb{R}^2)$; see Ibrahim, Masmoudi and Nakanishi [76] for details.

Lemma 2.5 *Suppose that $g : \mathbb{R} \to [0, +\infty)$ is a Borel function and define the functional G by $G(u) := \int_{\mathbb{R}^2} g(u(x))\mathrm{d}x$. Then for any $K > 0$ we have the following* (B) *and* (C)*:*

(B) *Boundedness: The following* (i) *and* (ii) *are equivalent.*

(i) $\limsup_{|t|\to+\infty} e^{-2|t|^2/K}|t|^2 g(t) < \infty$, $\limsup_{|t|\to 0} |t|^{-2} g(t) < \infty$.

(ii) *There exists a constant $C_{g,K} > 0$ such that*

$$u \in H^1(\mathbb{R}^2),\ \|\nabla u\|_2^2 \leq 2\pi K \Rightarrow \int_{\mathbb{R}^2} g(u)\mathrm{d}x \leq C_{g,K}\int_{\mathbb{R}^2} |u|^2 \mathrm{d}x.$$

(C) *Compactness: The following* (iii) *and* (iv) *are equivalent.*

(iii) $\limsup_{|t|\to+\infty} e^{-2|t|^2/K}|t|^2 g(t) = 0$, $\lim_{|t|\to 0} |t|^{-2} g(t) = 0$.

(iv) *For any radially symmetric sequence $\{u_n\} \subset H^1(\mathbb{R}^2)$ satisfying $\int_{\mathbb{R}^2} |\nabla u_n|^2 \mathrm{d}x \leq 2\pi K$ and weakly converging to some $u \in H^1(\mathbb{R}^2)$, we have $G(u_n) \to G(u)$.*

Now, we establish a relation between the attainability of A^∞ and the Trudinger-Moser inequality with the exact growth:

$$\sup_{\substack{u\in H^1(\mathbb{R}^2)\\ \|\nabla u\|_2 \leq 1}} \int_{\mathbb{R}^2} \frac{e^{4\pi|u|^2} - 1}{(1+|u|)^2}\mathrm{d}x \leq C\|u\|_2^2. \tag{2.39}$$

For this purpose, we introduce the Trudinger-Moser ratio

$$C_{\rm TM}^{L}(F) := \sup\left\{\frac{2}{\|u\|_2^2}\int_{\mathbb{R}^2} F(u){\rm d}x : u \in H^1(\mathbb{R}^2)\setminus\{0\}, \|\nabla u\|_2^2 \le L\right\}, \tag{2.40}$$

the Trudinger-Moser threshold

$$R(F) := \sup\left\{L > 0 : C_{\rm TM}^{L}(F) < +\infty\right\}, \tag{2.41}$$

and we denote by $C_{\rm TM}^{*}(F)$ the ratio at the threshold, that is,

$$C_{\rm TM}^{*}(F) = C_{\rm TM}^{R(F)}(F). \tag{2.42}$$

Using conditions (2.3) and (2.4), and applying Lemma 2.5, we derive that

$$R(F) = \frac{4\pi}{\alpha_0}. \tag{2.43}$$

If (F0) holds, to apply Schwarz symmetrization, we let

$$\tilde{f}(t) = \begin{cases} f(t), & \text{for all } t > 0, \\ -f(-t), & \text{for all } t \le 0. \end{cases} \tag{2.44}$$

Observe that $\tilde{f}$ satisfies the same conditions as f. Furthermore, by the maximum principle, solutions of $(\mathcal{K})_\infty$ with $\tilde{f}$ are also solutions of $(\mathcal{K})_\infty$ with f. Thus, without loss of generality, we can replace f by $\tilde{f}$, and we will always adopt the convention that f has been replaced by $\tilde{f}$; we keep however the same notation f in the following discussion of this section.

Lemma 2.6 *Assume that f satisfies* (F0)–(F3). *If*

$$A^\infty < \frac{2a\pi}{\alpha_0} + \frac{4b\pi^2}{\alpha_0^2},$$

then A^∞ is attained and $A^\infty = \Phi^\infty(u)$, where $u \in H_r^1(\mathbb{R}^2)$ is, under a suitable change of scale, a positive least-energy solution of equation $(\mathcal{K})_\infty$.

Proof. By Schwarz symmetrization and Lemma 2.3, there exists a sequence $\{u_n\} \subset \mathcal{P}_\infty \cap H_r^1(\mathbb{R}^2)$ satisfying

$$\frac{a}{2}\|\nabla u_n\|_2^2 + \frac{b}{4}\|\nabla u_n\|_2^4 \to A^\infty \text{ and } \|u_n\|_2 = 1. \tag{2.45}$$

By reflexivity, there exists $u \in H_r^1(\mathbb{R}^2)$ such that $u_n \rightharpoonup u$ in $H^1(\mathbb{R}^2)$.

Now, we prove that if

$$A^\infty < \frac{2a\pi}{\alpha_0} + \frac{4b\pi^2}{\alpha_0^2},$$

then A^∞ is attained. Note that

$$\begin{cases} \dfrac{at}{2} + \dfrac{bt^2}{4} < \dfrac{2a\pi}{\alpha_0} + \dfrac{4b\pi^2}{\alpha_0^2} \\ t \ge 0, \end{cases} \quad \Leftrightarrow \quad 0 \le t < R(F) = \frac{4\pi}{\alpha_0}. \tag{2.46}$$

Picking up $2/K > \alpha_0$ satisfying $\lim_{n\to\infty}\|\nabla u_n\|_2^2 \le 2\pi K$, then relation (2.4) yields

$$\lim_{|t|\to\infty} \frac{|t|^2 F(t)}{e^{2|t|^2/K}} = 0. \tag{2.47}$$

From (2.3), (2.47) and (C) of Lemma 2.5, we derive that

$$\lim_{n\to\infty}\int_{\mathbb{R}^2} F(u_n)\mathrm{d}x = \int_{\mathbb{R}^2} F(u)\mathrm{d}x. \tag{2.48}$$

Since $J^\infty(u_n) = 0$ and $\|u_n\|_2 = 1$, by (2.48), we have

$$0 < V_\infty = \lim_{n\to\infty} V_\infty\|u_n\|_2^2 = \lim_{n\to\infty}\int_{\mathbb{R}^2} F(u_n)\mathrm{d}x = \int_{\mathbb{R}^2} F(u)\mathrm{d}x, \tag{2.49}$$

which implies that $u \neq 0$ and $J^\infty(u) \leq 0$. By the weak lower semicontinuity of the norm and relation (2.48), we have

$$J^\infty(u) = V_\infty\|u\|_2^2 - \int_{\mathbb{R}^2} F(u)\mathrm{d}x \leq \lim_{n\to\infty}\left(V_\infty\|u_n\|_2^2 - \int_{\mathbb{R}^2} F(u_n)\mathrm{d}x\right) = 0 \tag{2.50}$$

and

$$0 < \frac{a}{2}\|\nabla u\|_2^2 + \frac{b}{4}\|\nabla u\|_2^4 \leq \lim_{n\to\infty}\left(\frac{a}{2}\|\nabla u_n\|_2^2 + \frac{b}{4}\|\nabla u_n\|_2^4\right) = A^\infty. \tag{2.51}$$

In order to prove that the infimum A^∞ is attained by u, it remains to show that $u \in \mathcal{P}_\infty$, namely $J^\infty(u) = 0$. Set

$$h(t) = J^\infty(tu) = t^2V_\infty\|u\|_2^2 - \int_{\mathbb{R}^2} F(tu)\mathrm{d}x.$$

Then $h(1) \leq 0$ by (2.50), and from (2.5) one can deduce that $h(t) > 0$ for $t > 0$ small enough. Consequently, there exists $t_0 \in (0,1]$ such that $J^\infty(t_0u) = 0$, namely $t_0u \in \mathcal{P}_\infty$. This together with (2.51) leads to

$$A^\infty \leq \frac{a}{2}\|\nabla(t_0u)\|_2^2 + \frac{b}{4}\|\nabla(t_0u)\|_2^4 \leq t_0^2A^\infty.$$

The above inequality and (2.51) show that $t_0 = 1$ and $\frac{a}{2}\|\nabla u\|_2^2 + \frac{b}{4}\|\nabla u\|_2^4 = A^\infty > 0$. Combining (2.49) with the fact that $J^\infty(u) = 0$, we have $\|u\|_2 = 1$. Applying Lemma 2.4, we have that the above u is a least-energy solution of $(\mathcal{K})_\infty$ under a suitable change of scale. Similarly as in the Proof of Theorem 1.3 of Figueiredo and Severo [55], we can derive that $u > 0$ in $\mathbb{R}^2$. The proof is complete. □

Lemma 2.7 *Assume that f satisfies conditions* (F0)–(F3). *Then the constrained minimization problem A^∞ associated to the functional Φ^∞ satisfies*

$$A^\infty < \frac{2a\pi}{\alpha_0} + \frac{4b\pi^2}{\alpha_0^2} \tag{2.52}$$

if $V_\infty < C^*_{\mathrm{TM}}(F)$, where $C^*_{\mathrm{TM}}(F)$ is given by (2.42).

Proof. We distinguish two cases: $C^*_{\mathrm{TM}}(F) < +\infty$ and $C^*_{\mathrm{TM}}(F) = +\infty$. In the case $C^*_{\mathrm{TM}}(F) < +\infty$, since $V_\infty < C^*_{\mathrm{TM}}(F)$, then $V_\infty < C^*_{\mathrm{TM}}(F) - \varepsilon_0$ for some $\varepsilon_0 > 0$. By the definition of $C^*_{\mathrm{TM}}(F)$, there exists some $u_0 \in H^1(\mathbb{R}^2)\setminus\{0\}$ such that

$$\|\nabla u_0\|_2^2 \leq R(F) \quad\text{and}\quad V_\infty < C^*_{\mathrm{TM}}(F) - \varepsilon_0 < \frac{2}{\|u_0\|_2^2}\int_{\mathbb{R}^2} F(u_0)\mathrm{d}x. \tag{2.53}$$

Then

$$J^\infty(u_0) = V_\infty\|u_0\|_2^2 - 2\int_{\mathbb{R}^2} F(u_0)\mathrm{d}x < 0. \tag{2.54}$$

Let $h(t) = J^\infty(tu_0)$ for $t > 0$. Since $h(1) < 0$ by (2.54), and $h(t) > 0$ for $t > 0$ small enough by (2.5), there exists $t_0 \in (0,1)$ such that $h(t_0) = J^\infty(t_0u_0) = 0$, namely $t_0u_0 \in \mathcal{P}^\infty$. Therefore, we have

$$A^\infty \leq \frac{a}{2}\|\nabla(t_0u_0)\|_2^2 + \frac{b}{4}\|\nabla(t_0u_0)\|_2^4 < \frac{a}{2}\|\nabla u_0\|_2^2 + \frac{b}{4}\|\nabla u_0\|_2^4 \leq \frac{2a\pi}{\alpha_0} + \frac{4b\pi^2}{\alpha_0^2},$$

which yields relation (2.52).

In the case $C_{\mathrm{TM}}^*(F) = +\infty$, for any $V_\infty > 0$, there exists $u_0 \in H^1(\mathbb{R}^2) \setminus \{0\}$ such that

$$\|\nabla u_0\|_2^2 \leq R(F) \ \text{ and } \ V_\infty\|u_0\|_2^2 < 2\int_{\mathbb{R}^2} F(u_0)\mathrm{d}x.$$

Hence we can repeat the same arguments as above to get the desired conclusion. □

2.2.2 Proof of Theorem 2.1 concluded

If $V_\infty < C_{\mathrm{TM}}^*(F)$, then Lemma 2.7 leads to

$$A^\infty < \frac{2a\pi}{\alpha_0} + \frac{4b\pi^2}{\alpha_0^2}.$$

Hence the assumptions of Lemma 2.6 are fulfilled, and we obtain the existence of a positive least-energy solution for equation $(\mathcal{K})_\infty$. Moreover, recalling (2.3) and in light of (B) of Lemma 2.5, we conclude that $C_{\mathrm{TM}}^*(F) = +\infty$ if and only if $\lim_{t\to+\infty} \frac{t^2F(t)}{e^{\alpha_0 t^2}} = +\infty$. □

2.2.3 Proof of Theorem 2.2

Besides the attainability of A^∞, Theorem 2.1 also shows that infimum A^∞ equals to the ground state m^∞. Proceeding to the proof of Theorem 2.2, we next investigate the interesting relation between the infimum A^∞ and the mountain pass level c^∞ defined by (2.9). Before this, we first verify that $\Phi^\infty(u)$ has a mountain pass geometry, in order to show that the mountain pass level c^∞ is well-defined. Indeed it has the following properties:

Lemma 2.8 *Assume that $V_\infty < C_{\mathrm{TM}}^*(F)$ and f satisfies* (F0)–(F3). *Then*

(i) *there exist $\rho_0 > 0$ and $\delta_0 > 0$ such that $\Phi^\infty(u) \geq \delta_0$ for all $\|u\| = \rho_0$;*

(ii) *there exists $u_0 \in H^1(\mathbb{R}^2)$ such that $\|u_0\| > \rho_0$ and $\Phi^\infty(u_0) < 0$.*

Proof. (i) By the Rellich embedding theorem, for $s \in [2,\infty)$, there exists $\gamma_s > 0$ such that

$$\|u\|_s \leq \gamma_s\|u\|, \quad \forall\, u \in H^1(\mathbb{R}^2). \tag{2.55}$$

By (F1) and (F2), we obtain for some constants $\alpha > 0$ and $C_1 > 0$

$$|F(t)| \leq \frac{V_\infty}{4}t^2 + C_1\left(e^{\alpha t^2} - 1\right)|t|^3, \quad \forall\, t \in \mathbb{R}. \tag{2.56}$$

In view of Lemma 2.1 (ii), we have

$$\int_{\mathbb{R}^2}\left(e^{2\alpha u^2} - 1\right)\mathrm{d}x = \int_{\mathbb{R}^2}\left(e^{2\alpha\|u\|^2(u/\|u\|)^2} - 1\right)\mathrm{d}x \leq \mathcal{C}(2\pi), \quad \forall\, \|u\| \leq \sqrt{\pi/\alpha}. \tag{2.57}$$

Next, from relations (2.56) and (2.57), we obtain

$$\begin{aligned}\int_{\mathbb{R}^2} F(u)\mathrm{d}x &\le \frac{V_\infty}{4}\|u\|_2^2 + C_1\int_{\mathbb{R}^2}\left(e^{\alpha u^2}-1\right)|u|^3\mathrm{d}x\\ &\le \frac{V_\infty}{4}\|u\|_2^2 + C_1\left[\int_{\mathbb{R}^2}\left(e^{2\alpha u^2}-1\right)\mathrm{d}x\right]^{1/2}\|u\|_6^3\\ &\le \frac{V_\infty}{4}\|u\|_2^2 + C_2\|u\|^3, \quad \forall\, \|u\| \le \sqrt{\pi/\alpha}. \end{aligned} \tag{2.58}$$

Hence, it follows from (2.8) and (2.58) that

$$\Phi^\infty(u) \ge \frac{a}{2}\|\nabla u\|_2^2 + \frac{V_\infty}{4}\|u\|_2^2 - C_2\|u\|^3, \quad \forall\, \|u\| \le \sqrt{\pi/\alpha}. \tag{2.59}$$

Therefore, there exist $\delta_0 > 0$ and $0 < \rho_0 < \sqrt{\pi/\alpha}$ such that

$$\Phi^\infty(u) \ge \delta_0, \quad \forall\, u \in S := \{u \in H^1(\mathbb{R}^2) : \|u\| = \rho_0\}. \tag{2.60}$$

(ii) This conclusion will be done in the proof of next lemma, see (2.71) below. □

Lemma 2.9 *Assume that* $V_\infty < C^*_{\mathrm{TM}}(F)$ *and* f *satisfies conditions* (F0)–(F3). *Then for any least-energy solution* $w(x)$ *of* $(\mathcal{K})_\infty$, *there exists a path* $\tilde{\gamma} \in \Gamma^\infty$ *such that* $w(x) \in \tilde{\gamma}([0,1])$ *and*

$$\max_{t\in[0,1]} \Phi^\infty(\tilde{\gamma}(t)) = \Phi^\infty(w).$$

Proof. Let w be a given least-energy solution of $(\mathcal{K})_\infty$ obtained in Theorem 2.1. We define a curve γ, constituted of the three pieces given by:

$$\gamma(\theta) = \begin{cases} \dfrac{\theta}{t_1} w_{t_1} & \text{if } \theta \in [0, t_1],\\ w_{[t_3(\theta - t_1) + (t_2-\theta)t_1]/(t_2-t_1)} & \text{if } \theta \in [t_1, t_2],\\ \dfrac{t_2(\theta - t_2) + t_3 - \theta}{t_3 - t_2} w_{t_3} & \text{if } \theta \in [t_2, t_3], \end{cases} \tag{2.61}$$

where $w_t(x) = w(x/t)$ and $0 < t_1 < 1 < t_2 < t_3$ are determined later.

It follows that $\gamma \in \mathcal{C}([0,1], H^1(\mathbb{R}^2))$. Since w is a weak solution of $(\mathcal{K})_\infty$, we have $\langle (\Phi^\infty)'(w), w\rangle = 0$, and so

$$\int_{\mathbb{R}^2} [f(w) - V_\infty w]\, w\mathrm{d}x = a\|\nabla w\|_2^2 + b\|\nabla w\|_2^4 > 0.$$

Then we can find $t_2 > 1$ such that

$$\int_{\mathbb{R}^2} [f(\xi w) - V_\infty \xi w]\, w\mathrm{d}x > 0, \quad \forall\, \xi \in [1, t_2]. \tag{2.62}$$

Set

$$\phi(t) = \frac{f(t)}{t} - V_\infty, \quad \forall\, t \in \mathbb{R}. \tag{2.63}$$

Then $\phi \in \mathcal{C}(\mathbb{R}, \mathbb{R})$, by (F1) and (F2). Moreover, relations (2.62) and (2.63) give

$$\int_{\mathbb{R}^2} \phi(\xi w) w^2 \mathrm{d}x > 0, \quad \forall\, \xi \in [1, t_2]. \tag{2.64}$$

Note that for any fixed $t > 0$,

$$\begin{aligned}\frac{\mathrm{d}}{\mathrm{d}\xi}\Phi^\infty(\xi w_t) &= \langle(\Phi^\infty)'(\xi w_t), w_t\rangle \\ &= \xi\left[a\|\nabla w_t\|_2^2 + \xi^2 b\|\nabla w_t\|_2^4 - \int_{\mathbb{R}^2}\phi(\xi w_t)w_t^2\mathrm{d}x\right] \\ &= \xi\left[a\|\nabla w\|_2^2 + \xi^2 b\|\nabla w\|_2^4 - t^2\int_{\mathbb{R}^2}\phi(\xi w)w^2\mathrm{d}x\right]. \qquad (2.65)\end{aligned}$$

Choosing $t_1 \in (0,1)$, we have

$$\begin{aligned}&a\|\nabla w\|_2^2 + \xi^2 b\|\nabla w\|_2^4 - t_1^2\int_{\mathbb{R}^2}\phi(\xi w)w^2\mathrm{d}x \\ &\geq a\|\nabla w\|_2^2 - t_1^2\int_{\mathbb{R}^2}\phi(\xi w)w^2\mathrm{d}x > 0, \quad \forall\, \xi \in [0,1]. \qquad (2.66)\end{aligned}$$

By (2.64), we can also choose $t_3 > t_2$ such that

$$\begin{aligned}&a\|\nabla w\|_2^2 + \xi^2 b\|\nabla w\|_2^4 - t_3^2\int_{\mathbb{R}^2}\phi(\xi w)w^2\mathrm{d}x \\ &\leq a\|\nabla w\|_2^2 + t_2^2 b\|\nabla w\|_2^4 - t_3^2\int_{\mathbb{R}^2}\phi(\xi w)w^2\mathrm{d}x \\ &\leq -\frac{2}{t_2^2-1}\left(a\|\nabla w\|_2^2 + b\|\nabla w\|_2^4\right), \quad \forall\, \xi \in [1,t_2]. \qquad (2.67)\end{aligned}$$

Thus, we can see by (2.66) that the function $\Phi^\infty\left(\frac{\theta}{t_1}w_{t_1}\right)$ is increasing on $\theta \in [0,t_1]$ and takes its maximum at $\theta = t_1$, namely

$$\Phi^\infty(\gamma(\theta)) = \Phi^\infty\left(\frac{\theta}{t_1}w_{t_1}\right) \leq \Phi^\infty(w_{t_1}), \quad \forall\, \theta \in [0,t_1]. \qquad (2.68)$$

Since $J^\infty(w) = \int_{\mathbb{R}^2}\left[V_\infty w^2 - 2F(w)\right]\mathrm{d}x = 0$ by the Pohozaev-type identity (see Lemma 2.2), we have

$$\begin{aligned}\Phi^\infty(w_t) &= \frac{a}{2}\|\nabla w\|_2^2 + \frac{b}{4}\|\nabla w\|_2^4 + \frac{t^2}{2}\int_{\mathbb{R}^2}\left[V_\infty w^2 - 2F(w)\right]\mathrm{d}x \\ &= \frac{a}{2}\|\nabla w\|_2^2 + \frac{b}{4}\|\nabla w\|_2^4 = \Phi^\infty(w) = m^\infty, \quad \forall\, t > 0. \qquad (2.69)\end{aligned}$$

By (2.65) and (2.67), we deduce that $\Phi^\infty(\xi w_{t_3})$ is decreasing on $\xi \in [1,t_2]$. Noticing that

$$\frac{t_2(\theta - t_2) + t_3 - \theta}{t_3 - t_2} \in [1,t_2] \;\Leftrightarrow\; \theta \in [t_2,t_3],$$

we then get that $\Phi^\infty\left(\frac{t_2(\theta-t_2)+t_3-\theta}{t_3-t_2}w_{t_3}\right)$ is decreasing on $\theta \in [t_2,t_3]$. Therefore,

$$\Phi^\infty(\gamma(\theta)) \leq \Phi^\infty(\gamma(t_2)) = \Phi^\infty(w_{t_3}), \quad \forall\, \theta \in [t_2,t_3]. \qquad (2.70)$$

Moreover, relation (2.67) yields

$$\begin{aligned}\Phi^\infty(\gamma(t_3)) &= \Phi^\infty(t_2 w_{t_3}) = \Phi^\infty(w_{t_3}) + \int_1^{t_2}\frac{\mathrm{d}}{\mathrm{d}\xi}\Phi^\infty(\xi w_{t_3})\mathrm{d}\xi \\ &\leq \frac{a}{2}\|\nabla w\|_2^2 + \frac{b}{4}\|\nabla w\|_2^4 - \int_1^{t_2}\frac{2\xi}{t_2^2-1}\left(a\|\nabla w\|_2^2 + b\|\nabla w\|_2^4\right)\mathrm{d}\xi \\ &= -\frac{a}{2}\|\nabla w\|_2^2 - \frac{3b}{4}\|\nabla w\|_2^4 < 0. \qquad (2.71)\end{aligned}$$

Combining (2.68), (2.69) and (2.70), we have

$$\Phi^\infty(\gamma(\theta)) \le \Phi^\infty(w) = m^\infty, \quad \forall\, \theta \in [0, t_3]. \tag{2.72}$$

Let $\tilde{\gamma}(\theta) = \gamma(t_3\theta)$ for all $\theta \in [0,1]$. Since $\tilde{\gamma} \in \Gamma^\infty$ due to (2.71), it follows from (2.72) that

$$\max_{t\in[0,t_3]} \Phi^\infty(\gamma(t)) = \Phi^\infty(w) = m^\infty, \tag{2.73}$$

where the definition of Γ^∞ is given by (2.10). The proof is completed. □

From the definition of c^∞, as a corollary to Lemma 2.9, we have the following result.

Corollary 2.2 *Assume that* $V_\infty < C^*_{\mathrm{TM}}(F)$ *and* f *satisfies* (F0)–(F3). *Then* $c^\infty \le m^\infty$.

Lemma 2.10 *Assume that* $V_\infty < C^*_{\mathrm{TM}}(F)$ *and* f *satisfies* (F0)–(F3). *Then* $c^\infty \ge A^\infty$.

Proof. To prove that $c^\infty \ge A^\infty = \inf_{u\in\mathcal{P}_\infty} \Phi^\infty(u)$, it suffices to show that

$$\gamma([0,1]) \cap \mathcal{P}_\infty \ne \emptyset \text{ for all } \gamma \in \Gamma^\infty. \tag{2.74}$$

The proof of (2.74) follows the same lines as in Lemma 4.1 of Jeanjean and Tanaka [80], so we omit it here. □

Proof of Theorem 2.2. In view of Theorem 2.1, we know that $A^\infty = m^\infty$. Therefore, Theorem 2.2 follows directly from Lemmas 2.9 and 2.10. □

2.3 Non-autonomous case with trapping potential: ground state solutions

In this section, we are concerned with the ground state solutions for $(\mathcal{K})$ with the trapping potential, that is, provided that V satisfies condition (V1).

Arguing as in the proof of Lemma 2.8, we can verify that $\Phi(u)$ has a mountain pass geometry. Applying the mountain pass theorem, we know that Φ possesses a Cerami sequence, reads as follows.

Lemma 2.11 *Assume that* $V \in \mathcal{C}(\mathbb{R}^2, [V_0, V_\infty])$ *and* f *satisfies* (F1), (F2) *and* (F4). *Then there exists a sequence* $\{u_n\} \subset H^1(\mathbb{R}^2)$ *such that*

$$\Phi(u_n) \to c, \qquad \|\Phi'(u_n)\|(1 + \|u_n\|) \to 0, \tag{2.75}$$

where

$$c = \inf_{\gamma\in\Gamma} \max_{t\in[0,1]} \Phi(\gamma(t)) \tag{2.76}$$

and

$$\Gamma = \left\{\gamma \in \mathcal{C}([0,1], H^1(\mathbb{R}^2)) : \gamma(0) = 0, \Phi(\gamma(1)) < 0\right\}. \tag{2.77}$$

Lemma 2.12 *Assume that* $V \in \mathcal{C}(\mathbb{R}^2, [V_0, V_\infty])$ *with* $V_\infty < C^*_{\mathrm{TM}}(F)$ *and* f *satisfies* (F1), (F2) *and* (F4). *Then*

$$c \le c^\infty < \frac{2\pi a}{\alpha_0} + \frac{4\pi^2 b}{\alpha_0^2},$$

where the definitions of c *and* c^∞ *are given by* (2.76) *and* (2.9).

Proof. First, we prove that $c \le c^\infty$. Since $\Phi(\gamma(1)) \le \Phi^\infty(\gamma(1)) < 0$ for any $\gamma \in \Gamma^\infty$ due to $V(x) \le V_\infty$ for all $x \in \mathbb{R}^2$, we have $\Gamma^\infty \subset \Gamma$. Then for any $\gamma \in \Gamma^\infty$, we obtain

$$\max_{t\in[0,1]} \Phi^\infty(\gamma(t)) \ge \max_{t\in[0,1]} \Phi(\gamma(t)) \ge \inf_{\gamma\in\Gamma} \max_{t\in[0,1]} \Phi(\gamma(t)) = c.$$

This relation, together with the arbitrariness of γ, yields

$$c^\infty = \inf_{\gamma\in\Gamma^\infty} \max_{t\in[0,1]} \Phi^\infty(\gamma(t)) \ge c.$$

Next, we prove that $c^\infty < \frac{2\pi a}{\alpha_0} + \frac{4\pi^2 b}{\alpha_0^2}$. Without loss of generality, we just consider the case $C^*_{\mathrm{TM}}(F) < +\infty$. Since $V_\infty < C^*_{\mathrm{TM}}(F)$, then $V_\infty + 2\varepsilon_0 < C^*_{\mathrm{TM}}(F)$ for some $\varepsilon_0 > 0$. In view of the definition of $C^*_{\mathrm{TM}}(F)$, there exists $\hat{u} \in H^1(\mathbb{R}^2)$ with $\|\nabla \hat{u}\|_2^2 \le \frac{4\pi}{\alpha_0}$ satisfying

$$(V_\infty + \varepsilon_0)\,\|\hat{u}\|_2^2 < 2\int_{\mathbb{R}^2} F(\hat{u})\mathrm{d}x. \tag{2.78}$$

This shows that $J^\infty(\hat{u}) < 0$, where the definition of J^∞ is given by (2.15).

Let $h(s) := J^\infty(s\hat{u})$ for $s > 0$. Since $h(1) < 0$ and $h(s) > 0$ for $s > 0$ small enough by (F2), then there exists $s_0 \in (0,1)$ satisfying $h(s_0) = 0$. Therefore, for $\tilde{u} := s_0\hat{u}$, we have

$$(V_\infty + \varepsilon_0)\,\|\tilde{u}\|_2^2 = 2\int_{\mathbb{R}^2} F(\tilde{u})\mathrm{d}x. \tag{2.79}$$

By (2.8) and (2.79), we get

$$\begin{aligned}
\Phi^\infty(\tilde{u}_t) &= \frac{a}{2}\|\nabla\tilde{u}\|_2^2 + \frac{b}{4}\|\nabla\tilde{u}\|_2^4 + \frac{t^2}{2}V_\infty\|\tilde{u}\|_2^2 - t^2\int_{\mathbb{R}^2} F(\tilde{u})\mathrm{d}x \\
&= \frac{a}{2}\|\nabla\tilde{u}\|_2^2 + \frac{b}{4}\|\nabla\tilde{u}\|_2^4 - \frac{\varepsilon_0 t^2}{2}\|\tilde{u}\|_2^2 \qquad (2.80)\\
&\le \frac{as_0^2}{2}\|\nabla\hat{u}\|_2^2 + \frac{bs_0^4}{4}\|\nabla\hat{u}\|_2^4 < \frac{2a\pi}{\alpha_0} + \frac{4b\pi^2}{\alpha_0^2}, \qquad \forall\, t > 0, \qquad (2.81)
\end{aligned}$$

where the last inequality follows from the fact that $\|\nabla\hat{u}\|_2^2 \le \frac{4\pi}{\alpha_0}$ and $s_0 \in (0,1)$. Using (F1), (F2) and Lemma 2.1 i), we deduce that there exists a constant $M_1 > 0$ independent of $\xi \in [0,1]$ such that

$$\int_{\mathbb{R}^2} \left|\frac{f(\xi\tilde{u})}{\xi\tilde{u}}\tilde{u}^2\right| \mathrm{d}x \le M_1, \qquad \forall\, \xi \in [0,1]. \tag{2.82}$$

Let $\tilde{u}_t(x) := \tilde{u}(x/t)$ for $t > 0$. Then it follows from (2.82) that

$$\begin{aligned}
\frac{d}{d\xi}\Phi^\infty(\xi\tilde{u}_t) &= \langle(\Phi^\infty)'(\xi\tilde{u}_t), \tilde{u}_t\rangle \\
&= \xi\left[a\|\nabla\tilde{u}\|_2^2 + \xi^2 b\|\nabla\tilde{u}\|_2^4 + t^2\left(V_\infty\|\tilde{u}\|_2^2 - \int_{\mathbb{R}^2}\frac{f(\xi\tilde{u})}{\xi\tilde{u}}\tilde{u}^2\mathrm{d}x\right)\right] \\
&\ge \xi\left[\|\nabla\tilde{u}\|_2^2 + t^2\left(V_\infty\|\tilde{u}\|_2^2 - M_1\right)\right], \qquad \forall\, t > 0,\ \xi \in [0,1]. \qquad (2.83)
\end{aligned}$$

Since $\tilde{u} \ne 0$, we can choose $t_0 \in (0,1)$ such that

$$\|\nabla\tilde{u}\|_2^2 + t_0^2\left(V_\infty\|\tilde{u}\|_2^2 - M_1\right) > 0. \tag{2.84}$$

Using (2.80), we know that there exists $T > 1$ such that $\Phi^\infty(\tilde{u}_T) < 0$. Let

$$\gamma^*(t) = \begin{cases} t_0^{-1}t\tilde{u}_{t_0}, & 0 \le t \le t_0, \\ \tilde{u}_{t_0+(T-t_0)(t-t_0)/(1-t_0)}, & t_0 \le t \le 1. \end{cases} \tag{2.85}$$

It follows that $\gamma^* \in \Gamma^\infty$, where Γ^∞ is defined by (2.10). Note that relations (2.83) and (2.84) show that $\Phi^\infty\left(t_0^{-1}t\tilde{u}_{t_0}\right)$ is increasing on $t \in [0, t_0]$. Hence it follows from (2.8) and (2.81) that

$$\Phi^\infty\left(t_0^{-1}t\tilde{u}_{t_0}\right) \le \Phi^\infty\left(\tilde{u}_{t_0}\right) < \frac{2a\pi}{\alpha_0} + \frac{4b\pi^2}{\alpha_0^2}, \quad \forall\, 0 \le t \le t_0. \tag{2.86}$$

From the definition of c^∞, (2.85) and (2.86), we derive that

$$c^\infty \le \max_{t\in[0,1]} \Phi^\infty(\gamma^*(t)) < \frac{2a\pi}{\alpha_0} + \frac{4b\pi^2}{\alpha_0^2}. \tag{2.87}$$

This completes the proof. □

Note that (F5) implies the following inequality:

$$\begin{aligned}
&\frac{1-t^4}{4}f(u)u + F(tu) - F(u) + \frac{(1-t^2)^2}{4}V_0u^2 \\
&= \int_t^1 \left[\frac{f(u)-V_0u}{u^3} - \frac{f(su)-V_0(su)}{(su)^3}\right] s^3u^4 \mathrm{d}s \\
&\ge 0, \ \ \forall\, u \ne 0,\ t \ge 0.
\end{aligned} \tag{2.88}$$

Letting $t = 0$ in (2.88), we have

$$\frac{1}{4}f(u)u - F(u) + \frac{1}{4}V_0u^2 \ge 0, \ \ \forall\, u \in \mathbb{R}. \tag{2.89}$$

It follows that the following properties are fulfilled.

Lemma 2.13 *Assume that* (V1), (F1), (F2) *and* (F5) *hold. Then*

$$\Phi(u) \ge \Phi(tu) + \frac{1-t^4}{4}\langle \Phi'(u), u\rangle + \frac{(1-t^2)^2a}{4}\|\nabla u\|_2^2, \quad \forall\, u \in H^1(\mathbb{R}^2),\ t \ge 0. \tag{2.90}$$

Corollary 2.3 *Assume that* (V1), (F1), (F2) *and* (F5) *hold. Then*

$$\Phi(u) \ge \max_{t\ge 0} \Phi(tu), \quad \forall\, u \in \mathcal{N}, \tag{2.91}$$

where $\mathcal{N}$ *is defined by* (2.13).

Lemma 2.14 *Assume that* (V1), (F1), (F2) *and* (F5) *hold. Then for any* $u \in H^1(\mathbb{R}^2)\setminus\{0\}$, *there exists* $t_u > 0$ *such that* $t_u u \in \mathcal{N}$.

Lemma 2.15 *Assume that* (V1), (F1), (F2) *and* (F5) *hold. Then*

$$c = c_N = \inf_{u\in\mathcal{N}} \max_{t\ge 0} \Phi(tu). \tag{2.92}$$

Lemma 2.16 *Assume that conditions* (V1), (F1), (F2), (F4) *and* (F5) *hold. Then any sequence satisfying relation* (2.75) *is bounded in* $H^1(\mathbb{R}^2)$.

Proof. Using (F4), by straightforward computation, there exists $R_0 > \beta_0$ such that

$$f(t)t - 8F(t) \ge 0, \quad \forall\, |t| \ge R_0. \tag{2.93}$$

By (2.6), (2.7), (2.75), (2.89) and (2.93), we have

$$
\begin{aligned}
c + o(1) &= \Phi(u_n) + o(1) = \Phi(u_n) - \frac{1}{4}\langle \Phi'(u_n), u_n\rangle \\
&= \frac{a}{4}\|\nabla u_n\|_2^2 + \frac{1}{4}\int_{\mathbb{R}^2} V(x)u_n^2 \mathrm{d}x + \int_{\mathbb{R}^2}\left[\frac{1}{4}f(u_n)u_n - F(u_n)\right]\mathrm{d}x \qquad (2.94)\\
&\geq \frac{a}{4}\|\nabla u_n\|_2^2 + \frac{1}{8}\int_{|u_n|\geq R_0} f(u_n)u_n \mathrm{d}x. \qquad (2.95)
\end{aligned}
$$

In view of (2.95), to prove the boundedness of $\{\|u_n\|\}$, it suffices to show the boundedness of $\{\|u_n\|_2\}$. To this end, arguing by contradiction, we assume that $\|u_n\|_2 \to \infty$ as $n \to \infty$. Let

$$
t_n = \left[\frac{2(c+1)}{V_0\|u_n\|_2^2}\right]^{1/4}. \qquad (2.96)
$$

Then $t_n \to 0$ as $n \to \infty$. Note that (2.89) implies that

$$
\frac{F(t) - \frac{1}{2}V_0 t^2}{t^4} \text{ is nondecreasing on } t \in (-\infty, 0) \text{ and } t \in (0, +\infty). \qquad (2.97)
$$

This gives

$$
F(tu_n) - \frac{1}{2}V_0 t^2 u_n^2 \leq t^4\left[F(u_n) - \frac{1}{2}V_0 u_n^2\right], \quad \forall\, t \in (0,1), n \in \mathbb{N}. \qquad (2.98)
$$

Noticing that $2t_n^4 \leq t_n^2 < 1$ for large enough $n \in \mathbb{N}$, from (F2), (2.94) and (2.98), we deduce that for large enough $n \in \mathbb{N}$,

$$
\begin{aligned}
\int_{\mathbb{R}^2} F(t_n u_n)\mathrm{d}x &= \int_{|u_n|\leq R_0} F(t_n u_n)\mathrm{d}x + \int_{|u_n|>R_0} F(t_n u_n)\mathrm{d}x \\
&\leq \int_{|u_n|\leq R_0} F(t_n u_n)\mathrm{d}x + \frac{V_0}{2}t_n^2(1-t_n^2)\int_{|u_n|>R_0} u_n^2\mathrm{d}x + t_n^4\int_{|u_n|>R_0} F(u_n)\mathrm{d}x \\
&\leq \frac{V_0}{4}\int_{|u_n|\leq R_0} |t_n u_n|^2\mathrm{d}x + \frac{V_0}{2}t_n^2(1-t_n^2)\int_{|u_n|>R_0} u_n^2\mathrm{d}x + \frac{t_n^4}{8}\int_{|u_n|>R_0} f(u_n)u_n\mathrm{d}x \\
&\leq \frac{V_0}{4}\int_{|u_n|\leq R_0} |t_n u_n|^2\mathrm{d}x + \frac{V_0}{2}t_n^2(1-t_n^2)\int_{|u_n|>R_0} u_n^2\mathrm{d}x + (c+1)t_n^4
\end{aligned}
$$

and

$$
\begin{aligned}
\int_{\mathbb{R}^2} F(t_n u_n)\mathrm{d}x &\leq \frac{V_0}{4}t_n^2\int_{|u_n|\leq R_0} u_n^2\mathrm{d}x + \frac{V_0}{2}t_n^2\int_{|u_n|>R_0} u_n^2\mathrm{d}x + \frac{V_0}{2}t_n^4\int_{|u_n|\leq R_0} u_n^2\mathrm{d}x \\
&\quad - \frac{V_0}{2}t_n^4\|u_n\|_2^2 + (c+1)t_n^4 \\
&\leq \frac{V_0}{2}t_n^2\int_{|u_n|\leq R_0} u_n^2\mathrm{d}x + \frac{V_0}{2}t_n^2\int_{|u_n|>R_0} u_n^2\mathrm{d}x - \frac{V_0}{2}t_n^4\|u_n\|_2^2 + (c+1)t_n^4 \\
&= \frac{V_0}{2}t_n^2\|u_n\|_2^2 - (c+1) + \frac{2(c+1)^2}{V_0\|u_n\|_2^2}. \qquad (2.99)
\end{aligned}
$$

By (2.6), (2.7), (2.75), (2.90), (2.94) and (2.99), we have

$$
\begin{aligned}
c+o(1) &= \Phi(u_n)+o(1) \geq \Phi(t_n u_n)+o(1) \\
&= \frac{at_n^2}{2}\|\nabla u_n\|_2^2 + \frac{bt_n^4}{4}\|\nabla u_n\|_2^4 + \frac{t_n^2}{2}\int_{\mathbb{R}^2} V(x)u_n^2\mathrm{d}x - \int_{\mathbb{R}^2} F(t_n u_n)\mathrm{d}x \\
&\geq \frac{t_n^2}{2}\int_{\mathbb{R}^2} V(x)u_n^2\mathrm{d}x - \frac{V_0}{2}t_n^2\|u_n\|_2^2 + (c+1) - \frac{2(c+1)^2}{V_0\|u_n\|_2^2} + o(1) \\
&\geq c+1+o(1).
\end{aligned}
$$

This contradiction shows the boundedness of $\{\|u_n\|_2\}$. We conclude that $\{u_n\}$ is bounded in $H^1(\mathbb{R}^2)$. □

As a direct consequence of [41, Lemma 2.1], we can get the following property.

Lemma 2.17 *Assume that* (F1) *and* (F2) *hold. Let $u_n \rightharpoonup \bar{u}$ in $H^1(\mathbb{R}^2)$ and*

$$\int_{\mathbb{R}^2} |f(u_n)u_n|\mathrm{d}x \leq K_0 \tag{2.100}$$

for some constant $K_0 > 0$. Then for any $\phi \in \mathcal{C}_0^\infty(\mathbb{R}^2)$

$$\lim_{n\to\infty}\int_{\mathbb{R}^2} f(u_n)\phi\mathrm{d}x = \int_{\mathbb{R}^2} f(\bar{u})\phi\mathrm{d}x. \tag{2.101}$$

Similarly to the proof of [145, Lemma 2.13], we can get the following lemma.

Lemma 2.18 *Assume that* (V1), (F1), (F2) *and* (F5) *hold. If $\bar{u} \in \mathcal{N}$ and $\Phi(\bar{u}) = c_N$, then $\bar{u}$ is a critical point of Φ, namely $\Phi'(\bar{u}) = 0$.*

2.3.1 Proof of Theorem 2.3

Applying Lemmas 2.11 and 2.16, we deduce that there exists a sequence $\{u_n\} \subset H^1(\mathbb{R}^2)$ satisfying (2.75) and $\|u_n\| \leq C_7$. Then relations (2.7) and (2.75) yield

$$\int_{\mathbb{R}^2} f(u_n)u_n\mathrm{d}x \leq C_8. \tag{2.102}$$

Thus, passing to a subsequence if necessary, we can assume that $u_n \rightharpoonup \bar{u}$ in $H^1(\mathbb{R}^2)$, $u_n \to \bar{u}$ in $L^s_{\mathrm{loc}}(\mathbb{R}^2)$ for $s \in [1,\infty)$ and $u_n \to \bar{u}$ a.e. in $\mathbb{R}^2$.

We distinguish the following two cases: (i) $\bar{u} \neq 0$; (ii) $\bar{u} = 0$.

Case (i): $\bar{u} \neq 0$. We assume that $l = \lim_{n\to\infty}\|\nabla u_n\|_2$ up to a subsequence. Then $\|\nabla\bar{u}\|_2 \leq l$. Applying Lemma 2.17, we get

$$\lim_{n\to\infty}\int_{\mathbb{R}^2} f(u_n)\varphi\mathrm{d}x = \int_{\mathbb{R}^2} f(\bar{u})\varphi\mathrm{d}x, \quad \forall\, \varphi \in \mathcal{C}_0^\infty(\mathbb{R}^2). \tag{2.103}$$

Note that there exists $\{\varphi_n\} \subset \mathcal{C}_0^\infty(\mathbb{R}^2)$ such that $\|\varphi_n - \bar{u}\| = o(1)$ since $C_0^\infty(\mathbb{R}^2)$ is dense in $H^1(\mathbb{R}^2)$. This, together with (2.103), (F1) and Lemma 2.1 i), gives

$$
\begin{aligned}
\left|\int_{\mathbb{R}^2} f(\bar{u})(\varphi_n - \bar{u})\mathrm{d}x\right| &\leq \int_{\mathbb{R}^2} |f(\bar{u})||\varphi_n - \bar{u}|\mathrm{d}x \\
&\leq \int_{\mathbb{R}^2} \left(|\bar{u}| + C_9 e^{2\alpha_0\bar{u}^2}\right)|\varphi_n - \bar{u}|\mathrm{d}x \\
&\leq \|\bar{u}\|_2\|\varphi_n - \bar{u}\|_2 + C_9\left(\int_{\mathbb{R}^2} e^{4\alpha_0\bar{u}^2}\mathrm{d}x\right)^{\frac{1}{2}}\|\varphi_n - \bar{u}\|_2 \\
&= o(1).
\end{aligned} \tag{2.104}
$$

By (2.7), (2.75) and (2.104), we have

$$\langle \Phi'(\bar{u}), \bar{u}\rangle + b(l^2 - \|\nabla \bar{u}\|_2^2)\int_{\mathbb{R}^2} \nabla \bar{u}\cdot\nabla \bar{u}\mathrm{d}x = \lim_{n\to\infty}\langle \Phi'(u_n), \bar{u}\rangle = 0,$$

which, together with $l^2 - \|\nabla \bar{u}\|_2^2 \geq 0$, implies that $\langle \Phi'(\bar{u}), \bar{u}\rangle \leq 0$.

Note that $\langle \Phi'(t\bar{u}), t\bar{u}\rangle > 0$ for small $t > 0$ by (F1) and (F2). Then there exists $\bar{t} \in (0, 1]$ such that

$$\langle \Phi'(\bar{t}\bar{u}), \bar{t}\bar{u}\rangle = 0 \quad \text{and} \quad \Phi(\bar{t}\bar{u}) \geq c_N. \tag{2.105}$$

Using (F5), we obtain

$$f(tu)tu \leq f(u)ut^4 + V_0(1-t^2)(tu)^2, \quad \forall\, 0 \leq t \leq 1,\ u \in \mathbb{R}. \tag{2.106}$$

Note that (2.88) implies

$$F(tu) \geq \frac{t^4-1}{4}f(u)u + F(u) - \frac{1-2t^2+t^4}{4}V_0u^2, \quad \forall\, t \geq 0,\ u \in \mathbb{R}. \tag{2.107}$$

Combining (2.106) with (2.107), we have

$$\frac{1}{4}f(tu)tu - F(tu) + \frac{V_0}{4}(tu)^2 \leq \frac{1}{4}f(u)u - F(u) + \frac{V_0}{4}u^2, \quad \forall\, 0 \leq t \leq 1,\ u \in \mathbb{R}. \tag{2.108}$$

Since $\bar{t} \in (0, 1]$, from (2.75), (2.105), lemma 2.15 and Fatou's lemma, it follows that

$$\begin{aligned}
c_N &\leq \Phi(\bar{t}\bar{u}) - \frac{1}{4}\langle \Phi'(\bar{t}\bar{u}), \bar{t}\bar{u}\rangle \\
&= \frac{\bar{t}^2}{4}\int_{\mathbb{R}^2}\left[a|\nabla \bar{u}|^2 + (V(x)-V_0)\bar{u}^2\right]\mathrm{d}x + \int_{\mathbb{R}^2}\left[\frac{1}{4}f(\bar{t}\bar{u})\bar{t}\bar{u} - F(\bar{t}\bar{u}) + \frac{V_0}{4}(\bar{t}\bar{u})^2\right]\mathrm{d}x \\
&\leq \frac{1}{4}\int_{\mathbb{R}^2}\left[a|\nabla \bar{u}|^2 + (V(x)-V_0)\bar{u}^2\right]\mathrm{d}x + \int_{\mathbb{R}^2}\left[\frac{1}{4}f(\bar{u})\bar{u} - F(\bar{u}) + \frac{V_0}{4}\bar{u}^2\right]\mathrm{d}x \\
&= \Phi(\bar{u}) - \frac{1}{4}\langle \Phi'(\bar{u}), \bar{u}\rangle \\
&\leq \lim_{n\to\infty}\left[\Phi(u_n) - \frac{1}{4}\langle \Phi'(u_n), u_n\rangle\right] = c = c_N.
\end{aligned}$$

This implies that $\bar{t} = 1$ and $u_n \to \bar{u}$ in $H^1(\mathbb{R}^2)$. Hence, $\Phi'(\bar{u}) = 0$ and $\Phi(\bar{u}) = c_N = c$.

Case (ii): $\bar{u} = 0$. Then $u_n \rightharpoonup 0$ in $H^1(\mathbb{R}^2)$. Moreover, we have

$$\Phi^\infty(u_n) \to c, \quad \|(\Phi^\infty)'(u_n)\|(1 + \|u_n\|) \to 0. \tag{2.109}$$

We claim that

$$\delta := \limsup_{n\to\infty}\sup_{y\in\mathbb{R}^2}\int_{B_1(y)}|u_n|^2\mathrm{d}x > 0. \tag{2.110}$$

Otherwise, if $\delta = 0$, then by Lions' concentration-compactness principle [153, Lemma 1.21], $u_n \to 0$ in $L^s(\mathbb{R}^2)$ for $s > 2$. Arguing as in the proof of (4.20) in Chen and Tang [32] and using (F2) and (F4), we can get

$$\int_{\mathbb{R}^2} F(u_n)\mathrm{d}x = o(1). \tag{2.111}$$

In view of Theorem 2.2, Lemma 2.12 and (2.52), we have

$$c \leq c^\infty = A^\infty < \frac{2\pi a}{\alpha_0} + \frac{4\pi^2 b}{\alpha_0^2}. \tag{2.112}$$

From (2.75), (2.111) and (2.112), it follows that

$$\frac{a}{2}\|\nabla u_n\|_2^2+\frac{1}{2}\int_{\mathbb{R}^2}V(x)u_n^2\mathrm{d}x+\frac{b}{4}\|\nabla u_n\|_2^4+o(1)=c\le c^\infty=A^\infty$$
$$=\frac{2a\pi}{\alpha_0}(1-4\bar{\varepsilon})+\frac{4b\pi^2}{\alpha_0^2}(1-4\bar{\varepsilon})^2, \quad (2.113)$$

where $0<\bar{\varepsilon}<\frac{1}{4}$.

Using (2.113), we obtain

$$\|\nabla u_n\|_2^2\le\frac{4\pi}{\alpha_0}(1-4\bar{\varepsilon}). \quad (2.114)$$

Choosing $q\in(1,2)$ satisfying

$$\frac{(1+\bar{\varepsilon})(1-3\bar{\varepsilon})q}{1-\bar{\varepsilon}}<1, \quad (2.115)$$

by (F1), we have

$$|f(t)|^q\le C_8\left[e^{\alpha_0(1+\bar{\varepsilon})qt^2}-1\right], \quad \forall\ |t|\ge 1. \quad (2.116)$$

From (2.114), (2.115), (2.116) and Lemma 2.1 ii), we deduce that

$$\int_{|u_n|\ge 1}|f(u_n)|^q\mathrm{d}x\le C_8\int_{\mathbb{R}^2}\left[e^{\alpha_0(1+\bar{\varepsilon})qu_n^2}-1\right]\mathrm{d}x$$
$$=C_8\int_{\mathbb{R}^2}\left[e^{\alpha_0(1+\bar{\varepsilon})q(\|\nabla u_n\|_2^2+4\pi\bar{\varepsilon}/\alpha_0)\frac{u_n^2}{(\|\nabla u_n\|_2^2+4\pi\bar{\varepsilon}/\alpha_0)}}-1\right]\mathrm{d}x\le C_9. \quad (2.117)$$

Let $q':=q/(q-1)>2$. Then (2.117) and the Hölder's inequality yield

$$\int_{|u_n|\ge 1}f(u_n)u_n\mathrm{d}x\le\left[\int_{|u_n|\ge 1}|f(u_n)|^q\mathrm{d}x\right]^{1/q}\|u_n\|_{q'}=o(1). \quad (2.118)$$

From (2.6), (2.7), (2.75), (2.111) and (2.118), it follows that

$$c+o(1)=\Phi(u_n)-\frac{1}{2}\langle\Phi'(u_n),u_n\rangle$$
$$=-\frac{b}{4}\|\nabla u_n\|_2^4+\int_{\mathbb{R}^2}\left[\frac{1}{2}f(u_n)u_n-F(u_n)\right]\mathrm{d}x\le o(1). \quad (2.119)$$

This contradiction shows that the claim (2.110) holds, namely $\delta>0$. Going if necessary to a subsequence, we can assume the existence of $k_n\in\mathbb{Z}^2$ such that

$$\int_{B_2(k_n)}|u_n|^2\mathrm{d}x>\frac{\delta}{2}>0.$$

Let $v_n(x)=u_n(x+k_n)$. Then

$$\int_{B_2(0)}|v_n|^2\mathrm{d}x>\frac{\delta}{2}>0. \quad (2.120)$$

Moreover, relation (2.109) gives

$$\Phi^\infty(v_n)\to c, \quad \|(\Phi^\infty)'(v_n)\|(1+\|v_n\|)\to 0. \quad (2.121)$$

Passing to a subsequence, we have $v_n \rightharpoonup \bar{v}$ in $H^1(\mathbb{R}^2)$, $v_n \to \bar{v}$ in $L^s_{\mathrm{loc}}(\mathbb{R}^2)$ for $s > 2$ and $v_n \to \bar{v}$ a.e. in $\mathbb{R}^2$. Then (2.120) yields $\bar{v} \neq 0$. In view of Theorem 2.1 and Theorem 2.2, we obtain

$$c^\infty = m^\infty = \inf\left\{\Phi^\infty(u) : u \in H^1(\mathbb{R}^2)\setminus\{0\},\ \langle(\Phi^\infty)'(u), u\rangle = 0\right\}. \tag{2.122}$$

As in the proof of (2.105), using (2.122), we can derive that there exists $\tilde{t} \in (0,1]$ such that

$$\langle(\Phi^\infty)'(\tilde{t}\bar{v}), \tilde{t}\bar{v}\rangle = 0 \quad \text{and} \quad \Phi^\infty(\tilde{t}\bar{v}) \geq c^\infty. \tag{2.123}$$

Since $\tilde{t} \in (0,1]$, from (2.121), (2.123), Lemma 2.12 and Fatou's lemma, we deduce

$$\begin{aligned}
c^\infty &\leq \Phi^\infty(\tilde{t}\bar{v}) - \frac{1}{4}\langle(\Phi^\infty)'(\tilde{t}\bar{v}), \tilde{t}\bar{v}\rangle \\
&= \frac{\tilde{t}^2}{4}\int_{\mathbb{R}^2}\left[a|\nabla\bar{v}|^2 + (V_\infty - V_0)\bar{v}^2\right]\mathrm{d}x + \int_{\mathbb{R}^2}\left[\frac{1}{4}f(\tilde{t}\bar{v})\tilde{t}\bar{v} - F(\tilde{t}\bar{v}) + \frac{V_0}{4}(\tilde{t}\bar{v})^2\right]\mathrm{d}x \\
&\leq \frac{1}{4}\int_{\mathbb{R}^2}\left[a|\nabla\bar{u}|^2 + (V_\infty - V_0)\bar{u}^2\right]\mathrm{d}x + \int_{\mathbb{R}^2}\left[\frac{1}{4}f(\bar{v})\bar{v} - F(\bar{v}) + \frac{V_0}{4}\bar{v}^2\right]\mathrm{d}x \\
&= \Phi^\infty(\bar{v}) - \frac{1}{4}\langle(\Phi^\infty)'(\bar{v}), \bar{v}\rangle \\
&\leq \lim_{n\to\infty}\left[\Phi^\infty(v_n) - \frac{1}{4}\langle(\Phi^\infty)'(v_n), v_n\rangle\right] = c = c_N \leq c^\infty,
\end{aligned}$$

which implies that $\tilde{t} = 1$, $v_n \to \bar{v}$ in $H^1(\mathbb{R}^2)$, $\langle(\Phi^\infty)'(\bar{v}), \bar{v}\rangle = 0$ and $\Phi^\infty(\bar{v}) = c = c_N$. As in Corollary 2.3, we have $c = c_N = \Phi^\infty(\bar{v}) \geq \max_{t\geq 0}\Phi^\infty(t\bar{v})$. Since $\bar{v} \neq 0$, by Lemma 2.14, there exists $\hat{t} > 0$ such that $\hat{t}\bar{v} \in \mathcal{N}$, and so $\Phi(\hat{t}\bar{v}) \geq c_N = c$. Therefore, it follows that

$$c = \Phi^\infty(\bar{v}) \geq \Phi^\infty(\hat{t}\bar{v}) = \Phi(\hat{t}\bar{v}) + \frac{\hat{t}^2}{2}\int_{\mathbb{R}^2}[V_\infty - V(x)]\bar{v}^2\mathrm{d}x \geq \Phi(\hat{t}\bar{v}) \geq c_N = c,$$

which yields $\Phi(\hat{t}\bar{v}) = c_N = c$. Applying Lemma 2.18, we get $\Phi'(\hat{t}\bar{v}) = 0$. This completes the proof of Theorem 2.3. □

By adapting the argument of Case ii) in the proof of Theorem 2.3, we can conclude Corollary 2.1.

2.4 Problem $(\mathcal{K})$ with radial potential

We are now concerned with the existence of nontrivial solutions for problem $(\mathcal{K})$ with radial potential V satisfying (V2). This is essentially achieved by restricting the working space in $H^1_r(\mathbb{R}^2)$. We recall that the embedding $H^1_r(\mathbb{R}^2) \hookrightarrow L^s(\mathbb{R}^2)$ is compact for any $s > 2$. Thus, if u is a critical point of Φ restricted to $H^1_r(\mathbb{R}^2)$, then u is a critical point of Φ on $H^1(\mathbb{R}^2)$.

With similar arguments as in the proof of Lemma 2.11, there exists a sequence $\{u_n\} \subset H^1_r(\mathbb{R}^2)$ satisfying

$$\Phi(u_n) \to c_r < \frac{2\pi a}{\alpha_0} + \frac{4\pi^2 b}{\alpha_0^2}, \qquad \|\Phi'(u_n)\|_{(H^1_r(\mathbb{R}^2))^*}(1 + \|u_n\|) \to 0, \tag{2.124}$$

$$c_r = \inf_{\gamma\in\Gamma_r}\max_{t\in[0,1]}\Phi(\gamma(t)), \tag{2.125}$$

where

$$\Gamma_r = \left\{\gamma \in \mathcal{C}([0,1], H^1_r(\mathbb{R}^2)) : \gamma(0) = 0, \Phi(\gamma(1)) < 0\right\}. \tag{2.126}$$

Lemma 2.19 *Assume that conditions* (V2) *and* (F1), (F2), (F3′), (F4) *and* (F6) *hold. Then any sequence satisfying* (2.124) *is bounded in* $H^1(\mathbb{R}^2)$.

Proof. Arguing as in (2.93)–(2.95), we establish that

$$\|\nabla u_n\|_2^2 \le C_1,\ \ 8\int_{|u_n|\ge R_0} F(u_n)\mathrm{d}x \le \int_{|u_n|\ge R_0} f(u_n)u_n\mathrm{d}x \le C_2. \tag{2.127}$$

In view of (2.127), to prove the boundedness of $\{\|u_n\|\}$, it remains to show the boundedness of $\{\|u_n\|_2\}$. To this end, arguing by contradiction, suppose that $\|u_n\|_2 \to +\infty$. Let $v_n := \frac{u_n}{\|u_n\|_2}$,

$$D_1 := \left\{t \in [-R_0, R_0] : \frac{f(t)}{t} < V_0\right\},\quad D_2 := \left\{t \in [-R_0, R_0] : \frac{f(t)}{t} \ge V_0\right\} \tag{2.128}$$

and

$$\mathcal{F}(t) := \frac{1}{2}f(t)t - F(t),\quad \forall\, t \in \mathbb{R}. \tag{2.129}$$

Using (F3′), we obtain

$$\mathcal{F}(t) \ge C_4,\quad \forall\, t \in D_2. \tag{2.130}$$

From (F1), (F2) and (2.130), we have

$$|f(t)|^2 \le C_5|t| \le C_6|t|\mathcal{F}(t),\quad \forall\, t \in D_2. \tag{2.131}$$

By (2.124), we get

$$c + o(1) = \Phi(u_n) + o(1) = \Phi(u_n) - \frac{1}{2}\langle \Phi'(u_n), u_n\rangle \ge -\frac{b}{4}\|\nabla u_n\|_2^4 + \int_{|u_n|\le R_0} \mathcal{F}(u_n)\mathrm{d}x,$$

which, together with (2.127), yields

$$\int_{|u_n|\le R_0} \mathcal{F}(u_n)\mathrm{d}x \le C_7. \tag{2.132}$$

Note that $\{\|v_n\|\}$ is bounded due to (2.127). Thus, it follows from (2.7), (2.124) and (2.132) that

$$\begin{aligned} V_0 &\le \lim_{n\to\infty}\left[\frac{1}{\|u_n\|_2^2}\int_{\mathbb{R}^2}\left[a|\nabla u_n|^2 + V(x)u_n^2\right]\mathrm{d}x + \frac{b\|\nabla u_n\|_2^4}{\|u_n\|_2^2}\right] \\ &= \lim_{n\to\infty}\left[\frac{1}{\|u_n\|_2^2}\int_{|u_n|\le R_0} f(u_n)u_n\mathrm{d}x + \frac{1}{\|u_n\|_2^2}\int_{|u_n|>R_0} f(u_n)u_n\mathrm{d}x\right], \end{aligned}$$

hence

$$\begin{aligned} V_0 &\le \limsup_{n\to\infty}\int_{D_1}\frac{f(u_n)}{u_n}v_n^2\mathrm{d}x + \limsup_{n\to\infty}\frac{1}{\|u_n\|_2}\int_{D_2}\frac{|f(u_n)|}{|u_n|^{1/2}}|v_n|^{3/2}\mathrm{d}x \\ &< V_0 + \limsup_{n\to\infty}\frac{\sqrt{C_6}}{\|u_n\|_2}\left[\int_{|u_n|\le R_0}\mathcal{F}(u_n)\mathrm{d}x\right]^{1/2}\|v_n\|_3^{3/2} = V_0. \end{aligned}$$

This contradiction shows the boundedness of $\{\|u_n\|_2\}$. Hence, $\{u_n\}$ is bounded in $H^1(\mathbb{R}^2)$ and the proof is now complete. □

2.4.1 Proof of Theorem 2.4

In view of Lemma 2.19, there exists a sequence $\{u_n\} \subset H_r^1(\mathbb{R}^2)$ satisfying (2.124) and $\|u_n\| \leq C_8$. Thus, up to a subsequence, we can assume that $u_n \rightharpoonup \bar{u}$ in $H_r^1(\mathbb{R}^2)$, $u_n \to \bar{u}$ in $L^s(\mathbb{R}^2)$ for $s > 2$ and $u_n \to \bar{u}$ a.e. in $\mathbb{R}^2$. Arguing as in the proof of relation (4.27) in Chen and Tang [32], we obtain

$$\lim_{n\to\infty} \int_{\mathbb{R}^2} F(u_n)\mathrm{d}x = \int_{\mathbb{R}^2} F(\bar{u})\mathrm{d}x. \tag{2.133}$$

Step 1. We prove that, up to a subsequence,

$$L := \lim_{n\to\infty} \|\nabla u_n\|_2^2 < \frac{4\pi}{\alpha_0} + \|\nabla \bar{u}\|_2^2. \tag{2.134}$$

Arguing by contradiction, we assume that $L \geq \frac{4\pi}{\alpha_0} + \|\nabla \bar{u}\|_2^2$. By (2.7) and (2.75), we get

$$0 = \lim_{n\to\infty} \langle \Phi'(u_n), u_n \rangle = aL + bL^2 + \lim_{n\to\infty} \int_{\mathbb{R}^2} \left[V(x)u_n^2 - f(u_n)u_n\right] \mathrm{d}x \tag{2.135}$$

and

$$\begin{aligned}
0 &= \lim_{n\to\infty} \langle \Phi'(u_n), \varphi \rangle \\
&= \lim_{n\to\infty} \left\{ \int_{\mathbb{R}^2} [a\nabla u_n \cdot \nabla\varphi + V(x)u_n\varphi]\,\mathrm{d}x + bL\int_{\mathbb{R}^2} \nabla u_n \cdot \nabla\varphi \mathrm{d}x - \int_{\mathbb{R}^2} f(u_n)\varphi \mathrm{d}x \right\} \\
&= \int_{\mathbb{R}^2} [a\nabla \bar{u} \cdot \nabla\varphi + V(x)\bar{u}\varphi]\,\mathrm{d}x + bL\int_{\mathbb{R}^2} \nabla \bar{u} \cdot \nabla\varphi \mathrm{d}x - \int_{\mathbb{R}^2} f(\bar{u})\varphi \mathrm{d}x, \quad \forall\, \varphi \in \mathcal{C}_0^\infty(\mathbb{R}^2).
\end{aligned}$$

Combining this relation with (2.104), the definition of L and the density of $\mathcal{C}_0^\infty(\mathbb{R}^2)$ in $H_r^1(\mathbb{R}^2)$, we obtain

$$0 = \lim_{n\to\infty} \langle \Phi'(u_n), \bar{u} \rangle = a\|\nabla \bar{u}\|_2^2 + \int_{\mathbb{R}^2} V(x)\bar{u}^2 \mathrm{d}x + bL\|\nabla \bar{u}\|_2^2 - \int_{\mathbb{R}^2} f(\bar{u})\bar{u}\mathrm{d}x. \tag{2.136}$$

If $L \geq \frac{4\pi}{\alpha_0} + \|\nabla \bar{u}\|_2^2$, from (2.75), (2.112), (2.135), (2.136) and Fatou's lemma, we then deduce

$$\begin{aligned}
\frac{2\pi a}{\alpha_0} + \frac{4\pi^2 b}{\alpha_0^2} &> c^\infty \geq c = \lim_{n\to\infty} \left[\Phi(u_n) - \frac{1}{4}\langle \Phi'(u_n), \bar{u}\rangle\right] \\
&= \frac{a}{2}L + \frac{b}{4}L^2 - \int_{\mathbb{R}^2} F(\bar{u})\mathrm{d}x + \frac{1}{2}\lim_{n\to\infty}\int_{\mathbb{R}^2} V(x)u_n^2 \mathrm{d}x \\
&\quad - \frac{a}{4}\|\nabla \bar{u}\|_2^2 + \frac{1}{4}\int_{\mathbb{R}^2} V(x)\bar{u}^2\mathrm{d}x + \frac{b}{4}L\|\nabla \bar{u}\|_2^2 + \frac{1}{4}\int_{\mathbb{R}^2} f(\bar{u})\bar{u}\mathrm{d}x \\
&= \frac{a}{4}L + \frac{a+bL}{4}(L - \|\nabla \bar{u}\|_2^2) + \frac{1}{2}\lim_{n\to\infty}\int_{\mathbb{R}^2} V(x)\left(u_n^2 - \bar{u}^2\right)\mathrm{d}x \\
&\quad + \frac{1}{4}\int_{\mathbb{R}^2} [f(\bar{u})\bar{u} - 4F(\bar{u}) + V(x)\bar{u}^2]\mathrm{d}x \geq \frac{2\pi a}{\alpha_0} + \frac{4\pi^2 b}{\alpha_0^2} + \left(a + \frac{4\pi b}{\alpha_0^2}\right)\frac{\|\nabla \bar{u}\|_2^2}{4},
\end{aligned}$$

a contradiction, which completes the proof of Step 1.

Step 2. We prove that $\Phi(\bar{u}) = c$ and $\Phi'(\bar{u}) = 0$.

From Step 1, we know that there exists $\hat{\varepsilon} > 0$ satisfying

$$\lim_{n\to\infty} \left[\|\nabla(u_n - \bar{u})\|_2^2\right] = \lim_{n\to\infty} \left[\|\nabla u_n\|_2^2 - \|\nabla \bar{u}\|_2^2\right] = \frac{4\pi(1-3\hat{\varepsilon})}{\alpha_0}. \tag{2.137}$$

Choosing $q \in (1,2)$ satisfying

$$\frac{(1+\hat{\varepsilon})^2(1-3\hat{\varepsilon})q^2}{1-\hat{\varepsilon}} < 1, \tag{2.138}$$

by (2.137), the Young inequality and Lemma 2.1, we have

$$\begin{aligned}\int_{\mathbb{R}^2}\left(e^{\alpha_0(1+\hat{\varepsilon})qu_n^2}-1\right)\mathrm{d}x &\le \int_{\mathbb{R}^2}\left(e^{\alpha_0(1+\hat{\varepsilon})^2\hat{\varepsilon}^{-1}q\bar{u}^2}e^{\alpha_0(1+\hat{\varepsilon})^2q(u_n-\bar{u})^2}-1\right)\mathrm{d}x\\ &\le \frac{(q-1)}{q}\int_{\mathbb{R}^2}\left(e^{\alpha_0(1+\hat{\varepsilon})^2\hat{\varepsilon}^{-1}q^2(q-1)^{-1}\bar{u}^2}-1\right)\mathrm{d}x\\ &\quad+\frac{1}{q}\int_{\mathbb{R}^2}\left(e^{\alpha_0(1+\hat{\varepsilon})^2q^2(u_n-\bar{u})^2}-1\right)\mathrm{d}x \le C_9. \end{aligned}\tag{2.139}$$

Using (F1) and (F2), for any $\varepsilon > 0$, there exists $C_\varepsilon > 0$ such that

$$|f(t)| \le \varepsilon|t| + C_\varepsilon\left(e^{\alpha_0(1+\hat{\varepsilon})t^2}-1\right), \quad \forall\, t \in \mathbb{R}. \tag{2.140}$$

Let $q' := q/(q-1) > 2$. Since $u_n \to u$ in $L^{q'}(\mathbb{R}^2)$, from (2.139), (2.140) and the Hölder inequality, we deduce that

$$\begin{aligned}\int_{\mathbb{R}^2} f(u_n)(u_n-\bar{u})\mathrm{d}x &\le \varepsilon\|u_n-\bar{u}\|_2^2 + C_9\int_{\mathbb{R}^2}\left(e^{\alpha u_n^2}-1\right)|u_n-\bar{u}|\mathrm{d}x\\ &\le \varepsilon\left(C_1+\|\bar{u}\|_2^2\right)+C_9\left[\int_{\mathbb{R}^2}\left(e^{\alpha_0(1+\hat{\varepsilon})u_n^2}-1\right)^q\mathrm{d}x\right]^{1/q}\|u_n-\bar{u}\|_{q'}\\ &\le \varepsilon\left(C_1+\|\bar{u}\|_2^2\right)+C_{10}\left[\int_{\mathbb{R}^2}\left(e^{\alpha_0(1+\hat{\varepsilon})qu_n^2}-1\right)\mathrm{d}x\right]^{1/q}\|u_n-\bar{u}\|_{q'}\\ &\le \varepsilon\left(C_1+\|\bar{u}\|_2^2\right)+C_{11}\|u_n-\bar{u}\|_{q'} \le \varepsilon\left(C_1+\|\bar{u}\|_2^2\right)+o(1), \end{aligned}\tag{2.141}$$

which, together with the arbitrariness of ε, yields

$$\lim_{n\to\infty}\int_{\mathbb{R}^2} f(u_n)(u_n-\bar{u})\mathrm{d}x = 0. \tag{2.142}$$

Therefore, it follows from (2.7), (2.75), (2.142) that

$$\begin{aligned} o(1) = \langle\Phi'(u_n), u_n-\bar{u}\rangle &= \int_{\mathbb{R}^2}\left[a\nabla u_n\cdot\nabla(u_n-\bar{u})+V(x)u_n(u_n-\bar{u})\right]\mathrm{d}x\\ &\quad+b\|\nabla u_n\|_2^2\int_{\mathbb{R}^2}\nabla u_n\cdot\nabla(u_n-\bar{u})\mathrm{d}x-\int_{\mathbb{R}^2}f(u_n)(u_n-\bar{u})\mathrm{d}x\\ &\ge a\|\nabla(u_n-\bar{u})\|_2^2+V_0\|u_n-\bar{u}\|_2^2+b\|\nabla u_n\|_2^2\|\nabla(u_n-\bar{u})\|_2^2+o(1)\\ &\ge \min\{a,V_0\}\|u_n-\bar{u}\|^2+o(1), \end{aligned}$$

which implies that $u_n \to \bar{u}$ in $H_r^1(\mathbb{R}^2)$.

Finally, the continuity of Φ and Φ' leads to $\Phi(\bar{u}) = c$ and $\Phi'(\bar{u})\big|_{H_r^1(\mathbb{R}^2)} = 0$. From Theorem 1.28 of Willem [153], we conclude that u is a critical point of Φ on $H^1(\mathbb{R}^2)$. □

2.5 Historical comments

Problem $(\mathcal{K})$ has a profound physical meaning, which was proposed firstly by Kirchhoff [88] in the case where $\mathbb{R}^2$ is replaced by the bounded domain $\Omega \subset \mathbb{R}$. Nonlocal equations of this type model the vibration of elastic strings by considering the effect of the changes in the length of strings. As is customary in quantum mechanics applications, the unknown u is the probability density function of a particle trapped inside a trapping potential well, traditionally modeled by the potential $V(x)$.

After the pioneering contributions of Lions [99] and Pohozaev [128], the following Kirchhoff-type problem

$$\begin{cases} -\left(a+b\int_{\mathbb{R}^N}|\nabla u|^2\mathrm{d}x\right)\Delta u+V(x)u=f(u) & \text{in } \mathbb{R}^N,\\ u\in H^1(\mathbb{R}^N) \end{cases} \tag{2.143}$$

has been studied intensively in many papers, where $a, b > 0$, $N \geq 2$, $V \in \mathcal{C}(\mathbb{R}^N, \mathbb{R})$ and $f \in \mathcal{C}(\mathbb{R}, \mathbb{R})$. By variational methods, a number of important results of the existence and multiplicity of solutions for (2.143) were established under various conditions on V and f, especially when $N \geq 3$. In higher dimensions corresponding to the case $N \geq 3$, the nonlinearity is usually required to have polynomial growth, and the notation of criticality is associated to the sharp Sobolev embedding $H^1(\mathbb{R}^N) \hookrightarrow L^{2^*}(\mathbb{R}^N)$ with $2^* := 2N/(N-2)$. In this framework, we refer to Pucci and Saldi [130] who established existence and multiplicity properties of non-negative entire solutions to stationary Kirchhoff eigenvalue problems driven by a general nonlocal integro-differential operator.

Coming to the case $N = 2$, much faster exponential growth is allowed for the nonlinearity and the Trudinger-Moser inequality replaces the sharp Sobolev inequality used for $N \geq 3$. In this chapter focused on the dimension $N = 2$ when the nonlinearity exhibits the critical exponential growth, which is more complicated than the case $N \geq 3$.

The first result on planar Kirchhoff equation with critical exponential growth is due to Figueiredo and Severo [55]. In this setting, based on the mountain pass theorem, they proved that the following Kirchhoff equation on the bounded domain $\Omega \subset \mathbb{R}^2$

$$\begin{cases} -\left(a+b\displaystyle\int_{\Omega}|\nabla u|^2\mathrm{d}x\right)\Delta u=f(u) & \text{in } \Omega,\\ u=0, & \text{on } \partial\Omega \end{cases} \tag{2.144}$$

has a positive ground state solution where $f \in \mathcal{C}(\mathbb{R}, \mathbb{R})$ satisfies (F1), (F2) and the following assumptions:

(F0) $f(t) = 0$ for all $t \leq 0$;

(F3′) $\lim_{t\to+\infty}\frac{tf(t)}{e^{\alpha_0 t^2}} > \frac{2}{\alpha_0 d^2}\left(a+\frac{4\pi b}{\alpha_0}\right)$, where d is the radius of the largest open ball contained in Ω;

(F4′) $f(t) > 0$ for all $t > 0$, and there exist $\hat{M}_0 > 0$ and $\hat{\beta}_0 > 0$ such that

$$F(t) \leq \hat{M}_0 f(t), \quad \forall\, t \geq \hat{\beta}_0;$$

(F5′) $\frac{f(t)}{t^3}$ is increasing on $(0, \infty)$.

Note that if $\alpha = 4\pi$, the Trudinger-Moser inequality (2.2) gives rise to the possible failure of compactness of the associated energy functional. In order to restore the compactness property, Figueiredo and Severo proved that the mountain pass level is less than the threshold $\frac{2a\pi}{\alpha_0} + \frac{4b\pi^2}{\alpha_0^2}$ under which Palais-Smale condition holds with the help of (F3′). Their arguments followed some ideas introduced by de Figueiredo, Miyagaki, and Ruf [41] in a paper concerned with the solvability of the elliptic type problem (2.144) with $b = 0$.

Next, assuming that the monotonicity condition (F5′) is replaced by the following Ambrosetti-Rabinowitz condition:

(F6′) $f(t)t \geq 4F(t) \geq 0$ for all $t \geq 0$,

Naimen and Tarsi [122] obtained the existence of positive solutions to problem (2.144).

The monotonicity condition (F5′) plays an important role in using a Nehari-type argument, and the Ambrosetti-Rabinowitz condition (F6′) can help proving the boundedness of (PS) sequences. The above results were improved and generalized by Chen, Tang, and Wei [34], by weakening (F3′), (F5′) and (F6′) to the following conditions:

(F3″) $\liminf_{t\to+\infty} \frac{t^2F(t)}{e^{\alpha_0 t^2}} > \frac{1}{e\alpha_0^2 d^2}\left(a + \frac{4\pi b}{\alpha_0}\right)$;

(F5″) $\frac{f(t)-a\lambda_1 t}{t^3}$ is non-decreasing on $(0,\infty)$, where λ_1 denotes the first eigenvalue of the Laplace operator $-\Delta$ with a Dirichlet boundary condition;

(F6″) $f(t)t - 4F(t) + \lambda_1 t^2 \geq 0$ for all $t \geq 0$.

Kirchhoff equations with more general nonlocal coefficient were also considered by Figueiredo and Severo [55], provided that $\left(a + b\|\nabla u\|_2^2\right)$ is replaced by more general continuous function $m(\|\nabla u\|_2^2)$ in (2.143). Assumptions (F3″), (F4′) and (F5″) (or (F6″)) are crucial in these works. For instance, condition (F3″) was used to yield the threshold of the mountain pass level. Also, by using (F4′) and (F5″) (or (F6″)) it can be shown the convergence of $(\mathrm{PS})_c$ sequences in $H_0^1(\Omega)$ provided that the mountain pass level lies below the threshold, that is, if $c < \frac{2a\pi}{\alpha_0} + \frac{4b\pi^2}{\alpha_0^2}$.

However, the methods used in papers [34, 55, 122] seem difficult to apply for Kirchhoff equation in $\mathbb{R}^2$ since they depend heavily on the compactness of the embeddings $H_0^1(\Omega) \hookrightarrow L^q(\Omega)$ for $q \geq 2$. Until now, almost all of the works dealing with planar Kirchhoff equations are set in bounded domains $\Omega \subset \mathbb{R}^2$, and there is no result available on the existence of nontrivial solutions for Kirchhoff equation $(\mathcal{K})$ with the critical exponential growth in $\mathbb{R}^2$, which is the focus of the present chapter.

2.6 Final remarks, perspectives, and open problems

This chapter is concerned with the rigorous mathematical analysis of a class of (autonomous and non-autonomous) planar Kirchhoff equations with critical exponential growth and trapping potential. *Why this combination?* Because it creates a mathematically optimal testbed:

(a) *Kirchhoff*: nonlocality in the source term;

(b) *critical exponential growth*: maximal growth in two-dimensions;

(c) *trapping potential*: confinement and compactness.

In summary, planar Kirchhoff systems with critical exponential growth (Trudinger-Moser type) and trapping potential offer deep insights into how nonlocality, criticality, and confinement interact in low-dimensional nonlinear systems. Physical patterns of this type include Bose-Einstein condensates in two-dimensional traps, Kirchhoff-type dispersion from nonlocal nonlinearities (thermal, molecular), and exponential response in photo-reactive materials.

The content of this chapter is based on the results established in [29]. In the local case, related properties can be found in [26, 30], while the case of planar Schrödinger-Newton systems with critical Trudinger-Moser growth was treated in [109, 110].

Some interesting open problems associated with the analysis developed in this chapter are stated in what follows.

We first propose the case where the nonlocal term becomes *degenerate*, potentially causing a change in the nature of the equation. In this new setting, the geometry of the associated energy functional changes dramatically.

Open Problem 1. Study the autonomous and non-autonomous versions of the degenerate Kirchhoff equation

$$\begin{cases} -\left(\displaystyle\int_{\mathbb{R}^2} |\nabla u|^{2\gamma} \mathrm{d}x\right) \Delta u + V(x)u = f(u) & \text{in } \mathbb{R}^2, \\ u \in H^1(\mathbb{R}^2), \end{cases}$$

where $\gamma > 0$, f has a critical exponential growth, and the potential V satisfies similar assumptions as in this chapter.

The next open problem concerns the case where the nonlinearity exceeds the critical exponential growth.

Open Problem 2. Can we establish the existence of least-energy or ground state solutions for problems $(\mathcal{K})_\infty$ or $(\mathcal{K})$ if $f(u) \sim e^{\alpha u^2} u^p$, where $p > 1$?

Another interesting open problem is concerned with the case of singular potentials.

Open Problem 3. Study problem $(\mathcal{K})$ on a bounded domain $\Omega \subset \mathbb{R}^2$ with a Kirchhoff term and critical exponential nonlinearity if $0 \in \Omega$ and with a singular potential (for instance, if $V(x) = \lambda/|x|^\gamma$ with $\gamma > 0$.

The case where the Laplace operator in problem $(\mathcal{K})$ is replaced by the fractional Laplacian $(-\Delta)^s$ with $s \in (0,1)$ is unexplored.

Open Problem 4. Study the fractional Kirchhoff equation

$$\begin{cases} \left(a + \displaystyle\int_{\mathbb{R}^2} |\nabla u|^2 \mathrm{d}x\right) (-\Delta)^s u + V(x)u = f(u) & \text{in } \mathbb{R}^2, \\ u \in H^1(\mathbb{R}^2), \end{cases}$$

where f has a critical exponential growth, and the potential V satisfies similar assumptions as in this chapter.

In the above problem, the combination of fractional nonlocal operators and the Kirchhoff nonlocality is not sufficiently studied, especially in the two-dimensional case with exponential growth. The appropriate Trudinger-Moser inequality for fractional Sobolev spaces should be explored.

We conclude this section with a class of fourth-order Kirchhoff equations, which appear in the von Kármán plate theory, piezoelectric laminates, capacitive MEMS devices, viscoelastic plates, etc.

Open Problem 5. Study the fourth-order Kirchhoff problem

$$\begin{cases} \Delta^2 u - \left(a + b \int_{\mathbb{R}^2} |\nabla u|^2 \mathrm{d}x\right) \Delta u + V(x)u = f(u) & \text{in } \mathbb{R}^2, \\ u \in H^1(\mathbb{R}^2), \end{cases}$$

where f has a critical growth as it is described by Adams' inequality.

The analysis of Kirchhoff-type equations with such nonlinearities is wide open.

2.7 Glossary

Trapping Potential: It is a function that creates an effective energy well to confine particles or waves within a localized region, preventing them from escaping to infinity. This concept is fundamental in studies of quantum mechanics, Bose-Einstein condensates, Kirchhoff problems, and nonlinear wave equations.

Critical Exponential Growth: A function $f(u)$ has critical exponential growth if its behavior as $u \to \infty$ to $e^{\alpha u^2}$ for some $\alpha > 0$. This is the maximal growth allowed by the two-dimensional Sobolev embedding. A growth faster than $e^{\alpha u^2}$ would make standard variational formulations ill-posed, as the energy functional might not be well-defined on the Sobolev space.

Autonomous Kirchhoff Equation: A nonlinear equation where the coefficients do not depend explicitly on the independent variables. It extends the classical D'Alembert wave equation by including a nonlocal term depending on the energy of the solution.

Non-autonomous Kirchhoff Equation: This class extends the classical Kirchhoff model by allowing explicit dependence on the independent variable in the coefficients or nonlinear terms. Non-autonomous Kirchhoff equations are more realistic for modeling media with inhomogeneities.

Mountain Pass: This notion refers both to the geometric configuration of the functional and to the resulting saddle point critical point whose existence is guaranteed by the mountain pass theorem.

Part II

Normalized Solutions

3

Non-autonomous Kirchhoff problems

Why normalized solutions? There are several reasons. Firstly, for their physical necessity, since prescribing the mass is often required to define a specific physical system, for instance because the gravitational interaction strength depends directly on the mass. Secondly, normalized stationary states define naturally the eigenvalue problem, which is a nonlinear eigenvalue problem constrained by mass normalization. Thirdly, problems of this type yield a key prediction. Indeed, solving this kind of problem for different masses reveals the mass-radius relationship, a critical signature used to test the model against astrophysical observations (e.g., for fuzzy dark matter). Part II of this monograph is devoted to the mathematical analysis of *normalized solutions* for several classes of Kirchhoff equations.

The non-autonomous case moves Kirchhoff equations from model problems to realistic physical patterns where spatial inhomogeneity is the rule, not the exception. The analysis developed in this chapter covers both the subcritical and supercritical abstract settings. This chapter is concerned with the existence of *normalized solutions* (also called *solutions with prescribed mass*) for the following class of Kirchhoff equations:

$$\begin{cases} -\left(a+b\displaystyle\int_{\mathbb{R}^3}|\nabla u|^2\mathrm{d}x\right)\Delta u-\lambda u=K(x)f(u) \text{ in } \mathbb{R}^3; \\ u\in H^1(\mathbb{R}^3), \end{cases}$$

where $a, b > 0$, λ is a parameter that appears as a Lagrange multiplier, $K \in \mathcal{C}(\mathbb{R}^3, \mathbb{R}^+)$ with $0 < \lim_{|y|\to\infty} K(y) \le \inf_{\mathbb{R}^3} K$, and $f \in \mathcal{C}(\mathbb{R}, \mathbb{R})$ satisfies general L^2-supercritical or L^2-subcritical conditions. We introduce some new analytical techniques in order to exclude the vanishing and the dichotomy cases of minimizing sequences due to the presence of the potential K and the lack of the homogeneity of the nonlinearity f.

The main challenge in the analysis developed in this chapter consists in the combined effects of several terms, namely:

- (i) *Kirchhoff nonlinearity*: the nonlocal term $\left(\int_{\mathbb{R}^3}|\nabla u|^2\mathrm{d}x\right)\Delta u$ makes the energy functional more complex;
- (ii) *Mass constraint*: the problem is set on the L^2-sphere $\mathcal{S}_c = \{u \in H^1(\mathbb{R}^3) : \|u\|_2^2 = c\}$, turning the parameter λ into an unknown;
- (iii) *Non-autonomous nonlinearity*: the x-dependence in the reaction term prevents the use of standard radial symmetry and concentration-compactness techniques in a straightforward way;
- (iv) *Lack of compactness*: due to the unbounded domain $\mathbb{R}^3$, even the sub-critical case faces loss of compactness, which is more severe in the super-critical regime.

A cornerstone in this analysis is to study the manifold where the Pohozaev identity holds. This set contains all critical points and helps bound the energy from below. Another feature is to use scaling arguments to transfer between different mass levels and to relate the Kirchhoff equation to a suitable limit problem.

DOI: 10.1201/9781003797159-3

3.1 Statement of the problem

This chapter deals with the existence of normalized solutions to the following non-autonomous Kirchhoff equation:

$$\begin{cases} -\left(a+b\int_{\mathbb{R}^3}|\nabla u|^2\mathrm{d}x\right)\Delta u-\lambda u=K(x)f(u) \quad \text{in } \mathbb{R}^3; \\ u\in H^1(\mathbb{R}^3), \end{cases} \tag{3.1}$$

where a, b are positive real numbers, λ is unknown and will appear as a Lagrange multiplier, $K \in \mathcal{C}(\mathbb{R}^3, \mathbb{R}^+)$ is a potential, and $f \in \mathcal{C}(\mathbb{R}, \mathbb{R})$.

Problem (3.1) is related to the stationary analogue of the Kirchhoff equation

$$u_{tt}-\left(a+b\int_{\mathbb{R}^3}|\nabla u|^2\mathrm{d}x\right)\Delta u=g(x,t), \tag{3.2}$$

Set $F(u)=\int_0^u f(t)\mathrm{d}t$. Solutions of problem (3.1) with $\|u\|_2^2=c>0$ can be obtained by looking for critical points of the following functional

$$I(u)=\frac{a}{2}\int_{\mathbb{R}^3}|\nabla u|^2\mathrm{d}x+\frac{b}{4}\left(\int_{\mathbb{R}^3}|\nabla u|^2\mathrm{d}x\right)^2-\int_{\mathbb{R}^3}K(x)F(u)\mathrm{d}x \tag{3.3}$$

on the manifold

$$\mathcal{S}_c=\left\{u\in H^1(\mathbb{R}^3):\|u\|_2^2=c\right\}. \tag{3.4}$$

where $F(u)=\int_0^u f(t)\mathrm{d}t$. In this case, the parameter $\lambda \in \mathbb{R}$ cannot be fixed but instead it appears as a Lagrange multiplier, and each critical point $u_c \in \mathcal{S}_c$ of $I|_{\mathcal{S}_c}$, corresponds to a Lagrange multiplier $\lambda_c \in \mathbb{R}$ such that (u_c, λ_c) solves (weakly) problem (3.1). In particular, if $u_c \in \mathcal{S}_c$ is a solution of the constrained minimization problem

$$\sigma(c):=\inf_{u\in\mathcal{S}_c} I(u), \tag{3.5}$$

then there exists $\lambda_c \in \mathbb{R}$ such that $I'(u_c)=\lambda_c u_c$, that is, the pair (u_c, λ_c) is a solution of problem (3.1).

3.1.1 Existence of solutions with prescribed mass

Let us first consider the L^2-*supercritical* case. Consider the submanifold

$$\mathcal{M}_c:=\left\{u\in\mathcal{S}_c: J(u):=\frac{\mathrm{d}}{\mathrm{d}t}I(u^t)\Big|_{t=1}=0\right\}. \tag{3.6}$$

Then $\mathcal{M}_c$ is a natural constraint of $I|_{\mathcal{S}_c}$ by using the Lagrange multiplier method, where

$$u^t(x):=t^{3/2}u(tx), \qquad \forall\, t>0, u\in H^1(\mathbb{R}^3), \tag{3.7}$$

and $u^t \in \mathcal{S}_c$ for all $t>0$ if $u \in \mathcal{S}_c$.

We establish the existence of a critical point for the energy functional I on $\mathcal{S}_c$ by considering the constrained minimization problem

$$m(c):=\inf_{u\in\mathcal{M}_c} I(u). \tag{3.8}$$

To this end, we introduce the following assumptions:

(F1) $f \in \mathcal{C}(\mathbb{R}, \mathbb{R})$, there exists $\mu \in (\frac{14}{3}, 6)$ such that $0 \leq f(t)t \leq \mu F(t)$ for all $t \in \mathbb{R}$, and $\text{meas}\{t \in \mathbb{R} : \mu F(t) - f(t)t = 0\} = 0$;

(F2) there exists $\theta \in (2, \frac{14}{3})$ such that $\lim_{|t| \to 0} \frac{F(t)}{|t|^\theta} = 0$ and $\lim_{|t| \to \infty} \frac{F(t)}{|t|^{14/3}} = +\infty$;

(F3) the mapping $t \mapsto [f(t)t - \theta F(t)]/|t|^{11/3}t$ is nondecreasing on $(-\infty, 0)$ and $(0, +\infty)$;

(K1) $K \in \mathcal{C}(\mathbb{R}^3, \mathbb{R}^+)$ and $0 < K_\infty := \lim_{|y| \to \infty} K(y) \leq K(x)$ for all $x \in \mathbb{R}^3$;

(K2) $K \in \mathcal{C}^1(\mathbb{R}^3, \mathbb{R}^+)$, $(6 - \mu)K(x) + 2\nabla K(x) \cdot x \geq 0$ for all $x \in \mathbb{R}^3$, and $3(\theta - 2)K(tx) - 2\nabla K(tx) \cdot (tx)$ is nonincreasing on $t \in (0, \infty)$ for every $x \in \mathbb{R}^3$.

The following result establishes an existence property in the L^2-supercritical case.

Theorem 3.1 *Assume that* (K1), (K2) *and* (F1)–(F3) *hold. Then for any* $c > 0$, *problem* (3.1) *has a solution* $(\bar{u}_c, \lambda_c) \in \mathcal{S}_c \times \mathbb{R}^-$ *such that*

$$I(\bar{u}_c) = \inf_{u \in \mathcal{M}_c} I(u) = \inf_{u \in \mathcal{S}_c} \max_{t > 0} I(u^t) > 0.$$

Next, in the L^2-*subcritical* case, we find a global minimizer and a local minimizer of I which are critical points of $I|_{\mathcal{S}_c}$. This is achieved by solving the minimization problem (3.5) and constructing a geometry of local minima for I, respectively. To this end, in addition to (K1), we introduce the following assumptions:

(F4) $f \in \mathcal{C}(\mathbb{R}, \mathbb{R})$, $f(t) = o(t)$ as $t \to 0$ and there exist constants $\mathcal{C} > 0$ and $p \in (\frac{10}{3}, \frac{14}{3})$ such that $|f(t)| \leq \mathcal{C}(1 + |t|^{p-1})$;

(F5) there exists $\mu_0 \in (2, \frac{14}{3})$ such that $f(t)t \geq \mu_0 F(t) > 0$ for all $t \in \mathbb{R} \setminus \{0\}$;

(F6) there exists $q_0 \in (2, \frac{10}{3})$ such that $\lim_{|t| \to 0} \frac{F(t)}{|t|^{q_0}} > 0$;

(F6′) $\lim_{|t| \to 0} \frac{F(t)}{|t|^{10/3}} = 0$;

(K3) the mapping $t \mapsto t^{(\mu_0 - 2)/2} K(tx)$ is nondecreasing on $(0, \infty)$ for every $x \in \mathbb{R}^3$.

Let

$$c_* := \inf \{c \in (0, +\infty) : \sigma(c) < 0\}. \tag{3.9}$$

Theorem 3.2 *Assume that* K *and* f *satisfy* (K1), (K3), (F4) *and* (F5).

(i) *If* (F6) *holds, then* $c_* = 0$ *and* I *admits a critical point* u_c *on* $\mathcal{S}_c$, *which is a negative global minimum of* I *when* $c > 0$.

(ii) *If* (F6′) *holds, then* $c_* > 0$, *and there exists* $c_0 \in (0, c_*)$ *such that* I *admits a critical point* u_c *on* $\mathcal{S}_c$ *which is a local minimum of* I *when* $c \in (c_0, c_*)$, *but* u_c *is a global minimum of* I *when* $c \in [c_*, +\infty)$. *In particular,*

$$I(u_c) \begin{cases} > 0, & \text{if } c \in (c_0, c_*); \\ = 0, & \text{if } c = c_*; \\ < 0, & \text{if } c \in [c_*, \infty). \end{cases} \tag{3.10}$$

Moreover, for the above critical point u_c, *there is a Lagrange multiplier* $\lambda_c \in \mathbb{R}$ *such that* (u_c, λ_c) *is a solution of problem* (3.1).

When $K \in \mathcal{C}(\mathbb{R}^3, \mathbb{R}^+)$ is bounded, f satisfies (F1) and (F2) (or (F4)), a standard argument shows that $I \in \mathcal{C}^1(H^1(\mathbb{R}^3), \mathbb{R})$.

Let us now define the *limit equation* associated to problem (3.1) by

$$\begin{cases} -\left(a + b\displaystyle\int_{\mathbb{R}^3} |\nabla u|^2 \mathrm{d}x\right) \Delta u - \lambda u = K_\infty f(u) \quad \text{in } \mathbb{R}^3; \\ u \in H^1(\mathbb{R}^3). \end{cases} \tag{3.11}$$

Corresponding to (3.3), (3.5), (3.6) and (3.8), we define

$$I^\infty(u) = \frac{a}{2}\int_{\mathbb{R}^3} |\nabla u|^2 \mathrm{d}x + \frac{b}{4}\left(\int_{\mathbb{R}^3} |\nabla u|^2 \mathrm{d}x\right)^2 - \int_{\mathbb{R}^3} K_\infty F(u) \mathrm{d}x, \tag{3.12}$$

$$\sigma^\infty(c) = \inf_{u \in \mathcal{S}_c} I^\infty(u), \tag{3.13}$$

$$\mathcal{M}_c^\infty = \left\{ u \in \mathcal{S}_c : J^\infty(u) := \frac{\mathrm{d}}{\mathrm{d}t} I^\infty(u^t)\Big|_{t=1} = 0 \right\} \tag{3.14}$$

and

$$m^\infty(c) = \inf_{u \in \mathcal{M}^\infty} I^\infty(u). \tag{3.15}$$

Remark that all above conclusions on problem (3.1) are also true for the limit equation (3.11), since $K(x) \equiv K_\infty$ satisfies (K1)–(K3).

Throughout this chapter we make use of the following notations:

- $H^1(\mathbb{R}^3)$ denotes the usual Sobolev space equipped with the inner product and norm

$$(u, v) = \int_{\mathbb{R}^3} (\nabla u \cdot \nabla v + uv) \mathrm{d}x, \quad \|u\| = (u, u)^{1/2}, \quad \forall\, u, v \in H^1(\mathbb{R}^3);$$

- $L^s(\mathbb{R}^3)$ $(1 \le s < \infty)$ denotes the Lebesgue space with the norm $\|u\|_s = \left(\int_{\mathbb{R}^3} |u|^s \mathrm{d}x\right)^{1/s}$;
- For any $u \in H^1(\mathbb{R}^3)$, $u^t(x) := t^{3/2}u(tx)$ and $u_t(x) := t^{1/2}u(x/t)$;
- For any $x \in \mathbb{R}^3$ and $r > 0$, $B_r(x) := \{y \in \mathbb{R}^3 : |y - x| < r\}$;
- $S = \inf_{u \in \mathcal{D}^{1,2}(\mathbb{R}^3)\setminus\{0\}} \|\nabla u\|_2^2 / \|u\|_6^2$;
- $C_1, C_2, \ldots$ denote positive constants possibly different in different places.

3.1.2 Basic ideas in the main proofs

For the L^2-supercritical case, some key arguments are stated in what follows.

(i) When $K(x) \equiv K_\infty$, Ye [158] showed that the mapping

$$c \mapsto m(c) \text{ is strictly decreasing on } (0, +\infty) \tag{3.16}$$

by using the translation invariance of I and the homogeneity of f. In such a way there are excluded the vanishing and the dichotomy cases of the minimizing sequence $\{u_n\}$ for $m(c) = \inf_{u \in \mathcal{M}_c} I(u)$ in applying the concentration-compactness principle. However, the approach used in [158] is valid only for autonomous equations and it does not work anymore for problem (3.1) with $K \neq$ constant and more general f. By establishing some new inequalities, we prove that $m(c)$ is nonincreasing and $m(c) > m(\tilde{c})$ for any $\tilde{c} > c$ provided $m(c)$ is attained. To bypass the difficulty caused by the lack of compactness of Sobolev embedding $H^1(\mathbb{R}^3) \hookrightarrow L^s(\mathbb{R}^3)$ for $2 \le s < 6$, we compare the constrained minimum $m(c)$ with the one of the limit equation (3.11) (that is, problem (3.1) with $K(x) = K_\infty$), and by using some subtle analysis we prove that $u_n \to \bar{u}$ in $H^1(\mathbb{R}^3)$ (after a translation and extraction of a subsequence if necessary) as long as $m(c) < m^\infty(c)$, and $\bar{u} \in \mathcal{S}(c)$ is a minimizer of $m(c)$, where the definition of $m^\infty(c)$ is given by relation (3.15).

(ii) To verify that $\mathcal{M}_c$ is a natural constraint on $\mathcal{S}_c$, we use a combination of the deformation lemma, some new inequalities and an intermediary theorem for continuous functions, other than the mountain pass theorem on $\mathcal{S}_c$ and the Lagrange multiplier method used in [155] and [158], respectively.

For the L^2-subcritical case, some basic ideas are described in what follows.

(i) The key step to prove that $\sigma(c) = \inf_{\mathcal{S}_c} I$ is achieved is to obtain the subadditivity inequality

$$\sigma(c) < \sigma(\alpha) + \sigma(c - \alpha), \qquad \forall\, 0 < \alpha < c. \tag{3.17}$$

To this end, Ye in [158] used the scaling $t \mapsto u(t^{-2/3}x)$. But this kind of scaling is not suitable in our case, excepting the case when $K(x)$ is a positive constant. Instead, we present another scaling $t \mapsto t^{1/2}u(x/t)$, and we succeed to prove that (3.17) still holds under assumptions of Theorem 3.2.

(ii) Compared with [158], it is more complicated to find a local minimizer of I on $\mathcal{S}_c$ which is a critical point of $I|_{\mathcal{S}_c}$ because $K(x)$ is variable and f is nonhomogeneous.

3.2 The L^2-supercritical case

This section is devoted to the proof of Theorem 3.1.

Lemma 3.1 *Assume that conditions* (K1) *and* (K2) *hold. Then $K(tx)$ is nonincreasing on $t \in (0, \infty)$ for every $x \in \mathbb{R}^3$.*

Proof. For any fixed $x \in \mathbb{R}^3$, we define the following function on $(0, \infty)$:

$$\varphi(t) = \frac{3(\theta - 2)}{2} K(tx) - \nabla K(tx) \cdot (tx). \tag{3.18}$$

Note that for every $x \in \mathbb{R}^3$,

$$K(tx) = t^{3(\theta-2)/2} \int_t^\infty \frac{\varphi(s)}{s^{3\theta/2-4}} \mathrm{d}s. \tag{3.19}$$

For any $0 < t_1 < t_2 < \infty$, by (K1), (K2) and (3.19), we have

$$\begin{aligned}
K(t_2x) - K(t_1x) &= t_2^{3(\theta-2)/2} \int_{t_2}^\infty \frac{\varphi(s)}{s^{3\theta/2-4}} \mathrm{d}s - t_1^{3(\theta-2)/2} \int_{t_1}^\infty \frac{\varphi(s)}{s^{3\theta/2-4}} \mathrm{d}s \\
&= \left(t_2^{3(\theta-2)/2} - t_1^{3(\theta-2)/2}\right) \int_{t_2}^\infty \frac{\varphi(s)}{s^{3\theta/2-4}} \mathrm{d}s - t_1^{3(\theta-2)/2} \int_{t_1}^{t_2} \frac{\varphi(s)}{s^{3\theta/2-4}} \mathrm{d}s \\
&\le \left(t_2^{3(\theta-2)/2} - t_1^{3(\theta-2)/2}\right) \varphi(t_2) \int_{t_2}^\infty \frac{1}{s^{3\theta/2-4}} \mathrm{d}s - t_1^{3(\theta-2)/2} \varphi(t_2) \int_{t_1}^{t_2} \frac{1}{s^{3\theta/2-4}} \mathrm{d}s \\
&= \varphi(t_2) \left[t_2^{3(\theta-2)/2} \int_{t_2}^\infty \frac{1}{s^{3\theta/2-4}} \mathrm{d}s - t_1^{3(\theta-2)/2} \int_{t_1}^\infty \frac{1}{s^{3\theta/2-4}} \mathrm{d}s \right] = 0.
\end{aligned}$$

This completes the proof. □

Lemma 3.2 *Assume that hypotheses* (K1) *and* (K2) *hold. Then for all $x \in \mathbb{R}^3$ and $t > 0$,*

$$h_0(x,t) := t^{\frac{3(\theta-2)}{2}}[K(t^{-1}x) - K(x)] - \frac{2\left(1 - t^{\frac{3(\theta-2)}{2}}\right)}{3(\theta-2)}\nabla K(x) \cdot x \geq 0, \tag{3.20}$$

$$t \mapsto K(tx) \text{ is nonincreasing on } (0,\infty) \text{ for every } x \in \mathbb{R}^3, \tag{3.21}$$

and

$$|\nabla K(x) \cdot x| \to 0 \text{ as } |x| \to \infty. \tag{3.22}$$

Proof. For any $x \in \mathbb{R}^3$, by (K2), we have

$$\begin{aligned}\frac{\mathrm{d}}{\mathrm{d}t}h_0(x,t) &= \frac{t^{\frac{3(\theta-2)}{2}-1}}{2}\left[3(\theta-2)K(t^{-1}x) - 2\nabla K(t^{-1}x)\cdot(t^{-1}x) - 3(\theta-2)K(x) + 2\nabla K(x)\cdot x\right] \\ &\begin{cases} \geq 0, & t \geq 1, \\ \leq 0, & 0 < t < 1, \end{cases}\end{aligned}$$

which implies that $h_0(x,t) \geq h_0(x,1) = 0$ for all $x \in \mathbb{R}^3$ and $t > 0$, hence (3.20) holds.

By (3.20) and the continuity of $h_0(x,\cdot)$, we obtain

$$\lim_{t\to 0} h_0(x,t) = -\frac{2}{3(\theta-2)}\nabla K(x) \cdot x \geq 0, \quad \forall\, x \in \mathbb{R}^3, \tag{3.23}$$

which leads to (3.21). Since $h_0(x,2) \geq 0$ for all $x \in \mathbb{R}^3$, we have

$$0 \leq -\nabla K(x) \cdot x \leq \frac{2^{\frac{3\theta-8}{2}}3(\theta-2)[K(x/2) - K(x)]}{2^{\frac{3(\theta-2)}{2}} - 1}, \quad \forall\, x \in \mathbb{R}^3.$$

Hence, relation (3.22) holds by letting $|x| \to \infty$ in the above inequality. □

Lemma 3.3 *Assume that* (F1)–(F3) *hold. Then for all $t > 0$ and $\tau \in \mathbb{R}$,*

$$h_1(t,\tau) := \frac{2\left(1 - t^{7-3\theta/2}\right)}{14 - 3\theta}[f(\tau)\tau - \theta F(\tau)] - \frac{2}{3}F(\tau) + \frac{2}{3}t^{-3\theta/2}F(t^{3/2}\tau) \geq 0 \tag{3.24}$$

and

$$\frac{F(t)}{|t|^{11/3}t} \text{ is nondecreasing on both } (-\infty,0) \text{ and } (0,+\infty). \tag{3.25}$$

Proof. For any $\tau \in \mathbb{R}$, by (F3), we have

$$\begin{aligned}\frac{\mathrm{d}}{\mathrm{d}t}h_1(t,\tau) &= t^{6-3\theta/2}|\tau|^{14/3}\left[\frac{f(t^{3/2}\tau)t^{3/2}\tau - \theta F(t^{3/2}\tau)}{|t^{3/2}\tau|^{14/3}} - \frac{f(\tau)\tau - \theta F(\tau)}{|\tau|^{14/3}}\right] \\ &\begin{cases} \geq 0, & t \geq 1, \\ \leq 0, & 0 < t < 1, \end{cases}\end{aligned}$$

which implies that $h_1(t,\tau) \geq h_1(1,\tau) = 0$ for all $t > 0$ and $\tau \in \mathbb{R}$, that is, inequality (3.24) holds. By (F3) and (3.24), we obtain

$$h_1(0,\tau) := \lim_{|t|\to 0} h(t,\tau) = \frac{2}{14-3\theta}\left[f(\tau)\tau - \frac{14}{3}F(\tau)\right] \geq 0, \quad \forall\, \tau \in \mathbb{R}. \tag{3.26}$$

From (3.26), we derive

$$\frac{\mathrm{d}}{\mathrm{d}t}\frac{F(t)}{|t|^{11/3}t} = \frac{3}{2|t|^7}\left[f(t)t - \frac{14}{3}F(t)\right] \geq 0.$$

This shows that relation (3.25) holds. □

Lemma 3.4 *Assume that conditions* (K1), (K2) *and* (F1)–(F3) *hold. Then*

$$\begin{aligned} h_2(x,t,\tau) &:= t^{-3}K(t^{-1}x)F(t^{3/2}\tau) - K(x)F(\tau) + \frac{3(1-t^4)}{8}K(x)[f(\tau)\tau - 2F(\tau)] \\ &\quad - \frac{1-t^4}{4}\nabla K(x)\cdot xF(\tau) \\ &\geq 0, \quad \forall\, x\in\mathbb{R}^3,\ t>0,\ \tau\in\mathbb{R}. \end{aligned} \tag{3.27}$$

Proof. For any $x\in\mathbb{R}^3$ and $\tau\in\mathbb{R}$, by (K1), (K2), (F1), (F3), (3.21) and (3.25), we have

$$\begin{aligned} \frac{\mathrm{d}}{\mathrm{d}t}h_2(x,t,\tau) &= -\frac{1}{t^4}\nabla K(t^{-1}x)\cdot(t^{-1}x)F(t^{3/2}\tau)+\frac{3}{2t^4}K(t^{-1}x)\left[f(t^{3/2}\tau)t^{3/2}\tau-2F(t^{3/2}\tau)\right] \\ &- \frac{3t^3}{2}K(x)[f(\tau)\tau-2F(\tau)]+t^3\nabla K(x)\cdot xF(\tau) \\ &= \frac{3}{2t^4}K(t^{-1}x)\left[f(t^{3/2}\tau)t^{3/2}\tau-\theta F(t^{3/2}\tau)\right] \\ &+ \frac{1}{2t^4}\left[3(\theta-2)K(t^{-1}x)-2\nabla K(t^{-1}x)\cdot(t^{-1}x)\right]F(t^{3/2}\tau) \\ &- \frac{3t^3}{2}K(x)[f(\tau)\tau-\theta F(\tau)]-\frac{t^3}{2}\left[3(\theta-2)K(x)-2\nabla K(x)\cdot x\right]F(\tau) \end{aligned}$$

It follows that

$$\begin{aligned} \frac{\mathrm{d}}{\mathrm{d}t}h_2(x,t,\tau) &= \frac{t^3|\tau|^7}{2}\left\{3K(t^{-1}x)\frac{f(t^{3/2}\tau)t^{3/2}\tau-\theta F(t^{3/2}\tau)}{|t^{3/2}\tau|^{14/3}}-3K(x)\frac{f(\tau)\tau-\theta F(\tau)}{|\tau|^{14/3}}\right. \\ &+ \left[3(\theta-2)K(t^{-1}x)-2\nabla K(t^{-1}x)\cdot(t^{-1}x)\right]\frac{F(t^{3/2}\tau)}{|t^{3/2}\tau|^{14/3}} \\ &- \left.\left[3(\theta-2)K(x)-2\nabla K(x)\cdot x\right]\frac{F(\tau)}{|\tau|^{14/3}}\right\}, \end{aligned}$$

hence

$$\frac{\mathrm{d}}{\mathrm{d}t}h_2(x,t,\tau)\begin{cases}\geq 0, & t\geq 1,\\ \leq 0, & 0<t<1,\end{cases}$$

This implies that $h_2(x,t,\tau)\geq h_2(x,1,\tau)=0$ for all $x\in\mathbb{R}^3$, $t>0$ and $\tau\in\mathbb{R}^3$, that is, relation (3.27) holds. □

By the scaling (3.7), we obtain

$$I(u^t)=\frac{at^2}{2}\int_{\mathbb{R}^3}|\nabla u|^2\mathrm{d}x+\frac{bt^4}{4}\left(\int_{\mathbb{R}^3}|\nabla u|^2\mathrm{d}x\right)^2-t^{-3}\int_{\mathbb{R}^3}K(t^{-1}x)F(t^{3/2}u)\mathrm{d}x \tag{3.28}$$

and

$$I^\infty(u^t)=\frac{at^2}{2}\int_{\mathbb{R}^3}|\nabla u|^2\mathrm{d}x+\frac{bt^4}{4}\left(\int_{\mathbb{R}^3}|\nabla u|^2\mathrm{d}x\right)^2-t^{-3}\int_{\mathbb{R}^3}K_\infty F(t^{3/2}u)\mathrm{d}x. \tag{3.29}$$

Noting that $J(u)=\frac{\mathrm{d}}{\mathrm{d}t}I(u^t)\Big|_{t=1}$ and $J^\infty(u):=\frac{\mathrm{d}}{\mathrm{d}t}I^\infty(u^t)\Big|_{t=1}$, it follows from (3.28) and (3.29) that

$$J(u)=a\|\nabla u\|_2^2+b\|\nabla u\|_2^4-\frac{3}{2}\int_{\mathbb{R}^3}K(x)\left[f(u)u-2F(u)\right]\mathrm{d}x+\int_{\mathbb{R}^3}\nabla K(x)\cdot xF(u)\mathrm{d}x \tag{3.30}$$

and

$$J^{\infty}(u)=a\|\nabla u\|_2^2+b\|\nabla u\|_2^4-\frac{3}{2}\int_{\mathbb{R}^3}K_{\infty}\left[f(u)u-2F(u)\right]\mathrm{d}x. \tag{3.31}$$

The following result establishes a useful energy estimate.

Lemma 3.5 *Assume that* (K1), (K2) *and* (F1)–(F3) *hold. Then*

$$I(u)\geq I\left(u^t\right)+\frac{1-t^4}{4}J(u)+\frac{a(1-t^2)^2}{4}\|\nabla u\|_2^2,\qquad \forall\ u\in H^1(\mathbb{R}^3),\ t>0. \tag{3.32}$$

Proof. By (3.3), (3.24), (3.25), (3.27), (3.28) and (3.30), we have

$$\begin{aligned}
I(u)-I\left(u^t\right) &= \frac{a(1-t^2)}{2}\|\nabla u\|_2^2+\frac{b(1-t^4)}{4}\|\nabla u\|_2^4\\
&+ \int_{\mathbb{R}^3}\left[t^{-3}K(t^{-1}x)F\left(t^{3/2}u\right)-K(x)F(u)\right]\mathrm{d}x\\
&= \frac{1-t^4}{4}\left\{a\|\nabla u\|_2^2+b\|\nabla u\|_2^4-\frac{3}{2}\int_{\mathbb{R}^3}K(x)\left[f(u)u-2F(u)\right]\mathrm{d}x\right.\\
&+ \left.\int_{\mathbb{R}^3}\nabla K(x)\cdot xF(u)\mathrm{d}x\right\}+\frac{a(1-t^2)^2}{4}\|\nabla u\|_2^2\\
&+ \int_{\mathbb{R}^3}\left\{t^{-3}K(t^{-1}x)F(t^{3/2}u)-K(x)F(u)+\frac{3(1-t^4)}{8}K(x)[f(u)u-2F(u)]\right.\\
&- \left.\frac{1-t^4}{4}\nabla K(x)\cdot xF(u)\right\}\mathrm{d}x.
\end{aligned}$$

This shows that relation (3.32) holds. □

A direct consequence of this result is the following property.

Corollary 3.1 *Assume that* (K1), (K2) *and* (F1)–(F3) *hold. Then*

$$I(u)=\max_{t>0}I\left(u^t\right),\qquad \forall\ u\in\mathcal{M}_c. \tag{3.33}$$

Lemma 3.6 *Assume that* (K1), (K2) *and* (F1)–(F3) *hold. Then for any* $u\in H^1(\mathbb{R}^3)\setminus\{0\}$, *there exists a unique* $t_u>0$ *such that* $u^{t_u}\in\mathcal{M}_c$.

Proof. Let $u\in H^1(\mathbb{R}^3)\setminus\{0\}$ be fixed and define the function $\zeta(t):=I\left(u^t\right)$ on $(0,\infty)$. By relations (3.28) and (3.30), we have

$$\begin{aligned}
\zeta'(t)=0 \quad &\Leftrightarrow\ at\|\nabla u\|_2^2+bt^3\|\nabla u\|_2^4-\frac{3}{2t^4}\int_{\mathbb{R}^3}K(t^{-1}x)\left[f(t^{3/2}u)t^{3/2}u-2F(t^{3/2}u)\right]\mathrm{d}x\\
&\quad +\frac{1}{t^4}\int_{\mathbb{R}^3}\nabla K(t^{-1}x)\cdot(t^{-1}x)F(t^{3/2}u)\mathrm{d}x=0\\
&\Leftrightarrow\ \frac{1}{t}J\left(u^t\right)=0\ \Leftrightarrow\ u^t\in\mathcal{M}_c.
\end{aligned} \tag{3.34}$$

Note that (3.21) and (3.25) lead to

$$K(t^{-1}x)F(t^{3/2}\tau)\leq t^7K(x)F(\tau),\qquad \forall\ x\in\mathbb{R}^3,\ t\in(0,1),\ \tau\in\mathbb{R}. \tag{3.35}$$

From (3.28) and (3.35), we deduce that

$$I(u^t)\geq\frac{at^2}{2}\|\nabla u\|_2^2+\frac{bt^4}{4}\|\nabla u\|_2^4-t^4\int_{\mathbb{R}^3}K(x)F(u)\mathrm{d}x,\qquad \forall\ t\in(0,1), \tag{3.36}$$

which implies that $\zeta(t) > 0$ for $t > 0$ small. Moreover, by (K1), (K2), (F1), (F2) and (3.28), we obtain that $\lim_{t\to 0}\zeta(t) = 0$ and $\zeta(t) < 0$ for t large. Therefore, $\max_{t\in(0,\infty)}\zeta(t)$ is achieved at $t_u > 0$ so that $\zeta'(t_u) = 0$ and $u^{t_u} \in \mathcal{M}_c$.

Next, we claim that t_u is unique for any $u \in H^1(\mathbb{R}^3)\setminus\{0\}$. Otherwise, for any given $u \in H^1(\mathbb{R}^3)\setminus\{0\}$, there exist positive constants $t_1 \neq t_2$ such that $u^{t_1}, u^{t_2} \in \mathcal{M}_c$, that is, $J(u^{t_1}) = J(u^{t_2}) = 0$. Then (3.32) implies

$$I\left(u^{t_1}\right) > I\left(u^{t_2}\right) + \frac{t_1^4 - t_2^4}{t_1^4} J\left(u^{t_1}\right) = I\left(u^{t_2}\right) > I\left(u^{t_1}\right) + \frac{t_2^4 - t_1^4}{t_2^4} J\left(u^{t_2}\right) = I\left(u^{t_1}\right). \quad (3.37)$$

This contradiction shows that $t_u > 0$ is unique for any $u \in H^1(\mathbb{R}^3)\setminus\{0\}$. □

Combining Corollary 3.1 with Lemma 3.6, we have the following property.

Lemma 3.7 *Assume that* (K1), (K2) *and* (F1)–(F3) *hold. Then*

$$\inf_{u\in\mathcal{M}_c} I(u) = m(c) = \inf_{u\in\mathcal{S}_c}\max_{t>0} I\left(u^t\right).$$

Lemma 3.8 *Assume that conditions* (K1), (K2) *and* (F1)–(F3) *hold. Then*

1. *there exists* $\rho_0 > 0$ *such that* $\|\nabla u\|_2 \geq \rho_0, \ \forall\, u \in \mathcal{M}_c$;
2. $m(c) = \inf_{u\in\mathcal{M}_c} I(u) > 0$.

Proof. (i) By (F1), we observe that the mapping

$$\frac{F(t)}{|t|^{\mu-1}t} \text{ is nonincreasing on both } (-\infty, 0) \text{ and } (0, +\infty). \quad (3.38)$$

From (3.25) and (3.38), we derive that for any $s \in \mathbb{R}$,

$$\begin{cases} |t|^{\mu}F(s) \leq F(st) \leq |t|^{\frac{14}{3}}F(s), & \text{if } |t| \leq 1; \\ |t|^{\frac{14}{3}}F(s) \leq F(st) \leq |t|^{\mu}F(s), & \text{if } |t| \geq 1, \end{cases} \quad (3.39)$$

which implies that there is a constant $C_0 > 0$ such that

$$0 \leq F(t) \leq C_0\left(|t|^{\frac{14}{3}} + |t|^{\mu}\right), \qquad \forall\, t \in \mathbb{R}. \quad (3.40)$$

By Lemma B.1 (Gagliardo–Nirenberg inequality), we have

$$\|u\|_s^s \leq \mathcal{C}(s)\|\nabla u\|_2^{\frac{3(s-2)}{2}}\|u\|_2^{\frac{6-s}{2}}, \qquad \forall\, s \in (2, 6). \quad (3.41)$$

Since $J(u) = 0$, $\forall u \in \mathcal{M}_c$, by (K1), (K2), (3.30), (3.40), (3.41) and the Sobolev inequality, we deduce that

$$\begin{aligned} a\|\nabla u\|_2^2 &\leq a\|\nabla u\|_2^2 + b\|\nabla u\|_2^4 \\ &= 2\int_{\mathbb{R}^3} K(x)\left[f(u)u - 2F(u)\right]\mathrm{d}x - 2\int_{\mathbb{R}^3}\nabla K(x)\cdot xF(u)\mathrm{d}x \\ &\leq C_1\left(\|u\|_{\frac{14}{3}}^{\frac{14}{3}} + \|u\|_6^6\right) \leq C_2\|\nabla u\|_2^4\|u\|_2^{2/3} + C_1 S^{-3}\|\nabla u\|_2^6 \\ &= C_2 c^{2/3}\|\nabla u\|_2^4 + C_1 S^{-3}\|\nabla u\|_2^6, \qquad \forall\, u \in \mathcal{M}_c, \end{aligned} \quad (3.42)$$

which concludes the proof of (i).

(ii) By (i) and (3.32) with $t \to 0$, we have

$$I(u) = I(u) - \frac{1}{4}J(u) \geq \frac{a}{4}\|\nabla u\|_2^2 \geq \frac{a}{4}\rho_0^2, \qquad \forall\, u \in \mathcal{M}_c. \quad (3.43)$$

Hence, $m(c) = \inf_{u\in\mathcal{M}_c} I(u) > 0$. □

Lemma 3.9 *Assume that* (K1), (K2) *and* (F1)–(F3) *hold. Then $m(c)$ is nonincreasing on $(0,\infty)$. In particular, if $m(c)$ is achieved, then $m(c) > m(\tilde{c})$ for any $\tilde{c} > c$.*

Proof. For any $c_2 > c_1 > 0$, there exists $\{u_n\} \subset \mathcal{M}_{c_1}$ such that

$$I(u_n) < m(c_1) + \frac{1}{n}.$$

Let $\xi = \sqrt{c_2/c_1} \in (1,\infty)$ and $v_n(x) = \xi^{-1/2}u_n(\xi^{-1}x)$. Then $\|v_n\|_2^2 = c_2$ and $\|\nabla v_n\|_2 = \|\nabla u_n\|_2$. By Lemma 3.6, there exists $t_n > 0$ such that $(v_n)^{t_n} \in \mathcal{M}_{c_2}$. Note that $(6-\mu)K(x) + \nabla K(x)\cdot x \geq 0$ for all $x \in \mathbb{R}^3$ implies

$$t \mapsto t^{\frac{6-\mu}{2}}K(tx) \text{ is nondecreasing on } (0,+\infty) \text{ for every } x \in \mathbb{R}^3. \tag{3.44}$$

Then it follows from (K2), (3.28), (3.32), (3.38) and (3.44) that

$$\begin{aligned}
& m(c_2) \leq I\left((v_n)^{t_n}\right) \\
= \; & \frac{at_n^2}{2}\|\nabla u_n\|_2^2 + \frac{bt_n^4}{4}\|\nabla u_n\|_2^4 - t_n^{-3}\xi^3\int_{\mathbb{R}^3} K(t_n^{-1}\xi x)F(t_n^{3/2}\xi^{-1/2}u_n)\mathrm{d}x \\
\leq \; & I\left((u_n)^{t_n}\right) + t_n^{-3}\int_{\mathbb{R}^3}\left[K(t_n^{-1}x)F(t_n^{3/2}u_n) - \xi^{\frac{6-\mu}{2}}K(t_n^{-1}\xi x)\xi^{\frac{\mu}{2}}F(t_n^{3/2}\xi^{-1/2}u_n)\right]\mathrm{d}x \\
\leq \; & I(u_n) - \frac{a(1-t_n^2)^2}{4}\|\nabla u_n\|_2^2 < m(c_1) + \frac{1}{n},
\end{aligned}$$

which shows that $m(c_2) \leq m(c_1)$ by letting $n \to \infty$.

Next, we assume that $m(c)$ is achieved, that is, there exists $u \in \mathcal{M}_c$ such that $I(u) = m(c)$. For any given $\tilde{c} > c$. Let $\tilde{\xi} = \tilde{c}/c \in (1,\infty)$ and $v(x) = \tilde{\xi}^{-1/2}u(\tilde{\xi}^{-1}x)$. Then $\|v\|_2^2 = \tilde{c}$ and $\|\nabla v\|_2 = \|\nabla u\|_2$. By Lemma 3.6, there exists $\tilde{t} > 0$ such that $v^{\tilde{t}} \in \mathcal{M}_{\tilde{c}}$. Then it follows from (K2), (3.28), (3.32), (3.38) and (3.44) that

$$\begin{aligned}
m(\tilde{c}) \; & \leq \; I(v^{\tilde{t}}) = \frac{a\tilde{t}^2}{2}\|\nabla u\|_2^2 + \frac{b\tilde{t}^4}{4}\|\nabla u\|_2^4 - \tilde{t}^{-3}\tilde{\xi}^3\int_{\mathbb{R}^3} K(\tilde{t}^{-1}\tilde{\xi}x)F(\tilde{t}^{3/2}\tilde{\xi}^{-1/2}u)\mathrm{d}x \\
& \leq \; I\left(u^{\tilde{t}}\right) + \tilde{t}^{-3}\int_{\mathbb{R}^3}\left[K(\tilde{t}^{-1}x)F(\tilde{t}^{3/2}u) - \tilde{\xi}^{\frac{6-\mu}{2}}K(\tilde{t}^{-1}\tilde{\xi}x)\tilde{\xi}^{\frac{\mu}{2}}F(\tilde{t}^{3/2}\tilde{\xi}^{-1/2}u)\right]\mathrm{d}x \\
& \leq \; I(u) - \frac{a(1-\tilde{t}^2)^2}{4}\|\nabla u\|_2^2 < m(c).
\end{aligned}$$

The proof is completed. □

Remark 3.1 *Lemmas 3.1–3.5 provide technical ingredients allowing generalization of the previous results to a new scenario when $K(x)$ is variable and $f(u)$ has a more general structure. But it is difficult to prove that $m(c)$ is strictly decreasing on $(0,\infty)$ in the same way as in the proof of* [158, Lemma 4.5], *because it is still unknown that the sequence $\{t_n\}$, involved in the proof of Lemma 3.9, is bounded from below for more general nonlinearity f.*

Lemma 3.10 *Assume that* (K1), (K2) *and* (F1)–(F3) *hold. Then $m(c) \leq m^\infty(c)$.*

Proof. In view of Lemmas 3.6 and 3.8, we have $\mathcal{M}_c^\infty \neq \emptyset$ and $m^\infty(c) > 0$. Assume by contradiction that $m(c) > m^\infty(c)$. Let $\varepsilon := m(c) - m^\infty(c)$. Then there exists u_ε^∞ such that

$$u_\varepsilon^\infty \in \mathcal{M}_c^\infty \quad \text{and} \quad m^\infty(c) + \frac{\varepsilon}{2} > I^\infty(u_\varepsilon^\infty). \tag{3.45}$$

In view of Lemma 3.6, there exists $t_\varepsilon > 0$ such that $(u_\varepsilon^\infty)^{t_\varepsilon} \in \mathcal{M}_c$. Since $K_\infty \le K(x)$ for all $\mathbb{R}^3$, it follows from (3.3), (3.12), (3.45) and Corollary 3.1 that

$$m^\infty(c) + \frac{\varepsilon}{2} > I^\infty(u_\varepsilon^\infty) \ge I^\infty\left((u_\varepsilon^\infty)^{t_\varepsilon}\right) \ge I\left((u_\varepsilon^\infty)^{t_\varepsilon}\right) \ge m(c).$$

This contradiction shows that $m(c) \le m^\infty(c)$. □

Similarly to [153], we have the following Brezis-Lieb type lemma.

Lemma 3.11 *Assume that* (K1), (K2), (F1) *and* (F2) *hold. If* $u_n \rightharpoonup \bar{u}$ *in* $H^1(\mathbb{R}^3)$, *then*

$$I(u_n) = I(\bar{u}) + I(u_n - \bar{u}) + \frac{b}{2}\|\nabla \bar{u}\|_2^2 \|\nabla(u_n - \bar{u})\|_2^2 + o(1) \tag{3.46}$$

and

$$J(u_n) = J(\bar{u}) + J(u_n - \bar{u}) + b\|\nabla \bar{u}\|_2^2 \|\nabla(u_n - \bar{u})\|_2^2 + o(1). \tag{3.47}$$

Lemma 3.12 *Assume that* (K1), (K2) *and* (F1)–(F3) *hold. Then* $m(c)$ *is achieved.*

Proof. In view of Lemmas 3.6 and 3.8, we have $\mathcal{M}_c \ne \emptyset$ and $m(c) > 0$. Let $\{u_n\} \subset \mathcal{M}_c$ be such that $I(u_n) \to m(c)$. Since $J(u_n) = 0$, then using (3.32) with $t \to 0$, it follows that

$$m(c) + o(1) = I(u_n) \ge \frac{a}{4}\|\nabla u_n\|_2^2, \tag{3.48}$$

which, together with $\|u_n\|_2^2 = c$, implies that $\{u_n\}$ is bounded in $H^1(\mathbb{R}^3)$. Passing to a subsequence, we have $u_n \rightharpoonup \bar{u}$ in $H^1(\mathbb{R}^3)$. Then $u_n \to \bar{u}$ in $L^s_{\text{loc}}(\mathbb{R}^3)$ for $2 \le s < 6$ and $u_n \to \bar{u}$ a.e. in $\mathbb{R}^3$. There are two possible cases: i) $\bar{u} = 0$ and ii) $\bar{u} \ne 0$.

Case i) $\bar{u} = 0$, namely $u_n \rightharpoonup 0$ in $H^1(\mathbb{R}^3)$. Then $u_n \to 0$ in $L^s_{\text{loc}}(\mathbb{R}^3)$ for $2 \le s < 6$ and $u_n \to 0$ a.e. in $\mathbb{R}^3$. By (K1) and (3.22), we observe that

$$\lim_{n\to\infty} \int_{\mathbb{R}^3} [K_\infty - K(x)] F(u_n) \mathrm{d}x = \lim_{n\to\infty} \int_{\mathbb{R}^3} \nabla K(x) \cdot x F(u_n) \mathrm{d}x = 0. \tag{3.49}$$

From (3.3), (3.12), (3.30), (3.31) and (3.49), one can get

$$I^\infty(u_n) \to m(c), \qquad J^\infty(u_n) \to 0. \tag{3.50}$$

From (3.31), (3.50), Lemma 3.6 and Lemma 3.7 (i), we obtain

$$a\rho_0^2 \le a\|\nabla u_n\|_2^2 + b\|\nabla u_n\|_2^4 = \frac{3}{2}\int_{\mathbb{R}^3} K_\infty \left[f(u_n)u_n - 2F(u_n)\right] \mathrm{d}x. \tag{3.51}$$

Using (3.40), (3.51) and Lions' concentration-compactness principle [153, Lemma 1.21], we prove that there exist $\delta > 0$ and $\{y_n\} \subset \mathbb{R}^3$ such that $\int_{B_1(y_n)} |u_n|^2 \mathrm{d}x > \delta$. Let $\hat{u}_n(x) = u_n(x + y_n)$. Then we have $\|\hat{u}_n\| = \|u_n\|$ and

$$J^\infty(\hat{u}_n) = o(1), \qquad I^\infty(\hat{u}_n) \to m(c), \qquad \int_{B_1(0)} |\hat{u}_n|^2 \mathrm{d}x > \delta. \tag{3.52}$$

Therefore, there exists $\hat{u} \in H^1(\mathbb{R}^3) \setminus \{0\}$ such that, passing to a subsequence,

$$\begin{cases} \hat{u}_n \rightharpoonup \hat{u}, & \text{in } H^1(\mathbb{R}^3); \\ \hat{u}_n \to \hat{u}, & \text{in } L^s_{\text{loc}}(\mathbb{R}^3), \ \forall\, s \in [1, 6); \\ \hat{u}_n \to \hat{u}, & \text{a.e. on } \mathbb{R}^3. \end{cases} \tag{3.53}$$

Let $w_n = \hat{u}_n - \hat{u}$. Then (3.53) and Lemma 3.11 yield

$$\|\hat{u}\|_2^2 =: \hat{c} \le c, \qquad \|w_n\|_2^2 =: \hat{c}_n \le c \text{ for large } n \in \mathbb{N}, \tag{3.54}$$

$$I^\infty(\hat{u}_n) = I^\infty(\hat{u}) + I^\infty(w_n) + \frac{b}{2}\|\nabla \hat{u}\|_2^2 \|\nabla w_n\|_2^2 + o(1), \tag{3.55}$$

and

$$J^\infty(\hat{u}_n) = J^\infty(\hat{u}) + J^\infty(w_n) + b\|\nabla \hat{u}\|_2^2 \|\nabla w_n\|_2^2 + o(1). \tag{3.56}$$

Let

$$\begin{aligned}\Psi^\infty(u) &:= I^\infty(u) - \frac{1}{4}J^\infty(u)\\ &= \frac{a}{4}\|\nabla u\|_2^2 + \frac{1}{8}\int_{\mathbb{R}^3} K_\infty[3f(u)u - 14F(u)]\mathrm{d}x, \qquad \forall\, u \in H^1(\mathbb{R}^3). \end{aligned}\tag{3.57}$$

By (3.26), one has $\Psi^\infty(u) > 0$ for all $u \in H^1(\mathbb{R}^3) \setminus \{0\}$. Moreover, it follows from (3.52), (3.55), (3.56) and (3.57) that

$$\Psi^\infty(w_n) \le m(c) - \Psi^\infty(\hat{u}) + o(1), \qquad J^\infty(w_n) \le -J^\infty(\hat{u}) + o(1). \tag{3.58}$$

If there exists a subsequence $\{w_{n_i}\}$ of $\{w_n\}$ such that $w_{n_i} = 0$, then it follows from (3.54), Lemmas 3.9 and 3.10 that

$$m^\infty(\hat{c}) \le I^\infty(\hat{u}) = m(c) \le m(\hat{c}) \le m^\infty(\hat{c}), \qquad J^\infty(\hat{u}) = 0,$$

which, together with $m^\infty(c) \le m^\infty(\hat{c}) \le I^\infty(\hat{u}) = m(c) \le m^\infty(c)$, implies

$$I^\infty(\hat{u}) = m^\infty(\hat{c}) = m(\hat{c}) = m(c) = m^\infty(c), \qquad J^\infty(\hat{u}) = 0. \tag{3.59}$$

Next, we assume that $w_n \ne 0$. We claim that $J^\infty(\hat{u}) \le 0$. Otherwise, if $J^\infty(\hat{u}) > 0$, then (3.58) implies $J^\infty(w_n) < 0$ for large n. In view of Lemma 3.6, there exists $t_n > 0$ such that $(w_n)^{t_n} \in \mathcal{M}^\infty_{\hat{c}_n}$. Then it follows from (3.12), (3.31), (3.32), (3.57), (3.58), Lemmas 3.9 and 3.10 that

$$\begin{aligned} m(c) - \Psi^\infty(\hat{u}) + o(1) &\ge \Psi^\infty(w_n) = I^\infty(w_n) - \frac{1}{4}J^\infty(w_n)\\ &\ge I^\infty\left((w_n)^{t_n}\right) - \frac{t_n^4}{4}J^\infty(w_n) \ge m^\infty(\hat{c}_n) - \frac{t_n^4}{4}J^\infty(w_n)\\ &\ge m^\infty(c) + o(1) \ge m(c) + o(1), \end{aligned}$$

which is impossible due to $\Psi^\infty(\hat{u}) > 0$. This shows that $J^\infty(\hat{u}) \le 0$. In view of Lemma 3.6, there exists $t_\infty > 0$ such that $\hat{u}^{t_\infty} \in \mathcal{M}^\infty_{\hat{c}}$. From (3.12), (3.31), (3.32), (3.52), (3.57), the weak semicontinuity of the norm, Fatou's lemma, Lemmas 3.9 and 3.10, we obtain

$$\begin{aligned} m(c) &= \lim_{n\to\infty}\left[I^\infty(\hat{u}_n) - \frac{1}{4}J^\infty(\hat{u}_n)\right] = \lim_{n\to\infty}\Psi^\infty(\hat{u}_n) \ge \Psi^\infty(\hat{u})\\ &= I^\infty(\hat{u}) - \frac{1}{4}J^\infty(\hat{u}) \ge I^\infty\left(\hat{u}^{t_\infty}\right) - \frac{t_\infty^4}{4}J^\infty(\hat{u})\\ &\ge m^\infty(\hat{c}) - \frac{t_\infty^4}{4}J^\infty(\hat{u}) \ge m(\hat{c}) \ge m(c), \end{aligned}$$

which implies that (3.59) holds for $w_n \ne 0$. From (3.54), (3.59), Lemmas 3.9 and 3.10, we then derive

$$m^\infty(c) \le m^\infty(\hat{c}) = I^\infty(\hat{u}) = m(\hat{c}) = m(c) \le m^\infty(c), \qquad J^\infty(\hat{u}) = 0,$$

which implies (3.59) holds for $w_n \neq 0$. Thus, $m^\infty(\hat{c})$ is achieved at $\hat{u}$. In view of Lemma 3.9, we deduce that $\|\hat{u}\|_2^2 = \hat{c} = c$ due to $m^\infty(\hat{c}) = m^\infty(c)$. By Lemma 3.6, there exists $\hat{t} > 0$ such that $\hat{u}^{\hat{t}} \in \mathcal{M}_c$. Then it follows from (3.28), (3.32), (3.38), (3.44) and (3.59) that

$$\begin{aligned} m(c) &\leq I\left(\hat{u}^{\hat{t}}\right) \leq I^\infty\left(\hat{u}^{\hat{t}}\right) \\ &\leq I^\infty(\hat{u}) - \frac{a(1-\hat{t}^2)^2}{4}\|\nabla \hat{u}\|_2^2 = m(c) - \frac{a(1-\hat{t}^2)^2}{4}\|\nabla \hat{u}\|_2^2, \end{aligned}$$

which implies that $\hat{u} \in \mathcal{M}_c$ and $I(\hat{u}) = m(c)$. Hence, $m(c)$ is achieved at $\hat{u} \in \mathcal{M}_c$.

Case ii) $\bar{u} \neq 0$. Let $v_n = u_n - \bar{u}$. Then Lemma 3.11 yields

$$\|\bar{u}\|_2^2 := \bar{c} \leq c, \quad \|v_n\|_2^2 := c_n \leq c \text{ for large } n \in \mathbb{N}, \tag{3.60}$$

$$I(u_n) = I(\bar{u}) + I(v_n) + \frac{b}{2}\|\nabla \bar{u}\|_2^2 \|\nabla v_n\|_2^2 + o(1), \tag{3.61}$$

and

$$J(u_n) = J(\bar{u}) + J(v_n) + b\|\nabla \bar{u}\|_2^2 \|\nabla v_n\|_2^2 + o(1). \tag{3.62}$$

Let

$$\begin{aligned} \Psi(u) &:= I(u) - \frac{1}{4}J(u) \\ &= \frac{a}{4}\|\nabla u\|_2^2 + \frac{1}{8}\int_{\mathbb{R}^3} K(x)[3f(u)u - 14F(u)]\mathrm{d}x \\ &- \frac{1}{4}\int_{\mathbb{R}^3} \nabla K(x) \cdot xF(u)\mathrm{d}x, \quad \forall\, u \in H^1(\mathbb{R}^3). \end{aligned} \tag{3.63}$$

By (3.23) and (3.26), one has $\Psi(u) > 0$ for all $u \in H^1(\mathbb{R}^3) \setminus \{0\}$. Similarly to the proof of (3.58), we deduce that

$$\Psi(v_n) \leq m(c) - \Psi(\bar{u}) + o(1), \quad J(v_n) \leq -J(\bar{u}) + o(1). \tag{3.64}$$

If there exists a subsequence $\{v_{n_i}\}$ of $\{v_n\}$ such that $v_{n_i} = 0$, then it follows from Lemma 3.9 that

$$m(\bar{c}) \leq I(\bar{u}) = m(c) \leq m(\bar{c}), \quad J(\bar{u}) = 0,$$

which implies

$$I(\bar{u}) = m(c) = m(\bar{c}), \quad J(\bar{u}) = 0. \tag{3.65}$$

Next, we prove that (3.65) holds for $v_n \neq 0$. To this end, we assume that $v_n \neq 0$. We claim that $J(\bar{u}) \leq 0$. Otherwise $J(\bar{u}) > 0$, then (3.64) implies $J(v_n) < 0$ for large n. In view of Lemma 3.6, there exists $t_n > 0$ such that $(v_n)^{t_n} \in \mathcal{M}_{c_n}$. From (3.32), (3.63) and (3.64), we obtain

$$\begin{aligned} m(c) - \Psi(\bar{u}) + o(1) &\geq \Psi(v_n) = I(v_n) - \frac{1}{4}J(v_n) \\ &\geq I\left((v_n)^{t_n}\right) - \frac{t_n^4}{4}J(v_n) \geq m(\bar{c}_n) \geq m(c) + o(1), \end{aligned}$$

which is impossible due to $\Psi(\bar{u}) > 0$. This shows that $J(\bar{u}) \leq 0$. In view of Lemma 3.6, there exists $\tilde{t} > 0$ such that $\bar{u}^{\tilde{t}} \in \mathcal{M}_{\bar{c}}$. Then it follows from (3.32), (3.63), the weak semicontinuity of norm, Fatou's lemma and Lemma 3.10 that

$$\begin{aligned} m(c) &= \lim_{n\to\infty}\left[I(u_n) - \frac{1}{4}J(u_n)\right] = \lim_{n\to\infty}\Psi(u_n) \\ &\geq \Psi(\bar{u}) = I(\bar{u}) - \frac{1}{4}J(\bar{u}) \geq I\left(\bar{u}^{\tilde{t}}\right) - \frac{\tilde{t}^4}{4}J(\bar{u}) \geq m(\bar{c}) \geq m(c), \end{aligned}$$

which implies (3.65) holds for $v_n \neq 0$. This shows that $m(\bar{c})$ is achieved at $\bar{u} \in \mathcal{M}_{\bar{c}}$. In view of Lemma 3.9, we have $\|\bar{u}\|_2^2 = \bar{c} = c$ due to $m(c) = m(\bar{c})$. By Lemma 3.6, there exists $\bar{t} > 0$ such that $\bar{u}^{\bar{t}} \in \mathcal{M}_c$. Then it follows from (3.28), (3.32), (3.38), (3.44) and (3.65) that

$$\begin{aligned} m(c) &\leq I\left(\bar{u}^{\bar{t}}\right) = \frac{a\bar{t}^2}{2}\|\nabla \bar{u}\|_2^2 + \frac{b\bar{t}^4}{4}\|\nabla \bar{u}\|_2^4 - \bar{t}^{-3}\int_{\mathbb{R}^3} K(\bar{t}^{-1}x)F(\bar{t}^{3/2}\bar{u})\mathrm{d}x \\ &= I\left(\bar{u}^{\bar{t}}\right) + \bar{t}^{-3}\int_{\mathbb{R}^3}\left[K(\bar{t}^{-1}x)F(\bar{t}^{3/2}\bar{u}) - K(\bar{t}^{-1}x)F(\bar{t}^{3/2}\bar{u})\right]\mathrm{d}x \\ &\leq I\left(\bar{u}^{\bar{t}}\right) \leq I(\bar{u}) - \frac{a(1-\bar{t}^2)^2}{4}\|\nabla \bar{u}\|_2^2 = m(c) - \frac{a(1-\bar{t}^2)^2}{4}\|\nabla \bar{u}\|_2^2, \end{aligned}$$

which implies that $\bar{u} \in \mathcal{M}_c$ and $I(\bar{u}) = m(c)$. Hence, $m(c)$ is achieved at $\bar{u} \in \mathcal{M}_c$. □

Lemma 3.13 *Assume that* (K1), (K2) *and* (F1)–(F3) *hold. If* $\bar{u} \in \mathcal{M}_c$ *and* $I(\bar{u}) = m(c)$, *then* $\bar{u}$ *is a critical point of* $I\big|_{\mathcal{S}_c}$.

Proof. Assume that $I|'_{\mathcal{S}_c}(\bar{u}) \neq 0$. Then there exist $\delta > 0$ and $\varrho > 0$ such that

$$u \in \mathcal{S}_c, \quad \|u - \bar{u}\| \leq 3\delta \Rightarrow \|I|'_{\mathcal{S}_c}(u)\| \geq \varrho. \tag{3.66}$$

By a standard argument, we deduce that

$$\lim_{t\to 1}\left\|\bar{u}^t - \bar{u}\right\| = 0. \tag{3.67}$$

Thus, there exists $\delta_1 \in (0, 1/4)$ such that

$$|t-1| < \delta_1 \Rightarrow \left\|\bar{u}^t - \bar{u}\right\| < \delta. \tag{3.68}$$

In view of (3.32), we have

$$I\left(\bar{u}^t\right) \leq I(\bar{u}) - \frac{a(1-t^2)^2}{4}\|\nabla \bar{u}\|_2^2 = m(c) - \frac{a(1-t^2)^2}{4}\|\nabla \bar{u}\|_2^2, \quad \forall\, t > 0. \tag{3.69}$$

From (F3), (3.21), (3.30), (3.34) and (3.35), we deduce that there exist $T_1 \in (0,1)$ and $T_2 \in (1,\infty)$ such that

$$J\left(\bar{u}^{T_1}\right) > 0, \quad J\left(\bar{u}^{T_2}\right) < 0. \tag{3.70}$$

Let $\varepsilon := \min\{a(1-T_1^2)^2\|\nabla \bar{u}\|_2^2/12, a(1-T_2^2)^2\|\nabla \bar{u}\|_2^2/12, 1, \varrho\delta/8\}$ and $\mathcal{S} := B(\bar{u},\delta) \cap \mathcal{S}_c$. By [153, Lemma 2.3], there exists a deformation $\eta \in \mathcal{C}([0,1] \times \mathcal{S}_c, \mathcal{S}_c)$ such that

(i) $\eta(1,u) = u$ if $I(u) < m(c) - 2\varepsilon$ or $I(u) > m(c) + 2\varepsilon$;

(ii) $\eta\left(1, I^{m(c)+\varepsilon} \cap \mathcal{S}\right) \subset I^{m(c)-\varepsilon}$;

(iii) $I(\eta(1,u)) \leq I(u)$, $\forall\, u \in \mathcal{S}_c$;

(iv) $\eta(1,u)$ is a homeomorphism of $\mathcal{S}_c$.

By Corollary 3.1, $I\left(\bar{u}^t\right) \leq I(\bar{u}) = m(c)$ for $t > 0$, then it follows from (3.68) and (ii) that

$$I\left(\eta\left(1, \bar{u}^t\right)\right) \leq m - \varepsilon, \quad \forall\, t > 0, \ |t-1| < \delta_1. \tag{3.71}$$

On the other hand, by (iii) and relation (3.69), one has

$$\begin{aligned} I\left(\eta\left(1, \bar{u}^t\right)\right) &\leq I\left(\bar{u}^t\right) \leq m - \frac{a(1-t^2)^2}{4}\|\nabla \bar{u}\|_2^2 \\ &\leq m - \delta_2\|\nabla \bar{u}\|_2^2, \quad \forall\, t > 0, \ |t-1| \geq \delta_1, \end{aligned} \tag{3.72}$$

where

$$\delta_2 := \frac{a}{4}\min\left\{(1-T_1^2)^2, (1-T_1^2)^2\right\} > 0.$$

Combining (3.71) with (3.72), we have

$$\max_{t\in[T_1,T_2]} I\left(\eta\left(1,\bar{u}^t\right)\right) < m(c). \tag{3.73}$$

Define $\Psi_0(t) := J\left(\eta\left(1,\bar{u}^t\right)\right)$ for $t>0$. It follows from (3.69) and i) that $\eta(1,\bar{u}^t)=\bar{u}^t$ for $t=T_1$ and $t=T_2$, which, together with (3.70), implies

$$\Psi_0(T_1) = J\left(\bar{u}^{T_1}\right) > 0, \qquad \Psi_0(T_2) = J\left(\bar{u}^{T_1}\right) < 0.$$

Since $\Psi_0(t)$ is continuous on $(0,\infty)$, then we have that $\eta\left(1,\bar{u}^t\right)\cap\mathcal{M}_c \neq \emptyset$ for some $t_0 \in [T_1,T_2]$, contradicting to the definition of $m(c)$. □

Lemma 3.14 *Assume that* (K1), (K2) *and* (F1)–(F3) *hold. If* $\bar{u}\in\mathcal{S}_c$ *is a critical point of* $I\big|_{\mathcal{S}_c}$, *then* $J(\bar{u})=0$, *and there exists* $\lambda_c<0$ *such that* $I'(\bar{u})-\lambda_c u = 0$.

Proof. Since $(I|_{\mathcal{S}_c})'(\bar{u})=0$, there exists $\lambda_c\in\mathbb{R}$ such that $I'(\bar{u})-\lambda_c\bar{u}=0$, and so

$$\langle I'(\bar{u})-\lambda_c\bar{u},u\rangle = a\|\nabla\bar{u}\|_2^2 + b\|\nabla\bar{u}\|_2^4 - \int_{\mathbb{R}^3} K(x)f(\bar{u})\bar{u}\mathrm{d}x - \lambda_c\|\bar{u}\|_2^2 = 0. \tag{3.74}$$

Moreover, $\bar{u}$ satisfies the following Pohozaev identity:

$$\mathcal{P}(\bar{u}) := \frac{a}{2}\|\nabla\bar{u}\|_2^2 + \frac{b}{2}\|\nabla\bar{u}\|_2^4 - \int_{\mathbb{R}^3}[3K(x)+\nabla K(x)\cdot x]F(\bar{u})\mathrm{d}x - \frac{3\lambda_c}{2}\|\bar{u}\|_2^2 = 0. \tag{3.75}$$

Relations (3.30), (3.74) and (3.75) yield

$$J(\bar{u}) = \frac{3}{2}\langle I'(\bar{u})-\lambda_c\bar{u},\bar{u}\rangle - \mathcal{P}(\bar{u}) = 0.$$

Noticing that $\|\bar{u}\|_2^2=c$, it follows from (K2), (F1), (3.74) and (3.75) that

$$\begin{aligned}
2\lambda c &= \int_{\mathbb{R}^3}\left\{K(x)f(\bar{u})\bar{u} - [6K(x)+2\nabla K(x)\cdot x]F(\bar{u})\right\}\mathrm{d}x \\
&= \int_{\mathbb{R}^3}\left\{K(x)[f(\bar{u})\bar{u}-\mu F(\bar{u})] - [(6-\mu)K(x)+2\nabla K(x)\cdot x]F(\bar{u})\right\}\mathrm{d}x < 0,
\end{aligned}$$

and so $\lambda_c<0$. This completes the proof. □

3.2.1 Proof of Theorem 3.1

In view of Lemmas 3.7, 3.8 and 3.12–3.14, for each $c>0$ there exists $\bar{u}_c\in\mathcal{M}_c$ such that

$$I(\bar{u}_c) = m(c) = \inf_{u\in\mathcal{S}_c}\max_{t>0} I((\bar{u}_c)^t) > 0, \qquad I'(\bar{u}_c) = 0,$$

and there exists Lagrange multiplier $\lambda_c\in\mathbb{R}^-$ such that $(\bar{u}_c,\lambda_c)$ is a solution of (3.1). □

3.3 The L^2-subcritical case

In this section, we give the proof of Theorem 3.2.

3.3.1 Global minimizers on the constraint $\mathcal{S}_c$

We start with a qualitative property of the constrained minimization problem.

Lemma 3.15 *Assume that* (K1) *and* (F4) *hold. Then*

(i) *for any* $c>0$, $\sigma(c)=\inf_{u\in\mathcal{S}_c} I(u)$ *is well defined and* $\sigma(c)\le 0$;

(ii) *for any* $c>0$, *then* $\sigma(c)<0$ *if* (F6) *holds;*

(iii) *there exists* $\mathcal{C}_0>0$ *such that* $\sigma(c)<0$ *for any* $c>\mathcal{C}_0$ *if* (F5) *holds.*

Proof. (i) Using (F4), for any $\varepsilon>0$, there exists $C_\varepsilon>0$ such that

$$|f(t)t|+|F(t)|\le \varepsilon|t|^2+C_\varepsilon|t|^p,\qquad \forall\, t\in\mathbb{R}. \tag{3.76}$$

By (3.3), (3.41) and (3.76), we obtain for all $u\in\mathcal{S}_c$

$$\begin{aligned} I(u) &\ge \frac{a}{2}\|\nabla u\|_2^2+\frac{b}{4}\|\nabla u\|_2^4-\sup_{\mathbb{R}^3}K\left(\varepsilon\|u\|_2^2+C_\varepsilon\|u\|_p^p\right)\\ &\ge \frac{a}{2}\|\nabla u\|_2^2+\frac{b}{4}\|\nabla u\|_2^4-\sup_{\mathbb{R}^3}K\left(\varepsilon\|u\|_2^2+C_\varepsilon\mathcal{C}(p)\|\nabla u\|_2^{\frac{3(p-2)}{2}}\|u\|_2^{\frac{6-p}{2}}\right), \end{aligned} \tag{3.77}$$

which, together with $0<3(p-2)/2<4$, shows that I is bounded from below on $\mathcal{S}_c$ for any $c>0$, hence $\sigma(c)$ is well defined. Noting that $u^t\in\mathcal{S}_c$ for all $u\in\mathcal{S}_c$, from (3.28) and (3.76), we deduce that $I(u^t)\to 0$ as $t\to 0$, and so $\sigma(c)\le 0$ for any $c>0$.

(ii) By (F4) and (F6), there exist $\delta_0,\varrho_0>0$ such that

$$|F(t)|\ge \delta_0|t|^{q_0},\qquad \forall\ |t|\le\varrho_0. \tag{3.78}$$

For any $c>0$, we choose a function $u_0\in\mathcal{C}_0^\infty(\mathbb{R}^3,[-\varrho_0,\varrho_0])$ satisfying $\|u_0\|_2^2=c$. Then it follows from (3.28) and (3.78) that

$$I(u_0^t)\le \frac{at^2}{2}\|\nabla u_0\|_2^2+\frac{bt^4}{4}\|\nabla u_0\|_2^4-K_\infty\delta_0 t^{3(q_0-2)/2}\|u_0\|_{q_0}^{q_0},\qquad \forall\ 0<t\le 1. \tag{3.79}$$

Since $0<3(q_0-2)/2<2$, relation (3.79) implies that $I({u_0}^t)<0$ for small $t\in(0,1)$. Jointly with the fact that $\|{u_0}^t\|_2=\|u_0\|_2$, we obtain $\sigma(c)\le\inf_{t\in(0,1]}I({u_0}^t)<0$.

(iii) Set the following scaling

$$u_t(x):=t^{1/2}u(x/t),\qquad \forall\ u\in H^1(\mathbb{R}^3),\ t>0. \tag{3.80}$$

Note that (F5) implies

$$\frac{F(t)}{|t|^{\mu_0-1}t}\ \text{is nondecreasing on } (-\infty,0) \text{ and } (0,\infty). \tag{3.81}$$

For any $u\in\mathcal{S}_1$, by (3.3), (3.80) and (3.81), we have for all $t>1$

$$\begin{aligned} I(u_t) &= \frac{at^2}{2}\|\nabla u\|_2^2+\frac{bt^4}{4}\|\nabla u\|_2^4-t^3\int_{\mathbb{R}^3}K(tx)F\left(t^{1/2}u\right)\mathrm{d}x\\ &\le \frac{at^2}{2}\|\nabla u\|_2^2+\frac{bt^4}{4}\|\nabla u\|_2^4-K_\infty t^{3+\mu_0/2}\int_{\mathbb{R}^3}F(u)\mathrm{d}x, \end{aligned} \tag{3.82}$$

which, together with $3+\mu_0/2>4$, implies that $I(u_t)\to-\infty$ as $t\to+\infty$. Since $\|u_t\|_2^2=t^4\|u\|_2^2=t^4$ for $u\in\mathcal{S}_1$ and $t>0$, there exists $\mathcal{C}_0>0$ such that $\sigma(c)<0$ for any $c>\mathcal{C}_0$. $\square$

Noticing that Lemma 3.15 implies

$$\{c \in (0,+\infty) \colon \sigma(c) < 0\} \neq \emptyset, \tag{3.83}$$

we deduce that

$$c_* = \inf\{c \in (0,+\infty) \colon \sigma(c) < 0\}$$

is well-defined.

Lemma 3.16 *Assume that* (K1) *and* (F4) *hold. Then the following properties are fulfilled:*

(i) *if* (F6) *holds, then* $c_* = 0$*;*

(ii) *if* (F5) *and* (F6′) *hold, then* $c_* \in (0,+\infty)$*, moreover,* $\sigma(c) = 0$ *for any* $c \in (0, c_*]$ *and* $\sigma(c) < 0$ *for any* $c > c_*$*.*

Proof. (i) Obviously, property (i) follows directly from Lemma 3.15 (ii).
(ii) We first prove that $c^* > 0$. From (F4), (F6′) and (3.41), we obtain

$$\begin{aligned}\int_{\mathbb{R}^3} F(u)\mathrm{d}x &\le C_3\left(\|u\|_{\frac{10}{3}}^{\frac{10}{3}} + \|u\|_p^p\right)\\ &\le C_4\left(\|\nabla u\|_2^2\|u\|_2^{\frac{4}{3}} + \|\nabla u\|_2^{\frac{3(p-2)}{2}}\|u\|_2^{\frac{6-p}{2}}\right), \qquad \forall\, u \in H^1(\mathbb{R}^3).\end{aligned} \tag{3.84}$$

By straightforward computation, there exists a constant $\epsilon > 0$ small enough such that

$$\epsilon^{\frac{2}{3}} \le \frac{a}{4C_4} \quad \text{and} \quad \epsilon^{\frac{6-p}{14-3p}} \le C_4^{\frac{4}{3p-14}}\frac{4a}{14-3p}\left(\frac{4b}{3p-10}\right)^{\frac{3p-10}{14-3p}}. \tag{3.85}$$

For any $u \in \mathcal{S}_\epsilon$, namely $\|u\|_2^2 = \epsilon$, it follows from (3.3), (3.84) and (3.85) that

$$\begin{aligned}I(u) &\ge \frac{a}{2}\|\nabla u\|_2^2 + \frac{b}{4}\|\nabla u\|_2^4 - C_4\left(\|u\|_2^{\frac{4}{3}}\|\nabla u\|_2^2 + \|u\|_2^{\frac{6-p}{2}}\|\nabla u\|_2^{\frac{3(p-2)}{2}}\right)\\ &\ge \frac{1}{4}\|\nabla u\|_2^2\left(a + b\|\nabla u\|_2^2 - C_4\epsilon^{\frac{6-p}{4}}\|\nabla u\|_2^{\frac{3p-10}{2}}\right).\end{aligned} \tag{3.86}$$

Next, combining Young's inequality with relation (3.85), we have

$$\begin{aligned}C_4\epsilon^{\frac{6-p}{4}}\|\nabla u\|_2^{\frac{3p-10}{2}} &= \left(\frac{4b}{3p-10}\right)^{\frac{3p-10}{4}}\|\nabla u\|_2^{\frac{3p-10}{2}} \cdot \left(\frac{3p-10}{4b}\right)^{\frac{3p-10}{4}} C_4\epsilon^{\frac{6-p}{4}}\\ &\le b\|\nabla u\|_2^2 + C_4^{\frac{4}{14-3p}}\frac{14-3p}{4}\left(\frac{3p-10}{4b}\right)^{\frac{3p-10}{14-3p}}\epsilon^{\frac{6-p}{14-3p}}\\ &\le b\|\nabla u\|_2^2 + a,\end{aligned} \tag{3.87}$$

which, together with (3.86), leads to $I(u) \ge 0$ for $u \in \mathcal{S}_\epsilon$. On the other hand, by Lemma 3.15 (i), we have $\sigma(c) \le 0$ for $c > 0$. This shows that $c_* = \inf\{c \in (0,+\infty), \sigma(c) < 0\} > 0$. Moreover, from the definition of c_* and Lemma 3.15 (i), we deduce that $\sigma(c) = 0$ for any $c \in (0, c_*]$ and $\sigma(c) < 0$ for any $c > c_*$. □

Lemma 3.17 *Assume that* (K1), (K3), (F4), (F5) *and either of* (F6) *and* (F6′) *hold. Then*

(i) *for any* $c > 0$*,* $\sigma(c)$ *is continuous;*

(ii) *for any* $c > c_*$*,*

$$\sigma(c) < \sigma(\alpha) + \sigma(c-\alpha), \qquad \forall\, 0 < \alpha < c. \tag{3.88}$$

Proof. (i) For any $c > 0$, let $c_n > 0$ and $c_n \to c$. For every $n \in \mathbb{N}$, let $u_n \in \mathcal{S}_{c_n}$ such that $I(u_n) < \sigma(c_n) + \frac{1}{n} \leq \frac{1}{n}$. Then (3.77) implies that $\{u_n\}$ is bounded in $H^1(\mathbb{R}^3)$ and

$$\sigma(c) \leq I\left(\sqrt{\frac{c}{c_n}}u_n\right) = I(u_n) + o(1) \leq \sigma(c_n) + o(1). \tag{3.89}$$

On the other hand, given a minimizing sequence $\{v_n\} \subset \mathcal{S}_c$ for I, we have

$$\sigma(c_n) \leq I\left(\sqrt{\frac{c_n}{c}}v_n\right) \leq I(v_n) + o(1) = \sigma(c) + o(1),$$

which, jointly to (3.89), gives $\lim_{n\to\infty} \sigma(c_n) = \sigma(c)$.

(ii) Note that (3.3) and (3.81) lead to

$$\begin{aligned} I(u_t) &= \frac{at^2}{2}\|\nabla u\|_2^2 + \frac{bt^4}{4}\|\nabla u\|_2^4 - t^3\int_{\mathbb{R}^3} K(tx)F\left(t^{1/2}u\right)\mathrm{d}x \\ &\leq \frac{at^2}{2}\|\nabla u\|_2^2 + \frac{bt^4}{4}\|\nabla u\|_2^4 - t^4\int_{\mathbb{R}^3} t^{(\mu_0-2)/2}K(tx)F(u)\mathrm{d}x \\ &\leq t^4 I(u) + \frac{at^2(1-t^2)}{2}\|\nabla u\|_2^2 \;<\; t^4 I(u), \qquad \forall\, u \in \mathcal{S}_c,\ c > 0,\ t > 1, \end{aligned} \tag{3.90}$$

where the definition of u_t is given by (3.80). Let $\{u_n\} \subset \mathcal{S}_c$ be such that $I(u_n) \to \sigma(c)$ for any $c > 0$. Since $\|(u_n)_t\|_2^2 = t^4\|u_n\|_2^2 = t^4 c$ for all $t > 0$, it follows from (3.90) that

$$\begin{aligned} \sigma(t^4 c) &\leq I((u_n)_t) \leq t^4 I(u_n) - \frac{at^2(t^2-1)}{2}\|\nabla u_n\|_2^2 \\ &\leq t^4\sigma(c) + o(1), \qquad \forall\, t > 1, \end{aligned}$$

which implies

$$\sigma(tc) \leq t\sigma(c), \qquad t > 1,\ c > 0. \tag{3.91}$$

Let $\{u_n\} \subset \mathcal{S}_c$ be such that $I(u_n) \to \sigma(c)$ for any $c > c^*$.

We claim that exists a constant $\rho_0 > 0$ such that

$$\liminf_{n\to\infty} \|\nabla u_n\|_2 > \rho_0. \tag{3.92}$$

Otherwise, if (3.92) is not true, then up to a subsequence, $\|\nabla u_n\|_2 \to 0$, and so (3.77) yields $0 > \sigma(c) = \lim_{n\to\infty} I(u_n) = 0$. This contradiction shows that (3.92) holds.

Since $\|(u_n)_t\|_2^2 = t^4\|u_n\|_2^2 = t^4 c$ for all $t > 0$, it follows from (3.90) and (3.92) that

$$\begin{aligned} \sigma(t^4 c) &\leq I((u_n)_t) \leq t^4 I(u_n) - \frac{at^2(t^2-1)}{2}\|\nabla u_n\|_2^2 \\ &\leq t^4\sigma(c) - \frac{at^2(t^2-1)}{2}\rho_0^2 + o(1), \qquad \forall\, t > 1, \end{aligned}$$

which implies

$$\sigma(tc) < t\sigma(c), \qquad t > 1,\ c > c^*. \tag{3.93}$$

If $\alpha > c^*$ and $c - \alpha > c^*$, we then conclude from (3.93) that for any $c > c^*$,

$$\sigma(c) = \frac{\alpha}{c}\sigma(c) + \frac{c-\alpha}{c}\sigma(c) < \sigma(\alpha) + \sigma(c-\alpha), \qquad \forall\, 0 < \alpha < c.$$

If $\alpha \leq c^*$ or $c - \alpha \leq c^*$, then Lemma 3.16 (ii) implies that $\sigma(c-\alpha) = 0$ or $\sigma(\alpha) = 0$, and we deduce easily from (3.93) that (3.88) holds for any $c > c^*$. This completes the proof. □

Lemma 3.18 *Assume that* (K1), (K3), (F4) *and* (F5) *hold. Then the following properties hold:*

(i) $\sigma(c) \leq \sigma^\infty(c)$ *for any* $c > 0$;

(ii) *if* (F6) *holds, then* $\sigma(c)$ *has a minimizer for any* $c > 0$;

(iii) *if* (F6′) *holds, then* $\sigma(c)$ *has a minimizer for any* $c \geq c_*$ *and* $\sigma(c_*) = 0$.

Proof. (i) Let $\{u_n\} \subset \mathcal{S}_c$ be such that $I^\infty(u_n) \to \sigma^\infty(c)$ for any $c > 0$. Since $K_\infty \leq K(x)$ for all $x \in \mathbb{R}^3$, it follows from (3.3) and (3.12) that

$$\sigma(c) \leq I(u_n) \leq I^\infty(u_n) = \sigma^\infty(c) + o(1),$$

which implies that $\sigma(c) \leq \sigma^\infty(c)$ for any $c > 0$.

(ii) In view of Lemma 3.15 (ii), we have $\sigma(c) < 0$ for any $c > 0$. Let $\{u_n\} \subset \mathcal{S}_c$ be such that $I(u_n) \to \sigma(c)$ for any $c > 0$. Then (3.77) implies that $\{u_n\}$ is bounded in $H^1(\mathbb{R}^3)$. Thus, we can assume that for some $\bar{u} \in H^1(\mathbb{R}^3)$ and up to a subsequence, $u_n \rightharpoonup \bar{u}$ in $H^1(\mathbb{R}^3)$. Here, we distinguish two cases: a) $\bar{u} \neq 0$ and b) $\bar{u} = 0$.

Case a): $\bar{u} \neq 0$. Then $u_n \rightharpoonup \bar{u}$ in $H^1(\mathbb{R}^3)$, $u_n \to \bar{u}$ in $L^s_{\text{loc}}(\mathbb{R}^3)$ for $2 \leq s < 6$ and $u_n \to \bar{u}$ a.e. in $\mathbb{R}^3$. By Lemmas 3.11 and 3.17, we have

$$\begin{aligned} \sigma(c) &= \lim_{n\to\infty} I(u_n) = I(\bar{u}) + \lim_{n\to\infty}\left[I(u_n - \bar{u}) + \frac{b}{2}\|\nabla\bar{u}\|_2^2\|\nabla(u_n - \bar{u})\|_2^2\right] \\ &\geq \sigma(\|\bar{u}\|_2^2) + \lim_{n\to\infty}\sigma(\|u_n - \bar{u}\|_2^2) = \sigma(\|\bar{u}\|_2^2) + \sigma(c - \|\bar{u}\|_2^2). \end{aligned} \tag{3.94}$$

If $\|\bar{u}\|_2^2 < c$, then (3.94) and Lemma 3.17 (ii) imply

$$\sigma(c) \geq \sigma(\|\bar{u}\|_2^2) + \sigma(c - \|\bar{u}\|_2^2) > \sigma(c),$$

which is impossible. This shows that $\|\bar{u}\|_2^2 = c = \|u_n\|_2^2$. Moreover, it follows from (3.94) that $u_n \to \bar{u}$ in $H^1(\mathbb{R}^3)$, and so $\sigma(c) = \lim_{n\to\infty} I(u_n) = I(\bar{u})$. Hence, $\bar{u}$ is a minimizer of $\sigma(c)$ for any $c > 0$ if $\bar{u} \neq 0$.

Case b): $\bar{u} = 0$, that is, $u_n \rightharpoonup 0$ in $H^1(\mathbb{R}^3)$. Then $u_n \to 0$ in $L^s_{\text{loc}}(\mathbb{R}^3)$ for $2 \leq s < 6$ and $u_n \to 0$ a.e. in $\mathbb{R}^3$. By (K1), we deduce that

$$\int_{\mathbb{R}^3}[K(x) - K_\infty]F(u_n)\mathrm{d}x = o(1). \tag{3.95}$$

Then (3.3), (3.12) and (3.95) imply

$$I^\infty(u_n) \to \sigma(c). \tag{3.96}$$

We claim that

$$\delta := \limsup_{n\to\infty}\sup_{y\in\mathbb{R}^3}\int_{B_1(y)}|u_n|^2\mathrm{d}x > 0. \tag{3.97}$$

In fact, if δ=0, then by Lions' concentration-compactness principle [153, Lemma 1.21], we have that $u_n \to 0$ in $L^s(\mathbb{R}^3)$ for $2 < s < 6$, and so (3.76) implies that $\int_{\mathbb{R}^3} F(u_n)\mathrm{d}x \to 0$. Then by (3.3), we have

$$0 > \sigma(c) = \lim_{n\to\infty} I(u_n) = \lim_{n\to\infty}\left(\frac{a}{2}\|\nabla u_n\|_2^2 + \frac{b}{4}\|\nabla u_n\|_2^4\right) \geq 0,$$

which is impossible. Hence, we have $\delta > 0$, and there exists $\{y_n\} \subset \mathbb{R}^3$ such that

$$\int_{B_1(y_n)} |u_n|^2 \mathrm{d}x \geq \frac{\delta}{2}. \tag{3.98}$$

Let $\hat{u}_n(x) = u_n(x + y_n)$. Then (3.96) leads to

$$\hat{u}_n \in \mathcal{S}_c, \qquad I^\infty(\hat{u}_n) \to \sigma(c). \tag{3.99}$$

In view of (3.98), we may assume that there exists $\hat{u} \in H^1(\mathbb{R}^3) \setminus \{0\}$ such that, passing to a subsequence,

$$\begin{cases} \hat{u}_n \rightharpoonup \hat{u}, & \text{in } H^1(\mathbb{R}^3); \\ \hat{u}_n \to \hat{u}, & \text{in } L^s_{\text{loc}}(\mathbb{R}^3), \ \forall\, s \in [1,6); \\ \hat{u}_n \to \hat{u}, & \text{a.e. on } \mathbb{R}^3. \end{cases} \tag{3.100}$$

Then it follows from (3.99), (3.100), (i), Lemmas 3.11 and 3.17 that

$$\begin{aligned} \sigma(c) &= \lim_{n\to\infty} I^\infty(\hat{u}_n) \\ &= I^\infty(\hat{u}) + \lim_{n\to\infty} \left[I^\infty(\hat{u}_n - \hat{u}) + \frac{b}{2}\|\nabla \hat{u}\|_2^2 \|\nabla(\hat{u}_n - \hat{u})\|_2^2 \right] \\ &\geq \sigma^\infty(\|\hat{u}\|_2^2) + \lim_{n\to\infty} \sigma^\infty(\|\hat{u}_n - \hat{u}\|_2^2) \\ &= \sigma^\infty(\|\hat{u}\|_2^2) + \sigma^\infty(c - \|\hat{u}\|_2^2) \geq \sigma(\|\hat{u}\|_2^2) + \sigma(c - \|\hat{u}\|_2^2), \end{aligned} \tag{3.101}$$

If $\|\hat{u}\|_2^2 < c$, then (3.101) and Lemma 3.17 (ii) imply

$$\sigma(c) \geq \sigma(\|\hat{u}\|_2^2) + \sigma(c - \|\hat{u}\|_2^2) > \sigma(c),$$

which is impossible. Hence, $\|\hat{u}\|_2^2 = c$, moreover, it follows from (3.101) that

$$\hat{u}_n \to \hat{u} \text{ in } H^1(\mathbb{R}^3), \quad \sigma(c) \leq I(\hat{u}) \leq I^\infty(\hat{u}) = \lim_{n\to\infty} I^\infty(\hat{u}_n) = \sigma(c). \tag{3.102}$$

This shows that $\hat{u}$ is a minimizer of $\sigma(c)$ for any $c > 0$.

(iii) In view of Lemma 3.15 (iii), we have $\sigma(c) < 0$ for any $c > c_*$. Arguing as in the proof of (ii), we deduce that $\sigma(c)$ is attained for any $c > c_*$. Let $c_n = c_* + \frac{1}{n}$. Note that Lemma 3.16 (ii), leads to $\sigma(c_n) < 0$ for every $n \in \mathbb{N}$. Arguing as in the proof of (ii), there exists $\{u_n\} \subset \mathcal{S}_{c_n}$ such that

$$I(u_n) = \sigma(c_n) < 0 \ \text{ for every } n \in \mathbb{N}. \tag{3.103}$$

By the definition of c_* and Lemma 3.17 (i), we have $I(u_n) = \sigma(c_n) \to \sigma(c_*) = 0$. Then (3.77) implies that $\{u_n\}$ is bounded in $H^1(\mathbb{R}^3)$. We then may assume that there exists $\bar{u} \in H^1(\mathbb{R}^3)$ such that, up to a subsequence, $u_n \rightharpoonup \bar{u}$ in $H^1(\mathbb{R}^3)$. To prove that $\sigma(c_*)$ is attained, we need to distinguish two cases: a) $\bar{u} \neq 0$ and b) $\bar{u} = 0$.

Case a): $\bar{u} \neq 0$. By Lemmas 3.11, 3.16 and 3.17, we have

$$\begin{aligned} 0 &= \sigma(c^*) = \lim_{n\to\infty} I(u_n) = I(\bar{u}) + \lim_{n\to\infty} \left[I(u_n - \bar{u}) + \frac{b}{2}\|\nabla \bar{u}\|_2^2 \|\nabla(u_n - \bar{u})\|_2^2 \right] \\ &\geq \sigma(\|\bar{u}\|_2^2) + \lim_{n\to\infty} \sigma(\|u_n - \bar{u}\|_2^2) \\ &= \sigma(\|\bar{u}\|_2^2) + \sigma(c^* - \|\bar{u}\|_2^2) = 0, \end{aligned}$$

which implies

$$\lim_{n\to\infty} \|\nabla(u_n - \bar{u})\|_2 = 0, \tag{3.104}$$

and so a standard argument shows that $I(\bar{u}) = \lim_{n\to\infty} I(u_n) = \sigma(c^*) = 0$.

To prove that $\bar{u}$ is a minimizer of $\sigma(c^*)$, it suffices to show that $\|\bar{u}\|_2^2 = c^*$. By contradiction, let us assume that $\|\bar{u}\|_2^2 < c^*$. Let

$$t^* = \left(\frac{c^*}{\|\bar{u}\|_2^2}\right)^{\frac{1}{4}}.$$

Then $t^* > 1$, and $\|\bar{u}_{t^*}\|_2^2 = (t^*)^4\|\bar{u}\|_2^2 = c^*$ by the scaling (3.80). Thus, it follows from (3.90) that

$$0 = \sigma(c^*) \le I(\bar{u}_{t^*}) \le (t^*)^4 I(\bar{u}) + \frac{a(t^*)^2[1-(t^*)^2]}{2}\|\nabla\bar{u}\|_2^2 < (t^*)^4 I(\bar{u}) = 0. \tag{3.105}$$

This contradiction shows that $\|\bar{u}\|_2^2 = c^*$. Hence, $\bar{u} \in \mathcal{S}_{c^*}$ and $I(\bar{u}) = \sigma(c^*) = 0$.

Case b): $\bar{u} = 0$. By (3.12) and (3.95), one has $I^\infty(u_n) \to 0$. We claim that (3.97) holds. In fact, if $\delta = 0$, by Lions' concentration-compactness principle [153, Lemma 1.21], one has $u_n \to 0$ in $L^s(\mathbb{R}^3)$ for $2 < s < 6$. By (F4) and (F6′), for any $\varepsilon > 0$, there exists $C_\varepsilon > 0$ such that

$$|F(t)| \le \varepsilon|t|^{\frac{10}{3}} + C_\varepsilon|t|^p, \qquad \forall\, t \in \mathbb{R}. \tag{3.106}$$

Then (3.106) implies $\int_{\mathbb{R}^3} F(u_n)dx \to 0$. Jointly with $I(u_n) \to 0$, we have $\|\nabla u_n\|_2 \to 0$. From (3.3), (3.41) and (3.106), we deduce

$$I(u_n) \ge \frac{1}{4}\|\nabla u_n\|_2^2\left(2a + b\|\nabla u_n\|_2^2 - 4\varepsilon\|u_n\|_2^{\frac{4}{3}} - 4C_\varepsilon\|u_n\|_2^{\frac{6-p}{2}}\|\nabla u_n\|_2^{\frac{3p-10}{2}}\right), \qquad \forall\, n \in \mathbb{N},$$

which, together with $3p - 10 > 0$ and arbitrariness of ε, implies that $I(u_n) \ge 0$ for large $n \in \mathbb{N}$. This contradicts (3.103), and thus (3.97) holds. Let $\hat{u}_n(x) = u_n(x + y_n)$. From (3.97), there exists $\hat{u} \in H^1(\mathbb{R}^3) \setminus \{0\}$ such that, passing to a subsequence, $\hat{u}_n \rightharpoonup \hat{u}$ in $H^1(\mathbb{R}^3)$. Similarly to the proof of (3.101), we have

$$\begin{aligned}
0 &= \sigma(c^*) = \lim_{n\to\infty} I^\infty(\hat{u}_n) \\
&= I^\infty(\hat{u}) + \lim_{n\to\infty}\left[I^\infty(\hat{u}_n - \hat{u}) + \frac{b}{2}\|\nabla\hat{u}\|_2^2\|\nabla(\hat{u}_n - \hat{u})\|_2^2\right] \\
&\ge \sigma^\infty(\|\hat{u}\|_2^2) + \lim_{n\to\infty}\sigma^\infty(\|\hat{u}_n - \hat{u}\|_2^2) \\
&= \sigma^\infty(\|\hat{u}\|_2^2) + \sigma^\infty(c^* - \|\hat{u}\|_2^2) \ge \sigma(\|\hat{u}\|_2^2) + \sigma(c^* - \|\hat{u}\|_2^2) = 0,
\end{aligned} \tag{3.107}$$

which implies

$$\lim_{n\to\infty}\|\nabla(\hat{u}_n - \hat{u})\|_2 = 0.$$

Proceeding as in the proof of Case a), we conclude that $\hat{u} \in \mathcal{S}_{c^*}$ and $I(\hat{u}) = \sigma(c^*) = 0$. □

3.3.2 Local minimizers on the manifold

We now look for a local minimizer of I on the constraint $\mathcal{S}_c$, which is a critical point of $I|_{\mathcal{S}_c}$.

For $k > 0$, set

$$\mathcal{S}_c(k) := \left\{u \in \mathcal{S}_c : \|\nabla u\|_2^2 = k\right\}.$$

For any $0 < c < c_*$ and $u \in \mathcal{S}_c$, it follows from (3.3), (3.41) and (3.106) that

$$I(u) \ge \frac{1}{4}\|\nabla u\|_2^2\left(2a + b\|\nabla u\|_2^2 - 4\varepsilon c_*^{\frac{2}{3}} - 4C_\varepsilon c_*^{\frac{6-p}{4}}\|\nabla u\|_2^{\frac{3p-10}{2}}\right). \tag{3.108}$$

Since $3p - 10 > 0$ and ε is arbitrary, there exists $k_0 > 0$ independent of c such that

$$I^\infty(u) \geq I(u) \geq \frac{ak}{4} > 0, \quad \forall\, u \in \mathcal{S}_c(k),\ 0 < k \leq k_0. \tag{3.109}$$

In view of Lemma 3.15 (i), we have

$$\bar{\sigma}(c) := \inf_{u\in\mathcal{S}_c\setminus\bigcup_{0<k\leq k_0}\mathcal{S}_c(k)} I(u) \geq \inf_{u\in\mathcal{S}_c} I(u) = \sigma(c) > -\infty, \quad \forall\, c > 0. \tag{3.110}$$

Lemma 3.19 *Assume that* (K1), (K3), (F4), (F5) *and* (F6′) *hold. Then* $\bar{\sigma}(c)$ *is continuous on* $(0, +\infty)$ *and for any* $c > 0$,

$$\bar{\sigma}(c) < \bar{\sigma}(\alpha) + \bar{\sigma}(c-\alpha), \quad \forall\, 0 < \alpha < c. \tag{3.111}$$

Proof. Similarly to the proof of Lemma 3.17, we deduce that $\bar{\sigma}(c)$ is continuous for any $c > 0$. Let $\{u_n\} \subset \mathcal{S}_c \setminus \bigcup_{0<k\leq k_0}\mathcal{S}_c(k)$ be such that $I(u_n) \to \bar{\sigma}(c)$ for any $c > 0$. Then (3.77) implies that $\{u_n\}$ is bounded in $H^1(\mathbb{R}^3)$. Noting that $\|\nabla u_n\|_2^2 > k_0$ and $\|(u_n)_t\|_2^2 = t^4\|u_n\|_2^2 = t^4 c$ for all $n \in \mathbb{N}$ and $t > 0$, it follows from (3.90) that

$$\begin{aligned}\bar{\sigma}(t^4 c) &\leq I((u_n)_t) \leq t^4 I(u_n) - \frac{at^2(t^2-1)}{2}\|\nabla u_n\|_2^2\\ &\leq t^4\bar{\sigma}(c) - \frac{at^2(t^2-1)}{2}k_0^2 + o(1), \quad \forall\, t > 1,\ c > 0,\end{aligned}$$

which implies

$$\bar{\sigma}(tc) < t\bar{\sigma}(c), \quad t > 1,\ c > 0. \tag{3.112}$$

Then it follows from (3.112) that

$$\bar{\sigma}(c) = \frac{\alpha}{c}\bar{\sigma}(c) + \frac{c-\alpha}{c}\bar{\sigma}(c) < \bar{\sigma}(\alpha) + \bar{\sigma}(c-\alpha), \quad \forall\, 0 < \alpha < c.$$

This completes the proof. □

The following property is concerned with the study of the geometry of local minima.

Lemma 3.20 *Assume that* (K1), (K3), (F4), (F5) *and* (F6′) *hold. Then there exists* $0 < c_0 < c_*$ *such that*

$$0 < \bar{\sigma}(c) < \frac{a}{4}k_0 \leq \inf_{u\in\mathcal{S}_c(k_0)} I(u), \quad \forall\, c \in (c_0, c_*). \tag{3.113}$$

Proof. By the definition of c_* (see (3.9)), Lemma 3.15 (i) and (3.110), one has

$$\bar{\sigma}(c) \geq \sigma(c) = 0, \quad \forall\, 0 < c < c_*. \tag{3.114}$$

We first prove that $\bar{\sigma}(c) > 0$ for all $0 < c < c_*$. If not, then $\bar{\sigma}(\bar{c}) = 0$ for some $\bar{c} \in (0, c_*)$, and so there exists $\{u_n\} \subset \mathcal{S}_{\bar{c}}\setminus\bigcup_{0<k\leq k_0}\mathcal{S}_{\bar{c}}(k)$ such that $I(u_n) \to 0$ and $\|\nabla u_n\|_2^2 > k_0$. Since $\|(u_n)_t\|_2^2 = t^4\|u_n\|_2^2 = t^4\bar{c}$ by the scaling (3.80), it follows from (3.90) that

$$\begin{aligned}\bar{\sigma}(t^4\bar{c}) &\leq \lim_{n\to\infty} I((u_n)_t) \leq \lim_{n\to\infty}\left[t^4 I(u_n) - \frac{at^2(t^2-1)}{2}\|\nabla u_n\|_2^2\right]\\ &\leq -\frac{ak_0t^2(t^2-1)}{2} < 0, \quad \forall\, 1 < t < \left(\frac{c_*}{\bar{c}}\right)^{1/4},\end{aligned}$$

which contradicts (3.114). Hence, $\bar{\sigma}(c) > 0$ for all $0 < c < c_*$.

Next, we prove that there exists $c_0 > 0$ such that the last inequality of (3.113) holds. In view of Lemma 3.18 (iii), there exists $u_* \in \mathcal{S}_{c_*}$ such that $I(u_*) = \sigma(c_*) = 0$. From the continuity of $I(tu_*)$ on $t \in (0, \infty)$, we deduce that there exists a constant $t_0 \in (0, 1)$ sufficiently close to 1 such that

$$I(tu_*) < \frac{a}{4}k_0, \qquad \forall\, t_0 \le t \le 1, \tag{3.115}$$

which, together with (3.109), implies

$$tu_* \in \mathcal{S}_{tc_*} \setminus \bigcup_{0<k\le k_0} \mathcal{S}_{tc_*}(k), \qquad \forall\, t_0 \le t \le 1. \tag{3.116}$$

Letting $c_0 = t_0 c_*$, it follows from (3.109), (3.115) and (3.116) that for any $c \in (c_0, c_*)$, there exists $v_* \in \mathcal{S}_c \setminus \bigcup_{0<k\le k_0} \mathcal{S}_c(k)$ such that

$$\bar{\sigma}(c) \le I(v_*) < \frac{a}{4}k_0 \le \inf_{u\in\mathcal{S}_c(k_0)} I(u).$$

The proof is completed. □

Define

$$\bar{\sigma}^\infty(c) := \inf_{u\in\mathcal{S}_c\setminus\bigcup_{0<k\le k_0}\mathcal{S}_c(k)} I^\infty(u) \ge \inf_{u\in\mathcal{S}_c} I^\infty(u) = \sigma^\infty(c) > -\infty, \qquad \forall\, c > 0. \tag{3.117}$$

Lemma 3.21 *Assume that* (K1), (K3), (F4), (F5) *and* (F6′) *hold. Then*

(i) $\bar{\sigma}(c) \le \bar{\sigma}^\infty(c)$ *for any* $c > 0$*;*

(ii) *for any* $c \in (c_0, c_*)$, (3.1) *admits a solution* $(\bar{u}_c, \bar{\lambda}_c) \in \left(\mathcal{S}_c \setminus \bigcup_{0<k<k_0} \mathcal{S}_c(k)\right)\times\mathbb{R}$ *such that* $I(\bar{u}_c) = \bar{\sigma}(c) > 0$.

Proof. (i) Similarly to the proof of Lemma 3.18 (i), we get $\bar{\sigma}(c) \le \bar{\sigma}^\infty(c)$ for any $c > 0$.

(ii) By Ekeland's variational principle, there exists a sequence $\{u_n\} \subset \mathcal{S}_c\setminus\bigcup_{0<k\le k_0}\mathcal{S}_c(k)$ such that

$$\begin{cases} \bar{\sigma}(c) \le I(u_n) \le \bar{\sigma}(c) + \frac{1}{n}, \\ I(v) \ge I(u_n) - \frac{1}{n}\|u_n - v\|, \qquad \forall\, v \in \mathcal{S}_c \setminus \bigcup_{0<k\le k_0} \mathcal{S}_c(k). \end{cases} \tag{3.118}$$

Then (3.77) implies that $\{u_n\}$ is bounded in $H^1(\mathbb{R}^3)$. We then may assume that for some $\bar{u}_c \in H^1(\mathbb{R}^3)$ such that up to a subsequence, $u_n \rightharpoonup \bar{u}_c$ in $H^1(\mathbb{R}^3)$, $u_n \to \bar{u}_c$ in $L^s_{\mathrm{loc}}(\mathbb{R}^3)$ for $2 \le s < 6$ and $u_n \to \bar{u}_c$ a.e. in $\mathbb{R}^3$. For brevity, we denote $\bar{u}_c$ by $\bar{u}$. We complete our proof in two steps as follows.

Step 1. We prove that if $\bar{u} \ne 0$, then $I(\bar{u}) = \bar{\sigma}(c)$ and there exists $\bar{\lambda} \in \mathbb{R}$ such that $I'(\bar{u}) - \bar{\lambda}u = 0$ for any $c \in (c_0, c_*)$.

Similarly to the proof of relation (3.17) in [158], we have

$$|\langle I'(u_n) - \lambda_n u_n, \varphi\rangle| \le \frac{C}{n}, \qquad \forall\, \varphi \in \mathcal{C}_0^\infty(\mathbb{R}^3), \tag{3.119}$$

where $\lambda_n = -\frac{\langle I'(u_n), u_n\rangle}{c}$ and $C > 0$ is a constant independent of n. Since $\{\lambda_n\}$ is bounded, we may assume that up to a subsequence, $\lambda_n \to \bar{\lambda}$ for some $\bar{\lambda} \in \mathbb{R}$. Now, we claim that

$$\text{up to a subsequence } \|\nabla(u_n - \bar{u})\|_2 \to 0 \text{ if } \bar{u} \ne 0. \tag{3.120}$$

Assume by contradiction that $\liminf_{n\to\infty} \|\nabla(u_n - \bar{u})\|_2^2 > 0$. Then it follows from the scaling (3.80) that there exists $T_0 > 1$ such that

$$\|\nabla\bar{u}_t\|_2^2 = t^2\|\nabla\bar{u}\|_2^2 > k_0, \quad \|\nabla(u_n - \bar{u})_t\|_2^2 = t^2\|\nabla(u_n - \bar{u})\|_2^2 > k_0, \quad \forall\, t > T_0. \tag{3.121}$$

By (3.90), (3.118) and Lemma 3.11, we have

$$\begin{aligned}
\bar{\sigma}(c)+o(1) &= I(u_n)=I(\bar{u})+I(u_n-\bar{u})+\frac{b}{2}\|\nabla\bar{u}\|_2^2\|\nabla(u_n-\bar{u})\|_2^2 \\
&\geq \frac{1}{t^4}\left[I(\bar{u}_t)+I((u_n-\bar{u})_t)\right]+\frac{at^2(t^2-1)}{2t^4}\left[\|\nabla\bar{u}\|_2^2+\|\nabla(u_n-\bar{u})\|_2^2\right] \\
&\geq \frac{1}{t^4}\left[\bar{\sigma}(t^4\|\bar{u}\|_2^2)+\bar{\sigma}(t^4c-t^4\|\bar{u}\|_2^2)\right]+\frac{at^2(t^2-1)}{2t^4}\|\nabla u_n\|_2^2+o(1) \\
&\geq \frac{1}{t^4}\bar{\sigma}(t^4c)+\frac{at^2(t^2-1)}{2t^4}k_0+o(1), \quad \forall\, t\geq T_0.
\end{aligned} \tag{3.122}$$

From (3.77), we deduce

$$\liminf_{t\to+\infty}\frac{\bar{\sigma}(t^4c)}{t^4}\geq 0. \tag{3.123}$$

By (3.123), there exists $T_1\geq T_0$ such that

$$\frac{1}{T_1^4}\bar{\sigma}(T_1^4c)\geq -\frac{ak_0}{16}, \qquad \frac{aT_1^2(T_1^2-1)}{2t^4}k_0\geq\frac{3ak_0}{8}. \tag{3.124}$$

Then (3.113), (3.122) and (3.124) lead to

$$\frac{ak_0}{4}>\bar{\sigma}(c)\geq\frac{1}{T_1^4}\bar{\sigma}(T_1^4c)+\frac{aT_1^2(T_1^2-1)}{2t^4}k_0\geq\frac{5ak_0}{16}. \tag{3.125}$$

This contradiction shows that relation (3.120) holds. Thus, from (F4), (F5), (3.118), (3.119) and (3.120), we obtain

$$\bar{\sigma}(c)=\lim_{n\to\infty}I(u_n)=I(\bar{u}) \tag{3.126}$$

and for all $\varphi\in\mathcal{C}_0^\infty(\mathbb{R}^3)$

$$\begin{aligned}
0 &= \lim_{n\to\infty}\langle I'(u_n)-\lambda_n u_n,\varphi\rangle \\
&= \lim_{n\to\infty}\left[a\int_{\mathbb{R}^3}\nabla u_n\cdot\nabla\varphi \mathrm{d}x+b\|\nabla u_n\|_2^2\int_{\mathbb{R}^3}\nabla u_n\cdot\nabla\varphi \mathrm{d}x\right. \\
&\quad \left.-\int_{\mathbb{R}^3}K(x)f(u_n)\varphi \mathrm{d}x-\lambda_n\int_{\mathbb{R}^3}u_n\varphi \mathrm{d}x\right] \\
&= a\int_{\mathbb{R}^3}\nabla\bar{u}\cdot\nabla\varphi \mathrm{d}x+b\|\nabla\bar{u}\|_2^2\int_{\mathbb{R}^3}\nabla\bar{u}\cdot\nabla\varphi \mathrm{d}x-\int_{\mathbb{R}^3}K(x)f(\bar{u})\varphi \mathrm{d}x-\bar{\lambda}\int_{\mathbb{R}^3}\bar{u}\varphi \mathrm{d}x \\
&= \langle I'(\bar{u})-\bar{\lambda}\bar{u},\varphi\rangle.
\end{aligned} \tag{3.127}$$

By Fatou's lemma, one has $\|\bar{u}\|_2^2\leq\liminf_{n\to\infty}\|u_n\|_2^2=c$. In view of (3.126) and (3.127), to finish the proof of Step 1, it suffices to show that $\|\bar{u}\|_2^2=c$.

Suppose by contradiction that $\|\bar{u}\|_2^2<c$. As in [158, (3.14)], for any $\varepsilon>0$, there exists $\delta=\delta(\varepsilon)>0$ such that if $u\in\bigcup_{k_0-\delta\leq k\leq k_0+\delta}\mathcal{S}_c(k)$, then

$$I(u)\geq\inf_{u\in\mathcal{S}_c(k_0)}I(u)-\varepsilon,$$

and so

$$u_n\in\mathcal{S}_c\setminus\bigcup_{0\leq k\leq k_0+\delta}\mathcal{S}_c(k),\ \text{for } n \text{ large enough}, \tag{3.128}$$

that is, $\{u_n\}$ stays away from the boundary. Jointly with (3.120), we have $\|\nabla \bar{u}\|_2^2 > k_0$. Let $\bar{t} := (c/\|\bar{u}\|_2^2)^{1/4} > 1$. By the scaling (3.80), one has $\|\bar{u}_{\bar{t}}\|_2^2 = \bar{t}^4\|\bar{u}\|_2^2$ and $\|\nabla \bar{u}_{\bar{t}}\|_2^2 = \bar{t}^2\|\nabla \bar{u}\|_2^2 > k_0$. Then it follows from (3.90) and (3.126) that

$$\bar{t}^4\bar{\sigma}(c) = \bar{t}^4 I(\bar{u}) \geq I(\bar{u}_{\bar{t}}) + \frac{a\bar{t}^2(\bar{t}^2-1)}{2}\|\nabla \bar{u}\|_2^2 \geq \bar{\sigma}(c) + \frac{a\bar{t}^2(\bar{t}^2-1)}{2}k_0, \tag{3.129}$$

which implies

$$\bar{\sigma}(c) \geq \frac{a\sqrt{c}k_0}{2(\|\bar{u}\| + \sqrt{c})} > \frac{ak_0}{4}, \quad \forall\, c \in (c_0, c_*).$$

This contradicts (3.113). Hence, we have $\|\bar{u}\|_2^2 = c$, moreover, (3.127) implies that $(\bar{u}, \bar{\lambda}) \in \left(\mathcal{S}_c \setminus \bigcup_{0<k\leq k_0} \mathcal{S}_c(k)\right) \times \mathbb{R}$ solves (3.1). The proof of Step 1 is completed.

Step 2. We prove that $\bar{u} \neq 0$.

Assume by contradiction that $\bar{u} = 0$, that is, $u_n \rightharpoonup 0$ in $H^1(\mathbb{R}^3)$. Then $u_n \to 0$ in $L^s_{\text{loc}}(\mathbb{R}^3)$ for $2 \leq s < 6$ and $u_n \to 0$ a.e. in $\mathbb{R}^3$. Then (3.3), (3.12), (3.95) and (3.118) imply

$$I^\infty(u_n) \to \bar{\sigma}(c). \tag{3.130}$$

We first prove that $\{u_n\}$ is not vanishing, that is, relation (3.97) holds. Otherwise, if $\{u_n\}$ is vanishing, by Lions' concentration-compactness principle, we obtain that $u_n \to 0$ in $L^s(\mathbb{R}^3)$ for $2 < s < 6$, and so (3.106) implies that $\int_{\mathbb{R}^3} F(u_n)\mathrm{d}x \to 0$. Since $\{u_n\} \subset \mathcal{S}_c \setminus \bigcup_{0<k\leq k_0} \mathcal{S}_c(k)$, by (3.3), (3.113) and (3.118), we have

$$\frac{ak_0}{2} + \frac{bk_0^2}{4} \leq \lim_{n\to\infty}\left(\frac{a}{2}\|\nabla u_n\|_2^2 + \frac{b}{4}\|\nabla u_n\|_2^4\right) = \lim_{n\to\infty} I(u_n) = \bar{\sigma}(c) < \frac{ak_0}{4},$$

which is impossible. Hence, relation (3.97) holds, and so there exists $\{y_n\} \subset \mathbb{R}^3$ such that (3.98) holds. Let $\tilde{u}_n(x) = u_n(x + y_n)$. Then (3.130) leads to

$$\tilde{u}_n \in \mathcal{S}_c \setminus \bigcup_{0<k\leq k_0} \mathcal{S}_c(k), \quad I^\infty(\tilde{u}_n) \to \bar{\sigma}(c). \tag{3.131}$$

By reflexivity, there exists $\tilde{u} \in H^1(\mathbb{R}^3) \setminus \{0\}$ such that, passing to a subsequence,

$$\begin{cases} \tilde{u}_n \rightharpoonup \tilde{u}, & \text{in } H^1(\mathbb{R}^3); \\ \tilde{u}_n \to \tilde{u}, & \text{in } L^s_{\text{loc}}(\mathbb{R}^3),\ \forall\, s \in [1, 6); \\ \tilde{u}_n \to \tilde{u}, & \text{a.e. on } \mathbb{R}^3. \end{cases} \tag{3.132}$$

We also have

$$|\langle (I^\infty)'(\tilde{u}_n) - \lambda_n \tilde{u}_n, \varphi\rangle| \leq \frac{C}{n}, \quad \forall\, \varphi \in \mathcal{C}_0^\infty(\mathbb{R}^3), \tag{3.133}$$

where λ_n and $C > 0$ are the same as that of (3.119). Arguing as in the proofs of (3.120), (3.126) and (3.127), we obtain

$$\text{up to a subsequence } \|\nabla(\tilde{u}_n - \tilde{u})\|_2 \to 0 \tag{3.134}$$

and

$$\bar{\sigma}(c) = \lim_{n\to\infty} I^\infty(\tilde{u}_n) = I^\infty(\tilde{u}). \tag{3.135}$$

Moreover, as in the proof of Step 1, we have $\|\tilde{u}\|_2^2 = c$ and $\|\nabla \tilde{u}\|_2^2 > k_0$. Hence, $\tilde{u} \in \mathcal{S}_c \setminus \bigcup_{0<k\leq k_0} \mathcal{S}_c(k)$. Since $K(x) \geq (\not\equiv) K_\infty$ and $\|\tilde{u}\|_2^2 = c > 0$, it follows that there exist $r > 0$ and $\bar{x}_1, \bar{x}_2 \in \mathbb{R}^3$ such that

$$K(x) - K_\infty > 0, \quad \forall\, |x - \bar{x}_1| < r, \quad |\tilde{u}(x)| > 0, \quad \forall\, |x - \bar{x}_2| < r.$$

Let $\hat{u}(x) = \tilde{u}(x - \bar{x}_1 + \bar{x}_2)$. Then $\|\hat{u}\|_2^2 = c$, $\|\nabla \hat{u}\|_2^2 > k_0$ and $I^\infty(\hat{u}) = I^\infty(\tilde{u})$, which together with (3.3), (3.12), implies that

$$\bar{\sigma}(c) = I^\infty(\hat{u}) > I(\hat{u}) \geq \bar{\sigma}(c). \tag{3.136}$$

This contradiction shows that $\bar{u} \neq 0$. □

3.3.3 Proof of Theorem 3.2

Note that if u_c is a critical point of $I|_{\mathcal{S}_c}$, then there exists $\lambda_c \in \mathbb{R}$ such that $I'(u_c) - \lambda_c u_c = 0$. From Lemmas 3.16, 3.18, and 3.21, we conclude that the conclusions of Theorem 3.2 hold. □

3.4 Historical comments

The Kirchhoff equation has been introduced for the first time in 1883 by Kirchhoff [88] in dimension 1, without forcing term and with Dirichlet boundary conditions, in order to describe the transversal free vibrations of a clamped string in which the dependence of the tension on the deformation cannot be neglected. This is a quasi-linear partial differential equation, namely the nonlinear part of the equation contains as many derivatives as the linear differential operator. The Kirchhoff equation is an extension of the classical D'Alembert wave equation for free vibrations of elastic strings. Kirchhoff's model takes into account the changes in length of the string produced by transverse vibrations.

From the mathematical point of view, problem (3.1) is nonlocal due to the appearance of the term $\int_{\mathbb{R}^3} |\nabla u|^2 \mathrm{d}x$ indicates that (3.1) is not a pointwise identity. This kind of problem has been paid much attention after the pioneering work of Lions [99], in which an abstract functional analysis framework was introduced.

Nowadays, since physicists are interested in normalized solutions, mathematical researchers began to focus on solutions having a prescribed L^2-norm, that is, solutions which satisfy $\|u\|_2^2 = c > 0$ for a priori given c. To the best of our knowledge, the study of solutions with prescribed norm was initiated by Jeanjean [78] in the framework of semilinear elliptic equations. In this chapter, we studied in the existence of solutions with L^2-prescribed norm and their qualitative properties in the framework of nonlocal Kirchhoff problems. Our analysis includes both the L^2-supercritical case and the L^2-subcritical growth.

To the best of our knowledge, solutions of problem (3.1) having a prescribed L^2-norm have been studied mainly in the case where $K(x) \equiv 1$. Let us introduce and review the few known results in this respect. Ye [158] studied the existence and non-existence of normalized solutions to the following special form of problem (3.1):

$$\begin{cases} -\left(a + b\displaystyle\int_{\mathbb{R}^3} |\nabla u|^2 \mathrm{d}x\right) \Delta u - \lambda u = |u|^{p-2}u \quad \text{in } \mathbb{R}^3; \\ u \in H^1(\mathbb{R}^3) \end{cases} \tag{3.137}$$

for $p \in (2, 6)$, and showed that $p = \frac{14}{3}$ is a L^2-critical exponent for problem (3.137), that is, for any given $c > 0$,

$$\begin{cases} \sigma(c) \in (-\infty, 0], & \text{if } p \in (2, \frac{14}{3}); \\ \sigma(c) = -\infty, & \text{if } p \in (\frac{14}{3}, 6). \end{cases}$$

More precisely, for any $p \in (2, \frac{14}{3})$, Ye [158] obtained the sharp existence of global constraint minimizers for (3.137) by solving the minimization problem (3.5). If $p \in (\frac{10}{3}, \frac{14}{3})$, Ye [158] found a local minimizer which is also a critical point of $I|_{\mathcal{S}_c}$ by constructing a geometry of local minima for I. For the case $p \in (\frac{14}{3}, 6)$, to look for a critical point of $I|_{\mathcal{S}_c}$, Ye [158] took a minimum on a suitable submanifold

$$\mathcal{M}_c := \left\{ u \in \mathcal{S}_c : J(u) := \frac{\mathrm{d}}{\mathrm{d}t} I(u^t)\Big|_{t=1} = 0 \right\},$$

and showed that $\mathcal{M}_c$ is a natural constraint of $I|_{\mathcal{S}_c}$ by using the Lagrange multiplier method, where

$$u^t(x) := t^{3/2} u(tx), \qquad \forall\, t > 0, u \in H^1(\mathbb{R}^3),$$

and $u^t \in \mathcal{S}_c$ for all $t > 0$ if $u \in \mathcal{S}_c$. Note that this method relies heavily on the power-type nonlinearity $f(u) = |u|^{p-2}u$ with $p \in (\frac{14}{3}, 6)$. Xie and Chen [155] generalized this special case to general nonlinearities f satisfying $\lim_{|t|\to\infty} \frac{F(t)}{|t|^{14/3}} = +\infty$. Xie and Chen [155] proved the existence of normalized solutions for problem (3.1) with $K(x) \equiv 1$ under additional growth and monotonicity conditions. For the L^2-critical case, Ye [157] proved that problem (3.137) with $p = \frac{14}{3}$ has a solution (u_c, λ_c) which satisfies $\|u_c\|_2^2 = c > c^*$ for some $c^* > 0$. Additionally, it was considered in [157, 158] the existence of normalized solutions for one-dimensional and two-dimensional autonomous Kirchhoff type equations with power-type nonlinearity.

3.5 Final remarks, perspectives, and open problems

This chapter is concerned with the qualitative analysis of solutions with prescribed mass of nonautonomous Kirchhoff equations in both sub- and super-critical cases. The analysis developed in this chapter shows how the Kirchhoff nonlocality influences the existence of normalized solutions compared to the classical Schrödinger case. A feature of the problem studied in this context is given by the presence of the nonlocal term $b(\int_{\mathbb{R}^3} |\nabla u|^2 \mathrm{d}x)\Delta u$ together with the L^2-norm constraint, which creates a competition between mass and Kirchhoff effects, complicating the geometry of the energy functional.

The content of this chapter is based on the results established in [27]. Related results on normalized solutions of Kirchhoff equations with generic double-behavior nonlinearities have been obtained by Cai and Rădulescu [17], while further properties of constraint minimizers of mass critical fractional Kirchhoff equations have been studied by Liu, Rădulescu and Yuan [103]. Recently, Chen and Tang [33] studied normalized solutions for Kirchhoff equations with Sobolev critical exponent and mixed nonlinearities.

Some interesting open problems associated with the analysis developed in this chapter are stated in what follows.

The first problem proposes to refine the analysis in the case of a non-autonomous potential associated with the linear term.

Open Problem 1. Study the following Kirchhoff equation

$$\begin{cases} -\left(a + b\displaystyle\int_{\mathbb{R}^3} |\nabla u|^2 \mathrm{d}x\right)\Delta u - V(x)u = K(x)f(u) & \text{in } \mathbb{R}^3; \\ u \in H^1(\mathbb{R}^3) \end{cases} \tag{3.138}$$

in the following cases:

(i) V is a confining potential, that is, $V(x) \to +\infty$ as $|x| \to \infty$;

(ii) V is a singular potential, that is, $V(x) = |x|^{-m}$ with $0 < m < 2$. In this case, the interplay between the nonlocal Kirchhoff term, the singular attractive potential, and the L^2-norm constraint is completely unexplored.

Another exciting directions concerns the case of trapping potentials and nodal solutions.

Open Problem 2. Assume that V is a trapping potential (or another type of potential, as described previously). Can one find normalized solutions of problem (3.138) with a prescribed number of nodes (sign changes)? The main difficulty is the coupling between the nonlocal term and the nodal structure, which breaks the natural symmetric decreasing rearrangement arguments.

The next open problem proposes a related analysis in the L^2-critical case, which is notoriously difficult under normalization.

Open Problem 3. Assume that the nonlinearity is focusing (attractive) like $f(u) \sim |u|^{\theta-1}$. This chapter explicitly treats subcritical ($\theta < \frac{14}{3}$) and supercritical ($\theta > \frac{14}{3}$) cases but avoids the L^2-critical case, that is, if $\theta = \frac{14}{3}$. In this case, does a nontrivial potential V help restore existence for certain masses c? Can one determine a sharp threshold mass c^* (depending on V)?

3.6 Glossary

Mass Subcritical for Kirchhoff Equations: In dimension N, this means that the nonlinearity grows slower than $|u|^{1+\frac{8}{N}}$, which is more restrictive than the usual $|u|^{1+\frac{4}{N}}$ for standard Schrödinger equations. This is because the Kirchhoff term $\|\nabla u\|_2^4$ changes the scaling balance when working with fixed L^2-norm.

Mass Supercritical for Kirchhoff Equations: In dimension N, this means that the nonlinearity grows faster than $|u|^{1+\frac{8}{N}}$. This means that the nonlinearity overwhelms the Kirchhoff stiffness at large amplitudes, creating a fundamentally different variational geometry requiring topological methods rather than direct minimization.

Threshold Mass: It is a critical prescribed mass where the Kirchhoff equation's solution structure changes qualitatively, mainly because of the balance between the nonlocal stiffness term and the focusing nonlinearity's strength. The reason of the existence of this value is that the Kirchhoff term cannot compete with the focusing nonlinearity when the mass is too large.

4

Autonomous Kirchhoff equations with Sobolev critical exponent

This chapter tackles the challenges of a Kirchhoff problem with double criticality (mass and Sobolev) in the context of the existence of solutions with prescribed mass. The analysis developed in what follows demonstrates how the nonlocal term interacts with the L^2-constraint and the critical reaction to create a rich and complex parameter space for existence of solutions. We highlight a qualitative difference from the classical semilinear Schrödinger equation (local setting) due to the nonlocal term's dependence on the gradient norm, which changes the scaling properties of the equation.

We investigate the existence and multiplicity properties of the normalized solutions to the following Kirchhoff equation with Sobolev critical growth

$$\begin{cases} -\left(a+b\int_{\mathbb{R}^3}|\nabla u|^2dx\right)\Delta u+\lambda u=\mu|u|^{p-2}u+|u|^4u & \text{in } \mathbb{R}^3,\\ u>0,\ \int_{\mathbb{R}^3}|u|^2dx=c^2, & \text{in } \mathbb{R}^3, \end{cases}$$

where a, b, c, $\mu>0$ and $4<p<6$. It is highlighted how the existence of solutions depends on the interplay between several parameters: the mass c, the exponent p, and the coefficients a, b and μ. The nonlocal coefficient $b>0$ is not just a perturbation but it qualitatively changes the existence theory compared to the semilinear case ($b=0$). In particular, b contributes to the formation of the threshold c^*.

We consider both the L^2-subcritical and the L^2-supercritical cases. The concentration-compactness principle and its variant for the constraint manifold are employed to handle the lack of compactness. In the L^2-subcritical case, by combining the truncation method, the concentration-compactness principle and genus theory, we obtain the multiplicity of normalized solutions for problem (P). For this mass sub-critical case, the solution is found as a global minimizer, and existence is generally proved for all $c>0$. The novelty here is overcoming the lack of compactness from the critical exponent in this minimization framework.

In the L^2-supercritical case, by using a fiber map and the concentration-compactness principle, we obtain a couple of normalized solutions for problem (P), as well as their asymptotic behavior. Crucial steps in the arguments involve proving that the mountain-pass energy level stays below a certain threshold, ensuring that the associated Palais-Smale sequence does not "vanish" and that its weak limit is nontrivial. This often involves comparing with the optimal constant in the Sobolev inequality. For the mass super-critical and critical case, the low mass regime $c<c^*$ is identified as the necessary and sufficient condition for the existence of the mountain-pass solution. This c^*-phenomenon is a direct and novel consequence of the interplay between the Kirchhoff term and the Sobolev critical exponent.

A feature of the study developed in this chapter for the three-dimensional case is that it unifies the sub- and super-critical regimes. *Why this case is fundamental?* Because the autonomous Kirchhoff equation with Sobolev critical exponent sits at the intersection of

DOI: 10.1201/9781003797159-4

three deep mathematical theories:

(i) *calculus of variations*: critical growth, lack of compactness;
(ii) *geometric analysis*: best constants, extremals, bubbling;
(iii) *nonlocal equation*: Kirchhoff-type nonlocality.

The basic abstract tools in this study are the following:

(a) *truncation method and fiber mapping*: to analyze the energy levels;
(b) *concentration-compactness principle*: for handling the lack of compactness;
(c) *genus theory*: to establish the multiplicity of solutions.

4.1 Statement of the problem

Consider the Kirchhoff equation

$$\begin{cases} -\left(a+b\displaystyle\int_{\mathbb{R}^3}|\nabla u|^2dx\right)\Delta u+\lambda u=\mu|u|^{p-2}u+|u|^4u & \text{in } \mathbb{R}^3,\\ u>0,\ \displaystyle\int_{\mathbb{R}^3}|u|^2dx=c^2, & \text{in } \mathbb{R}^3, \end{cases} \tag{P}$$

where $a,\ b,\ c,\ \mu>0$ and $4<p<6$.

Set

$$H^1(\mathbb{R}^3)=\{u\in L^2(\mathbb{R}^3):|\nabla u|\in L^2(\mathbb{R}^3)\}$$

with the inner product

$$\langle u,v\rangle=\int_{\mathbb{R}^3}[a\nabla u\cdot\nabla v+uv]dx$$

and the norm

$$\|u\|=\langle u,u\rangle^{1/2}.$$

We are now in a position to state the first two main results of this chapter. Essentially, these results establish the existence of normalized solutions for high perturbations of the forcing term. Our analysis covers both the L^2-subcritical and the L^2-supercritical cases. In the first setting, we obtain the multiplicity of the normalized solutions, while in the L^2-supercritical case, we find a couple of normalized solutions.

Theorem 4.1 *If $4<p<\frac{14}{3}$, for given $k\in\mathbb{N}$, then there exist $\alpha>0$ independent of k and $\mu_k:=\mu(k)$ such that problem (P) has at least k pairs $(u_j,\lambda_j)\in H^1(\mathbb{R}^3)\times\mathbb{R}$ of weak solutions for $\mu\geq\mu_k$ and $c\in\left(0,(\frac{\alpha}{\mu})^{\frac{2}{6-p}}\right]$ with $\int_{\mathbb{R}^3}|u_j|^2dx=c^2$, $\lambda_j>0$ for all $j\in[1,k]$.*

Theorem 4.2 *If $\frac{14}{3}<p<6$, then there exists $\mu^*=\mu^*(c)>0$ such that for $\mu\geq\mu^*$, problem (P) has a couple $(u_c,\lambda_c)\in H^1(\mathbb{R}^3)\times\mathbb{R}$ of weak solutions with $\int_{\mathbb{R}^3}|u_c|^2dx=c^2$ and $\lambda_c>0$.*

Problem (P) on the whole space $\mathbb{R}^3$ is invariant under translations, which leads to the lack of compactness. In order to overcome this phenomenon, we consider the function space $H^1_{rad}(\mathbb{R}^3)$ of radial functions as the working space, where

$$H^1_{rad}(\mathbb{R}^3) := \{u \in H^1(\mathbb{R}^3) : \ u \text{ is radial}\}.$$

Then the embedding $H^1_{rad}(\mathbb{R}^3) \hookrightarrow L^t(\mathbb{R}^3)$ is compact for any $2 < t < 6$. In such a way, we consider the associated energy functional $I : H^1_{rad}(\mathbb{R}^3) \to \mathbb{R}$ given by

$$I(u) = \frac{1}{2}a \int_{\mathbb{R}^3} |\nabla u|^2 dx + \frac{b}{4}\left(\int_{\mathbb{R}^3} |\nabla u|^2 dx\right)^2 - \frac{\mu}{p}\int_{\mathbb{R}^3} |u|^p dx - \frac{1}{6}\int_{\mathbb{R}^3} |u|^6 dx,$$

restricted to the following sphere in $L^2(\mathbb{R}^3)$

$$S(c) := \{u \in H^1_{rad}(\mathbb{R}^3) : \|u\|_2 = c\}.$$

Next, we provide the asymptotic behavior of the solution u_c as $c \to +\infty$.

Theorem 4.3 *If u_c is the solution obtained in Theorem 4.2, then* $\lim\limits_{c\to+\infty} I(u_c) = 0.$

The Sobolev critical exponent also leads to the lack of compactness. Even the embedding of the radially symmetric space $H^1_{rad}(\mathbb{R}^3)$ into $L^6(\mathbb{R}^3)$ is not compact. Furthermore, the embedding $H^1_{rad}(\mathbb{R}^3) \hookrightarrow L^2(\mathbb{R}^3)$ is also not compact. Then, the weak limit of Palais-Smale sequences could leave the constrained manifold $S(c)$. Hence, we need to estimate finely the Lagrange multiplier, which is vital in obtaining compactness. With the aid of the concentration-compactness principle, we overcome the difficulty.

No matter $4 < p < \frac{14}{3}$ or $\frac{14}{3} < p < 6$, the energy $I(u)$ on the constrained manifold $S(c)$ is unbounded from below. Hence, it is unlikely to obtain a solution to problem (P) by minimizing method.

4.2 The mass subcritical case

Let us first recall the definition of genus. Let X be a Banach space and A be a subset of X. The set A is said to be symmetric if $u \in A$ implies that $-u \in A$. Denote by Σ the family of closed symmetric subsets A of X such that $0 \notin A$, that is,

$$\Sigma = \{A \subset X\backslash\{0\} : A \text{ is closed and symmetric with respect to the origin}\}.$$

For $A \in \Sigma$, define

$$\gamma(A) = \begin{cases} 0, & \text{if } A = \emptyset, \\ \inf\Big\{k \in \mathbb{N} : \exists \text{ an odd } \varphi \in C\big(A, \mathbb{R}^k\backslash\{0\}\big)\Big\}, & \\ +\infty, & \text{if no such odd map exists,} \end{cases}$$

and $\Sigma_k = \{A \in \Sigma : \gamma(A) \geq k\}$.

For $u \in S(c)$, by the Gagliardo–Nirenberg inequality and the Sobolev embedding theorem we have

$$\begin{aligned} I(u) =& \frac{1}{2}a\int_{\mathbb{R}^3}|\nabla u|^2dx + \frac{b}{4}\left(\int_{\mathbb{R}^3}|\nabla u|^2dx\right)^2 - \frac{\mu}{p}\int_{\mathbb{R}^3}|u|^pdx - \frac{1}{6}\int_{\mathbb{R}^3}|u|^6dx \\ \geq& \frac{1}{2}a\int_{\mathbb{R}^3}|\nabla u|^2dx + \frac{b}{4}\left(\int_{\mathbb{R}^3}|\nabla u|^2dx\right)^2 - \frac{\mu}{p}C_pc^{\frac{6-p}{2}}\|\nabla u\|_2^{\frac{3(p-2)}{2}} - \frac{1}{6S^3}\|\nabla u\|_2^6 \\ =:& M(\|\nabla u\|_2), \end{aligned}$$

where

$$M(t) = \frac{1}{2}at^2 + \frac{b}{4}t^4 - \frac{\mu}{p}C_pc^{\frac{6-p}{2}}t^{\frac{3(p-2)}{2}} - \frac{1}{6S^3}t^6.$$

Since $4 < p < \frac{14}{3}$, we obtain $\frac{3(p-2)}{2} < 4 < 6$. By straightforward computation, there exists $\alpha > 0$ such that as $\mu c^{\frac{6-p}{2}} \leq \alpha$, then the function $M(\cdot)$ attains its positive local maximum. More precisely, there exist two constants $0 < R_1 < R_2 < +\infty$ such that $M(\cdot) < 0$ in the intervals $(0, R_1)$ and $(R_2, +\infty)$, and $M(\cdot) > 0$ in the interval (R_1, R_2). Let $\tau(\cdot) \in C^\infty(\mathbb{R}^+, [0, 1])$ be a nonincreasing function such that $\tau(t) = 1$ for $t \leq R_1$ and $\tau(t) = 0$ for all $t \geq R_2$.

4.2.1 Proof of Theorem 4.1

Define the truncated functional

$$I_\tau(u) = \frac{1}{2}a\int_{\mathbb{R}^3}|\nabla u|^2dx + \frac{b}{4}\left(\int_{\mathbb{R}^3}|\nabla u|^2dx\right)^2 - \frac{\mu}{p}\int_{\mathbb{R}^3}|u|^pdx - \frac{\tau(\|\nabla u\|_2)}{6}\int_{\mathbb{R}^3}|u|^6dx.$$

For $u \in S(c)$, again by the Gagliardo–Nirenberg inequality and the Sobolev embedding theorem one has

$$\begin{aligned} I_\tau(u) \geq& \frac{1}{2}a\int_{\mathbb{R}^3}|\nabla u|^2dx + \frac{b}{4}\left(\int_{\mathbb{R}^3}|\nabla u|^2dx\right)^2 - \frac{\mu}{p}C_pc^{\frac{6-p}{2}}\|\nabla u\|_2^{\frac{3(p-2)}{2}} \\ &- \frac{\tau(\|\nabla u\|_2)}{6S^3}\|\nabla u\|_2^6 =: \hat{M}(\|\nabla u\|_2), \end{aligned}$$

where

$$\hat{M}(t) = \frac{1}{2}at^2 + \frac{b}{4}t^4 - \frac{\mu}{p}C_pc^{\frac{6-p}{2}}t^{\frac{3(p-2)}{2}} - \frac{\tau(t)}{6S^3}t^6.$$

Then the definition of $\tau(\cdot)$ implies that when $c \in \left(0, (\frac{\alpha}{\mu})^{\frac{2}{6-p}}\right]$, $\hat{M}(\cdot) < 0$ in the interval $(0, R_1)$ and $\hat{M}(\cdot) > 0$ in the interval $(R_1, +\infty)$.

In the following, we always assume that $c \in \left(0, (\frac{\alpha}{\mu})^{\frac{2}{6-p}}\right]$. Without loss of generality, we may assume that

$$\frac{1}{2}ar^2 + \frac{b}{4}r^4 - \frac{1}{6S^3}r^6 \geq 0 \text{ for } r \in [0, R_1] \text{ and } R_1^2 < a^{\frac{1}{2}}S^{\frac{3}{2}}. \tag{4.1}$$

Lemma 4.1 *The following properties are fulfilled:*

(i) $I_\tau \in C^1(H^1_{rad}(\mathbb{R}^3), \mathbb{R})$.

(ii) I_τ *is coercive and bounded from below on* $S(c)$. *Moreover, if* $I_\tau \leq 0$, *then* $\|\nabla u\|_2 \leq R_1$ *and* $I_\tau(u) = I(u)$.

(iii) $I_\tau|_{S(c)}$ *satisfies the* $(PS)_d$ *condition for all* $d < 0$.

Proof. The proofs of (i) and (ii) are straightforward. For (iii), let $\{u_n\}$ be a $(PS)_d$ sequence of $I_\tau|_{S(c)}$ with $d < 0$, i.e., $I_\tau(u_n) \to d < 0$ and $\|I_\tau|'_{S(c)}(u_n)\| \to 0$ as $n \to \infty$. By (ii), $\|\nabla u_n\|_2 \le R_1$ for large n, and $\{u_n\}$ is also a $(PS)_d$ sequence of $I|_{S(c)}$ with $d < 0$. Then, $\{u_n\}$ is bounded in $H^1_{rad}(\mathbb{R}^3)$. Hence, up to a subsequence, there exists $u \in H^1_{rad}(\mathbb{R}^3)$ such that $u_n \rightharpoonup u$ in $H^1_{rad}(\mathbb{R}^3)$ and $u_n \to u$ in $L^t(\mathbb{R}^3)$ for $2 < t < 6$ and $u_n(x) \to u(x)$ a.e. $x \in \mathbb{R}^3$. Since $4 < p < \frac{14}{3}$,

$$\lim_{n\to\infty} \int_{\mathbb{R}^3} |u_n|^p dx = \int_{\mathbb{R}^3} |u|^p dx.$$

We claim that $u \not\equiv 0$. Otherwise, $\lim_{n\to\infty} \int_{\mathbb{R}^3} |u_n|^p dx = 0$, and whence by (4.1) we have

$$\begin{aligned} 0 > d &= \lim_{n\to\infty} I(u_n) \\ &= \lim_{n\to\infty} \left[\frac{1}{2} a \int_{\mathbb{R}^3} |\nabla u_n|^2 dx + \frac{b}{4}\left(\int_{\mathbb{R}^3} |\nabla u_n|^2 dx\right)^2 - \frac{\mu}{p}\int_{\mathbb{R}^3} |u_n|^p dx - \frac{1}{6}\int_{\mathbb{R}^3} |u_n|^6 dx\right] \\ &\ge \lim_{n\to\infty} \left[\frac{1}{2} a\|\nabla u_n\|_2^2 + \frac{b}{4}\|\nabla u_n\|_2^4 - \frac{\mu}{p}\int_{\mathbb{R}^3} |u_n|^p dx - \frac{1}{6S^3}\|\nabla u_n\|_2^6\right] \\ &\ge -\frac{\mu}{p}\lim_{n\to\infty} \int_{\mathbb{R}^3} |u_n|^p dx = 0, \end{aligned}$$

a contradiction.

On the other hand, let $\Phi(v) := \frac{1}{2}\int_{\mathbb{R}^3} |v|^2 dx$, $\forall v \in H^1(\mathbb{R}^3)$, then $S(c) = \Phi^{-1}(\{\frac{c^2}{2}\})$. By Proposition 5.12 in Willem [153], there exists a sequence $\{\lambda_n\} \subset \mathbb{R}$ such that $\|I'(u_n) - \lambda_n \Phi'(u_n)\| \to 0$ as $n \to \infty$, which means that

$$-\left(a + b\int_{\mathbb{R}^3} |\nabla u_n|^2 dx\right)\Delta u_n - \mu|u_n|^{p-2}u_n - |u_n|^4 u_n = \lambda_n u_n + o(1) \text{ in } H^{-1}_{rad}(\mathbb{R}^3). \quad (4.2)$$

Therefore, for $\varphi \in H^1_{rad}(\mathbb{R}^3)$,

$$\begin{aligned} &\left(a + b\int_{\mathbb{R}^3} |\nabla u_n|^2 dx\right)\int_{\mathbb{R}^3} \nabla u_n \cdot \nabla\varphi dx - \mu\int_{\mathbb{R}^3} |u_n|^{p-2}u_n\varphi dx - \int_{\mathbb{R}^3} |u_n|^4 u_n\varphi dx \\ = \ &\lambda_n \int_{\mathbb{R}^3} u_n\varphi dx + o(1)\|\varphi\|. \end{aligned} \quad (4.3)$$

It follows that

$$\left(a + b\int_{\mathbb{R}^3} |\nabla u_n|^2 dx\right)\int_{\mathbb{R}^3} |\nabla u_n|^2 dx - \mu\int_{\mathbb{R}^3} |u_n|^p dx - \int_{\mathbb{R}^3} |u_n|^6 dx = \lambda_n c^2 + o(1).$$

The boundedness of $\{\|u_n\|\}$ yields that $\{\lambda_n\}$ is bounded in $\mathbb{R}$. Then, up to a subsequence, there exists $\lambda_c \in \mathbb{R}$ such that $\lambda_n \to \lambda_c$ as $n \to \infty$.

We now claim that $\lambda_c < 0$. Indeed, by the above identity and $I(u_n) \to d < 0$ as $n \to \infty$ combined with $p > 4$, we obtain

$$\lambda_n c^2 + o(1) < 4d + \frac{4-p}{p}\mu\int_{\mathbb{R}^3} |u_n|^p dx - \frac{1}{3}\int_{\mathbb{R}^3} |u_n|^6 dx + o(1) \le 4d + o(1).$$

Letting $n \to \infty$, we have $\lambda_c < 0$.

In the sequel, we prove that $u_n \to u$ in $L^6(\mathbb{R}^3)$ by using the concentration-compactness principle. In fact, there exist two positive measures μ, $\nu \in \mathcal{M}(\mathbb{R}^3)$ such that $|\nabla u_n|^2 \rightharpoonup \mu$ and $|u_n|^6 \rightharpoonup \nu$ in $\mathcal{M}(\mathbb{R}^3)$ as $n \to \infty$, and for an at most countable index set J, we have

$$\begin{cases} \nu = |u|^6 + \Sigma_{j\in J}\nu_j\delta_{x_j}, & \nu_j > 0, \\ \mu \ge |\nabla u|^2 + \Sigma_{j\in J}\mu_j\delta_{x_j}, & \mu_j > 0, \\ S\nu_j^{\frac{1}{3}} \le \mu_j, & \forall j \in J. \end{cases}$$

Suppose that J is nonempty but finite. Let $\varphi_\rho \in C^\infty(\mathbb{R}^3)$ be a cut-off function with $\varphi_\rho(x) = 1$ for $|x-x_j| \le \frac{1}{2}\rho$, $\varphi_\rho(x) = 0$ for $|x-x_j| \ge \rho$ and $|\nabla\varphi_\rho| \le \frac{1}{\sqrt{\rho}}$. By the boundedness of $\{u_n\}$ in $H^1_{rad}(\mathbb{R}^3)$ we know that $\{\varphi_\rho u_n\}$ is also bounded in $H^1_{rad}(\mathbb{R}^3)$. Therefore

$$\begin{aligned} o(1) &= \langle I'(u_n), \varphi_\rho u_n\rangle \\ &= \left(a + b\int_{\mathbb{R}^3} |\nabla u_n|^2 dx\right)\int_{\mathbb{R}^3} \nabla u_n \cdot \nabla(\varphi_\rho u_n)dx - \mu\int_{\mathbb{R}^3} \varphi_\rho |u_n|^p dx \\ &- \int_{\mathbb{R}^3} \varphi_\rho |u_n|^6 dx. \end{aligned}$$

But

$$\int_{\mathbb{R}^3} \nabla u_n \cdot \nabla(\varphi_\rho u_n)dx = \int_{\mathbb{R}^3} |\nabla u_n|^2 \varphi_\rho dx + \int_{\mathbb{R}^3} u_n \nabla u_n \cdot \nabla\varphi_\rho dx.$$

By Hölder's inequality we derive that

$$\begin{aligned} \left|\int_{\mathbb{R}^3} u_n \nabla u_n \cdot \nabla\varphi_\rho dx\right| &\le \left(\int_{\mathbb{R}^3} u_n^2 |\nabla\varphi_\rho|^2 dx\right)^{\frac{1}{2}} \cdot \left(\int_{\mathbb{R}^3} |\nabla u_n|^2 dx\right)^{\frac{1}{2}} \\ &\le C\left(\int_{\mathbb{R}^3} u_n^2 |\nabla\varphi_\rho|^2 dx\right)^{\frac{1}{2}} \\ &= C\left(\int_{B_\rho(x_j)\setminus B_{\frac{\rho}{2}}(x_j)} u_n^2 |\nabla\varphi_\rho|^2 dx\right)^{\frac{1}{2}}. \end{aligned} \tag{4.4}$$

We notice that

$$\begin{aligned} &\left|\int_{B_\rho(x_j)\setminus B_{\frac{\rho}{2}}(x_j)} u_n^2 |\nabla\varphi_\rho|^2 dx - \int_{B_\rho(x_j)\setminus B_{\frac{\rho}{2}}(x_j)} u^2 |\nabla\varphi_\rho|^2 dx\right| \\ \le\ & \frac{1}{\rho}\int_{B_\rho(x_j)\setminus B_{\frac{\rho}{2}}(x_j)} |u_n + u||u_n - u| dx \\ \le\ & \frac{1}{\rho}\left(\int_{\mathbb{R}^3} |u_n - u|^p dx\right)^{\frac{1}{p}} \left(\int_{\mathbb{R}^3} |u_n + u|^6 dx\right)^{\frac{1}{6}} \left(\int_{B_\rho(x_j)} dx\right)^{\frac{5}{6}-\frac{1}{p}} \\ \le\ & C\rho^{3(\frac{5}{6}-\frac{1}{p})-1}\left(\int_{\mathbb{R}^3} |u_n - u|^p dx\right)^{\frac{1}{p}}. \end{aligned}$$

By the absolute continuity of the Lebesgue integral, we obtain

$$\begin{aligned} \lim_{\rho\to 0}\lim_{n\to\infty}\int_{B_\rho(x_j)\setminus B_{\frac{\rho}{2}}(x_j)} u_n^2 |\nabla\varphi_\rho|^2 dx &= \lim_{\rho\to 0}\int_{B_\rho(x_j)\setminus B_{\frac{\rho}{2}}(x_j)} u^2 |\nabla\varphi_\rho|^2 dx \\ &\le \lim_{\rho\to 0}\left[\frac{1}{\rho}\left(\int_{B_\rho(x_j)} |u|^6 dx\right)^{\frac{1}{3}} \left(\int_{B_\rho(x_j)} dx\right)^{\frac{2}{3}}\right] = 0. \end{aligned}$$

By (4.4), we obtain $\lim_{\rho\to 0}\lim_{n\to\infty}\int_{\mathbb{R}^3} u_n \nabla u_n \cdot \nabla\varphi_\rho dx = 0$. Again by the absolute continuity of the Lebesgue integral we have

$$\lim_{\rho\to 0}\lim_{n\to\infty}\int_{\mathbb{R}^3} |u_n|^p \varphi_\rho dx = \lim_{\rho\to 0}\int_{\mathbb{R}^3} |u|^p \varphi_\rho dx = 0.$$

All together with the above estimates, we get that

$$\begin{aligned} \lim_{\rho\to 0}\left[\int_{\mathbb{R}^3} |u|^6 \varphi_\rho dx + \Sigma_{j\in J}\int_{\mathbb{R}^3} \nu_j \delta_{x_j} \varphi_\rho dx\right] &= \lim_{\rho\to 0}\int_{\mathbb{R}^3} \varphi_\rho d\nu = \lim_{\rho\to 0}\lim_{n\to\infty}\int_{\mathbb{R}^3} |u_n|^6 \varphi_\rho dx \\ &= \lim_{\rho\to 0}\lim_{n\to\infty}\left(a+b\int_{\mathbb{R}^3} |\nabla u_n|^2 dx\right)\int_{\mathbb{R}^3} |\nabla u_n|^2 \varphi_\rho dx \\ &\ge a\lim_{\rho\to 0}\lim_{n\to\infty}\int_{\mathbb{R}^3} |\nabla u_n|^2 \varphi_\rho dx = a\lim_{\rho\to 0}\int_{\mathbb{R}^3} \varphi_\rho d\mu \\ &\ge a\lim_{\rho\to 0}\left[\int_{\mathbb{R}^3} |\nabla u|^2 \varphi_\rho dx + \Sigma_{j\in J}\int_{\mathbb{R}^3} \mu_j \delta_{x_j} \varphi_\rho dx\right]. \end{aligned}$$

Hence, $\nu_j \geq a\mu_j$. Then, $\mu_j \geq S\nu_j^{\frac{1}{3}} \geq S(a\mu_j)^{\frac{1}{3}}$, i.e., $\mu_j \geq a^{\frac{1}{2}} S^{\frac{3}{2}}$, so

$$\begin{aligned} R_1^2 &\geq \limsup_{n\to\infty} \|\nabla u_n\|_2^2 \geq \limsup_{n\to\infty} \int_{\mathbb{R}^3} |\nabla u_n|^2 \varphi_\rho dx = \int_{\mathbb{R}^3} \varphi_\rho d\mu \\ &\geq \int_{\mathbb{R}^3} |\nabla u|^2 \varphi_\rho dx + \Sigma_{j\in J} \int_{\mathbb{R}^3} \mu_j \delta_{x_j} \varphi_\rho dx \geq \mu_j \geq a^{\frac{1}{2}} S^{\frac{3}{2}}, \end{aligned}$$

a contradiction with (4.1). Thus, $J = \emptyset$. Then,

$$\lim_{n\to\infty} \int_{\mathbb{R}^3} |u_n|^6 \varphi_\rho dx = \int_{\mathbb{R}^3} \varphi_\rho d\nu = \int_{\mathbb{R}^3} |u|^6 \varphi_\rho dx,$$

so $u_n \to u$ in $L^6_{loc}(\mathbb{R}^3)$.

On the other hand, since $u_n \in H^1_{rad}(\mathbb{R}^3)$ and $\{u_n\}$ is bounded in $H^1(\mathbb{R}^3)$, $|u_n(x)| \leq \frac{\|u_n\|}{|x|} \leq \frac{C}{|x|}$ for every $|x| \geq 1$. So $|u_n(x)|^6 \leq \frac{C}{|x|^6}$ for every $|x| \geq 1$. Since $\frac{1}{|\cdot|^6} \in L^1(\mathbb{R}^3 \backslash B_R(0)$ and $u_n(x) \to u(x)$ a.e. on $\mathbb{R}^3\backslash B_R(0)$, by the Lebesgue dominated convergence theorem, $u_n \to u$ in $L^6(\mathbb{R}^3\backslash B_R(0))$.

Consequently, $u_n \to u$ in $L^6(\mathbb{R}^3)$. Thus, up to a subsequence if necessary, set $B := \lim_{n\to\infty} \|\nabla u_n\|_2^2 \geq 0$. Then $0 < \|\nabla u\|_2^2 \leq B$. By (4.3),

$$(a+bB)\int_{\mathbb{R}^3} \nabla u \cdot \nabla \varphi dx - \mu \int_{\mathbb{R}^3} |u|^{p-2} u\varphi dx - \int_{\mathbb{R}^3} |u|^4 u\varphi dx = \lambda_c \int_{\mathbb{R}^3} u\varphi dx.$$

It follows that

$$\begin{aligned} &\lim_{n\to\infty} \left[\left(a + b\int_{\mathbb{R}^3} |\nabla u_n|^2 dx\right) \int_{\mathbb{R}^3} |\nabla u_n|^2 dx - \lambda_n \|u_n\|_2^2 \right] \\ &= \lim_{n\to\infty} \left[\mu \|u_n\|_p^p + \|u_n\|_6^6\right] \\ &= \mu\|u\|_p^p + \|u\|_6^6 = \left(a + bB\right) \int_{\mathbb{R}^3} |\nabla u|^2 dx - \lambda_c \|u\|_2^2. \end{aligned}$$

By $\lambda_c < 0$ we deduce that

$$\lim_{n\to\infty} -\lambda_c \|u_n\|_2^2 = -\lambda_c \|u\|_2^2, \quad a \lim_{n\to\infty} \int_{\mathbb{R}^3} |\nabla u_n|^2 dx = a \int_{\mathbb{R}^3} |\nabla u|^2 dx.$$

Hence, $u_n \to u$ in $H^1_{rad}(\mathbb{R}^3)$ and $\|u\|_2 = c$. □

For $\varepsilon > 0$, set

$$I_\tau^{-\varepsilon} = \{u \in H^1_{rad}(\mathbb{R}^3) \cap S(c) : I_\tau(u) \leq -\varepsilon\} \subset H^1_{rad}(\mathbb{R}^3).$$

Since I_τ is even and continuous on $H^1_{rad}(\mathbb{R}^3)$, it follows that $I_\tau^{-\varepsilon}$ is closed and symmetric. Then the following property holds.

Lemma 4.2 *Given $n \in \mathbb{N}$, there exist $\varepsilon_n := \varepsilon(n) > 0$ and $\mu_n := \mu(n) > 0$ such that $\gamma(I_\tau^{-\varepsilon}) \geq n$ for $0 < \varepsilon \leq \varepsilon_n$ and $\mu \geq \mu_n$.*

Set

$$\Sigma_k := \{D \subset H^1_{rad}(\mathbb{R}^3) \cap S(c) : D \text{ is closed and symmetric, } \gamma(D) \geq k\},$$

and $d_k := \inf_{D\in\Sigma_k} \sup_{u\in D} I_\tau(u) > -\infty$ for all $k \in \mathbb{N}$ by Lemma 4.1 (*ii*) and

$$K_d := \{u \in H^1_{rad}(\mathbb{R}^3) \cap S(c) : I'_\tau(u) = 0, I_\tau(u) = d\}.$$

Lemma 4.3 *If $d = d_k = d_{k+1} = \cdots = d_{k+r}$, then $\gamma(K_d) \geq r+1$. In particular, I_τ has at least $r+1$ nontrivial critical points.*

Proof. For $\varepsilon > 0$, we observe that $I_\tau^{-\varepsilon} \in \Sigma$. For any $k \in \mathbb{N}$, by the previous lemma there exist $\varepsilon_k = \varepsilon(k) > 0$ and $\mu_k = \mu(k) > 0$ such that if $0 < \varepsilon \leq \varepsilon_k$ and $\mu \geq \mu_k$, we get $\gamma(I_\tau^{-\varepsilon}) \geq k$. Then $I_\tau^{-\varepsilon_k} \in \Sigma_k$, and $d_k \leq \sup\limits_{u \in I_\tau^{-\varepsilon_k}} I_\tau(u) = -\varepsilon_k < 0$. Suppose that $0 > d = d_k = d_{k+1} = \cdots = d_{k+r}$, then Lemma 4.1 *(iii)* implies that I_τ satisfies the $(PS)_d$ condition. Hence, K_d is a compact set. We conclude that $I_\tau|_{S(c)}$ possesses at least $r+1$ critical points. □

Proof of Theorem 4.1 concluded. By Lemma 4.1 *(ii)*, the critical points of I_τ founded in Lemma 4.3 are the critical points of I. So, Theorem 4.1 is proved. □

4.3 The mass supercritical case

In this section, we consider the case $\frac{14}{3} < p < 6$. At this time, $\frac{3(p-2)}{2} > 4$. It follows that the truncated functional I_τ is still unbounded from below on $S(c)$. Therefore, we cannot use the truncation technique to study problem (P) when $\frac{14}{3} < p < 6$. In this setting, we define $f(t) = \mu|t|^{p-2}t + |t|^4 t$ for all $t \in \mathbb{R}$, and introduce the following auxiliary functional

$$\mathcal{I} : S(c) \times \mathbb{R} \to \mathbb{R}, \quad (u, \tau) \mapsto I(\tau * u),$$

where $(\tau * u)(x) := e^{\frac{3}{2}\tau} u(e^\tau x)$. Then

$$\int_{\mathbb{R}^3} |\nabla(\tau * u)|^2 dx = e^{2\tau} \int_{\mathbb{R}^3} |\nabla u|^2 dx$$

and

$$\int_{\mathbb{R}^3} |\tau * u|^q dx = e^{\frac{q-2}{2}3\tau} \int_{\mathbb{R}^3} |u|^q dx, \quad \forall q \in [2, 6].$$

We have

$$\begin{aligned}
\mathcal{I}(u,\tau) &= I(\tau * u) = I\big(e^{\frac{3}{2}\tau} u(e^\tau x)\big) \\
&= \frac{1}{2} a \int_{\mathbb{R}^3} \left|\nabla(\tau * u)\right|^2 dx + \frac{b}{4}\Big(\int_{\mathbb{R}^3} \left|\nabla(\tau * u)\right|^2 dx\Big)^2 - \int_{\mathbb{R}^3} F(\tau * u) dx \\
&= \frac{1}{2} a e^{2\tau} \int_{\mathbb{R}^3} |\nabla u|^2 dx + \frac{b}{4} e^{4\tau}\Big(\int_{\mathbb{R}^3} |\nabla u|^2 dx\Big)^2 - e^{-3\tau} \int_{\mathbb{R}^3} F\Big(e^{\frac{3\tau}{2}} u(x)\Big) dx \\
&= \frac{1}{2} a e^{2\tau} \int_{\mathbb{R}^3} |\nabla u|^2 dx + \frac{b}{4} e^{4\tau}\Big(\int_{\mathbb{R}^3} |\nabla u|^2 dx\Big)^2 - \frac{\mu}{p} \cdot e^{\frac{3(p-2)}{2}\tau} \int_{\mathbb{R}^3} |u|^p dx \\
&\quad - \frac{1}{6} \cdot e^{6\tau} \int_{\mathbb{R}^3} |u|^6 dx.
\end{aligned}$$

The above estimates imply the following asymptotic properties.

Lemma 4.4 *([162], Lemma 3.1) Let $u \in S(c)$ be arbitrary but fixed. Then*

(i) $\displaystyle\int_{\mathbb{R}^3} \left|\nabla(\tau * u)\right|^2 dx \to 0$ *and* $\mathcal{I}(u,\tau) \to 0$ *as* $\tau \to -\infty$.

(ii) $\displaystyle\int_{\mathbb{R}^3} \left|\nabla(\tau * u)\right|^2 dx \to +\infty$ *and* $\mathcal{I}(u,\tau) \to -\infty$ *as* $\tau \to +\infty$.

With the aid of the Gagliardo–Nirenberg inequality we can obtain the next property; see Lemma 3.2 of Zhang and Han [162] for more details.

Lemma 4.5 *There exists $K(c) > 0$ sufficiently small such that*

$$I(u) > 0 \quad \textit{for } u \in A \textit{ and } 0 < \sup_{u\in A} I(u) < \inf_{u\in B} I(u),$$

where

$$A := \left\{u \in S(c) : \int_{\mathbb{R}^3} |\nabla u|^2 dx \leq K(c)\right\} \quad \textit{and} \quad B := \left\{u \in S(c) : \int_{\mathbb{R}^3} |\nabla u|^2 dx = 2K(c)\right\}.$$

As a consequence of Lemmas 4.4 and 4.5, we observe that for fixed $u_0 \in S(c)$, there exist two constants τ_1, τ_2 satisfying $\tau_1 < 0 < \tau_2$ such that

$$\int_{\mathbb{R}^3} |\nabla u_1|^2 dx < \frac{K(c)}{2}, \ \int_{\mathbb{R}^3} |\nabla u_2|^2 dx > 2K(c)$$

and

$$I(u_1) > 0, \ I(u_2) < 0,$$

where $u_1 := \tau_1 * u_0 \in S(c)$ and $u_2 := \tau_2 * u_0 \in S(c)$.

In the following, define the mountain pass level $\gamma_\mu(c)$ by

$$\gamma_\mu(c) := \inf_{g\in\Gamma} \max_{t\in[0,1]} I(g(t)),$$

where

$$\Gamma := \{g \in C([0,1], S(c)) : g(0) = u_1, g(1) = u_2\}.$$

Then for any $g \in \Gamma$,

$$\max_{t\in[0,1]} I(g(t)) > \max\{I(u_1), I(u_2)\}.$$

It follows that $\gamma_\mu(c) > 0$. Then $\gamma_\mu(c)$ satisfies the following property.

Lemma 4.6 $\lim\limits_{\mu\to+\infty} \gamma_\mu(c) = 0.$

Proof. Taking $g_0(t) := \big[(1-t)\tau_1 + t\tau_2\big] * u_0 \in \Gamma$, then

$$\begin{aligned}
0 \quad &< \gamma_\mu(c) \leq \max_{t\in[0,1]} I(g_0(t)) = \max_{t\in[0,1]} I\Big(\big[(1-t)\tau_1 + t\tau_2\big] * u_0\Big) \\
&= \max_{t\in[0,1]} \Big\{\frac{1}{2} a e^{2[(1-t)\tau_1+t\tau_2]} \int_{\mathbb{R}^3} |\nabla u_0|^2 dx + \frac{b}{4} e^{4[(1-t)\tau_1+t\tau_2]} \Big(\int_{\mathbb{R}^3} |\nabla u_0|^2 dx\Big)^2 \\
&-\tfrac{\mu}{p} \cdot e^{\frac{3(p-2)}{2}[(1-t)\tau_1+t\tau_2]} \textstyle\int_{\mathbb{R}^3} |u_0|^p dx - \frac{1}{6} \cdot e^{6[(1-t)\tau_1+t\tau_2]} \int_{\mathbb{R}^3} |u_0|^6 dx\Big\}.
\end{aligned}$$

It follows that

$$\begin{aligned}
\gamma_\mu(c) \quad &\leq \max_{r\geq 0} \Big\{\frac{1}{2} a r^2 \int_{\mathbb{R}^3} |\nabla u_0|^2 dx + \frac{b}{4} r^4 \Big(\int_{\mathbb{R}^3} |\nabla u_0|^2 dx\Big)^2 - \frac{\mu}{p} \cdot r^{\frac{3(p-2)}{2}} \int_{\mathbb{R}^3} |u_0|^p dx \\
&-\frac{1}{6} r^6 \int_{\mathbb{R}^3} |u_0|^6 dx\Big\} \\
&\leq \max_{r\geq 0} \Big\{\frac{1}{2} a r^2 \int_{\mathbb{R}^3} |\nabla u_0|^2 dx + \frac{b}{4} r^4 \Big(\int_{\mathbb{R}^3} |\nabla u_0|^2 dx\Big)^2 - \frac{\mu}{p} \cdot r^{\frac{3(p-2)}{2}} \int_{\mathbb{R}^3} |u_0|^p dx\Big\} \\
&=: A.
\end{aligned}$$

We distinguish the following two cases.

Case 1. We have $r^2 \int_{\mathbb{R}^3} |\nabla u_0|^2 dx \geq 1$. Then

$$\begin{aligned} A &\leq \max_{r\geq 0} \Big\{ \frac{1}{2} a r^4 \left(\int_{\mathbb{R}^3} |\nabla u_0|^2 dx \right)^2 + \frac{b}{4} r^4 \left(\int_{\mathbb{R}^3} |\nabla u_0|^2 dx \right)^2 - \frac{\mu}{p} \cdot r^{\frac{3(p-2)}{2}} \int_{\mathbb{R}^3} |u_0|^p dx \Big\} \\ &\leq C \left(\frac{1}{\mu} \right)^{\frac{8}{3p-14}} \to 0 \text{ (as } \mu \to +\infty). \end{aligned}$$

Case 2. We have $r^2 \int_{\mathbb{R}^3} |\nabla u_0|^2 dx < 1$. Then

$$\begin{aligned} A &\leq \max_{r\geq 0} \Big\{ \frac{1}{2} a r^2 \int_{\mathbb{R}^3} |\nabla u_0|^2 dx + \frac{b}{4} r^2 \int_{\mathbb{R}^3} |\nabla u_0|^2 dx - \frac{\mu}{p} \cdot r^{\frac{3(p-2)}{2}} \int_{\mathbb{R}^3} |u_0|^p dx \Big\} \\ &\leq C \left(\frac{1}{\mu} \right)^{\frac{4}{3p-10}} \to 0 \text{ (as } \mu \to +\infty). \end{aligned}$$

The proof is now complete. □

By Proposition 3.5 in [162], there exists a sequence $\{u_n\} \subset S(c)$ satisfying

$$I(u_n) \to \gamma_\mu(c), \ \|I'|_{S(c)}(u_n)\| \to 0 \text{ and } Q(u_n) \to 0$$

as $n \to \infty$, where

$$Q(u_n) = a \int_{\mathbb{R}^3} |\nabla u_n|^2 dx + b \left(\int_{\mathbb{R}^3} |\nabla u_n|^2 dx \right)^2 + 3 \int_{\mathbb{R}^3} F(u_n) dx - \frac{3}{2} \int_{\mathbb{R}^3} f(u_n) u_n dx.$$

We deduce that $\{u_n\}$ is bounded in $H^1_{rad}(\mathbb{R}^3)$.

Set $\Phi(v) := \frac{1}{2} \int_{\mathbb{R}^3} |v|^2 dx$, $\forall v \in H^1(\mathbb{R}^3)$. Then $S(c) = \Phi^{-1}\big(\{\frac{c^2}{2}\}\big)$. Thus, there exists a sequence $\{\lambda_n\} \subset \mathbb{R}$ such that

$$\|I'(u_n) - \lambda_n \Phi'(u_n)\| \to 0$$

as $n \to \infty$, which implies that

$$-\left(a + b \int_{\mathbb{R}^3} |\nabla u_n|^2 dx \right) \Delta u_n - f(u_n) = \lambda_n u_n + o(1) \text{ in } H^{-1}_{rad}(\mathbb{R}^3). \tag{4.5}$$

We conclude that for all $\varphi \in H^1_{rad}(\mathbb{R}^3)$,

$$\left(a + b \int_{\mathbb{R}^3} |\nabla u_n|^2 dx \right) \int_{\mathbb{R}^3} \nabla u_n \cdot \nabla \varphi dx - \int_{\mathbb{R}^3} f(u_n) \varphi dx = \lambda_n \int_{\mathbb{R}^3} u_n \varphi dx + o(1)\|\varphi\|. \tag{4.6}$$

We can estimate λ_n as follows.

Lemma 4.7 *The sequence $\{\lambda_n\}$ is bounded in $\mathbb{R}$ and $\lambda_n = -\frac{1}{c^2} \cdot \frac{6-p}{2p} \mu \int_{\mathbb{R}^3} |u_n|^p dx + o(1)$.*

Proof. By relation (4.6) and since $u_n \in S(c)$, we obtain

$$\left(a + b \int_{\mathbb{R}^3} |\nabla u_n|^2 dx \right) \int_{\mathbb{R}^3} |\nabla u_n|^2 dx - \int_{\mathbb{R}^3} f(u_n) u_n dx = \lambda_n \int_{\mathbb{R}^3} |u_n|^2 dx + o(1) = \lambda_n c^2 + o(1),$$

which implies

$$\lambda_n = \frac{1}{c^2} \Big[\left(a + b \int_{\mathbb{R}^3} |\nabla u_n|^2 dx \right) \int_{\mathbb{R}^3} |\nabla u_n|^2 dx - \int_{\mathbb{R}^3} f(u_n) u_n dx \Big] + o(1).$$

By the boundedness of $\{u_n\}$ in $H^1_{rad}(\mathbb{R}^3)$, we know that $\{\lambda_n\}$ is bounded in $\mathbb{R}$. Moreover, combining with $Q(u_n) \to 0$ as $n \to \infty$ we see that

$$\begin{aligned}\lambda_n &= \frac{1}{c^2}\Big[\frac{3}{2}\int_{\mathbb{R}^3} f(u_n)u_n dx - 3\int_{\mathbb{R}^3} F(u_n)dx - \int_{\mathbb{R}^3} f(u_n)u_n dx\Big] + o(1)\\ &= \frac{1}{c^2}\Big[\frac{1}{2}\int_{\mathbb{R}^3} f(u_n)u_n dx - 3\int_{\mathbb{R}^3} F(u_n)dx\Big] + o(1)\\ &= \frac{1}{c^2}\Big[\frac{1}{2}\int_{\mathbb{R}^3} (\mu|u_n|^p + |u_n|^6)dx - 3\int_{\mathbb{R}^3} \left(\frac{\mu}{p}|u_n|^p + \frac{1}{6}|u_n|^6\right)dx\Big] + o(1)\\ &= -\frac{1}{c^2}\cdot\frac{6-p}{2p}\cdot\mu\int_{\mathbb{R}^3}|u_n|^p dx + o(1).\end{aligned}$$

From the boundedness of $\{u_n\}$ in $H^1_{rad}(\mathbb{R}^3)$, up to a subsequence, there exists $u \in H^1_{rad}(\mathbb{R}^3)$ such that $u_n \rightharpoonup u$ in $H^1_{rad}(\mathbb{R}^3)$ and $u_n \to u$ in $L^t(\mathbb{R}^3)$ for $2 < t < 6$ and $u_n(x) \to u(x)$ a.e. $x \in \mathbb{R}^3$. Since $\frac{14}{3} < p < 6$, then

$$\lim_{n\to\infty}\int_{\mathbb{R}^3}|u_n|^p dx = \int_{\mathbb{R}^3}|u|^p dx, \tag{4.7}$$

which completes the proof. □

Lemma 4.8 *There exists $\mu^* = \mu^*(c) > 0$ such that $u \neq 0$ for all $\mu \geq \mu^*$.*

Proof. Suppose that $u = 0$. Then taking into account relation (4.7) and Lemma 4.7 one has $\lim_{n\to\infty}\int_{\mathbb{R}^3}|u_n|^p dx = 0$ and $\lim_{n\to\infty}\lambda_n = 0$. Combining with (4.6), we get that

$$a\int_{\mathbb{R}^3}|\nabla u_n|^2 dx + b\left(\int_{\mathbb{R}^3}|\nabla u_n|^2 dx\right)^2 - \|u_n\|_6^6 = o(1).$$

Up to a subsequence, we have

$$a\int_{\mathbb{R}^3}|\nabla u_n|^2 dx + b\left(\int_{\mathbb{R}^3}|\nabla u_n|^2 dx\right)^2 \to l \geq 0$$

and

$$\|u_n\|_6^6 \to l$$

as $n \to \infty$.

If $l = 0$, we deduce from the expression of $I(u_n)$ that $\gamma_\mu(c) = 0$, which is a contradiction. Hence, $l > 0$. By the definition of S, we have

$$S \leq \frac{\int_{\mathbb{R}^3}|\nabla u_n|^2 dx}{\|u_n\|_6^2} \leq \frac{1}{a}\cdot\frac{a\int_{\mathbb{R}^3}|\nabla u_n|^2 dx + b\left(\int_{\mathbb{R}^3}|\nabla u_n|^2 dx\right)^2}{\|u_n\|_6^2} \to \frac{1}{a}\cdot l^{\frac{2}{3}}$$

as $n \to \infty$. It follows that $l \geq a^{\frac{3}{2}}S^{\frac{3}{2}}$. Consequently, relation (3.17) yields

$$\begin{aligned}\gamma_\mu(c) &= \lim_{n\to\infty} I(u_n) = \lim_{n\to\infty}\Big\{\frac{1}{2}a\int_{\mathbb{R}^3}|\nabla u_n|^2 dx + \frac{b}{4}\left(\int_{\mathbb{R}^3}|\nabla u_n|^2 dx\right)^2 - \frac{\mu}{p}\int_{\mathbb{R}^3}|u_n|^p dx\\ &\quad -\frac{1}{6}\int_{\mathbb{R}^3}|u_n|^6 dx\Big\} \geq \frac{1}{12}l \geq \frac{1}{12}a^{\frac{3}{2}}S^{\frac{3}{2}},\end{aligned}$$

a contradiction. This completes the proof. □

Subsequently, by virtue of the concentration-compactness principle, we can obtain the following lemma. The proof is similar to the proof of Lemma 4.1 (*iii*) and we omit it here.

Lemma 4.9 *$u_n \to u$ in $L^6(\mathbb{R}^3)$ for $\mu \geq \mu^*$.*

4.3.1 Proof of Theorem 4.2

Fix $\mu \geq \mu^*$. By Lemma 4.7, we may assume that $\lambda_n \to \lambda_c$ as $n \to \infty$. Combining with Lemma 4.8 and (3.17), we deduce that

$$\lim_{n\to\infty} \lambda_n = -\frac{1}{c^2} \cdot \frac{6-p}{2p}\mu \lim_{n\to\infty} \int_{\mathbb{R}^3} |u_n|^p dx = -\frac{1}{c^2} \cdot \frac{6-p}{2p}\mu \int_{\mathbb{R}^3} |u|^p dx < 0,$$

and so $\lambda_c < 0$. By relation (4.6) together with $\lambda_c < 0$, we obtain

$$\lim_{n\to\infty} a \int_{\mathbb{R}^3} |\nabla u_n|^2 dx = a \int_{\mathbb{R}^3} |\nabla u|^2 dx \text{ and } \lim_{n\to\infty} \|u_n\|_2^2 = \|u\|_2^2.$$

Hence, $u_n \to u$ in $H^1_{rad}(\mathbb{R}^3)$ and $\|u\|_2 = c$. The proof is now complete. □

4.4 Asymptotic analysis in the mass supercritical case

In this section, we study the asymptotic behavior when c tends to infinity and provided that $\frac{14}{3} < p < 6$. By Theorem 4.2, for any $c > 0$, there exists $\mu^* = \mu^*(c) > 0$ such that as $\mu \geq \mu^*$, problem (P) has a solution $(u_c, \lambda_c) \in H^1(\mathbb{R}^3) \times \mathbb{R}$ with $\int_{\mathbb{R}^3} |u_c|^2 dx = c^2$ and $\lambda_c > 0$. By the proof, we also know that $I(u_c) = \gamma_\mu(c)$.

4.4.1 Proof of Theorem 4.3

We start with some auxiliary properties.

Lemma 4.10 *For any $u \in H^1(\mathbb{R}^3)\backslash\{0\}$, the following properties hold:*

(i) there exists a unique $\tau(u) \in \mathbb{R}$ such that $Q(\tau(u) * u) = 0$;

(ii) $I(\tau(u) * u) > I(\tau * u)$ for any $\tau \neq \tau(u)$.

Proof. (i) We first observe that

$$\begin{aligned}\frac{d}{d\tau} I(\tau * u) &= a e^{2\tau} \int_{\mathbb{R}^3} |\nabla u|^2 dx + b e^{4\tau} \left(\int_{\mathbb{R}^3} |\nabla u|^2 dx \right)^2 - e^{6\tau} \int_{\mathbb{R}^3} |u|^6 dx \\ &\quad - \frac{3\mu(p-2)}{2p} e^{\frac{3(p-2)}{2}\tau} \int_{\mathbb{R}^3} |u|^p dx = Q(\tau * u).\end{aligned}$$

It follows from Lemma 4.4 that $I(\tau * u)$ reaches the global maximum at some $\tau(u) \in \mathbb{R}$, and so

$$Q(\tau(u) * u) = \frac{d}{d\tau} I(\tau(u) * u) = 0.$$

In the sequel, we prove that such $\tau(u)$ is unique. By contradiction, suppose that there exist $\tau_1(u) < \tau_2(u)$ such that $Q(\tau_i(u) * u) = 0$, where $i = 1,\ 2$. By straightforward computation, we obtain

$$a\left[\frac{1}{e^{2\tau_1(u)}} - \frac{1}{e^{2\tau_2(u)}}\right] \int_{\mathbb{R}^3} |\nabla u|^2 dx = \frac{3\mu(p-2)}{2p} \left[e^{\frac{3(p-2)}{2}\tau_1(u) - 4\tau_1(u)} - e^{\frac{3(p-2)}{2}\tau_2(u) - 4\tau_2(u)} \right] \int_{\mathbb{R}^3} |u|^p dx + \left[e^{2\tau_1(u)} - e^{2\tau_2(u)}\right] \int_{\mathbb{R}^3} |u|^6 dx,$$

which is a contradiction due to the fact that $\frac{3(p-2)}{2} > 4$.

(ii) This is a direct consequence of (i). □

Set

$$\mathcal{Q}(c) = \{u \in S(c) : Q(u) = 0\}.$$

Then the following property holds.

Lemma 4.11 $\gamma_\mu(c) \leq \inf_{u \in \mathcal{Q}(c)} I(u)$.

Proof. For any $u \in \mathcal{Q}(c)$, we have $u \in S(c)$ and $Q(u) = 0$. The above argument between Lemmas 4.5 and 4.6 implies that there exist $\tau_1(u) < 0 < \tau_2(u)$ such that

$$\tau_1(u) * u \in S(c) \text{ and } \tau_2(u) * u \in S(c).$$

Define

$$g(t) = [(1-t)\tau_1(u) + t\tau_2(u)] * u \in \Gamma, \ \forall t \in [0,1].$$

Then, by Lemma 4.10 we get that

$$\gamma_\mu(c) \leq \max_{t \in [0,1]} I(g(t)) = \max_{t \in [0,1]} I([(1-t)\tau_1(u) + t\tau_2(u)] * u) = I(u),$$

which is our conclusion. □

Lemma 4.12 $\gamma_\mu(c) \to 0$ *as* $c \to +\infty$.

Proof. Take $u \in S_1 \cap L^\infty(\mathbb{R}^N)$ and set $u_c := cu \in S(c)$ for any $c > 1$. By virtue of Lemma 4.10 (i), there exists a unique $\tau_1(c) \in \mathbb{R}$ such that $\tau_1(c) * u_c \in \mathcal{Q}(c)$. As a consequence, by Lemma 4.11 one has

$$0 < \gamma_\mu(c) \leq \inf_{u \in \mathcal{Q}(c)} I(u) \leq I(\tau_1(c) * u_c) \leq \frac{ac^2}{2} e^{2\tau_1(c)} \int_{\mathbb{R}^3} |\nabla u|^2 dx + \frac{b}{4} c^4 e^{4\tau_1(c)} \left(\int_{\mathbb{R}^3} |\nabla u|^2 dx \right)^2.$$

To conclude the proof, it suffices to show that

$$\lim_{c \to +\infty} c^2 e^{2\tau_1(c)} = 0.$$

As a matter of fact, by $Q(\tau_1(c) * u_c) = 0$ we know that

$$\begin{aligned} & ae^{2\tau_1(c)} c^2 \int_{\mathbb{R}^3} |\nabla u|^2 dx + be^{4\tau_1(c)} c^4 \left(\int_{\mathbb{R}^3} |\nabla u|^2 dx \right)^2 \\ &= \frac{3\mu(p-2)}{2p} e^{\frac{3(p-2)}{2}\tau_1(c)} c^p \int_{\mathbb{R}^3} |u|^p dx + e^{6\tau_1(c)} c^6 \int_{\mathbb{R}^3} |u|^6 dx, \end{aligned}$$

namely

$$\begin{aligned} & a \int_{\mathbb{R}^3} |\nabla u|^2 dx + be^{2\tau_1(c)} c^2 \left(\int_{\mathbb{R}^3} |\nabla u|^2 dx \right)^2 \\ &= \frac{3\mu(p-2)}{2p} e^{[\frac{3(p-2)}{2} - 2]\tau_1(c)} c^{p-2} \int_{\mathbb{R}^3} |u|^p dx + e^{4\tau_1(c)} c^4 \int_{\mathbb{R}^3} |u|^6 dx \\ &= \frac{3\mu(p-2)}{2p} [c^2 e^{2\tau_1(c)}]^{\frac{3p-10}{4}} c^{\frac{6-p}{2}} \int_{\mathbb{R}^3} |u|^p dx + e^{4\tau_1(c)} c^4 \int_{\mathbb{R}^3} |u|^6 dx. \end{aligned}$$

Since $\lim_{c \to +\infty} c^{\frac{6-p}{2}} = +\infty$, we conclude that $\lim_{c \to +\infty} c^2 e^{2\tau_1(c)} = 0$. □

The proof of Theorem 4.3 is now a direct consequence of Lemma 4.12.

4.5 Historical comments

Consider the Kirchhoff equation

$$-\left(a+b\int_{\mathbb{R}^3}|\nabla u|^2dx\right)\Delta u+\lambda u=f(u)\qquad \text{in } \mathbb{R}^3, \tag{4.8}$$

where $a,\ b>0$ and $\lambda\in\mathbb{R}$. The presence of the nonlocal term $\left(\int_{\mathbb{R}^3}|\nabla u|^2dx\right)\Delta u$ causes several technical difficulties that make the mathematical analysis of problem (4.8) particularly interesting. At the same time, the interest of studying Kirchhoff-type equations comes from the physical background of such problems. Usually, u denotes the displacement, $f(u)$ is the external force, b is the initial tension, while a is related to the intrinsic properties of the string, such as the Young modulus.

This first model of the Kirchhoff equation is without forcing term and with Dirichlet boundary conditions and it describes the transversal free vibrations of a clamped string in which the dependence of the tension on the deformation cannot be neglected. This is a quasilinear partial differential equation, namely the nonlinear part of the equation contains as many derivatives as the linear differential operator. The Kirchhoff equation is an extension of the classical d'Alembert wave equation for free vibrations of elastic strings. Kirchhoff's model takes into account the changes in length of the string produced by transverse vibrations.

There are two substantially different viewpoints in terms of the frequency λ in problem (4.8). One is to regard the frequency λ as a given constant. In this situation, solutions of problem (4.8) are critical points of the corresponding action functional on the working space. The other point of view is to regard the frequency λ as an unknown quantity to the problem (4.8). In this situation, it is natural to prescribe the value of the mass so that λ can be interpreted as a Lagrange multiplier. Nowadays, some physicists are very interested in the solutions satisfying

$$\int_{\mathbb{R}^3}|u|^2dx=c^2>0$$

for *a priori* given c, since the mass admits a clear physical meaning. For example, from a physical point of view, the mass $\|u\|_2^2$ may represent the number of particles of each component in Bose-Einstein condensates or the power supply in the nonlinear optics framework. In addition, such solutions can give a better insight of the dynamical properties, like orbital stability or instability, and can describe attractive Bose-Einstein condensates.

In order to study solutions of problem (4.8) satisfying the normalized condition $\int_{\mathbb{R}^3}|u|^2dx=c^2$, it suffices to consider the critical point of the functional

$$E(u)=\frac{1}{2}a\int_{\mathbb{R}^3}|\nabla u|^2dx+\frac{b}{4}\left(\int_{\mathbb{R}^3}|\nabla u|^2dx\right)^2-\int_{\mathbb{R}^3}F(u)dx$$

on the constrained manifold

$$S(c):=\left\{u\in H^1(\mathbb{R}^3):\int_{\mathbb{R}^3}|u|^2dx=c^2\right\}.$$

For instance, Guo, Zhang and Zou [65] studied the existence and blow-up behavior of the normalized solutions for the equation

$$-\left(a+b\int_{\mathbb{R}^N}|\nabla u|^2dx\right)\Delta u+V(x)u=\mu|u|^{p-2}u+\lambda u, \tag{4.9}$$

where $1 \leq N \leq 4$ and $2 < p < 2^*$. Ye [158] pointed out that $2 + \frac{8}{N}$ is the L^2-critical exponent for (4.9). In Ye [157] it is the existence of normalized solutions to problem (4.9) with $p = 2 + \frac{8}{N}$ and $V \equiv 0$. Chen, Rădulescu and Tang [27] established the existence of the normalized solutions for Kirchhoff-type equation

$$-\left(a + b \int_{\mathbb{R}^3} |\nabla u|^2 dx\right) \Delta u - \lambda u = K(x) f(u),$$

where $f \in C(\mathbb{R}, \mathbb{R})$ satisfies general L^2-supercritical or L^2-subcritical condition. When $\frac{14}{3} < p < 6$, Zhang and Han [162] considered the existence of normalized solutions for problem (P) by calculating the threshold of the mountain pass level. The approach developed in this chapter is different and easier than their method. Alternatively, we also consider the L^2-subcritical case. We point out that $p > 4$ only ensures the corresponding Lagrange multiplier is negative, see the proof of Lemma 4.1 *(iii)*.

4.6 Final remarks, perspectives, and open problems

This chapter establishes a foundation for Kirchhoff equations with Sobolev critical exponent in the framework of normalized solutions. The analysis is developed in the three-dimensional case but it can be extended with similar arguments to higher dimensions $N \geq 3$. The novelty in the study developed here is that it mixes the *nonlocal term* $a + b\|\nabla u\|_2^2$, the *Sobolev critical term* $|u|^{2^*-2}u$, and the *L^2-constraint*, with the exponent p in both subcritical and supercritical regimes in the L^2-sense.

The content of this chapter is based on the results established in [93]. Related properties on normalized solutions can be found in [69], [156] (for Choquard equations) and [31], [28] (for Schrödinger-type equations with critical exponential growth) [94] (for critical Schrödinger equations), and [97] (for Schrödinger-Bopp-Podolsky systems with logarithmic reaction).

Some interesting open problems associated with the analysis developed in this chapter are stated in what follows.

The first problem proposes to refine the analysis in the case of a general nonlocal term.

Open Problem 1. The analysis carried out in this chapter is done for the nonlocal term $M(s) = a + bs$. For more general M satisfying certain conditions, can one still get normalized solutions with Sobolev critical exponent? What conditions on M ensure that the mountain pass energy is below the critical threshold for compactness?

If $N = 2$, then the Sobolev critical exponent 2^* is $+\infty$. In this case, one can study Trudinger–Moser critical growth with Kirchhoff term and L^2-norm constraint. The next open problem highlights the case of low dimensions.

Open Problem 2. Study problem (P) if $N = 2$.

The following perspective generated by the problem studied in this chapter is concerned with orbital stability/instability of the normalized ground state.

Open Problem 3. In mass supercritical case with L^2-constraint, the ground state energy is often negative and mountain-pass type. In this case, the Kirchhoff term complicates the virial identity for stability analysis. It seems that the stability of normalized solutions for Kirchhoff equations with critical exponent problems is largely open.

We now propose an open problem in the nonlocal defocusing case for the Kirchhoff equation.

Open Problem 4. Physically, $b > 0$ is relevant in the expression of the nonlocal term M. Mathematically, if $b < 0$, the geometry changes drastically. Establish whether normalized solutions can exist only for small positive values of c.

The following research direction bridges Kirchhoff equations, critical exponent, and potential theory.

Open Problem 5. Study the following non-autonomous Kirchhoff equation under various hypotheses on the potential $V(x)$:

$$\begin{cases} -\left(a + b\displaystyle\int_{\mathbb{R}^3} |\nabla u|^2 dx\right) \Delta u + \lambda u = \mu |u|^{p-2}u + V(x)|u|^4 u & \text{in } \mathbb{R}^3, \\ u > 0, \displaystyle\int_{\mathbb{R}^3} |u|^2 dx = c^2, & \text{in } \mathbb{R}^3. \end{cases}$$

4.7 Glossary

Concentration-Compactness Principle: It is a fundamental tool in the calculus of variations and nonlinear PDEs, particularly for problems where compactness is lost due to invariance under non-compact symmetry groups (such as, translations, dilations). This principle systematically analyzes how sequences of functions can fail to be precompact in Sobolev embeddings and describes the possible defects of compactness.

Gagliardo–Nirenberg Inequality: This is a basic interpolation inequality in Sobolev spaces that controls intermediate derivatives by interpolating between lower-order and higher-order derivatives. In essence, it says that the L^p-norm of some derivative of u can be controlled by part of a higher derivative and part of the function itself.

Virial Identity: This is essentially the Pohozaev identity, which is a crucial integral identity derived by testing the equation with $x \cdot \nabla u$. This identity, combined with the equation itself, imposes a relation between the norm of a solution and its energy level, which is particularly restrictive in the critical case.

Nonlocal Defocusing Case: It corresponds to $b < 0$ in the nonlocal term. Mathematically, this means that the effective diffusion coefficient M decreases as the gradient norm increases. This can lead to loss of ellipticity if M becomes negative, causing ill-posedness.

5

Kirchhoff equations with double-behavior reaction

This chapter highlights the challenges of a Kirchhoff problem with mixed nonlinearities in the context of the existence of solutions with prescribed mass. This is a natural constraint in problems stemming from physics, such as standing waves in quantum mechanics. The analysis developed in this chapter bridges techniques from the well-studied theory of normalized solutions for semilinear equations to the more complex nonlocal Kirchhoff case, establishing optimal existence results for a broad and generic class of nonlinearities with a specific double growth behavior.

In this chapter, we investigate the existence of normalized solutions to the following Kirchhoff equation with double-behavior reaction

$$\begin{cases} -\left(a+b\displaystyle\int_{\mathbb{R}^3}|\nabla u|^2\,dx\right)\Delta u+\lambda u=f(u) \quad \text{in } \mathbb{R}^3,\\ \displaystyle\int_{\mathbb{R}^3}|u|^2\,dx=c^2,\\ (\lambda,u)\in\mathbb{R}\times H^1(\mathbb{R}^3),\end{cases}$$

where $a, b, c > 0$, and the parameter $\lambda \in \mathbb{R}$ is the frequency (or chemical potential). The reaction term f is a nonlinearity whose growth is superlinear but subcritical.

The difficulty generated by the nonlocal term $(\int_{\mathbb{R}^3}|\nabla u|^2\,dx)\Delta u$ prevents the direct application of standard variational methods for normalized solutions. The *double-behavior growth* refers to our focus on nonlinearities $f(u)$ that behave like:

(i) *L^2-subcritical at the origin and L^2-critical at infinity;*

(ii) *L^2-critical at the origin and L^2-supercritical at infinity.*

These hypotheses create a double-energy behavior: the associated energy functional is unbounded from below on the mass constraint, but exhibits a local minima geometry for certain mass levels.

The analysis is developed on the natural constraint set defined by the Pohozaev identity associated with the problem. This is a key technical tool to handle the nonlocality and the Lagrange multiplier. A major technical hurdle is overcoming potential lack of compactness due to the unboundedness of the domain (entire Euclidean space). This is achieved by using a *strict subadditivity inequality* for the energy, which is verified by the generic class of nonlinearities used in this study.

A feature of the analysis developed in this chapter is that it bridges the variational methods for normalized solutions (developed by Jeanjean, Bartsch, Soave, etc.) with the specific analysis required for Kirchhoff-type nonlocal operators. *Why this case is fundamental?* Because, in comparison with the semilinear case ($b = 0$), this problem addresses the following challenges:

(i) *scaling arguments failure*: standard scalings used for semilinear problems do not work in a proper way;

DOI: 10.1201/9781003797159-5

(ii) *interaction with mass constraint*: the nonlocality interacts with the L^2-constraint in a subtle way, making it difficult to apply established theories for normalized solutions (such as the Cazenave-Lions method or the Jeanjean mountain pass approach) without significant new ideas;

(iii) *lack of compactness*: the problem inherits the usual lack of compactness from working in the entire space and gains additional complications from the nonlocal term when analyzing minimizing sequences.

The basic abstract tools in this study are the following:

(a) *the Pohozaev manifold constraint*, in order to "tame" the nonlocal term;

(b) *fibering map analysis* to handle the mass constraint;

(c) *strict subadditivity*, which is the primary tool to prevent splitting of sequences and ensure compactness;

(d) *critical point theory on manifolds*,

all adapted to the intricate nonlocal structure of the Kirchhoff problem.

5.1 Statement of the problem

This chapter is concerned with solutions of the following Kirchhoff equation

$$\begin{cases} -\left(a+b\displaystyle\int_{\mathbb{R}^3}|\nabla u|^2\,dx\right)\Delta u+\lambda u=f(u) \quad \text{in } \mathbb{R}^3, \\ \displaystyle\int_{\mathbb{R}^3}|u|^2\,dx=c^2, \\ (\lambda,u)\in\mathbb{R}\times H^1(\mathbb{R}^3), \end{cases} \tag{5.1}$$

where $a,b,c>0$, the nonlinearity $f(u)$ corresponds to the L^2-subcritical, L^2-critical, L^2-supercritical, and Sobolev critical cases, which is more general than the sum of two powers.

A related problem was studied by Ye [158] who considered the Kirchhoff equation

$$-\left(a+b\int_{\mathbb{R}^3}|\nabla u|^2\,dx\right)\Delta u=\lambda u+|u|^{p-2}u \quad \text{in } \mathbb{R}^3, \tag{5.2}$$

where $p\in(2,6)$. Ye showed that for any given $c>0$, we have

$$\begin{cases} \sigma(c)\in(-\infty,0], & \text{if} \quad p\in(2,\bar{p})\,; \\ \sigma(c)=-\infty, & \text{if} \quad p\in(\bar{p},6)\,, \end{cases} \tag{5.3}$$

where

$$\bar{p}:=\frac{14}{3},\quad \sigma(c):=\inf_{u\in S_c}E(u),\quad S_c:=\left\{u\in H^1(\mathbb{R}^3):\int_{\mathbb{R}^3}|u|^2\,dx=c^2\right\}, \tag{5.4}$$

$$E(u):=\frac{a}{2}\int_{\mathbb{R}^3}|\nabla u|^2\,dx+\frac{b}{4}\left(\int_{\mathbb{R}^3}|\nabla u|^2\,dx\right)^2-\frac{1}{p}\int_{\mathbb{R}^3}|u|^p\,dx.$$

For the L^2-subcritical case, that is, if $p\in(2,\bar{p})$, Ye obtained the sharp existence of global constraint minimizers for (5.2) by solving the minimization problem (5.4). If $p\in(\tilde{p},\bar{p})$, Ye

TABLE 5.1
Classification of solutions for a Kirchhoff equation with power-type double-behavior reaction

p, q	α	Classifications of solutions
$2 < p < \tilde{p}$, $\bar{p} < q < 6$	$0 < \alpha < \min\{\alpha_*, \alpha^*\}$	A local minimizer
$2 < p < \tilde{p}$, $q = 6$	$0 < \alpha < \min\{\alpha_*, \alpha^*, \alpha^{**}\}$	A local minimizer
$\bar{p} < p < q < 6$	$\alpha > 0$	A mountain pass solution
$\bar{p} < p < 6$, $q = 6$	$\alpha > 0$	A mountain pass solution

also found a local minimizer which is a critical point of $E|_{S_c}$ by constructing a geometry of local minima for E, where $\tilde{p} := \frac{10}{3}$.

For the L^2-supercritical case when $p \in (\bar{p}, 6)$, since the minimization problem (5.4) is not available due to (5.3), in order to look for a critical point of $E|_{S_c}$, Ye [158] took the minimum on a suitable submanifold

$$\mathcal{M}_c := \left\{ u \in S_c : J(u) := \frac{\partial}{\partial t} E\left(u^t\right)\big|_{t=1} = 0 \right\}$$

and showed that $\mathcal{M}_c$ is a natural constraint of $E|_{S_c}$. This is done by using the Lagrange multiplier method, where

$$u^t(x) := t^{\frac{3}{2}} u(tx), \quad \forall t > 0,\, u \in H^1(\mathbb{R}^3),$$

and $u^t \in S_c$ for all $t > 0$ if $u \in S_c$.

For the L^2-critical case, that is, if $p = \bar{p}$, Ye [157] proved that equation (5.2) has a solution (u_c, λ_c) which satisfies $\left(\int_{\mathbb{R}^3} |u_c|^2 \, dx\right)^{\frac{1}{2}} = c > c^*$ for some $c^* > 0$. In another paper, Ye [159] analyzed the concentration behavior of normalized solutions. We also refer to Qi and Zou [132] who obtained the exact number and expressions of the positive normalized solutions for the nonlinear Kirchhoff equation (5.2). More generally, when $|u|^{p-2}u$ is replaced by $K(x)f(u)$, the existence and multiplicity of normalized solutions were studied by Chen, Rădulescu and Tang [27].

Li, Luo and Tang [92] were concerned with the existence and asymptotic behavior of Kirchhoff equation

$$-\left(a + b \int_{\mathbb{R}^3} |\nabla u|^2 \, dx\right) \Delta u = \lambda u + |u|^{p-2} u + \alpha |u|^{q-2} u \quad \text{in } \mathbb{R}^3$$

under the normalized constant $\int_{\mathbb{R}^3} u^2 \, dx = c^2$, where a, b, $c > 0$, $2 < q < \bar{p} < p \leq 6$ or $\bar{p} < q < p \leq 6$, $\alpha > 0$, and $\lambda \in \mathbb{R}$ appears as a Lagrange multiplier. They applied the methods introduced in Soave [140] to obtain the following results:

In Table 5.1, we explain some notations:

$$\alpha^* := \left[\frac{\frac{a}{2}\left(\frac{bq}{4\mathcal{C}_q^q}\right)^{\frac{2-p\gamma_p}{q\gamma_q-4}}}{c^{p(1-\gamma_p)+\frac{q(1-\gamma_q)(4-p\gamma_p)}{q\gamma_q-4}}} + \frac{\left(\frac{b}{4}\right)^{\frac{q\gamma_q-p\gamma_p}{q\gamma_q-4}}\left(\frac{q}{\mathcal{C}_q^q}\right)^{\frac{4-p\gamma_p}{q\gamma_q-4}}}{c^{p(1-\gamma_p)+\frac{q(1-\gamma_q)(4-p\gamma_p)}{q\gamma_q-4}}} \right] \frac{p\mathcal{C}_{p,q}}{\mathcal{C}_p^p};$$

$$\alpha_* := \left[\frac{p(q\gamma_q-4)b}{4(q\gamma_q-p\gamma_p)\mathcal{C}_p^p}\right]\left[\frac{q(4-p\gamma_p)b}{4(q\gamma_q-p\gamma_p)\mathcal{C}_q^q}\right]^{\frac{4-p\gamma_p}{q\gamma_q-4}} \frac{1}{c^{p(1-\gamma_p)+\frac{q(1-\gamma_q)(4-p\gamma_p)}{q\gamma_q-4}}};$$

$$\alpha^{**} := \frac{2(\frac{b}{\gamma_p})^{\frac{p\gamma_p}{4}}}{(6-p\gamma_p)\mathcal{C}_p^p}\left[\frac{12p}{4-p\gamma_p}\left(\frac{a\mathcal{S}\Lambda}{3}+\frac{b\mathcal{S}^2\Lambda^2}{12}\right)\right]^{1-\frac{p\gamma_p}{4}}\frac{1}{c^{p(1-\gamma_p)}},$$

$$\mathcal{C}_{p,q} := \left(\frac{8(4-p\gamma_p)}{q\gamma_q(q\gamma_q-2)(q\gamma_q-p\gamma_p)}\right)^{\frac{4-p\gamma_p}{q\gamma_q-4}} - \left(\frac{8(4-p\gamma_p)}{q\gamma_q(q\gamma_q-2)(q\gamma_q-p\gamma_p)}\right)^{\frac{q\gamma_q-p\gamma_p}{q\gamma_q-4}},$$

$$\mathcal{S} := \inf_{u\in D^{1,2}(\mathbb{R}^3)\setminus\{0\}}\frac{\int_{\mathbb{R}^3}|\nabla u|^2\,dx}{(\int_{\mathbb{R}^3}|u|^6\,dx)^{\frac{1}{3}}},\ \Lambda := \frac{b\mathcal{S}^2}{2}+\sqrt{a\mathcal{S}+\frac{b^2\mathcal{S}^4}{4}},$$

$$\mathcal{C}_q := \sup_{u\in H^1(\mathbb{R}^3)\setminus\{0\}}\frac{(\int_{\mathbb{R}^3}|u|^q\,dx)^{\frac{1}{q}}}{(\int_{\mathbb{R}^3}|\nabla u|^2\,dx)^{\frac{\gamma_q}{2}}(\int_{\mathbb{R}^3}u^2\,dx)^{\frac{1-\gamma_q}{2}}}.$$

In what follows, we extend these results to a general framework. More precisely, we address the existence of normalized solutions for Kirchhoff equations in the following cases in this chapter:

(i) f is L^2-subcritical at the origin and L^2-critical at infinity;

(ii) f is L^2-critical at the origin and L^2-supercritical at infinity.

The aim in this chapter is to provide a general class of nonlinearities and find the relationship between the solutions of problem (5.1) and of the following equation

$$\begin{cases} -\left(a+b\displaystyle\int_{\mathbb{R}^3}|\nabla u|^2\,dx\right)\Delta u+\lambda u=f(u) \quad \text{in } \mathbb{R}^3,\\ \displaystyle\int_{\mathbb{R}^3}|u|^2\,dx\le c^2,\\ (\lambda,u)\in\mathbb{R}\times H^1(\mathbb{R}^3). \end{cases} \tag{5.5}$$

In addition, we introduce the set

$$D_c := \left\{v\in H^1(\mathbb{R}^3) : \int_{\mathbb{R}^3}v^2\,dx\le c^2\right\}.$$

The idea of working with D_c was introduced by Bieganowski and Mederski [11] in the context of nonlinearities with L^2-critical or L^2-supercritical growth at the origin and L^2-supercritical and Sobolev-subcritical at infinity. The first advantage is that the weak limit of a sequence in the L^2-ball D_c still belongs to D_c, while this is not the case with the L^2-sphere S_c because the embedding $H^1(\mathbb{R}^3)\hookrightarrow L^2(\mathbb{R}^3)$ is not compact, not even studying radially symmetric functions. This makes it easier to obtain a minimizer of I over suitable subsets and gives additional information to use when we verify it belongs to S_c. The second advantage of working with D_c is that the Lagrange multipliers coming from this constraint are nonnegative because they are associated with a minimizer. This property is helpful when one has no other way to obtain information about the Lagrange multipliers.

Under standard assumptions about $f(u)$, Bieganowski and Mederski [11] proved that the solutions of (5.1) (resp., (5.5)) are the critical points of the energy functional

$$I : u\in H^1(\mathbb{R}^3)\mapsto \frac{a}{2}\int_{\mathbb{R}^3}|\nabla u|^2\,dx+\frac{b}{4}\left(\int_{\mathbb{R}^3}|\nabla u|^2\,dx\right)^2-\int_{\mathbb{R}^3}F(u)\,dx\in\mathbb{R}$$

restricted to the C^1 manifold S_c (resp., D_c) with the Lagrange multiplier $\lambda\in\mathbb{R}$, that is, they are critical points of the functional

$$u\in H^1(\mathbb{R}^3)\mapsto I(u)+\frac{\lambda}{2}\int_{\mathbb{R}^3}|u|^2\,dx\in\mathbb{R}$$

for some $\lambda \in \mathbb{R}$, where $F(t) = \int_0^t f(s)\,ds$. Moreover, every critical point of the above functional belongs to $W^{2,q}_{\text{loc}}(\mathbb{R}^3)$ for all $q < \infty$ and it satisfies both the following Pohozaev equality

$$a\int_{\mathbb{R}^3} |\nabla u|^2\,dx + b\left(\int_{\mathbb{R}^3} |\nabla u|^2\,dx\right)^2 = 6\int_{\mathbb{R}^3} F(u)\,dx - 3\lambda\int_{\mathbb{R}^3} |u|^2\,dx$$

and the Nehari identity

$$\langle I'(u), u\rangle + \lambda\int_{\mathbb{R}^3} |u|^2\,dx = 0.$$

By a linear combination of the two equalities above, every solution satisfies

$$P(u) := a\int_{\mathbb{R}^3} |\nabla u|^2\,dx + b\left(\int_{\mathbb{R}^3} |\nabla u|^2\,dx\right)^2 - \frac{3}{2}\int_{\mathbb{R}^3} H(u)\,dx = 0,$$

where $H(u) := \langle f(u), u\rangle - 2F(u)$ ($\langle\cdot,\cdot\rangle$ is the scalar product in $\mathbb{R}$).

The above arguments show that it is natural to introduce the constraint

$$\mathcal{P}(u) := \left\{u \in H^1(\mathbb{R}^3)\backslash\{0\} : P(u) = 0\right\},$$

which contains all the nontrivial solutions to (5.1) or (5.5) and does not depend on λ.

Observe that every nontrivial solution of (5.5) belongs to $\mathcal{P} \cap D$ and every nontrivial solution of problem (5.1) belongs to $\mathcal{P} \cap S_c \subset \mathcal{P} \cap D_c$.

Define

$$m(c) := \inf_{D_c} I.$$

Definition 5.1 *By a ground state solution of problem* (5.5) *we mean a nontrivial solution which minimizes the energy I among all the nontrivial solutions.*

In particular, if (λ, u) solves (5.5) and $I(u) = \inf_{\mathcal{P}\cap D_c} I$, then (λ, u) is a ground state solution.

Definition 5.2 *By a ground state solution of problem* (5.1) *we mean that* (λ, u) *solves* (5.1) *and* $I(u) = \inf_{\mathcal{P}\cap D_c} I$.

For the case (i), we assume the following hypotheses.

(A_0) $F \in C^1(\mathbb{R})$ and there exists $C > 0$ such that $|f(t)| = |\nabla F(t)| \le C(|t| + |t|^5)$ for every $t \in \mathbb{R}$.

(A_1) $\eta_\infty := \lim_{|t|\to+\infty} \frac{F(t)}{|t|^{\bar{p}}} < +\infty$.

(A_2) $\lim_{t\to 0} \frac{F(t)}{|t|^2} = 0$.

(A_3) $\eta_0 := \lim_{t\to 0} \frac{F(t)}{|t|^{\bar{p}}} > 0$.

We also need hypotheses about the radii c, namely

$$\frac{4}{b}\eta_\infty c^{\bar{p}(1-\bar{p})} C_{\bar{p}}^{\bar{p}} < 1, \tag{5.6}$$

$$\eta_0 > \max\left\{\frac{1}{2}, \frac{1}{2}\left(\frac{c^2}{\|\omega\|_2^2}\right)^{-\frac{2}{3}}\right\}, \tag{5.7}$$

where ω is the unique positive radial solution of (5.19). In particular, if $\eta_\infty = 0$ (resp., $\eta_0 = \infty$), then relation (5.6) (resp., (5.7)) is automatically satisfied. In addition, for c, $\tilde{R} > 0$, we set

$$D_c^{\tilde{R}} := \left\{u \in D_c : \|\nabla u\|_2 < \tilde{R}\right\} \quad \text{and} \quad m_c^{\tilde{R}} := \inf_{D_c^{\tilde{R}}} I.$$

For the case (ii), we assume that there exists $\kappa \in [0, +\infty)$ such that F is of the form

$$F(u) = \bar{F}(u) + \frac{\kappa}{6}|u|^6$$

for some $\bar{F} : \mathbb{R} \mapsto \mathbb{R}$. We set $\bar{f} = \nabla \bar{F}$, $\bar{H}(u) = \langle \bar{f}(u), u\rangle - 2\bar{F}(u)$, $\bar{h} = \nabla \bar{H}$, $h := \nabla H$, and consider the following assumptions:

- (B_0) $\bar{f}$ and $\bar{h}$ are continuous and there exists $\bar{C} > 0$ such that $|\bar{h}| \leq \bar{C}(|u| + |u|^5)$.
- (B_1) $\eta := \limsup_{t\to 0} \frac{\bar{F}(t)}{|t|^{\bar{p}}} < +\infty$.
- (B_2) If $\kappa = 0$, then $\lim_{|t|\to+\infty} \frac{\bar{F}(t)}{|t|^{\bar{p}}} = +\infty$; if $\kappa \in (0, +\infty)$, then $\liminf_{|t|\to+\infty} \frac{\bar{F}(t)}{|t|^{\bar{p}}} > 0$.
- (B_3) $\lim_{|t|\to+\infty} \frac{\bar{F}(t)}{|t|^6} = 0$.
- (B_4) $\bar{p} \cdot \bar{H}(u) \leq \langle \bar{h}(u), u\rangle$.
- (B_5) $\frac{8}{3}\bar{F} \leq \bar{H} \leq 4\bar{F}$.

We also need hypotheses about the radii c, namely

$$\frac{8}{b}\eta C_{\bar{p}}^{\bar{p}} c^{\bar{p}(1-\gamma_{\bar{p}})} < 1. \tag{5.8}$$

In what follows, if $f_1, f_2 : \mathbb{R} \mapsto \mathbb{R}$, then we write $f_1 \preceq f_2$ if and only if $f_1 \leq f_2$ and for every $\epsilon > 0$ there exists $t \in \mathbb{R}$, $|t| < \epsilon$ such that $f_1(t) < f_2(t)$.

Remark 5.1 *We observe that* $\lim_{t\to\infty} \frac{F(t)}{|t|^{\bar{p}}} > 0$ *if condition* (B_2) *holds. Moreover, if* F, H *satisfy* (B_1)–(B_5), *then* $\bar{F}, \bar{H} \geq 0$. *If* (B_0) *and* (B_5) *hold, then* I *and* P *are of class* C^1.

The main results of this chapter are the following.

Theorem 5.1 *Assume that conditions* (A_0), (A_2), (A_3) *hold and*

$$c^2 < 2\left(\frac{a\mathcal{S}}{6K_0}\right)^{\frac{3}{2}}, \tag{5.9}$$

where K_0 *is defined in* (5.12). *If* $c_n \to c$, $\tilde{R}_n \to R_1$ *(where* $R_1 > 0$ *is defined in Lemma* 5.1*), and* $u_n \in D_{c_n}^{\tilde{R}_n}$ *are such that* $I(u_n) \to m_c^{R_1}$, *then there exist* $u \in S_c \cap D_c^{R_1}$ *and* $\lambda_u > 0$ *such that* $u_n \to u$ *up to translations and up to subsequences,* $I(u) = m_c^{R_1} < 0$, *u has a constant sign, is nowhere zero, and* (λ_u, u) *is a solution of problem* (5.1).

Remark 5.2 *In view of Theorem 5.1, we give the existence of local minimizers in the case that the nonlinearity f is L^2-subcritical at the origin. In this case, we do not distinguish between nonlinear terms that have L^2-subcritical, L^2-critical, or L^2-supcritical growth at infinity. In fact, one of the purposes of this chapter is to understand what reasonably minimal hypotheses we need both for the various steps and for the main results.*

The following result establishes the existence of global minimizers with negative if the nonlinearity f is L^2-subcritical at the origin and L^2-critical at infinity.

Theorem 5.2 *Assume that conditions* (A_0)–(A_3), (5.6) *and* (5.7) *hold. Then there exists a solution* $(\lambda, u) \in (0,\infty) \times S_c$ *of problem* (5.1) *such that* $0 > I(u) = \inf_{D_c} I = \inf_{S_c} I$. *Furthermore, if f is locally Lipschitz continuous or F is even, then every minimizer of* $I|_{D_c}$ *has a constant sign and, up to a translation, is radial and radially monotone.*

Theorem 5.3 *Assume that* (B_0)–(B_5) *and* (5.8) *hold and*

$$\begin{aligned}\inf_{\mathcal{P}\cap D_c} I &< \frac{a\kappa^{-\frac{1}{2}}}{6}\left(b\kappa^{-\frac{1}{2}}\mathcal{S}^3 + \sqrt{(b\kappa^{-\frac{1}{2}}\mathcal{S}^3)^2 + 4a\mathcal{S}^3}\right) \\ &\quad + \frac{b\kappa^{-\frac{1}{2}}}{24}\left(b\kappa^{-\frac{1}{2}}\mathcal{S}^3 + \sqrt{(b\kappa^{-\frac{1}{2}}\mathcal{S}^3)^2 + 4a\mathcal{S}^3}\right)^2,\end{aligned} \tag{5.10}$$

if $\kappa \in (0,\infty)$.

(*i*) *Then there exists* $u \in \mathcal{P} \cap D_c$ *such that* $I(u) = \inf_{\mathcal{P}\cap D_c} I$. *In addition, u is a radial, nonnegative, and radially nonincreasing function of class* C^2, *provided that F is of the form*

$$F(u) = \hat{F}(u) + \beta|u|^{\zeta}, \tag{5.11}$$

where $\hat{F} : \mathbb{R} \mapsto [0,\infty)$ *is even,* $\bar{p} \le \zeta < 6$ *and* $\beta \ge 0$.

(*ii*) *If, moreover,* $(B_4, \preceq)$ *holds, then there exist* $\lambda \in (0,\infty)$ *and* $u \in \mathcal{P} \cap S_c$ *such that* (λ, u) *is a ground state solution of problem* (5.1).

Remark 5.3 *In view of Theorem* 5.3(*ii*), *we observe that the obtained ground state u of equation* (5.5) *belongs to* $\partial D_c = S_c$. *This means that u is a ground state solution of equation* (5.1), *too.*

In what follows, for $1 \le p \le \infty$, the norm in $L^p(\mathbb{R}^3)$ is denoted by $\|\cdot\|_p$ for simplicity. We assume that the Hilbert space $H^1(\mathbb{R}^3)$ is endowed with the standard norm $\|u\|^2 = \int_{\mathbb{R}^3}(|\nabla u|^2 + u^2)dx$.

5.2 Reaction L^2-subcritical at the origin and L^2-critical at infinity

In this section, we discuss the existence of solutions to problem (5.1) in the case that the nonlinearity f is L^2-subcritical at the origin. Let us start with the proof of the existence of negative-energy solutions by giving a proper estimate below of the energy functional I on D_c. In view of (A_0), we see that if $F(\xi) > 0$ for some $\xi > 0$ (which occurs if condition (A_3) holds), then

$$K_0 := \sup_{0 \neq t \in \mathbb{R}} \frac{F(t)}{t^2 + t^6} > 0 \tag{5.12}$$

is well defined. Here, we point out that whenever we mention K_0, we implicitly assume that $F > 0$ somewhere and $K_0 < +\infty$. So from (5.12) and the definition of $\mathcal{S}$, for $u \in D_c$,

$$\begin{aligned} I(u) &\geq \frac{a}{2}\|\nabla u\|_2^2 + \frac{b}{4}\|\nabla u\|_2^4 - K_0(\|u\|_2^2 + \|u\|_6^6) \\ &\geq \frac{a}{2}\|\nabla u\|_2^2 + \frac{b}{4}\|\nabla u\|_2^4 - K_0 c^2 - K_0 \mathcal{S}^{-3}\|\nabla u\|_2^6 \\ &\geq \frac{a}{2}\|\nabla u\|_2^2 - K_0 c^2 - K_0 \mathcal{S}^{-3}\|\nabla u\|_2^6, \end{aligned} \tag{5.13}$$

which implies that $I|_{D_c}$ is bounded below on bounded subsets in $H^1(\mathbb{R}^3)$.

Define the mapping $g : (0, +\infty) \times (0, +\infty) \mapsto \mathbb{R}$ by

$$g(c, t) := \frac{a}{2} - K_0 c^2 t^{-2} - K_0 \mathcal{S}^{-3} t^4.$$

It follows that

$$I(u) \geq g(c, \|\nabla u\|_2)\|\nabla u\|_2^2 \quad \text{for all } u \in D_c. \tag{5.14}$$

Lemma 5.1 *The following properties hold:*

(i) For every $c > 0$, the function $t \mapsto g(c, t)t^2$ has a unique critical point, which is a global maximizer.

(ii) If (5.9) holds, then there exist $0 < R_1 < R_2$ such that $g(c, R_1) = g(c, R_2) = 0$, $g(c, t) > 0$ for $t \in (R_1, R_2)$, and $g(c, t) < 0$ for $t \in (0, R_1) \cup (R_2, +\infty)$.

(iii) If $t > 0$ and $c_1 \geq c_2 > 0$, then for every $s \in \left[t\frac{c_2}{c_1}, t\right]$, we have $g(c_2, s) \geq g(c_1, t)$.

(iv) If c satisfies (5.9), then there exists $\epsilon > 0$ such that (5.13) is verified by every $c' \in (c - \epsilon, c + \epsilon)$ and the functions $c' \in (c - \epsilon, c + \epsilon) \mapsto R_i(c') \in (0 + \infty)$, $i \in \{1, 2\}$ with R_i defined in (ii), are invertible and of class C^1.

Proof. It follows from the geometry of $g(c, t)$ and straightforward computation that (i) and (ii) hold true. We also remark that $g(c_2, t) \geq g(c_1, t)$ for all $t > 0$. Furthermore,

$$g\left(c_2, \frac{c_2}{c_1}t\right) - g(c_1, t) = \frac{K_0}{\mathcal{S}^3}\left[1 - \left(\frac{c_2}{c_1}\right)^4\right] t^4 \geq 0.$$

Then by (i) and the regularity of $g(c_2, \cdot)$, property (iii) holds true. Finally, by the regularity of g, $\partial g(c, R_1) > 0$, $\partial_t g(c, R_2) < 0$ and the implicit function theorem, we obtain that (iv) holds true. □

We remark that if $u \in H^1(\mathbb{R}^3)\backslash\{0\}$ and $s > 0$, then $s * u := s^{\frac{3}{2}} u(s \cdot)$ satisfies

$$\|s * u\|_2 = \|u\|_2 \quad \text{and} \quad \|\nabla(s * u)\|_2 = s\|\nabla u\|_2.$$

The following property establishes that the infimum of I on $D_c^{\tilde{R}}$ is negative.

Lemma 5.2 *Assume that* (A_0) *and* (A_3) *hold. Then* $m_c^{\tilde{R}} \in (-\infty, 0)$ *for every* $c, \tilde{R} > 0$.

Proof. We first observe that the estimate given in (5.13) implies $m_c^{\tilde{R}} > -\infty$. Next, for all $u \in D_c \cap L^\infty(\mathbb{R}^3)\backslash\{0\}$ and $s > 0$,

$$I(s*u) = s^2\left(\frac{a}{2}\|\nabla u\|_2^2 + s^2\frac{b}{4}\|\nabla u\|_2^4 - \frac{1}{(s^{\frac{3}{2}})^{\bar{p}}}\int_{\mathbb{R}^3} F(s^{\frac{3}{2}}u)\,dx\right).$$

From (A_3), we have $F(s^{\frac{3}{2}}u) > 0$ a.e. in $\operatorname{supp} u$ for sufficiently small $s > 0$. So, using Fatou's lemma and (A_3) again, we observe that

$$\lim_{s\to 0^+} \frac{1}{(s^{\frac{3}{2}})^{\bar{p}}}\int_{\mathbb{R}^3} F(s^{\frac{3}{2}}u)\,dx = +\infty.$$

This means that $I(s*u) < 0$ for sufficiently small $s > 0$. Since $s*u \in D_c^{\tilde{R}}$ for small $s > 0$, we deduce that $m_c^{\tilde{R}} \in (-\infty, 0)$ for every $c, \tilde{R} > 0$. □

Next, we claim that $m_c^{R_1}$ cannot be achieved on $\partial D_c^{R_1}$. Indeed, it follows from Lemma 5.1(ii) and Lemma 5.2 that there exists $\epsilon > 0$ such that for any $s \in [R_1 - \epsilon, R_1]$,

$$0 \geq g(c,s) \geq \frac{m_c^{R_1}}{2R_1^2}.$$

Then combining (5.14) and Lemma 5.2, we have for any $u \in D_c$, it holds that $R_1 - \epsilon \leq \|\nabla u\|_2 \leq R_1$ and

$$I(u) \geq g(c, \|\nabla u\|_2)\|\nabla u\|_2^2 \geq R_1^2\frac{m_c^{R_1}}{2R_1^2} > m_c^{R_1}.$$

We now establish a very useful subadditivity property of $m_c^{R_1}$.

Lemma 5.3 *Assume that conditions* (A_0), (A_3) *and* (5.9) *hold. Then for every* $\bar{c} \in (0, c)$ *there holds*

$$m_c^{R_1} \leq m_{\bar{c}}^{R_1} + m_{\sqrt{c^2-\bar{c}^2}}^{R_1}.$$

Furthermore, the above inequality is strict if at least one between $m_c^{R_1}$ *and* $m_{\sqrt{c^2-\bar{c}^2}}^{R_1}$ *is achieved.*

Proof. We first show that for any $\sigma \in \left[1, \frac{c}{\bar{c}}\right]$,

$$m_{\sigma\bar{c}}^{R_1} = \sigma^2 m_{\bar{c}}^{R_1} \tag{5.15}$$

and the inequality is strict if $m_c^{R_1}$ is achieved and $\sigma > 1$. By relation (5.14), Lemma 5.2 and the definition of $m_c^{R_1}$, one has that for every $\epsilon > 0$ small enough, there exists $u \in D_c^{R_1}$ such that

$$g(\bar{c}, \|\nabla u\|_2)\|\nabla u\|_2^2 \leq I(u) \leq m_c^{R_1} + \epsilon < 0. \tag{5.16}$$

On the other hand, if $\|\nabla u\|_2 \geq R_1 \cdot \frac{\bar{c}}{c}$, then combining Lemma 5.1(iii) with $c_1 = c, c_2 = \bar{c}$, $t = R_1$ and $s = \|\nabla u\|_2$, we infer that

$$g(\bar{c}, \|\nabla u\|_2) \geq g(c, R_1) = 0.$$

This contradicts relation (5.16). Hence $u \in D_{\bar{c}}^{\bar{c}\cdot\frac{R_1}{c}}$.

Define $v(\cdot) := u\left(\frac{\cdot}{\sigma^{\frac{2}{3}}}\right) \in D_{\sigma\bar{c}}^{R_1}$ and

$$m_{\sigma\bar{c}}^{R_1} \leq I(v) = \sigma^2\left(\frac{a}{2\sigma^{\frac{4}{3}}}\|\nabla u\|_2^2 + \frac{b}{4\sigma^{\frac{2}{3}}}\|\nabla u\|_2^4 - \int_{\mathbb{R}^3} F(u)\,dx\right) \leq \sigma^2 I(u) \leq \sigma^2(m_c^{R_1} + \epsilon),$$

which implies (5.15) by the arbitrariness of ϵ.

Now, if $m_{\bar{c}}^{R_1}$ is achieved and $\sigma > 1$, we can take $u \in D_{\bar{c}}^{R_1}$ as the minimizer and similar to the above argument, we have

$$m_{\sigma\bar{c}}^{R_1} < \sigma^2 I(u) = \sigma^2 m_{\bar{c}}^{R_1}. \tag{5.17}$$

When $\bar{c}^2 > c^2 - \bar{c}^2$, we get

$$m_c^{R_1} \overset{(1)}{\leq} \frac{c^2}{\bar{c}^2} m_{\bar{c}}^{R_1} = m_{\bar{c}}^{R_1} + \frac{c^2 - \bar{c}^2}{\bar{c}^2} m_{\bar{c}}^{R_1} \overset{(2)}{\leq} m_{\bar{c}}^{R_1} + m_{\sqrt{c^2-\bar{c}^2}}^{R_1}.$$

Here, the inequality (1) (resp., (2)) is strict if $m_{\bar{c}}^{R_1}$ (resp., $m_{\sqrt{c^2-\bar{c}^2}}^{R_1}$) is achieved. Similarly, when $\bar{c}^2 < c^2 - \bar{c}^2$, we get

$$m_c^{R_0} \overset{(3)}{\leq} \frac{c^2}{c^2 - \bar{c}^2} m_{\sqrt{c^2-\bar{c}^2}}^{R_1} = m_{\sqrt{c^2-\bar{c}^2}}^{R_1} + \frac{\bar{c}^2}{c^2 - \bar{c}^2} m_{\sqrt{c^2-\bar{c}^2}}^{R_1} \overset{(4)}{\leq} m_{\sqrt{c^2-\bar{c}^2}}^{R_1} + m_{\bar{c}}^{R_1}.$$

Here, the inequality (3) (resp., (4)) is strict if $m_{\sqrt{c^2-\bar{c}^2}}^{R_1}$ (resp., $m_{\bar{c}}^{R_1}$) is achieved. In particular, if $\bar{c}^2 = c^2 - \bar{c}^2$ (that is, $\sqrt{2}\bar{c} = c$), by relation (5.15), it holds that $m_{\sqrt{2}\bar{c}}^{R_1} = 2m_{\bar{c}}^{R_1}$, while the inequality being strict if $m_{\bar{c}}^{R_1}$ is achieved follows from (5.17). Thus, the proof of Lemma 5.3 is completed. □

The following property gives the relative compactness in $L^p(\mathbb{R}^3)$ of minimizing sequences.

Lemma 5.4 *Assume that* (A_0), (A_2), (A_3) *and* (5.9) *hold. If* $c_n \to c$, $\tilde{R}_n \to R_1$ *and* $\{u_n\} \subset H^1(\mathbb{R}^3)$ *satisfies* $u_n \in D_{c_n}^{\tilde{R}_n}$ *for every* $n \in \mathbb{N}$, $I(u_n) \to m_c^{R_1}$, *then there exist* $\{y_n\} \subset \mathbb{R}^3$ *and* $\tilde{u} \in D_c \backslash \{0\}$ *such that* $\|\nabla \tilde{u}\|_2 \leq R_1$ *and* $u_n(\cdot + y_n) \to \tilde{u}$ *in* $L^p(\mathbb{R}^3)$ *for any* $p \in [2, 6)$.

Proof. We observe that since $\{u_n\}$ is bounded in $H^1(\mathbb{R}^3)$, in view of the interpolation inequality and the Sobolev embedding, it is enough to verify the statement for $p = 2$. Hence suppose by contradiction that

$$\lim_{n\to+\infty} \sup_{y\in\mathbb{R}^3} \int_{B(y,1)} |u_n|^2\, dx = 0.$$

Then by Lemma I.1 in Lions [100], we know that for any $q \in (2, 6)$, $\lim_{n\to+\infty} \|u_n\|_q = 0$.

Now, we fix $q \in (2, 6)$ and $\epsilon > 0$. By (C_1) and (5.12), there exists $C = C(q, \epsilon, K_0)$ such that for every $t \in \mathbb{R}$,

$$F(t) \leq \epsilon t^2 + C|t|^q + K_0 t^6.$$

Since $g(c, R_1) = 0$ for $\epsilon \ll 1$, by Lemma 5.2, it holds that

$$\begin{aligned}
0 > m_c^{R_1} &= \lim_{n\to+\infty} I(u_n) \geq \limsup_{n\to+\infty} \left(\frac{a}{2}\|\nabla u_n\|_2^2 + \frac{b}{4}\|\nabla u_n\|_2^4 - \epsilon\|u_n\|_2^2 - C\|u_n\|_q^q - K_0\|U_N\|_6^6 \right) \\
&\geq \left(\frac{a}{2} - \frac{K_0 R_1^4}{S^3} \right) \limsup_{n\to+\infty} \|\nabla u_n\|_2^2 + \limsup_{n\to+\infty} \|\nabla u_n\|_2^4 - \epsilon c^2 \\
&\geq \frac{K_0 c^2}{R_1^2} \limsup_{n\to+\infty} \|\nabla u_n\|_2^2 + \frac{b}{4} \limsup_{n\to+\infty} \|\nabla u_n\|_2^4 - \epsilon c^2 \geq -\epsilon c^2,
\end{aligned}$$

which is a contradiction. Thus, there exists $\{y_n\} \subset \mathbb{R}^3$ such that

$$\limsup_{n\to+\infty} \int_{B(0,1)} |u_n(\cdot + y_n)|^2\, dx > 0.$$

This means that there exists $\tilde{u} \in D_c \backslash \{0\}$ with $\|\nabla \tilde{u}\|_2 \le R_1$ such that $v_n(x) := u_n(x + y_n) - \tilde{u}(x)$. So, up to a subsequence, we have $v_n \rightharpoonup 0$ in $H^1(\mathbb{R}^3)$ and $v_n(x) \to 0$ for a.e. $x \in \mathbb{R}^3$.

Denote $\alpha_1 := \|\tilde{u}\|_2$ and assume by contradiction that $\alpha_2 := \lim_{n\to+\infty} \|v_n\|_2 > 0$ along a subsequence. In view of (C_0) and the Young's inequality, for every $\delta > 0$, there exists $C_\delta > 0$ such that for all $t, s \in \mathbb{R}$,

$$\begin{aligned}|F(t+s) - F(t)| &\le \int_0^1 |f(t+\tau s)||s|\,d\tau \le C \int_0^1 \left(|t+\tau s| + |t+\tau s|^5\right)|s|\,d\tau \\ &\le C\left(|t| + |s| + |t|^5 + |s|^5\right)|s| \le \delta(t^2 + t^6) + C_\delta(s^2 + s^6).\end{aligned}$$

Then using the Brezis-Lieb lemma in [14] and the weak convergence in $H^1(\mathbb{R}^3)$, we obtain that

$$\lim_{n\to+\infty} \left(\|u_n\|_2^2 - \|v_n\|_2^2\right) = \|\tilde{u}\|_2^2,$$

$$\lim_{n\to+\infty} \left(\|\nabla u_n\|_2^2 - \|\nabla v_n\|_2^2\right) = \|\nabla \tilde{u}\|_2^2$$

and

$$\lim_{n\to+\infty} (I(u_n) - J(v_n)) = J(\tilde{u}),$$

where

$$J(u) := \frac{a}{2}\|\nabla u\|_2^2 + \frac{bA}{4}\|\nabla u\|_2^2 - \int_{\mathbb{R}^3} F(u)\,dx,$$

and $A := \lim_{n\to+\infty} \|\nabla u_n\|_2^2$. In particular, we have $\alpha_2^2 \le c^2 - \alpha_1^2$ and $\|\nabla v_n\| < R_1$ for $n \gg 1$.

Next, we define for every n,

$$\tilde{v}_n := \begin{cases} v_n & \text{if } \|v_n\|_2 \le \alpha_2, \\ \alpha_2 \dfrac{v_n}{\|v_n\|_2} & \text{if } \|v_n\|_2 > \alpha_2. \end{cases}$$

Then $\tilde{v}_n \in D_{\alpha_2}^{R_1}$. Using (A_0), we obtain by straightforward computation that

$$\lim_{n\to+\infty} [J(v_n) - J(\tilde{v}_n)] = 0.$$

Thus, by Lemma 5.3 and the fact that $I(t), J(t)$ are continuous, we conclude that $m_c^{R_1}$ is nonincreasing in c. So, up to a subsequence,

$$\begin{aligned}m_c^{R_1} &= \lim_{n\to+\infty} I(u_n) = J(\tilde{u}) + \lim_{n\to+\infty} J(\tilde{v}_n) \ge I(\tilde{u}) + \lim_{n\to+\infty} I(\tilde{v}_n) \ge m_{\alpha_1}^{R_1} + m_{\alpha_2}^{R_1} \\ &\ge m_{\sqrt{\alpha_1^2+\alpha_2^2}}^{R_1} \ge m_c^{R_1}.\end{aligned}$$

This means that $m_{\alpha_1}^{R_1}$ is achieved at $\tilde{u}$ and using Lemma 5.8 again, we obtain

$$m_{\alpha_1}^{R_1} + m_{\alpha_2}^{R_1} > m_{\sqrt{\alpha_1^2+\alpha_2^2}}^{R_1}.$$

This is impossible. So, the proof of Lemma 5.4 is completed. □

5.2.1 Proof of Theorem 5.1: existence and asymptotics

Suppose that $\{c_n\} \subset (0, +\infty)$, $\{\tilde{R}_n\} \subset (0, +\infty)$, and $\{u_n\} \subset H^1(\mathbb{R}^3)$ satisfies $u_n \in D_{c_n}^{\tilde{R}_n}$ for every $n \in \mathbb{N}$, $I(u_n) \to m_c^{R_1}$. Then in view of Lemma 5.4, there exists $\tilde{u} \in D_c \backslash \{0\}$ with $\|\nabla \tilde{u}\|_2 \leq R_1$ such that replacing $u_n(\cdot + y_n)$ with u_n, we can assume that, up to a subsequence, $u_n \rightharpoonup \tilde{u}$ in $H^1(\mathbb{R}^3)$ and $u_n \to \tilde{u}$ and a.e. in $\mathbb{R}^3$. Moreover, $I(\tilde{u}) \geq m_c^{R_1}$.

Let $v_n := u_n - \tilde{u}$. By weak convergence, we have

$$\lim_{n \to +\infty} (\|\nabla u_n\|_2^2 - \|\nabla v_n\|_2^2) = \|\nabla \tilde{u}\|_2^2 > 0,$$

which implies that $\|\nabla v_n\|_2 < R_1$ for $n \gg 1$. Furthermore, by Theorem 2 of Brezis and Lieb [14], we have

$$m_c^{R_1} = \lim_{n \to +\infty} I(u_n) \geq \lim_{n \to +\infty} J(v_n) + J(\tilde{u}) \geq \lim_{n \to +\infty} I(v_n) + m_c^{R_1}.$$

Combining $g(c, R_1) = 0$, relation (5.12) and $\lim_{n \to +\infty} \|v_n\|_2 = 0$, we obtain

$$\begin{aligned} 0 &\geq \lim_{n \to +\infty} I(v_n) \geq \lim_{n \to +\infty} \left[\frac{a}{2}\|\nabla v_n\|_2^2 - K_0(\|v_n\|_2^2 + \|v_n\|_6^6)\right] \\ &\geq \left(\frac{a}{2} - \frac{K_0 R_1^4}{\mathcal{S}^3}\right) \lim_{n \to +\infty} \|\nabla v_n\|_2^2 = \frac{K_0 c^2}{R_1^2} \lim_{n \to +\infty} \|\nabla v_n\|_2^2 \geq 0. \end{aligned}$$

This means that $\lim_{n \to +\infty} \|\nabla v_n\|_2 = 0$, that is, $u_n \to \tilde{u}$ in $H^1(\mathbb{R}^3)$. Then it follows from Lemma 5.2 that $I(\tilde{u}) = \lim_{n \to +\infty} I(u_n) = m_c^{R_1} < 0$ and $\|\nabla \tilde{u}\|_2 < R_1$. So, there exists $\lambda_{\tilde{u}} \in \mathbb{R}$ such that

$$-\left(a + b\int_{\mathbb{R}^3} |\nabla \tilde{u}|^2 \, dx\right) \Delta \tilde{u} + \lambda_{\tilde{u}} \tilde{u} = f(\tilde{u}) \quad \text{in } \mathbb{R}^3.$$

In particular, if $\lambda_{\tilde{u}} \leq 0$, then by the Pohozaev identity, we have

$$a\int_{\mathbb{R}^3} |\nabla \tilde{u}|^2 \, dx + b\left(\int_{\mathbb{R}^3} |\nabla \tilde{u}|^2 \, dx\right)^2 = 6\int_{\mathbb{R}^3} F(\tilde{u}) \, dx - 3\lambda_{\tilde{u}} \int_{\mathbb{R}^3} |\tilde{u}|^2 \, dx \geq 6\int_{\mathbb{R}^3} F(\tilde{u}) \, dx,$$

which yields that

$$0 > I(\tilde{u}) \geq \frac{a}{3}\int_{\mathbb{R}^3} |\nabla \tilde{u}|^2 \, dx + \frac{b}{12}\left(\int_{\mathbb{R}^3} |\nabla \tilde{u}|^2 \, dx\right)^2.$$

This is a contradiction. Thus, $\lambda_{\tilde{u}} > 0$. We also have $\tilde{u} \in S_c$.

We now claim that $\tilde{u}$ has a constant sign. Denote $\tilde{u}_\pm := \max\{\pm\tilde{u}, 0\}$ and $c_\pm := \|\tilde{u}_\pm\|_2$. Then $\pm\tilde{u}_\pm \in D_{c_\pm}^{R_1} \subset D_c^{R_1}$. If $\|\tilde{u}_+\|_2 \cdot \|\tilde{u}_-\|_2 > 0$, then in virtue of Lemma 5.3 and $c^2 = c_+^2 + c_-^2$,

$$m_c^{R_1} = I(\tilde{u}) \geq I(\tilde{u}_+) + I(-\tilde{u}_-) \geq m_{c_+}^{R_1} + m_{c_-}^{R_1} \geq m_c^{R_1}.$$

Thus, $m_{c_+}^{R_1}$ is achieved at $\tilde{u}_+$ and $m_{c_-}^{R_1}$ is achieved at $-\tilde{u}_-$. Then using Lemma 5.3 again, we obtain $m_{c_+}^{R_1} + m_{c_-}^{R_1} > m_c^{R_1}$. This is impossible and the claim is true. In addition, since

$$-\left(a + b\int_{\mathbb{R}^3} |\nabla \tilde{u}|^2 \, dx\right) \Delta \tilde{u} + \left(\lambda_{\tilde{u}} + \max\left\{-\frac{f(\tilde{u})}{\tilde{u}}, 0\right\}\right) \tilde{u} = \max\left\{\frac{f(\tilde{u})}{\tilde{u}}, 0\right\} \tilde{u},$$

by the maximum principle in Gilbarg and Trudinger [63], it holds that $\tilde{u}$ does not vanish. The proof of Theorem 5.1 is now complete. □

5.2.2 Global minimizers with negative energy

This section is devoted to the proof of Theorem 5.2. We start with some auxiliary properties.

Lemma 5.5 *Assume that conditions* (A_0)*–*(A_2) *and* (5.6) *hold. Then* $I|_{D_c}$ *is coercive and bounded from below.*

Proof. By (A_1) and (A_2), for every $\epsilon > 0$ there exists $C_\epsilon > 0$ such that for any $u \in H^1(\mathbb{R}^3)$,

$$F(u) \leq C_\epsilon u^2 + (\epsilon + \eta_\infty)|u|^{\bar{p}}.$$

Combining Lemma $B.1$ and (5.6), we obtain that for every $u \in D_c$,

$$\begin{aligned} I(u) &\geq \frac{a}{2}\|\nabla u\|_2^2 + \frac{b}{4}\|\nabla u\|_2^4 - C_\epsilon\|u\|_2^2 + (\epsilon + \eta_\infty)\|u\|_{\bar{p}}^{\bar{p}} \\ &\geq \frac{b}{4}\|\nabla u\|_2^4 - C_\epsilon c^2 - (\epsilon + \eta_\infty)c^{\bar{p}(1-\gamma_{\bar{p}})}C_{\bar{p}}^{\bar{p}}\|\nabla u\|_2^4 \\ &= \left(\frac{b}{4} - (\epsilon + \eta_\infty)c^{\bar{p}(1-\gamma_{\bar{p}})}C_{\bar{p}}^{\bar{p}}\right)\|\nabla u\|_2^4 - C_\epsilon c^2. \end{aligned}$$

We conclude that Lemma 5.5 holds true for sufficiently small ϵ. □

We now recall that for $u \in H^1(\mathbb{R}^3)\backslash\{0\}$ and $s > 0$, we define $s * u(x) := s^{\frac{3}{2}}u(sx)$. In particular, we know that $\|u\|_2 = \|s * u\|_2^2$.

Lemma 5.6 *Assume that conditions* (A_0), (A_3) *and* (5.7) *hold. Then* $\inf_{D_c} I < 0$.

Proof. Given $u \in L^\infty(\mathbb{R}^3) \cap D_c\backslash\{0\}$, we have

$$I(s * u) = \frac{as^2}{2}\|\nabla u\|_2^2 + s^4\left(\frac{b}{4}\|\nabla u\|_2^4 - \int_{\mathbb{R}^3} \frac{F(s^{\frac{3}{2}}u)}{\left(s^{\frac{3}{2}}\right)^{\bar{p}}}\,dx\right).$$

In view of (A_3), it holds that $F(s^{\frac{3}{2}}u) \geq 0$ and

$$I(s * u) \leq s^2\left[\frac{a}{2}\|\nabla u\|_2^2 + \frac{b}{4}\|\nabla u\|_2^4 - \int_{\mathbb{R}^3} \frac{F(s^{\frac{3}{2}}u)}{\left(s^{\frac{3}{2}}\right)^{\bar{p}}}\,dx\right]$$

for sufficiently small $\epsilon > 0$. Hence if $\eta_\infty = +\infty$, then

$$\lim_{s\to 0^+} \int_{\mathbb{R}^3} \frac{F(s^{\frac{3}{2}}u)}{\left(s^{\frac{3}{2}}\right)^{\bar{p}}}\,dx = +\infty.$$

If $\eta_0 < +\infty$, then

$$\limsup_{s\to 0^+}\left(\frac{a}{2}\|\nabla u\|_2^2 + \frac{b}{4}\|\nabla u\|_2^4 - \int_{\mathbb{R}^3} \frac{F(s^{\frac{3}{2}}u)}{\left(s^{\frac{3}{2}}\right)^{\bar{p}}}\,dx\right) \leq \frac{a}{2}\|\nabla u\|_2^2 + \frac{b}{4}\|\nabla u\|_2^4 - \eta_0\|u\|_{\bar{p}}^{\bar{p}}.$$

So the statement holds true if $u \in D_c$ and

$$\frac{a}{2}\|\nabla u\|_2^2 + \frac{b}{4}\|\nabla u\|_2^4 < \eta_0\|u\|_{\bar{p}}^{\bar{p}}. \tag{5.18}$$

By [91], let $\omega \in H^1(\mathbb{R}^3) \cap L^\infty(\mathbb{R}^3)$ be the unique positive radial solution of the problem

$$-\left(a+b\int_{\mathbb{R}^3}|\nabla v|^2\,dx\right)\Delta v+v=|v|^{\bar{p}-2}v \quad \text{in } \mathbb{R}^3. \tag{5.19}$$

If we fix $u(x):=\omega\left(\frac{x}{t}\right)$ for some $t>0$, then it follows from (5.7) that $u\in D_c$ and relation (5.18) holds true if

$$\frac{1}{\sqrt{2\eta_0}}<t<\min\left\{\left(\frac{c^2}{\|\omega\|_2^2}\right)^{\frac{1}{3}},1\right\}. \tag{5.20}$$

This completes the proof of Lemma 5.6. □

Lemma 5.7 *Let $d_1, d_2>0$ and suppose that (A_0)–(A_2) hold. Then*

(i) $m(\sqrt{d_1^2+d_2^2})\le m(d_1)+m(d_2)$;

(ii) if $m(d_1)$ or $m(d_2)$ are attained at a nontrivial function, then

$$m\left(\sqrt{d_1^2+d_2^2}\right)<m(d_1)+m(d_2).$$

Proof. (i) Fix $\epsilon>0$. Then there exist $u_{d_1}\in D_{d_1}\cap C_0^\infty(\mathbb{R}^3)$ and $u_{d_2}\in D_{d_2}\cap C_0^\infty(\mathbb{R}^3)$ such that $I(u_{d_1})\le m(d_1)+\epsilon$ and $I(u_{d_2})\le m(d_2)+\epsilon$. Moreover, we assume that the supports of u_{d_1} and u_{d_2} are disjoint. Hence

$$I(u_{d_1}+u_{d_2})\le I(u_{d_1})+I(u_{d_2})$$

and

$$\|u_{d_1}+u_{d_2}\|_2^2=\|u_{d_1}\|_2^2+\|u_{d_2}\|_2^2\le d_1^2+d_2^2.$$

This implies that

$$m\left(\sqrt{d_1^2+d_2^2}\right)\le I(u_{d_1}+u_{d_2})\le m(d_1)+m(d_2)+2\epsilon.$$

(ii) Without loss of generality, we assume that $m(u_{d_1})$ is achieved. Then we let $u\in D_{d_1}\backslash\{0\}$ such that $I(u)=m(d_1)$. On the other hand, we let $\epsilon>0$ and $v\in D_{d_2}$ such that

$$I(v)\le m(d_2)+\frac{d_2^2}{d_1^2+d_2^2}\epsilon<m(d_2)+\frac{d_2^2}{d_1^2}\epsilon.$$

Then for every $s\ge 1$, we obtain

$$\begin{aligned} m(\sqrt{s}d_2)\le I(v(\cdot/s^{1/3}))&=s\left[\frac{a}{2s^{\frac{2}{3}}}\|\nabla v\|_2^2+\frac{b}{4s^{\frac{1}{3}}}\|\nabla v\|_2^4-\int_{\mathbb{R}^3}F(v)\,dx\right]\le sI(v)\\ &\le s\left[m(d_2)+\frac{d_2^2}{d_1^2+d_2^2}\epsilon\right]<s\left[m(d_2)+\frac{d_2^2}{d_1^2}\epsilon\right]. \end{aligned} \tag{5.21}$$

If $d_1\ge d_2$, then it follows from (5.21) that

$$m(d_1)=m\left(\frac{d_1}{d_2}d_2\right)<\frac{d_1^2}{d_2^2}\left(m(d_2)+\frac{d_2^2}{d_1^2}+\frac{d_2^2}{d_1^2}\epsilon\right)=\frac{d_1^2}{d_2^2}m(d_2)+\epsilon,$$

while if $d_1<d_2$, then it follows from (5.21) that

$$m\left(\sqrt{d_1^2+d_2^2}\right)=m\left(\sqrt{\frac{d_1^2+d_2^2}{d_2^2}}d_2\right)\le\frac{d_1^2+d_2^2}{d_2^2}\left(m(d_2)+\frac{d_2^2}{d_1^2+d_2^2}\epsilon\right)=\frac{d_1^2+d_2^2}{d_2^2}m(d_2)+\epsilon.$$

Thus by the arbitrariness of ϵ, we see that

$$\begin{aligned} m(d_1) &\le \frac{d_1^2}{d_2^2} m(d_2) \quad \text{if } d_1 \ge d_2, \\ m\left(\sqrt{d_1^2+d_2^2}\right) &\le \frac{d_1^2+d_2^2}{d_2^2} m(d_2) \quad \text{if } d_1 < d_2. \end{aligned} \tag{5.22}$$

Furthermore, arguing as in (5.21), we find that $m(\sqrt{s}d_1) < sm(d_1)$ for every $s > 1$. Then in view of (5.22), if $d_1 \ge d_2$, then

$$m\left(\sqrt{d_1^2+d_2^2}\right) < \frac{d_1^2+d_2^2}{d_1^2} m(d_1) = m(d_1) + \frac{d_2^2}{d_1^2} m(d_1) \le m(d_1) + m(d_2).$$

If $d_1 < d_2$, then

$$m\left(\sqrt{d_1^2+d_2^2}\right) \le \frac{d_1^2+d_2^2}{d_2^2} m(d_2) = \frac{d_1^2}{d_2^2} m(d_2) + m(d_2) < m(d_1) + m(d_2).$$

This completes the proof of Lemma 5.7(ii). □

Lemma 5.8 *Assume that conditions* (A_0)–(A_3), (5.6) *and* (5.7) *hold. Then every minimizing sequence for* $I|_{D_c}$ *is relatively compact in* $L^p(\mathbb{R}^3)$, *up to translations, for all* $2 < p < 6$.

Proof. Suppose that $\{u_n\} \subset D_c$ satisfies $\lim_{n\to+\infty} I(u_n) = m(c) = \inf_{D_c} I$. Then by Lemma 5.5, the sequence $\{u_n\}$ is bounded. Moreover, if

$$\lim_{n\to+\infty} \max_{y\in\mathbb{R}^3} \int_{B(y,1)} u_n^2 \, dx = 0,$$

then by Lemma I.1 in [100], we obtain that $u_n \to 0$ in $L^{\bar{p}}(\mathbb{R}^3)$, whence $\int_{\mathbb{R}^3} F(u_n)\, dx \to 0$ from (A_1) and (A_2). In addition, up to subsequence, $u_n \rightharpoonup 0$ in $H^1(\mathbb{R}^3)$ and $u_n \to 0$ a.e. in $\mathbb{R}^3$. Therefore, $\liminf_{n\to+\infty} I(u_n) \ge 0$, which contradicts Lemma 5.6. Thus, there exist $u \in D\backslash\{0\}$ and $y_n \in \mathbb{R}^3$ such that, up to a subsequence,

$$v_n := u_n(\cdot - y_n) - u \rightharpoonup 0 \quad \text{in } H^1(\mathbb{R}^3)$$

and $v_n \to 0$ a.e. in $\mathbb{R}^3$. From Theorem 1 of Brezis and Lieb [14], we get

$$\lim_{n\to+\infty} [J(u_n) - J(v_n)] = J(u),$$

where

$$J(u) := \frac{a}{2}\|\nabla u\|_2^2 + \frac{bK}{4}\|\nabla u\|_2^2 - \int_{\mathbb{R}^3} F(u)\, dx$$

and $K := \lim_{n\to+\infty} \|\nabla u_n\|_2^2$. If

$$\lim_{n\to+\infty} \max_{y\in\mathbb{R}^3} \int_{B(y,1)} v_n^2 \, dx = 0,$$

then using Lemma I.1 in [100] again, the statement holds true. If not, as before, there exist $v \in H^1(\mathbb{R}^3)\backslash\{0\}$ and $z_n \in \mathbb{R}^3$ such that, up to a subsequence, $w_n \to 0$ a.e. in $\mathbb{R}^3$,

$$w_n := v_n(\cdot - z_n) - v \rightharpoonup 0 \quad \text{in } H^1(\mathbb{R}^3),$$

$$\lim_{n\to+\infty} [J(v_n) - J(w_n)] = J(v), \quad \text{and} \quad \lim_{n\to+\infty} \left[\|u_n\|_2^2 - \|w_n\|_2^2\right] = \|u\|_2^2 + \|v\|_2^2.$$

Then

$$\liminf_{n\to+\infty} \|u_n\|_2^2 - \|u\|_2^2 - \|v\|_2^2 = \liminf_{n\to+\infty} \|w_n\|_2^2 =: d^2 \geq 0.$$

If $d > 0$, we let $\bar{w}_n := \frac{d}{\|w_n\|_2} w_n \in S_d$. It follows that

$$\lim_{n\to+\infty} [J(w_n) - J(\bar{w}_n)] = 0.$$

Using now Lemma 5.7(i), we obtain

$$\begin{aligned} m(c) &= \lim_{n\to+\infty} I(u_n) \geq J(u) + J(v) + \lim_{n\to+\infty} J(w_n) = J(u) + J(v) + \lim_{n\to+\infty} J(\bar{w}_n) \\ &\geq m(c_1) + m(c_2) + m(d) \geq m\left(\sqrt{c_1^2 + c_2^2 + d^2}\right) \geq m(c), \end{aligned}$$

where $c_1 := \|u\|_2$ and $c_2 := \|v\|_2$. This implies that the above inequalities are in fact equalities. In particular, since $m(c_1)$ and $m(c_2)$ are achieved, by Lemma 5.7(ii), we know that

$$m(c) \geq m(c_1) + m(c_2) + m(d) > m\left(\sqrt{c_1^2 + c_2^2 + d^2}\right) \geq m(c).$$

This is impossible.

If $d = 0$, then $w_n \to 0$ in $L^2(\mathbb{R}^3)$ and by (A_1) and (A_2), $\int_{\mathbb{R}^3} F(w_n)\,dx \to 0$ as $n \to +\infty$. Thus, $\liminf_{n\to+\infty} J(w_n) \geq 0$. It follows that

$$m(c) = \lim_{n\to+\infty} I(u_n) \geq J(u)+J(v)+ \lim_{n\to+\infty} J(w_n) \geq m(c_1)+m(c_2) \geq m\left(\sqrt{c_1^2 + c_2^2}\right) \geq m(c).$$

We can get a contradiction as before. The proof of Lemma 5.8 is now complete. □

Lemma 5.9 *Assume that conditions* (A_0)*–*(A_3)*,* (5.6) *and* (5.7) *hold. Then* $\inf_{D_c} I$ *is achieved.*

Proof. Suppose that $\{u_n\} \subset D_c$ satisfies $\lim_{n\to+\infty} I(u_n) = \inf_{D_c} I$. In view of Lemmas 5.5 and 5.8, there exists $u \in D$ such that, up to a subsequence and translations, $u_n \rightharpoonup u$ in $H^1(\mathbb{R}^3)$ and $u_n \to u$ in $L^p(\mathbb{R}^3)$ $(2 < p < 6)$, and $u_n \to u$ a.e. in $\mathbb{R}^3$ as $n \to +\infty$, which yields

$$\inf_{D_c} I = \lim_{n\to\infty} I(u_n) \geq I(u) \geq \inf_{D_c} I.$$

We conclude that $\inf_{D_c} I$ is achieved. □

Proof of Theorem 5.2. From Lemmas 5.6, 5.9, we see that there exists $u \in D_c$ such that $I(u) = \inf_{D_c} I < 0$. Then it follows from Proposition A.1 of Mederski and Schino [114] that there exists $\lambda \geq 0$ such that

$$-\left(a + b\int_{\mathbb{R}^3} |\nabla u|^2\,dx\right)\Delta u + \lambda u = f(u).$$

Assume by contradiction that $\lambda = 0$. By bootstrap regularity, $u \in W^{2,p}_{\text{loc}}(\mathbb{R}^3)$ for every $p < \infty$. Then u satisfies the following Pohozaev identity

$$a\int_{\mathbb{R}^3} |\nabla u|^2\,dx + b\left(\int_{\mathbb{R}^3} |\nabla u|^2\,dx\right)^2 = 6\int_{\mathbb{R}^3} F(u)\,dx,$$

which yields that

$$0 > I(u) = \frac{a}{3}\|\nabla u\|_2^2 + \frac{b}{12}\|\nabla u\|_2^4.$$

This contradiction implies that $(\lambda, u) \in (0, +\infty) \times H^1(\mathbb{R}^3)$ solves problem (5.1). If f is locally Lipschitz continuous, we can argue as in the proof of Theorem 1.4 in Jeanjean and Lu [79], otherwise, the result follows from the properties of the Schwarz rearrangements. □

5.3 Reaction L^2-critical at the origin and L^2-supercritical at infinity

In this section, we establish the existence of ground state solutions for problems (5.1) and (5.5) in the case that the nonlinearity f is L^2-critical at the origin and L^2-supercritical at infinity and we also give the relation between these ground states. First of all, by Lemma 2.1 of Bieganowski and Mederski [11], we have the following result.

Lemma 5.10 *Let $f_1, f_2 \in C(\mathbb{R})$ and suppose that there exists $C > 0$ such that $|f_1(u)| + |f_2(u)| \le C(|u|^2 + |u|^6)$ for every $u \in H^1(\mathbb{R}^3)$. Then $f_1 \preceq f_2$ if and only if $f_1 \le f_2$ and*

$$\int_{\mathbb{R}^3} [f_1(u) - f_2(u)]\, dx < 0$$

for every $u \in H^1(\mathbb{R}^3)\backslash\{0\}$.

In what follows, we always assume that condition (B_0) holds. Our purpose is to establish the relation between the energy functional I and $\mathcal{P} \cap D_c$.

Lemma 5.11 *Assume that conditions (B_1)–(B_3) and (B_5) hold. Then*

$$\inf\left\{\|\nabla u\|_2^2 : u \in \mathcal{P} \cap D_c\right\} > 0.$$

Proof. From (B_1)–(B_3), (B_5) and Lemma B.1 (Gagliardo–Nirenberg inequality), we infer that for every $\epsilon > 0$, there exists $C_\epsilon > 0$ such that for every $u \in \mathcal{P} \cap D_c$,

$$\begin{aligned} a\|\nabla u\|_2^2 + b\|\nabla u\|_2^4 &= \frac{3}{2}\int_{\mathbb{R}^3} H(u)\, dx \le 6\left(C_\epsilon \|u\|_6^6 + (\epsilon + \eta)\|u\|_{\bar{p}}^{\bar{p}}\right) \\ &\le 6\left(C_\epsilon C_6^6 \|\nabla u\|_2^6 + (\epsilon + \eta) C_{\bar{p}}^{\bar{p}} c^{\frac{4}{3}} \|\nabla u\|_2^4\right), \end{aligned}$$

which implies that

$$a\|\nabla u\|_2^2 \le 6\left(C_\epsilon C_6^6 \|\nabla u\|_2^6 + (\epsilon + \eta) C_{\bar{p}}^{\bar{p}} c^{\frac{4}{3}} \|\nabla u\|_2^4\right).$$

We conclude that $\inf\left\{\|\nabla u\|_2^2 : u \in \mathcal{P} \cap D_c\right\} > 0$. □

Now, recall that for $u \in H^1(\mathbb{R}^3)\backslash\{0\}$ and $s > 0$, we define the maps $(s * u)(x) := s^{\frac{3}{2}} u(sx)$ and $\psi(s) := I(s * u)$.

Lemma 5.12 *Assume that conditions (B_1)–(B_5) hold and let $u \in H^1(\mathbb{R}^3)\backslash\{0\}$ be such that*

$$\eta < \frac{b\|\nabla u\|_2^4}{8\|u\|_{\bar{p}}^{\bar{p}}}. \tag{5.23}$$

Then there exist $s_1 = s_1(u) > 0$ and $s_2 = s_2(u) \ge s_1$ such that every $s \in [s_1, s_2]$ is a global maximizer for ψ and ψ is increasing on $(0, s_1)$ and decreasing on $(s_2, +\infty)$. Moreover, $s * u \in \mathcal{P}$ if and only if $s \in [s_1, s_2]$, $P(s * u) > 0$ if and only if $s \in (0, s_1)$, and $P(s * u) < 0$ if and only if $s > s_2$. If $(B_4, \preceq)$ holds, then $s_1 = s_2$.

Proof. From (B_1), we see that

$$\psi(s) = \frac{as^2}{2}\int_{\mathbb{R}^3} |\nabla u|^2\, dx + \frac{bs^4}{4}\left(\int_{\mathbb{R}^3} |\nabla u|^2\, dx\right)^2 - \int_{\mathbb{R}^3} \frac{F(s^{\frac{3}{2}} u)}{s^3}\, dx \to 0$$

as $s \to 0^+$ and by (B$_2$), one has $\lim_{s\to\infty} \psi(s) = -\infty$. On the other hand, it follows from (B$_1$) and (B$_3$) that for every $\epsilon > 0$, there exists $C_\epsilon > 0$ such that

$$F(u) \leq (\epsilon + \eta)|u|^{\bar{p}} + C_\epsilon |u|^6.$$

Thus by (5.23), we have

$$\psi(s) \geq \frac{as^2}{2}\int_{\mathbb{R}^3} |\nabla u|^2\,dx + \frac{bs^4}{4}\left(\int_{\mathbb{R}^3} |\nabla u|^2\,dx\right)^2 - s^4(\eta+\epsilon)\int_{\mathbb{R}^3} |u|^{\bar{p}}\,dx - C_\epsilon s^6 \int_{\mathbb{R}^3} |u|^6\,dx > 0$$

for sufficiently small ϵ and s. Thus, there exists an interval $[s_1, s_2] \subset (0, +\infty)$ such that $\psi|_{[a,b]} = \max \psi$. Furthermore,

$$\psi'(s) = as\int_{\mathbb{R}^3} |\nabla u|^2\,dx + bs^3\left(\int_{\mathbb{R}^3} |\nabla u|^2\,dx\right)^2 - \frac{3}{2}\int_{\mathbb{R}^3} \frac{H(s^{3/2}u)}{s^5}\,dx$$

and the function

$$s \in (0, +\infty) \mapsto \int_{\mathbb{R}^3} \frac{H(s^{3/2}u)}{s^5}\,dx$$

is nondecreasing (resp. increasing) by (B$_4$) (resp. (B$_4$, $\preceq$) and Lemma 5.10) and tends to $+\infty$ as $s \to +\infty$ by (B$_2$) and (B$_5$). This implies that $\psi'(s) > 0$ if $s \in (0, s_1)$ and $\psi'(s) < 0$ if $s > s_2$ and that $s_1 = s_2$ if (B$_4$, $\preceq$) holds.

Finally, we obtain

$$s\psi'(s) = as^2\int_{\mathbb{R}^3} |\nabla u|^2\,dx + bs^4\left(\int_{\mathbb{R}^3} |\nabla u|^2\,dx\right)^2 - \frac{3}{2}\int_{\mathbb{R}^3} \frac{H(s^{3/2}u)}{s^3}\,dx = P(s * u).$$

We conclude that $s*u \in \mathcal{P}$ if and only if $s \in [s_1, s_2]$, $P(s*u) > 0$ if and only if $s \in (0, s_1)$, and $P(s * u) < 0$ if and only if $s > s_2$. The proof of Lemma 5.12 is completed. □

Remark 5.4 *We observe that* (5.8) *implies* (5.23) *provided that* $u \in D_c$. *Indeed, it follows from* (5.8) *and the Gagliardo–Nirenberg inequality that*

$$\eta\|u\|_{\bar{p}}^{\bar{p}} \leq \eta \cdot C_{\bar{p}}^{\bar{p}}\|\nabla u\|_2^4\|u\|_2^{\bar{p}(1-\gamma_{\bar{p}})} \leq \eta C_{\bar{p}}^{\bar{p}} c^{\bar{p}(1-\gamma_{\bar{p}})}\|\nabla u\|_2^4 < \frac{b}{8}\|\nabla u\|_2^4.$$

Lemma 5.13 *Assume that conditions* (B_1)*–*(B_5) *and* (5.8) *hold. Then* I *is coercive on* $\mathcal{P} \cap D_c$.

Proof. For all $u \in \mathcal{P}$, by (B$_5$), it holds that

$$I(u) = I(u) - \frac{1}{4}P(u) = \frac{a}{4}\int_{\mathbb{R}^3} |\nabla u|^2\,dx + \int_{\mathbb{R}^3} \frac{3}{8}H(u) - G(u)\,dx \geq 0.$$

Thus, I is nonnegative on $\mathcal{P} \cap D_c$.

In the following, suppose that $\{u_n\} \subset \mathcal{P} \cap D_c$ satisfies $\lim_{n\to+\infty} \|\nabla u_n\|_2 = +\infty$. Then we define

$$s_n := \|\nabla u_n\|_2^{-1} > 0 \quad \text{and} \quad v_n := s_n * u_n.$$

Note that when $s_n \to 0$, $\|v_n\|_2 = \|u\|_2 \leq c$ and $\|\nabla v_n\|_2^2 = 1$, which yields that $\{v_n\}$ is bounded in $H^1(\mathbb{R}^3)$. If we assume that

$$\limsup_{n\to+\infty} \max_{y\in\mathbb{R}^3} \int_{B(y,1)} |v_n|^2\,dx > 0,$$

then there exist $\{y_n\} \subset \mathbb{R}^3$ and $v \in H^1(\mathbb{R}^3)$ such that, up to a subsequence, $v_n(\cdot + y_n) \rightharpoonup v \neq 0$ in $H^1(\mathbb{R}^3)$ and $v_n(\cdot + y_n) \to v$ a.e. in $\mathbb{R}^3$. Thus, by (B_2),

$$0 \leq \frac{I(u_n)}{\|\nabla u_n\|_2^4} \leq \frac{a}{2\|\nabla u_n\|_2^2} + \frac{b}{4} - \int_{\mathbb{R}^3} \frac{F(u_n)}{\|\nabla u_n\|_2^4}\, dx = \frac{a}{2\|\nabla u_n\|_2^2} + \frac{b}{4} - s_n^7 \int_{\mathbb{R}^3} F(u_n(s_n x))\, dx$$

$$= \frac{a}{2\|\nabla u_n\|_2^2} + \frac{b}{4} - s_n^7 \int_{\mathbb{R}^3} F(s_n^{-\frac{3}{2}} v_n)\, dx = \frac{a}{2\|\nabla u_n\|_2^2} + \frac{b}{4} - \int_{\mathbb{R}^3} \frac{F(s_n^{-\frac{3}{2}} v_n)}{|s_n^{-\frac{3}{2}} v_n|^{\bar{p}}} |v_n|^{\bar{p}}\, dx$$

$$= \frac{a}{2\|\nabla u_n\|_2^2} + \frac{b}{4} - \int_{\mathbb{R}^3} \frac{F(s_n^{-\frac{3}{2}} v_n(x+y_n))}{|s_n^{-\frac{3}{2}} v_n(x+y_n)|^{\bar{p}}} |v_n(x+y_n)|^{\bar{p}}\, dx \to -\infty.$$

This contradiction implies that

$$\lim_{n\to+\infty} \max_{y\in\mathbb{R}^3} \int_{B(y,1)} |v_n|^2\, dx = 0.$$

Then from Lemma I.1 in Lions [100], we deduce that $v_n \to 0$ in $L^{\bar{p}}(\mathbb{R}^3)$. Since $s_n^{-1} * v_n = u_n \in \mathcal{P}$, it follows from Lemma 5.12 that for every $s > 0$,

$$I(u_n) = I(s_n^{-1} * v_n) \geq I(s * v_n) = \frac{as^2}{2} + \frac{bs^4}{4} - s^3 \int_{\mathbb{R}^3} F(s^{\frac{3}{2}} v_n(s\cdot))\, dx.$$

Since

$$\lim_{n\to+\infty} \int_{\mathbb{R}^3} F(s^{3/2} v_n(s\cdot))\, dx = 0,$$

we obtain that $\liminf_{n\to+\infty} I(u_n) \geq \frac{as^2}{2} + \frac{bs^4}{4}$ for every $s > 0$, that is, $\lim_{n\to+\infty} I(u_n) = +\infty$. The proof of Lemma 5.13 is now complete. □

Lemma 5.14 *Assume that conditions* (B_1)–(B_5) *and* (5.8) *hold. Then* $\rho := \inf_{\mathcal{P}\cap D_c} I > 0$.

Proof. From the Gagliardo–Nirenberg inequality and relation (5.8), we obtain that for every $\epsilon > 0$ there exists $C_\epsilon > 0$ such that

$$\int_{\mathbb{R}^3} F(u)\, dx \leq C_\epsilon C_6^6 \|\nabla u\|_2^6 + (\epsilon + \eta) C_{\bar{p}}^{\bar{p}} c^{\bar{p}(1-\gamma_{\bar{p}})} \|\nabla u\|_2^4$$
$$\leq \left(C_\epsilon C_6^6 \|\nabla u\|_2^4 + \epsilon C_{\bar{p}}^{\bar{p}} c^{\bar{p}(1-\gamma_{\bar{p}})} + \frac{b}{8} \right) \|\nabla u\|_2^4.$$

Choosing

$$\delta := \frac{1}{(\frac{8}{b} C_\epsilon C_6^6)^{\frac{1}{4}}} \quad \text{and} \quad \epsilon := \frac{b}{8} \cdot \frac{1}{C_{\bar{p}}^{\bar{p}} c^{\bar{p}(1-\gamma_{\bar{p}})}},$$

and since $\|\nabla u\|_2 \leq \delta$, we obtain

$$\int_{\mathbb{R}^3} F(u)\, dx \leq \frac{3b}{8} \|\nabla u\|_2^4,$$

which yields that $I(u) \geq \frac{b}{8}\|\nabla u\|_2^4$. Then taking $u \in \mathcal{P} \cap D_c$ such that $\|\nabla u\|_2 \leq \delta$, we obtain $I(u) \geq \frac{b}{8}\|\nabla u\|_2^4$.

Define

$$s := \frac{\delta}{\|\nabla u\|_2^4} \quad \text{and} \quad v := s * u.$$

Obviously, $\|v\|_2 = \|u\|_2 \le c$ and $\|\nabla v\|_2 = \delta$. From Lemma 5.12, we get

$$I(u) \ge I(v) \ge \frac{b}{8}\|\nabla v\|_2^4 = \frac{b}{8}\cdot \delta^4 > 0.$$

The proof of Lemma 5.14 is completed. □

Lemma 5.15 *Assume that* $\kappa \in (0, +\infty)$, *conditions* (B_1)–(B_5), (5.8) *hold, and there exist* $\bar{p} \le p \le 6$, $\bar{p} \le q < 6$ *such that*

$$\liminf_{|u|\to 0} \frac{\bar{F}(u)}{|u|^p} > 0 \quad \text{and} \quad \liminf_{|u|\to\infty} \frac{\bar{F}(u)}{|u|^q} > 0 \tag{5.24}$$

with $\max\{p,q\}/2 - \min\{p,q\} < -1$. *Then*

$$\begin{aligned} \rho &< \frac{a\kappa^{-\frac{1}{2}}}{3}\left(\frac{1}{2}(b\kappa^{-\frac{1}{2}}\mathcal{S}^3 + \sqrt{(b\kappa^{-\frac{1}{2}}\mathcal{S}^3)^2 + 4a\mathcal{S}^3})\right) \\ &+ \frac{b\kappa^{-\frac{1}{2}}}{12}\left(\frac{1}{2}(b\kappa^{-\frac{1}{2}}\mathcal{S}^3 + \sqrt{(b\kappa^{-\frac{1}{2}}\mathcal{S}^3)^2 + 4a\mathcal{S}^3})\right)^2. \end{aligned} \tag{5.25}$$

Proof. Consider the Kirchhoff equation

$$-\left(a + b\int_{\mathbb{R}^3} |\nabla u|^2\,dx\right)\Delta u = u^5 \quad \text{in } \mathbb{R}^3. \tag{5.26}$$

By straightforward computation, the function $\Phi_\epsilon(x) = K\phi_\epsilon =: K\frac{(3\epsilon^2)^{\frac{1}{4}}}{(\epsilon^2+|x|^2)^{\frac{1}{2}}}$, where $\epsilon > 0$, $K > 0$ and $K^2 = \frac{1}{2}\left(b\mathcal{S}^{\frac{3}{2}} + \sqrt{b^2\mathcal{S}^3 + 4a}\right)$ is a positive ground state solution of problem (5.26). Here, the best Sobolev constant $\mathcal{S}$ is achieved by ϕ_ϵ. Since $\Phi_\epsilon \notin L^2(\mathbb{R}^3)$, let $0 \le \varphi \in C_0^\infty(\mathbb{R}^3)$ radial such that $\varphi \equiv 1$ in B_1 and $\varphi \equiv 0$ in $\mathbb{R}^3 \backslash B_2$, where B_r stands for the closed ball centered at 0 of radius r. For every $\epsilon > 0$, define

$$u_\epsilon^\kappa := \kappa^{-\frac{1}{4}}\varphi\Phi_\epsilon \quad \text{and} \quad v_\epsilon^\kappa := \frac{c}{\|u_\epsilon^\kappa\|_2}u_\epsilon^\kappa \in D_c.$$

By Struwe [141, p. 179], we obtain that

$$\|\nabla(\varphi\Phi_\epsilon)\|_2^2 = K^2\mathcal{S}^{\frac{3}{2}} + O(\epsilon),$$

$$\|\varphi\Phi_\epsilon\|_6^2 = K^{\frac{1}{3}}\mathcal{S}^{\frac{1}{2}} + O(\epsilon^2),$$

$$\|\varphi\Phi_\epsilon\|_2^2 = \hat{C}\epsilon + O(\epsilon^2),$$

where $\hat{C}$ depends only on φ.

We now show that

$$\int_{\mathbb{R}^3} |\varphi\Phi_\epsilon|^i \chi_{\{\varphi\Phi_\epsilon \ge 1\}}\,dx \ge \tilde{C}\epsilon^{3-\frac{1}{2}i}$$

for some positive constant $\tilde{C}$ and sufficiently small $\epsilon > 0$, where $i \in \{p,q\}$ and χ_Γ stands for the characteristic function of Γ. Indeed, suppose that $|x|^2 \le K^2(3\epsilon^2)^{\frac{1}{2}} - \epsilon^2$. If ϵ is sufficiently small, then $x \in B_1$ and $\varphi(x)\Phi_\epsilon(x) = \Phi_\epsilon(x) \ge 1$. It follows that

$$\int_{\mathbb{R}^3} |\varphi\Phi_\epsilon|^i \chi_{\{\varphi\Phi_\epsilon\}\ge 1}\,dx \ge \int_{\left\{|x|\le \left(K^2(3\epsilon^2)^{\frac{1}{2}} - \epsilon^2\right)^{\frac{1}{2}}\right\}} |\Phi_\epsilon|^i\,dx = \tilde{C}\epsilon^{3-\frac{1}{2}i}.$$

Next, we define $s_\epsilon > 0$ such that $s_\epsilon * v_\epsilon^\kappa \in \mathcal{P}$. Based on the argument of Lemma 5.6, for every $\delta > 0$, there exists $C_\delta > 0$ not depending on ϵ and κ such that

$$\begin{aligned} as_\epsilon^2\|\nabla v_\epsilon^\kappa\|_2^2 + bs_\epsilon^4\|\nabla v_\epsilon^\kappa\|_2^4 &\le 6\left(s_\epsilon^4(\eta+\delta)\|v_\epsilon^\kappa\|_{\bar p}^{\bar p} + s_\epsilon^6 C_\delta\kappa\|v_\epsilon^\kappa\|_6^6\right) \\ &\le 6\left(s_\epsilon^4(\eta+\delta)C_{\bar p}^{\bar p}c^{\frac{4}{3}}\|\nabla v_\epsilon^\kappa\|_2^4 + s_\epsilon^6 C_\delta\kappa\|v_\epsilon^\kappa\|_6^6\right). \end{aligned}$$

Then by using (5.8), we obtain that

$$s_\epsilon^4 \ge \frac{a\|\nabla v_\epsilon^\kappa\|_2^2}{C_\delta\kappa\|v_\epsilon^\kappa\|_6^6} = \frac{a}{C_\delta\kappa}\cdot\frac{\|\nabla(\varphi\Phi_\epsilon)\|_2^2\cdot\|u_\epsilon^\kappa\|_2^4}{\|\varphi\Phi_\epsilon\|_6^6}.$$

In addition, $\bar F(u) > 0$ for $u \ne 0$. Then taking ϵ sufficiently small, we get

$$\begin{aligned} \rho \le I(s_\epsilon * v_\epsilon^\kappa) &\le -\int_{\mathbb{R}^3}\bar F(s_\epsilon * v_\epsilon^\kappa)\,dx \\ &+ \max_{s>0}\left[\frac{as^2}{2}\int_{\mathbb{R}^3}|\nabla v_\epsilon^\kappa|^2\,dx + \frac{bs^4}{4}\left(\int_{\mathbb{R}^3}|\nabla v_\epsilon^\kappa|^2\,dx\right)^2 - \frac{s^6}{6}\kappa\int_{\mathbb{R}^3}|v_\epsilon^\kappa|^6\,dx\right] \\ &\le -\int_{\mathbb{R}^3}\bar F(s_\epsilon * v_\epsilon^\kappa)\,dx + \frac{a^2}{2}t_*\|\nabla v_\epsilon^\kappa\|_2^2 + \frac{b}{4}t_*^2\|\nabla v_\epsilon^\kappa\|_2^4 - \frac{\kappa}{6}t_*^3\|v_\epsilon^\kappa\|_6^6 \\ &= -\int_{\mathbb{R}^3}\bar F(s_\epsilon * v_\epsilon^\kappa)\,dx + \frac{a\kappa^{-\frac{1}{2}}}{3}\left(\frac{1}{2}(b\kappa^{-\frac{1}{2}}\mathcal{S}^3 + \sqrt{(b\kappa^{-\frac{1}{2}}\mathcal{S}^3)^2 + 4a\mathcal{S}^3})\right) \\ &+ \frac{b\kappa^{-\frac{1}{2}}}{12}\left(\frac{1}{2}(b\kappa^{-\frac{1}{2}}\mathcal{S}^3 + \sqrt{(b\kappa^{-\frac{1}{2}}\mathcal{S}^3)^2 + 4a\mathcal{S}^3})\right)^2 + O(\epsilon), \end{aligned}$$

where

$$t_* := \frac{b\|\nabla v_\epsilon^\kappa\|_2^4 + \sqrt{b^2\|\nabla v_\epsilon^\kappa\|_2^8 + 4a\kappa\|v_\epsilon^\kappa\|_6^6\cdot\|\nabla v_\epsilon^\kappa\|_2^2}}{2\kappa\|v_\epsilon^\kappa\|_6^6}.$$

Now we estimate $\int_{\mathbb{R}^3}\bar F(s_\epsilon * v_\epsilon^\kappa)\,dx$ as $\epsilon \to 0^+$. Combining (5.24), (B_2) and (B_3), we find that there exists $C > 0$ such that $\bar F(u) \ge C|u|^p$ if $|u| \le 1$ and $\bar F(u) \ge C|u|^q$ if $|u| > 1$. Hence in virtue of $p, q < 6$, we obtain

$$\begin{aligned} \int_{\mathbb{R}^3}\bar F(s_\epsilon * v_\epsilon^\kappa)\,dx &\ge Cs_\epsilon^{3(\frac{p}{2}-1)}\int_{\mathbb{R}^3}|v_\epsilon^\kappa|^p\chi_{\left\{|(s_\epsilon^\kappa)^{\frac{3}{2}}v_\epsilon^\kappa|\le 1\right\}}\,dx + Cs_\epsilon^{3(\frac{q}{2}-1)}\int_{\mathbb{R}^3}|v_\epsilon^\kappa|^q\chi_{\left\{|(s_\epsilon^\kappa)^{\frac{3}{2}}v_\epsilon^\kappa|>1\right\}}\,dx \\ &\ge C'\|\varphi\Phi_\epsilon\|_2^{3(\frac{p}{2}-1)-p}\int_{\mathbb{R}^3}|\varphi\Phi_\epsilon|^p\chi_{\left\{|(s_\epsilon^\kappa)^{\frac{3}{2}}v_\epsilon^\kappa|\le 1\right\}}\,dx \\ &\quad + C'\|\varphi\Phi_\epsilon\|_2^{3(\frac{q}{2}-1)-q}\int_{\mathbb{R}^3}|\varphi\Phi_\epsilon|^q\chi_{\left\{|(s_\epsilon^\kappa)^{\frac{3}{2}}v_\epsilon^\kappa|>1\right\}}\,dx \\ &\ge C'\|\varphi\Phi_\epsilon\|_2^{(\frac{3}{2}-1)p-N}\int_{\mathbb{R}^3}|\varphi\Phi_\epsilon|^p\chi_{\left\{|(s_\epsilon^\kappa)^{\frac{3}{2}}v_\epsilon^\kappa|\le 1\right\}}\chi_{\{\varphi\Phi_\epsilon\ge 1\}}\,dx \\ &\quad + C'\|\varphi\Phi_\epsilon\|_2^{(\frac{3}{2}-1)q-N}\int_{\mathbb{R}^3}|\varphi\Phi_\epsilon|^q\chi_{\left\{|(s_\epsilon^\kappa)^{\frac{3}{2}}v_\epsilon^\kappa|>1\right\}}\chi_{\{\varphi\Phi_\epsilon\ge 1\}}\,dx \\ &\ge C'\min\left\{\|\varphi\Phi_\epsilon\|_2^{(\frac{3}{2}-1)p-N}, \|\varphi\Phi_\epsilon\|_2^{(\frac{3}{2}-1)q-N}\right\}\cdot\int_{\mathbb{R}^3}|\varphi\Phi_\epsilon|^{\min\{p,q\}}\chi_{\{\varphi\Phi_\epsilon\ge 1\}}\,dx \\ &\ge C''\|\varphi\Phi_\epsilon\|_2^{\frac{1}{2}\max\{p,q\}-3}\epsilon^{3-\frac{1}{2}\min\{p,q\}} \\ &\ge C''\frac{\epsilon^{3-\frac{\min\{p,q\}}{2}}}{(\hat C\epsilon)^{\frac{3}{2}-\frac{\max\{p,q\}}{4}} + O(\epsilon^{3-\frac{\max\{p,q\}}{2}})} \ge C'''\epsilon^{\frac{3+\frac{\max\{p,q\}}{2}-\frac{\min\{p,q\}}{2}}{2}}. \end{aligned}$$

Here, $C', C'', C''' > 0$ are constants and $0 < \frac{3+\frac{\max\{p,q\}}{2}-\frac{\min\{p,q\}}{2}}{2} < 1$. Then the following properties hold

$$O(\epsilon) - C'''\epsilon^{\frac{3+\frac{\max\{p,q\}}{2}-\frac{\min\{p,q\}}{2}}{2}} < 0$$

for sufficiently small ϵ and ρ is chosen so that

$$\rho < \kappa^{-\frac{1}{2}}\left[\frac{a}{3}\left(\frac{1}{2}(b\kappa^{-\frac{1}{2}}\mathcal{S}^3 + \sqrt{(b\kappa^{-\frac{1}{2}}\mathcal{S}^3)^2 + 4a\mathcal{S}^3})\right) + \frac{b}{12}\left(\frac{1}{2}(b\kappa^{-\frac{1}{2}}\mathcal{S}^3 + \sqrt{(b\kappa^{-\frac{1}{2}}\mathcal{S}^3)^2 + 4a\mathcal{S}^3})\right)^2\right].$$

The proof of Lemma 5.15 is completed. □

Comparing with (5.24), we give other sufficient conditions for (5.25) to hold.

Lemma 5.16 *Assume that conditions* (B_1)*–*(B_5) *hold and* $\kappa \in (0,\infty)$. *Then*

> *(i) If* $\eta = 0$ *and* $\lim_{t\to 0}\frac{\bar{F}(t)}{t^6} = \infty$, *then there exists* $c_0 > 0$ *such that* (5.25) *is satisfied provided that* $c > c_0$.
>
> *(ii) If relation* (5.8) *holds and* $\lim_{t\to+\infty}\frac{\bar{F}(t)}{|t|^{\bar{p}}} = +\infty$, *then there exists* $\kappa_0 > 0$ *such that* (5.25) *is satisfied provided that* $\kappa < \kappa_0$.

Proof. (i) We only need to show that $\rho \to 0$ as $c \to +\infty$. By contradiction, suppose that $c_n \to +\infty$. We take $u \in L^\infty(\mathbb{R}^3)$ such that $\|u\|_2 = 1$. In particular, we also assume that $c_n > 1$. Denote $u_n := c_n u$ satisfying $\|u_n\|_2 = c_n$. Then in view of Lemma 5.12, there exists $s_n > 0$ such that $v_n := s_n^{\frac{3}{2}} u_n(s_n\cdot) \in \mathcal{P}$. Note that $\|v_n\|_2 = \|u_n\|_2$. Then

$$\begin{aligned} 0 < \inf\{I(w) : w \in \mathcal{P}, \|w\|_2 \le c_n\} \le I(v_n) &\le \frac{a}{2}\|\nabla v_n\|_2^2 + \frac{b}{4}\|\nabla v_n\|_2^4 \\ &= \frac{a}{2}(s_n c_n)^2\|\nabla u\|_2^2 + \frac{b}{4}(s_n c_n)^4\|\nabla u\|_2^4. \end{aligned}$$

Thus, it is enough for us to prove that $s_n c_n \to 0$. We observe that

$$\begin{aligned} a(s_n c_n)^2\|\nabla u\|_2^2 + b(s_n c_n)^4\|\nabla u\|_2^4 = a\|\nabla v_n\|_2^2 + b\|\nabla v_n\|_2^4 &= \frac{3}{2}\int_{\mathbb{R}^3} H(v_n)\,dx \\ &= \frac{3}{2}s_n^{-3}\int_{\mathbb{R}^3} H(s_n^{\frac{3}{2}} c_n u)\,dx \end{aligned}$$

and

$$\frac{a}{(s_n c_n)^2}\|\nabla u\|_2^2 + b\|\nabla u\|_2^4 = \frac{3}{2}s_n^{-7}c_n^{-4}\int_{\mathbb{R}^3} H(s_n^{\frac{3}{2}} c_n u)\,dx = \frac{3}{2}c_n^{\frac{2}{3}}\int_{\mathbb{R}^3}\frac{H(s_n^{\frac{3}{2}} c_n u)}{|s_n^{\frac{3}{2}} c_n u|^{\bar{p}}}|u|^{\bar{p}}\,dx,$$

which implies that

$$\lim_{n\to+\infty}\int_{\mathbb{R}^3}\frac{H(s_n^{\frac{3}{2}} c_n u)}{|s_n^{\frac{3}{2}} c_n u|^{\bar{p}}}\,dx = 0.$$

This means that $s_n^{\frac{3}{2}} c_n \to 0$.

In what follows, we fix $\epsilon > 0$. In view of (B_5) and $\lim_{t\to 0}\frac{F(t)}{t^6} = \infty$, there holds

$$H(s) \ge \frac{4}{3}F(s) \ge \epsilon^{-1}s^6$$

for $|s|$ small enough. Then considering $u \in L^\infty(\mathbb{R}^3)$, for n large enough, we have

$$\frac{a}{(s_n c_n)^2}\|\nabla u\|_2^2 + b\|\nabla u\|_2^4 = \frac{3}{2}s_n^{-7}c_n^{-4}\int_{\mathbb{R}^3} H(s_n^{\frac{3}{2}}c_n u)\,dx \geq \epsilon^{-1}\frac{3}{2}s_n^{-7}c_n^{-4}|s_n^{\frac{3}{2}}c_n|^6\|u\|_6^6$$
$$= \epsilon^{-1}\frac{3}{2}(s_n c_n)^2\|u\|_6^6,$$

which implies that $s_n c_n \to 0$ as $n \to +\infty$.

(ii) Take any $u \in D_c\backslash\{0\}$ and observe that (5.23) holds. Based on Lemma 5.12 with $\kappa = 0$, there exists $\tilde{s} > 0$ such that $\tilde{s} * u \in \mathcal{P}$,

$$c \leq I(\tilde{s} * u) \leq \max_{s>0} I(s * u) \leq \max_{s>0}\left(\frac{as^2}{2}\|\nabla u\|_2^2 + \frac{bs^4}{4}\|\nabla u\|_2^4 - \int_{\mathbb{R}^3}\frac{\bar{F}(s^{\frac{3}{2}}u)}{s^3}\,dx\right), \quad (5.27)$$

and the latter expression of (5.27) is finite. Thus we choose $\kappa_0 > 0$ sufficiently small such that

$$\kappa^{-\frac{1}{2}}\left[\frac{a}{3}\left(\frac{1}{2}(b\kappa^{-\frac{1}{2}}\mathcal{S}^3 + \sqrt{(b\kappa^{-\frac{1}{2}}\mathcal{S}^3)^2 + 4a\mathcal{S}^3})\right) + \frac{b}{12}\left(\frac{1}{2}(b\kappa^{-\frac{1}{2}}\mathcal{S}^3 + \sqrt{(b\kappa^{-\frac{1}{2}}\mathcal{S}^3)^2 + 4a\mathcal{S}^3})\right)^2\right]$$

is greater than the right-hand side of (5.27). The proof is now complete. □

Remark 5.5 *We now provide explicit examples of nonlinearities that satisfy the assumptions of Lemma* 5.16. *Let $\epsilon > 0$ be small enough. If $\bar{f}(u) = \min\{|u|^{4-\epsilon}, |u|^{\frac{8}{3}}\}u$ and κ is not sufficiently small, then we can use Lemma* 5.16(*i*) *provided that c is large enough, but not* (*ii*). *On the other hand, if F is of the form* (5.11) *and $\bar{f}(u) = \min\{|u|^4, |u|^{\frac{8}{3}+\epsilon}\}u$, then if c is not large enough, we can use Lemma* 5.16(*ii*) *provided that κ is small enough, but not* (*i*).

Lemma 5.17 *Assume that conditions* (B_1)-(B_5) *and* (5.8) *hold and either $\kappa = 0$ or $\kappa \in (0, \infty)$ and relation* (5.25) *is satisfied. Then ρ is achieved.*

Proof. Suppose that $\{u_n\} \subset \mathcal{P} \cap D_c$ satisfies $\lim_{n\to+\infty} I(u_n) = \rho$. Then it follows from Lemma 5.13 that $\{||u_n||\}$ is bounded. Moreover, by Theorem 1.4 in Mederski [113], we find $\{\tilde{u}_i\}_{i=0}^\infty \subset H^1(\mathbb{R}^3)$ and $\{y_n^i\}_{i=0}^\infty \subset \mathbb{R}^3$ for any $n \geq 1$ such that $y_n^0 = 0$, $\lim_{n\to+\infty}|y_n^i - y_n^j| = 0$ if $i \neq j$, for every $G : \mathbb{R}^3 \mapsto \mathbb{R}$ of class C^1, there holds, up to a subsequence,

$$u_n(\cdot + y_n^i) \rightharpoonup \tilde{u}_i \quad \text{in } H^1(\mathbb{R}^3) \text{ as } n \to +\infty,$$

$$\lim_{n\to+\infty}\int_{\mathbb{R}^3}|\nabla u_n|^2\,dx = \sum_{j=0}^{i}\int_{\mathbb{R}^3}|\nabla\tilde{u}_j|^2\,dx + \lim_{n\to+\infty}\int_{\mathbb{R}^3}|\nabla v_n^i|^2\,dx,$$

$$\limsup_{n\to+\infty}\int_{\mathbb{R}^3}G(u_n)\,dx = \sum_{i=0}^{\infty}\int_{\mathbb{R}^3}G(\tilde{u}_i)\,dx,$$

where

$$v_n^i(x) := u_n(x) - \sum_{j=0}^{i}\tilde{u}_j(x - y_n^i).$$

Define $\mathcal{A} := \{i \geq 0 : \tilde{u}_i \neq 0\}$. Now we let $\kappa \in (0, \infty)$ such that relation (5.25) is satisfied. We claim that $\mathcal{A} \neq \emptyset$. Assume by contradiction that $\tilde{u}_i = 0$ for every $i \geq 0$. Then combining (B_1), (B_3), and (B_5), it holds that

$$\lim_{u\to 0}\frac{\bar{H}(u)}{u^2} = \lim_{|u|\to\infty}\frac{\bar{H}(u)}{u^6} = 0$$

and

$$a\int_{\mathbb{R}^3}|\nabla u_n|^2\,dx+b\left(\int_{\mathbb{R}^3}|\nabla u_n|^2\,dx\right)^2=\frac{3}{2}\int_{\mathbb{R}^3}H(u_n)\,dx=\frac{3}{2}\int_{\mathbb{R}^3}\bar{H}(u_n)\,dx+\kappa\int_{\mathbb{R}^3}u_n^6\,dx$$
$$=\kappa\int_{\mathbb{R}^3}u_n^6\,dx+o(1).$$

This means that

$$a\int_{\mathbb{R}^3}|\nabla u_n|^2\,dx+b\left(\int_{\mathbb{R}^3}|\nabla u_n|^2\,dx\right)^2\le\mathcal{S}^{-3}\kappa\left(\int_{\mathbb{R}^3}|\nabla u_n|^2\,dx\right)^3+o(1).$$

In addition, in view of Lemma 5.11, we know that $\lim_{n\to+\infty}\int_{\mathbb{R}^3}|\nabla u_n|^2\,dx>0$. Then

$$\rho=\lim_{n\to+\infty}I(u_n)=\lim_{n\to+\infty}\left(I(u_n)-\frac{1}{6}P(u_n)\right)$$
$$\ge\kappa^{-\frac{1}{2}}\left[\frac{a}{3}\left(\frac{1}{2}(b\kappa^{-\frac{1}{2}}\mathcal{S}^3+\sqrt{(b\kappa^{-\frac{1}{2}}\mathcal{S}^3)^2+4a\mathcal{S}^3})\right)+\frac{b}{12}\left(\frac{1}{2}(b\kappa^{-\frac{1}{2}}\mathcal{S}^3+\sqrt{(b\kappa^{-\frac{1}{2}}\mathcal{S}^3)^2+4a\mathcal{S}^3})\right)^2\right],$$

which contradicts the assumption (5.8). Thus, $\mathcal{A}\neq\emptyset$.

Next, we claim that for every $i\in\mathcal{A}$, there holds either $u_n(\cdot+y_n^i)\to\tilde{u}_i$ in $D^{1,2}(\mathbb{R}^3)$ or

$$a\int_{\mathbb{R}^3}|\nabla\tilde{u}_i|^2\,dx+b\left(\int_{\mathbb{R}^3}|\nabla\tilde{u}_i|^2\,dx\right)^2<\frac{3}{2}\int_{\mathbb{R}^3}H(\tilde{u}_i)\,dx.$$

Assume by contradiction that there exists $i\in\mathcal{A}$ such that, up to a subsequence, $v:=\lim_{n\to\infty}\int_{\mathbb{R}^3}|\nabla v_n|^2\,dx>0$ and

$$a\int_{\mathbb{R}^3}|\nabla\tilde{u}_i|^2\,dx+b\left(\int_{\mathbb{R}^3}|\nabla\tilde{u}_i|^2\,dx\right)^2\ge\frac{3}{2}\int_{\mathbb{R}^3}H(\tilde{u}_i)\,dx, \tag{5.28}$$

where $v_n:=u_n(\cdot+y_n^i)-\tilde{u}_i$. Using Vitali's convergence theorem, there holds

$$\begin{aligned}\int_{\mathbb{R}^3}[H(u_n)-H(v_n)]\,dx&=\int_{\mathbb{R}^3}\int_0^1-\frac{d}{ds}H(u_n-s\tilde{u}_i)\,dsdx=\int_{\mathbb{R}^3}\int_0^1h(u_n-s\tilde{u}_i)\tilde{u}_i\,dsdx\\&\to\int_0^1\int_{\mathbb{R}^3}h(\tilde{u}_i-s\tilde{u}_i)\tilde{u}_i\,dxds=\int_{\mathbb{R}^3}\int_0^1-\frac{d}{ds}H(\tilde{u}_i-s\tilde{u}_i)\,dsdx\\&=\int_{\mathbb{R}^3}H(\tilde{u}_i)\,dx\end{aligned} \tag{5.29}$$

as $n\to+\infty$. Then passing to a subsequence, we obtain

$$a\left(\int_{\mathbb{R}^3}|\nabla v_n|^2\,dx+\int_{\mathbb{R}^3}|\nabla\tilde{u}_i|^2\,dx\right)+b\left(\int_{\mathbb{R}^3}|\nabla v_n|^2\,dx\right)^2+b\left(\int_{\mathbb{R}^3}|\nabla\tilde{u}_i|^2\,dx\right)^2$$
$$+2b\int_{\mathbb{R}^3}|\nabla v_n|^2\,dx\cdot\int_{\mathbb{R}^3}|\nabla\tilde{u}_i|^2\,dx=\frac{3}{2}\left(\int_{\mathbb{R}^3}H(v_n)\,dx+\int_{\mathbb{R}^3}H(\tilde{u}_i)\,dx\right)+o(1).$$

Combining with (5.28), we obtain

$$a\int_{\mathbb{R}^3}|\nabla v_n|^2\,dx+b\left(\int_{\mathbb{R}^3}|\nabla v_n|^2\,dx\right)^2\le\frac{3}{2}\int_{\mathbb{R}^3}H(v_n)\,dx+o(1).\tag{5.30}$$

Now we define $\bar{R}_n>0$ such that $v_n(\bar{R}_n\cdot)\in\mathcal{P}$. We show that $\bar{R}_n\to 1$ as $n\to+\infty$. In particular, if

$$\frac{3}{2}\int_{\mathbb{R}^3}H(v_n)\,dx<a\int_{\mathbb{R}^3}|\nabla v_n|^2\,dx+b\left(\int_{\mathbb{R}^3}|\nabla v_n|^2\,dx\right)^2$$

holds for all n, then by (5.30) and since $v>0$, we infer that $\bar{R}_n\to 1$ as $n\to+\infty$. If, passing to a subsequence,

$$a\int_{\mathbb{R}^3}|\nabla v_n|^2\,dx+b\left(\int_{\mathbb{R}^3}|\nabla v_n|^2\,dx\right)^2\le\frac{3}{2}\int_{\mathbb{R}^3}H(v_n)\,dx$$

holds, then we deduce that $\bar{R}_n\ge 1$. We observe that

$$\lim_{n\to+\infty}\left(\|u_n\|_2^2-\|v_n\|_2^2\right)=\|\tilde{u}_i\|_2^2>0.$$

Then $v_n\in D_c$ and $v_n(\bar{R}_n\cdot)\in\mathcal{P}\cap D_c$ for a.e. n. Thus using (5.29), the Brezis-Lieb lemma and (B_5), we obtain that

$$\begin{aligned}
\rho&\le I(v_n(\bar{R}_n\cdot))=I(v_n(\bar{R}_n\cdot))-\frac{1}{4}P(v_n(\bar{R}_n\cdot))\\
&=\frac{1}{\bar{R}_n}\cdot\frac{a}{4}\int_{\mathbb{R}^3}|\nabla v_n|^2\,dx+\frac{1}{\bar{R}_n^3}\cdot\int_{\mathbb{R}^3}\frac{3}{8}H(v_n)-G(v_n)\,dx\\
&\le\frac{a}{4}\int_{\mathbb{R}^3}|\nabla v_n|^2\,dx+\int_{\mathbb{R}^3}\frac{3}{8}H(v_n)-G(v_n)\,dx\\
&\le\frac{a}{4}\int_{\mathbb{R}^3}|\nabla u_n|^2\,dx+\int_{\mathbb{R}^3}\frac{3}{8}H(u_n)-G(u_n)\,dx+o(1)\\
&=I(u_n)-\frac{1}{4}P(u_n)+o(1)=I(u_n)+o(1)=\rho+o(1),
\end{aligned}$$

which yields that $\bar{R}_n\to 1$. Hence

$$\begin{aligned}
a\int_{\mathbb{R}^3}|\nabla v_n|^2\,dx+b\left(\int_{\mathbb{R}^3}|\nabla v_n|^2\,dx\right)^2&=\frac{3}{2}\int_{\mathbb{R}^3}H(v_n)\,dx+o(1)=\kappa\int_{\mathbb{R}^3}v_n^6\,dx+o(1)\\
&\le\mathcal{S}^{-3}\kappa\left(\int_{\mathbb{R}^3}|\nabla v_n|^2\,dx\right)^3+o(1).
\end{aligned}$$

On the other hand, using the fact that

$$I(u_n)=J(v_n)+J(\tilde{u}_i)+o(1)$$

and

$$I(\tilde{u}_i)\ge I(\tilde{u}_i)-\frac{1}{4}P(\tilde{u}_i)+\frac{1}{4}P(\tilde{u}_i)\ge\int_{\mathbb{R}^3}\frac{3}{8}H(\tilde{u}_i)-P(\tilde{u}_i)\,dx\ge 0,$$

where

$$J(u):=\frac{a}{2}\|\nabla u\|_2^2+\frac{bA}{4}\|\nabla u\|_2^2-\int_{\mathbb{R}^3}F(u)\,dx,\quad A:=\lim_{n\to+\infty}\|\nabla u_n\|_2^2,$$

we deduce that

$$\begin{aligned}\rho = &\lim_{n\to+\infty} I(u_n) \geq I(\tilde{u}_i) + \lim_{n\to\infty} I(v_n)\\ \geq &\kappa^{-\frac{1}{2}}\left[\frac{a}{3}\left(\frac{1}{2}(b\kappa^{-\frac{1}{2}}\mathcal{S}^3+\sqrt{(b\kappa^{-\frac{1}{2}}\mathcal{S}^3)^2+4a\mathcal{S}^3})\right)+\frac{b}{12}\left(\frac{1}{2}(b\kappa^{-\frac{1}{2}}\mathcal{S}^3+\sqrt{(b\kappa^{-\frac{1}{2}}\mathcal{S}^3)^2+4a\mathcal{S}^3})\right)^2\right].\end{aligned}$$

This is impossible. Thus, for every $i \in \mathcal{A}$, there holds either $u_n(\cdot + y_n^i) \to \tilde{u}_i$ in $D^{1,2}(\mathbb{R}^3)$ or

$$a\int_{\mathbb{R}^3} |\nabla \tilde{u}_i|^2\,dx + b\left(\int_{\mathbb{R}^3} |\nabla \tilde{u}_i|^2\,dx\right)^2 < \frac{3}{2}\int_{\mathbb{R}^3} H(\tilde{u}_i)\,dx.$$

Let $i \in \mathcal{A}$. Based on the above conclusion, if

$$a\int_{\mathbb{R}^3} |\nabla \tilde{u}_i|^2\,dx + b\left(\int_{\mathbb{R}^3} |\nabla \tilde{u}_i|^2\,dx\right)^2 < \frac{3}{2}\int_{\mathbb{R}^3} H(\tilde{u}_i)\,dx,$$

then there exists $\bar{R} > 1$ such that $\tilde{u}_i(\bar{R}\cdot) \in \mathcal{P}$ and $\tilde{u}_i(\bar{R}\cdot) \in D_c$. So by Fatou's lemma and (B_5), we deduce that

$$\begin{aligned}\rho \leq I(\tilde{u}_i(\bar{R}\cdot)) = I(\tilde{u}(\bar{R}\cdot)) - \frac{1}{4}P(\tilde{u}_i(\bar{R}\cdot)) &= \frac{1}{\bar{R}}\cdot\frac{a}{4}\|\nabla\tilde{u}_i\|_2^2 + \frac{1}{R^3}\int_{\mathbb{R}^3}\frac{3}{8}H(\tilde{u}_i) - F(\tilde{u}_i)\,dx\\ &< \liminf_{n\to\infty}\left[\frac{a}{4}\|\nabla u_n\|_2^2 + \int_{\mathbb{R}^3}\frac{3}{8}H(u_n) - F(u_n)\,dx\right]\\ &= \liminf_{n\to\infty}\left[I(u_n) - \frac{1}{4}P(u_n)\right] = \liminf_{n\to\infty} I(u_n) = \rho,\end{aligned}$$

which is a contradiction. Hence $u_n(\cdot + y_n^i) \to \tilde{u}$ in $D^{1,2}(\mathbb{R}^3)$. Then combining the Brezis-Lieb lemma and the interpolation inequality, there holds

$$\begin{aligned}\int_{\mathbb{R}^3} H(u_n - \tilde{u})\,dx &\leq \check{C}\left(\|u_n - \tilde{u}\|_{\bar{p}}^{\bar{p}} + \|u_n - \tilde{u}\|_6^6\right)\\ &\leq \check{C}\left(\|u_n - \tilde{u}\|_2^{2\tau}\|u_n - \tilde{u}\|_6^{6(1-\tau)} + \|u_n - \tilde{u}\|_6^6\right) \to 0\end{aligned}$$

and $\int_{\mathbb{R}^3} H(u_n)\,dx \to \int_{\mathbb{R}^3} H(\tilde{u}_i)\,dx$ as $n \to \infty$ for some $\check{C} > 0$ and $\tau = \frac{6-\bar{p}}{4}$. This means that $\tilde{u}_i \in \mathcal{P} \cap D_c$. Moreover $I(\tilde{u}_i) = c$. When we consider the case $\kappa = 0$, similar to the above argument, we easily get the conclusion of Lemma 5.17. □

Lemma 5.18 *Assume that conditions* (B_1)*–*(B_5) *and* (5.8) *hold,* F *is of the form* (5.11)*, and either* $\kappa = 0$ *or* $\kappa \in (0, +\infty)$ *and* (5.25) *holds. Then* ρ *is achieved by a nonnegative and radially nonincreasing function.*

Proof. Let $\tilde{u} \in \mathcal{P} \cap D_c$ such that $I(\tilde{u}) = \rho$ is given by Lemma 5.17. Let u be the Schwarz rearrangement of $|\tilde{u}|$. In addition, let $s_1 = s_1(u)$ be defined by Lemma 5.12. In virtue of the properties of the Schwarz rearrangement, we have

$$P(1 * u) = P(u) \leq P(\tilde{u}) = 0,$$

which along with Lemma 5.12 yields that $s_1 \leq 1$. This means that $P(s_1 * \tilde{u}) \geq 0$.

Define $\varsigma := \frac{3}{2}(\zeta - 2) \geq 4$.

It follows that

$$\begin{aligned}
\rho \le I(s_1 * u) &= I(s_1 * u) - \frac{1}{\varsigma} P(s_1 * u) \\
&= s_1^2 \left(\frac{a}{2} - \frac{a}{\varsigma}\right) \int_{\mathbb{R}^3} |\nabla u|^2 \, dx + s_1^4 \left(\frac{b}{4} - \frac{b}{\varsigma}\right) \left(\int_{\mathbb{R}^3} |\nabla u|^2 \, dx\right)^2 \\
&\quad + \frac{1}{s_1^3} \int_{\mathbb{R}^3} \frac{3}{4\varsigma} \hat{H}(s_1^{\frac{3}{2}} u) - F(s_1^{\frac{3}{2}} u) \, dx - \frac{1}{s_1^3} \beta \left[1 - \frac{1}{\varsigma}(\zeta - 2)\right] \int_{\mathbb{R}^3} |s_1^{\frac{3}{2}} u|^{\varsigma} \, dx \\
&\le s_1^2 \left(\frac{a}{2} - \frac{a}{\varsigma}\right) \int_{\mathbb{R}^3} |\nabla \tilde{u}|^2 \, dx + s_1^4 \left(\frac{b}{4} - \frac{b}{\varsigma}\right) \left(\int_{\mathbb{R}^3} |\nabla \tilde{u}|^2 \, dx\right)^2 \\
&\quad + \frac{1}{s_1^3} \int_{\mathbb{R}^3} \frac{3}{4\varsigma} \hat{H}(s_1^{\frac{3}{2}} \tilde{u}) - F(s_1^{\frac{3}{2}} \tilde{u}) \, dx - \frac{1}{s_1^3} \beta \left[1 - \frac{1}{\varsigma}(\zeta - 2)\right] \int_{\mathbb{R}^3} |s_1^{\frac{3}{2}} \tilde{u}|^{\varsigma} \, dx \\
&= I(s_1 * \tilde{u}) - \frac{1}{\varsigma} P(s_1 * \tilde{u}) \le I(s_1 * \tilde{u}) \le I(\tilde{u}) = \rho,
\end{aligned}$$

which implies that $I(s_1 * u) = c$. This completes the proof. □

The following lemma concerns the sign of the Lagrange multiplier when the corresponding constraint is given by an inequality and the critical point of the critical point of restricted functional is the minimizer.

Lemma 5.19 *Let $\mathcal{H}$ be a real Hilbert space and $f, \phi, \psi \in C^1(\mathcal{H})$. Suppose that for every $x \in \phi^{-1}(0) \cap \psi^{-1}(0)$, the mapping*

$$(\phi'(x), \psi'(x)) : \mathcal{H} \mapsto \mathbb{R}^2$$

is surjective. If $\bar{x} \in \mathcal{H}$ minimizes f over

$$\{x \in \mathcal{H} : \phi(x) \le 0 \quad \text{and} \quad \psi(x) = 0\},$$

then there exist $\lambda \in [0, +\infty)$ and $\sigma \in \mathbb{R}$ such that

$$f'(\bar{x}) + \lambda \phi(\bar{x}) + \sigma \psi(\bar{x}) = 0.$$

Proof. We argue similarly as in the case $m = n = 1$ provided by Proposition $A.1$ in [114]. □

Lemma 5.20 *Assume that conditions (B_1)–(B_3), $(B_4, \preceq)$ and (5.8) hold. Then the following properties hold.*

(*i*) *Let $u \in \mathcal{P} \cap D_c$ such that $I(u) = \rho$ and u is radial. Then u is of class C^2.*

(*ii*) *If, in addition, F is of the form (5.11) and u is nonnegative, then $u \in \partial D_c$.*

Proof. (i) In view of Lemma 5.19, we set $f = I$, $\phi(u) = \|v\|_2^2 - c^2$, $\psi(u) = P(u)$, $u \in \mathcal{H} = H^1(\mathbb{R}^3)$. Then there exist $\lambda \in [0, +\infty)$ and $\sigma \in \mathbb{R}$ such that

$$-a(1 - 2\sigma)\Delta u - b(1 - 4\sigma) \left(\int_{\mathbb{R}^3} |\nabla u|^2 \, dx\right) \Delta u + \lambda u = \partial F(u) - \sigma \frac{3}{2} \partial H(u) \tag{5.31}$$

and u satisfies the Nehari identity

$$\begin{aligned}
a(1 - 2\sigma) \int_{\mathbb{R}^3} |\nabla u|^2 \, dx + b(1 - 4\sigma) \left(\int_{\mathbb{R}^3} |\nabla u|^2 \, dx\right)^2 + \lambda \int_{\mathbb{R}^3} u^2 \, dx \\
+ \int_{\mathbb{R}^3} \sigma \frac{3}{2} \langle h(u), u \rangle - \langle f(u), u \rangle \, dx = 0.
\end{aligned} \tag{5.32}$$

Then if $\sigma = \frac{1}{4}$, then using $(B_4, \preceq)$, (B_5) and (5.32), we obtain that

$$0 \geq \int_{\mathbb{R}^3} \frac{3}{8}\langle h(u), u\rangle - \langle f(u), u\rangle \, dx = \int_{\mathbb{R}^3} \frac{3}{8}\langle h(u), u\rangle - H(u) - 2F(u)\, dx$$
$$> \int_{\mathbb{R}^3} \frac{3}{4} H(u) - 2F(u)\, dx \geq 0,$$

which is a contradiction. Thus, $\sigma \neq \frac{1}{4}$ and u satisfies the Pohozaev identity

$$\begin{aligned} a(1-2\sigma)\int_{\mathbb{R}^3} |\nabla u|^2\, dx + b(1-4\sigma)\left(\int_{\mathbb{R}^3} |\nabla u|^2\, dx\right)^2 + 3\lambda \int_{\mathbb{R}^3} u^2\, dx \\ + 6\int_{\mathbb{R}^3} \sigma \cdot \frac{3}{2} H(u) - F(u)\, dx = 0. \end{aligned} \tag{5.33}$$

Combining (5.32) and (5.33), one has

$$a(1-2\sigma)\int_{\mathbb{R}^3} |\nabla u|^2\, dx + b(1-4\sigma)\left(\int_{\mathbb{R}^3} |\nabla u|^2\, dx\right)^2 + \frac{3}{2}\int_{\mathbb{R}^3} \sigma 3\left(\frac{1}{2}\langle h(u), u\rangle - H(u)\right)$$
$$- H(u)\, dx = 0.$$

This and the fact that $u \in \mathcal{P}$ imply that

$$(1-4\sigma)\int_{\mathbb{R}^3} H(u)\, dx + \frac{4a\sigma}{3}\int_{\mathbb{R}^3} |\nabla u|^2\, dx + \int_{\mathbb{R}^3} \sigma 3\left(\frac{1}{2}\langle h(u), u\rangle - H(u)\right) - H(u)\, dx = 0,$$

which yields

$$\sigma \int_{\mathbb{R}^3} \langle h(u), u\rangle - \bar{p}H(u) + \frac{4a}{3}|\nabla u|^2\, dx = 0.$$

Then by $(B_4, \preceq)$, we have $\sigma = 0$. Moreover, by Theorem 2.3 of Brezis and Lieb [15], we deduce that u is of class C^2.

(ii) First of all, we claim that $u \in \partial D_c$. Assume by contradiction that $\|u\|_2 < c$. Then $\lambda = 0$ and by relations (5.32), (5.33), we infer that

$$\int_{\mathbb{R}^3} [\langle f(u), u\rangle - 6F(u)]\, dx = 0. \tag{5.34}$$

In virtue of (B_5),

$$6F(u(x)) = \langle f(u(x)), u(x)\rangle \tag{5.35}$$

for every $x \in \mathbb{R}^3$. Since $\hat{F}$ satisfies (B_5), there holds $6\hat{F}(u(x)) \geq f(u(x))u(x)$. On the other hand, we observe that $6\beta|u(x)|^\zeta \geq \beta\zeta|u(x)|^\zeta$ by $\zeta < 6$. Thus, it follows from (5.35) that the above inequality is actually equality. Since $\zeta < 6$, we have $\beta = 0$ or $|u(x)|^\zeta = 0$ for every $x \in \mathbb{R}^3$. This means that

$$6\hat{F}(u(x)) = f(u(x))u(x)$$

for every $x \in \mathbb{R}^3$. Now since $u \in H^1(\mathbb{R}^3) \cap C^2$, there exists an open interval $\mathcal{I} \subset \mathbb{R}$ such that $0 \in \bar{\mathcal{I}}$ and $6\hat{F}(t) = f(t)t$ for $t \in \bar{\mathcal{I}}$. Then $\hat{F}(t) = \frac{\kappa}{6}t^6$ for $t \in \bar{\mathcal{I}}$ and u solves

$$-\left(a + b\int_{\mathbb{R}^3} |\nabla u|^2\, dx\right)\Delta u = \kappa u^5.$$

Since $u \geq 0$, then up to a scaling and translations, u is an Aubin-Talenti function, which is not L^2-integrable. So $u \in \partial D_c$. □

Proof of Theorem 5.3. The statement (i) of Theorem 5.3 follows from Lemmas 5.17 and 5.18. For Theorem 5.3(ii), using Lemma 5.20(i), we obtain that u is of class C^2. Combining Lemma 5.19, there exist $\lambda \in [0, \infty)$ and $\sigma \in \mathbb{R}$ such that (5.31) holds, which along with the proof of Lemma 5.20(i) implies that $\sigma = 0$. Moreover, by Lemma 5.20(ii) and the maximum principle [50, Lemma IX.V.1], the conclusion of Theorem 5.3(ii) holds true. □

5.4 Historical comments

The Kirchhoff equation (5.1) is related to the stationary analogue of the problem

$$\begin{cases} u_{tt} - \left(a + b\int_\Omega |\nabla u|^2\, dx\right)\Delta u = f(x,u), & (x,t) \in \Omega \times (0,+\infty), \\ u(x,t) = 0, & (x,t) \in \partial\Omega \times [0,+\infty). \end{cases} \tag{5.36}$$

Kirchhoff [88] introduced problem (5.36) as an extension of the D'Alembert wave equation

$$\rho\frac{\partial^2 u}{\partial t^2} - \left(\frac{\rho_0}{h} + \frac{E}{2L}\int_0^L \left|\frac{\partial u}{\partial x}\right|^2 dx\right)\frac{\partial^2 u}{\partial x^2} = f(x,u)$$

for free vibrations of elastic strings, where ρ denotes the mass density, u denotes the lateral displacement, ρ_0 is the initial axial tension, h denotes the cross section area, E is the Young modulus, L denotes the length of the string and f stands for the external force. Kirchhoff's model takes into account the changes in the length of the string produced by transverse vibrations. Some early classical studies of Kirchhoff equations can be seen in Bernstein [10] and Pohozaev [128].

Equation (5.36) received much attention only after Lions [99] introduced an abstract framework to the problem. Some interesting results can be found in [7, 22, 38]. In [7], Arosio et al. studied the Cauchy–Dirichlet type problem related to (5.36) in the Hadamard sense as a special case of an abstract second-order Cauchy problem in a Hilbert space. Cavalcanti et al. [22] considered the question of the existence and uniqueness of regular global solutions for the Kirchhoff–Carrier equation subject to nonlinear boundary dissipation without restriction on the initial data and obtained uniform decay rates by assuming a nonlinear feedback acting on the boundary. D'Ancona and Spagnolo [38] proved the existence of a global classical periodic solution for the degenerate Kirchhoff equation. They also explained the Kirchhoff equation as an example of a quasi-linear hyperbolic Cauchy problem that describes the transverse oscillations of a stretched string. In particular, problem (5.36) with $a = 0$ models a string with zero initial tension. This is called the *degenerate Kirchhoff equation*, see Iturriaga and Massa [77] and Pucci and Rădulescu [129].

5.5 Final remarks, perspectives, and open problems

This chapter deals with the study of normalized solutions for Kirchhoff-type equations with non-homogeneous nonlinearities exhibiting a double-behavior growth, for instance $f(u) \sim \mu_1|u|^{p-2}u$ near zero and $f(u) \sim \mu_2|u|^{2-2}u$ near infinity, with p, q in different ranges.

Why the case of a double-behavior reaction is important? Because this specific interplay between the growth of the reaction near origin and at infinity relative to the L^2-critical exponent is a central technical feature, which creates a competition between two different variational geometries. A major effort consists in handling the lack of compactness due to the Sobolev supercritical growth at infinity (which is still subcritical in the Sobolev sense) through the interplay with the mass constraint and the Kirchhoff term.

The content of this chapter is based on the results established in [19]. Related properties can be found in Autuori, Colasuonno and Pucci [8] (for polyharmonic systems), Bieganowski, Mederski and Schino [12] (for polyharmonic equations and curl-curl problems), Cai and

Rădulescu [17, 18] (for (p,q)-equations), Chen, Jin, Rădulescu and Zhou [25] (for Choquard equations), Jeanjean, Zhang and Zhong [81, 82] (for Schrödinger equations), and Liu and Rădulescu [101] (for quasilinear problems with nonlocal reaction).

The main perspectives generated by the analysis developed in this chapter are the following:

(i) *Existence and multiplicity properties:* It is highlighted a generic class of nonlinearities for which the problem admits multiple normalized solutions.

(ii) *Unification of scattering regimes:* The *double-behavior* of the nonlinearity creates a competition between a *focusing* effect at small masses and a *defocusing* effect at large masses, leading to richer structure. This nonstandard behavior of the reaction term allows to handle cases where the associated energy functional is:

– *coercive,* due to the growth near zero (in the L^2-subcritical case);

– *non-coercive,* due to the growth near infinity (in the L^2-supercritical case).

(iii) *Interaction with the Kirchhoff term:* The study clarifies how the nonlocal term $(\int_{\mathbb{R}^3} |\nabla u|^2 dx)\Delta u$ interacts with mass-preserving constraints. This acts as an additional feedback that influences the geometry of the energy functional, often making the compactness analysis (Palais-Smale condition) more delicate, especially at critical frequencies λ.

The open problems inspired by the analysis carried out in this chapter are primarily in the directions of critical exponents, generalized operators, stability analysis, transition from generic to explicit results, extension to different domains, and asymptotic analysis.

The first problem proposes to refine this analysis for the Sobolev criticality.

Open Problem 1. Study problem (5.1) when the growth at infinity of the nonlinearity reaches the Sobolev critical exponent. In this case, the compactness loss is more severe. Can one establish existence and multiplicity properties for generic reaction terms in this case? What are the possible blow-up scenarios?

The second open problem proposes to generalize the nonlocal structure.

Open Problem 2. (i) Study problem (5.1) for more general nonlocal terms, for instance of the type $M(\int_{\mathbb{R}^3} |\nabla u|^2 dx)\Delta u$ with a wider class of functions M.

(ii) What about fractional operators like

$$\left(a + b\int\int |u(x) - u(y)|^2/|x-y|^{N+2s}\right)(-\Delta)^s u?$$

A major research direction concerns dynamics and stability properties of the normalized ground state solutions. This is important because the Kirchhoff term makes the spectral analysis of the linearized operator around such solutions highly non-standard.

Open Problem 3. Study orbital stability/instability of solutions to problem (5.1).

The analysis developed in this chapter proves that *most* nonlinearities (in a topological sense) yield multiplicity of solutions. The following open problem proposes the transition from generic to explicit results.

Open Problem 4. Can we find verifiable, explicit criteria on f (beyond the double-behavior asymptotics) that guarantee multiplicity of solutions to problem (5.1)?

Next, we propose to develop this analysis in the case of bounded domains.

Open Problem 5. The entire Euclidean space offers translation invariance. The lack of translation invariance complicates the topological arguments for multiplicity.

(i) How do the results translate to *bounded* or *exterior* domains with Dirichlet/Neumann conditions?

(ii) In the case of *bounded* domains and for the pure mass-supercritical case, study the existence of a mountain-pass solution for *any* prescribed mass. Also, solving the pure supercritical case on bounded domains remains a significant challenge.

A very interesting open problem is concerned with the limiting case when the coefficient of the nonlocal term converges to zero, recovering the standard semilinear Schrödinger equation.

Open Problem 6. What is the behavior of the solutions found in this chapter as the Kirchhoff parameter b goes to 0? Does one recover the results in the semilinear case, and is the convergence singular?

5.6 Glossary

Pohozaev Manifold Constraint: It is a variational tool that encodes a necessary condition (Pohozaev identity) satisfied by all solutions. This method transforms an indefinite variational problem into a coercive minimization problem.

Fibering Map Analysis: It is a powerful technique used to construct and analyze constraint manifolds like the Nehari and Pohozaev manifolds. This method transforms the search for critical points in an infinite-dimensional space into a scalar optimization along parameterized rays.

Strict Subadditivity: For normalized solutions of Kirchhoff equations, this is the key compactness condition. Once strict subadditivity is established, the concentration-compactness technique yields existence of ground states and, with additional topological arguments, multiple normalized solutions.

Jeanjean Mountain Pass Technique: It is a groundbreaking method in critical point theory, specifically designed to find nontrivial solutions of nonlinear elliptic problems with strong lack of compactness, particularly those with indefinite nonlinearities. This technique is a sophisticated variant of the classical mountain pass theorem of Ambrosetti and Rabinowitz.

Cazenave-Lions Method: It is also known as the *concentration-compactness method via minimizing sequences* and is a fundamental technique for proving existence of ground states for elliptic equations. This method is particularly effective for problems with scaling invariance and lack of compactness.

Part III

Concentration Phenomena

6

Fractional p-Kirchhoff equations

Part III of this volume deals with concentration properties of solutions for several classes of nonlinear systems with Kirchhoff nonlocal term.

Why concentration properties of solutions?

Because this study gets to the heart of one of the main analytical challenges for the Kirchhoff equation. One of the reasons is that concentration properties of solutions are crucial for Kirchhoff problems because they determine whether the energy functional associated with the equation satisfies the Palais-Smale compactness condition. This property is the cornerstone of applying powerful variational methods to establish the existence of weak solutions.

In this chapter, we investigate the existence and concentration of positive solutions for the following class of fractional p-Kirchhoff problems:

$$\begin{cases} \left(\varepsilon^{sp}a + \varepsilon^{2sp-3}b\,[u]_{s,p}^{p}\right)(-\Delta)_p^s u + V(x)u^{p-1} = f(u) & \text{in } \mathbb{R}^3, \\ u \in W^{s,p}(\mathbb{R}^3), \quad u > 0 & \text{in } \mathbb{R}^3, \end{cases}$$

where ε is a small positive parameter, a and b are positive constants, $s \in (0,1)$ and $p \in (1,\infty)$ are such that $sp \in (\frac{3}{2}, 3)$, $(-\Delta)_p^s$ is the fractional p-Laplacian operator, $f : \mathbb{R} \to \mathbb{R}$ is a superlinear continuous function with subcritical growth and $V : \mathbb{R}^3 \to \mathbb{R}$ is a continuous potential having a local minimum. We also prove a multiplicity result and relate the number of positive solutions with the topology of the set where the potential V attains its minimum values. Finally, it is established an existence result when $f(u) = u^{q-1} + \gamma u^{r-1}$, where $\gamma > 0$ is sufficiently small, and the powers q and r satisfy $2p < q < p_s^* \leq r$. The main results extend and complement concentration properties known for simpler cases (like the classical Schrödinger equation or the fractional Laplacian) to this much more general and complex setting. The problem studied in this chapter models both nonlocal continuum mechanics with long-range interactions and patterns arising in fractional quantum mechanics and anomalous diffusion.

The analysis developed in what follows features two distinct nonlocal effects that interact:

(i) *Fractional p-Laplacian* $(-\Delta)_p^s$: this operator is itself nonlocal, as it depends on values of the function across the entire space, not just local gradients. The fractional p-Laplace operator is degenerate when $p > 2$ and singular when $1 < p < 2$. This affects the regularity theory and the linearization around solutions.

(ii) *Kirchhoff term* $\left(a + b\,[u]_{s,p}^{p}\right)$: this term accounts for nonlocal elasticity effects and it multiplies the main differential operator and depends on the global Gagliardo seminorm $[u]_{s,p}^{p}$ of the solution. In such a way it is introduced a second layer of nonlocality, in the sense that the coefficient of the equation itself depends on the overall energy of the solution.

DOI: 10.1201/9781003797159-6

Additionally, the presence of the potential V creates concentration properties of solutions. This potential represents a heterogeneous external field; for instance, an electrostatic potential in a quantum dot, or a density trap in Bose-Einstein condensates. The concentration result (as $\varepsilon \to 0$) describes the localization of states (like electrons or bosons) in low-potential regions, which is a fundamental phenomenon in quantum mechanics and condensed matter physics.

A hierarchy of the basic abstract tools used in this chapter is now described.

(a) *Foundation*: fractional Sobolev spaces and embeddings;

(b) *Existence framework*: mountain pass theory, maximum principle, Ljusternik-Schnirelmann category theory, and Nehari manifold analysis;

(c) *Localization technique*: penalization method with tail estimates;

(d) *Qualitative properties*: fractional regularity and decay estimates,

all adapted to the intricate nonlocal structure of the Kirchhoff problem.

6.1 Statement of the problem

Consider the following fractional p-Kirchhoff equation

$$\begin{cases} \left(\epsilon^{sp}a+\epsilon^{2sp-3}b\,[u]^p_{s,p}\right)(-\Delta)^s_p u+V(x)u^{p-1}=f(u) & \text{in } \mathbb{R}^3, \\ u\in W^{s,p}(\mathbb{R}^3), \quad u>0 & \text{in } \mathbb{R}^3, \end{cases} \tag{P_ϵ}$$

where $\epsilon>0$ is a small parameter, $a,b>0$ are constants, $s\in(0,1)$ and $p\in(1,\infty)$ are such that $sp\in(\frac{3}{2},3)$, $(-\Delta)^s_p$ is the fractional p-Laplacian operator which, up to normalization factors, may be defined for every function $u\in C^\infty_c(\mathbb{R}^3)$ as

$$(-\Delta)^s_p u(x)=2\lim_{r\to 0}\int_{\mathbb{R}^3\setminus\mathcal{B}_r(x)}\frac{|u(x)-u(y)|^{p-2}(u(x)-u(y))}{|x-y|^{3+sp}}dy \quad (x\in\mathbb{R}^3),$$

and $W^{s,p}(\mathbb{R}^3)$ is the fractional Sobolev space of functions $u\in L^p(\mathbb{R}^3)$ such that

$$[u]^p_{s,p}:=\int_{\mathbb{R}^3}\int_{\mathbb{R}^3}\frac{|u(x)-u(y)|^p}{|x-y|^{3+sp}}\,dxdy<\infty$$

endowed with the natural norm

$$\|u\|^p_{s,p}=[u]^p_{s,p}+|u|^p_p.$$

Assume that $V\in C^0(\mathbb{R}^3,\mathbb{R})$ satisfies the following assumptions:

(V_1) there exists $V_1>0$ such that $V_1:=\inf_{x\in\mathbb{R}^3}V(x)$;

(V_2) there exists an open bounded set $\Lambda\subset\mathbb{R}^3$ such that

$$0<V_0:=\inf_{\Lambda}V<\min_{\partial\Lambda}V.$$

We suppose that $f\in C^0(\mathbb{R},\mathbb{R})$ vanishes in $(-\infty,0)$, and fulfills the following conditions:

(f_1) $f(t)=o(t^{2p-1})$ as $t\to 0^+$,

(f_2) there exists $\nu \in (2p, p_s^*)$, with $p_s^* = \frac{3p}{3-sp}$, such that

$$\lim_{t\to\infty} \frac{f(t)}{t^{\nu-1}} = 0,$$

(f_3) there exists $\vartheta \in (2p, p_s^*)$ such that $0 < \vartheta F(t) := \vartheta \int_0^t f(\tau)\, d\tau \leq t f(t)$ for all $t > 0$,

(f_4) the map $t \mapsto \dfrac{f(t)}{t^{2p-1}}$ is increasing in $(0, \infty)$.

We emphasize that under the control of the fractional parameter $s \in (0,1)$, the condition $sp \in (\frac{3}{2}, 3)$ forces $p \in (\frac{3}{2}, \infty)$. In particular, the restriction $sp \in (\frac{3}{2}, 3)$ implies that $2p < p_s^*$, therefore condition (f_3) makes sense. A typical example of function which satisfies (f_1)–(f_4) is given by $f(t) = \sum_{i=1}^{k} \alpha_i t^{r_i - 1}$ with $\alpha_i > 0$ not all null and $2p < r_i < p_s^*$ for all $i \in \{1, \ldots, k\}$.

The following result establishes a connection between the global maximum points of solutions and the infimum of the potential V on Λ.

Theorem 6.1 *Assume that conditions* (V_1)–(V_2) *and* (f_1)–(f_4) *are fulfilled. Then, there exists* $\epsilon_0 > 0$ *such that, for all* $\epsilon \in (0, \epsilon_0)$, *problem* ($P_\epsilon$) *has a positive solution* u_ϵ. *Moreover, if* η_ϵ *denotes a global maximum point of* u_ϵ, *then* $\lim_{\epsilon\to 0} V(\eta_\epsilon) = V_0$, *and there exists* $C > 0$ *such that*

$$u_\epsilon(x) \leq \frac{C \epsilon^{3+sp}}{\epsilon^{3+sp} + |x - \eta_\epsilon|^{3+sp}} \quad \forall x \in \mathbb{R}^3.$$

The proof of Theorem 6.1 relies on suitable variational arguments. We first adapt in a suitable way a penalization argument which consists in modifying the nonlinearity f outside Λ, considering an auxiliary problem and then we check that, for $\epsilon > 0$ small enough, the solutions of the modified problem are solutions of the original one. In order to achieve this purpose, we look for critical points of the corresponding energy functional $\mathcal{I}_\epsilon$. The main difficulties that arise in the study of $\mathcal{I}_\epsilon$ are related to the presence of the Kirchhoff term $[u]_{s,p}^{p}(-\Delta)_p^s$ and the lack of compactness caused by the unboundedness of the domain $\mathbb{R}^3$. Indeed, in general, it is not trivial to prove that the weak limits of bounded Palais-Smale sequences are critical points of $\mathcal{I}_\epsilon$ when we consider Kirchhoff problems. Moreover, the non–Hilbertian structure of the fractional Sobolev spaces $W^{s,p}(\mathbb{R}^3)$ when $p \neq 2$ makes this analysis rather tough. To circumvent these difficulties, we develop some arguments which take care of the nonlocal character of the leading operator $(-\Delta)_p^s$ and that allow us to recover the compactness of the functional $\mathcal{I}_\epsilon$; see Lemma 6.4. After that, we show that the solution of the modified problem is also a solution of the original one by combining a Moser iteration argument [120] with the Hölder continuity result established for $(-\Delta)_p^s$. Finally, we establish a decay estimate for the solutions of problem (P_ϵ).

In the second part of this chapter, we deal with the multiplicity of positive solutions to problem (P_ϵ). In this case, we replace (V_2) by the following assumption:

(V_2') $V_0 = V_1$ and $M = \{x \in \Lambda : V(x) = V_0\} \neq \emptyset$.

We recall that if Y is a given closed set of a topological space X, we denote by $cat_Y(X)$ the Ljusternik-Schnirelmann category of Y in X, that is, the least number of closed and contractible sets in X which cover Y; see Willem [153] for more details.

Theorem 6.2 *Assume that conditions* (V_1)-(V_2') *and* (f_1)-(f_4) *hold. Then, for any* $\delta > 0$ *such that*

$$M_\delta = \{x \in \mathbb{R}^3 : \operatorname{dist}(x, M) \leq \delta\} \subset \Lambda,$$

there exists $\epsilon_\delta > 0$ *such that, for any* $\epsilon \in (0, \epsilon_\delta)$, *problem* (P_ϵ) *has at least* $cat_{M_\delta}(M)$ *positive solutions. Furthermore, if* u_ϵ *denotes one of these positive solutions and* $\eta_\epsilon \in \mathbb{R}^3$ *is a global maximum point of* u_ϵ, *then*

$$\lim_{\epsilon \to 0} V(\eta_\epsilon) = V_0.$$

In order to prove Theorem 6.2, we need to use some suitable variational and topological arguments. More precisely, to obtain multiple solutions, we study the modified functional $\mathcal{I}_\epsilon$ on the associated Nehari manifold $\mathcal{N}_\epsilon$. Anyway, due to the fact that f is only continuous, the Nehari manifold $\mathcal{N}_\epsilon$ is not differentiable and some well-known arguments for C^1-Nehari manifolds do not work in our situation. Next, we use a technique in which the main ingredient is to make fine comparisons between the category of some sublevel sets of the modified functional $\mathcal{I}_\epsilon$ and the category of the set M.

In the last part of this chapter, we consider the case when the nonlinearity f has a critical or supercritical growth. More precisely, we study the following nonlocal problem

$$\begin{cases} \left(\epsilon^{sp} a + \epsilon^{2sp-3} b\, [u]_{s,p}^p\right)(-\Delta)_p^s u + V(x) u^{p-1} = u^{q-1} + \gamma u^{r-1} & \text{in } \mathbb{R}^3, \\ u \in W^{s,p}(\mathbb{R}^3), \quad u > 0 & \text{in } \mathbb{R}^3, \end{cases} \tag{6.1}$$

where $\epsilon, \gamma > 0$ and the powers q and r are such that $2p < q < p_s^* \leq r$.

Theorem 6.3 *Assume that conditions* (V_1)-(V_2) *hold. Then, there exists* $\gamma_0 > 0$ *such that, for any* $\gamma \in (0, \gamma_0)$, *there exists* $\epsilon_\gamma > 0$ *such that, for any* $\epsilon \in (0, \epsilon_\gamma)$, *problem* (6.1) *has a positive solution* u_ϵ. *Moreover, if* η_ϵ *denotes a global maximum point of* u_ϵ, *then* $\lim_{\epsilon \to 0} V(\eta_\epsilon) = V_0$.

Remark 6.1 *If we assume* (V_1)-(V_2'), *then the conclusion of Theorem 6.2 holds true for* ϵ *and* γ *sufficiently small.*

Differently from the study of (P_ϵ), an additional difficulty arises in the analysis of problem (6.1). Indeed, when $r > p_s^*$, equation (6.1) becomes supercritical, and we cannot directly apply variational techniques because the corresponding energy functional is not well-defined on the fractional Sobolev space $W^{s,p}(\mathbb{R}^3)$. To overcome this hitch, we truncate the nonlinearity involving the critical or supercritical power, and we consider a new subcritical problem for which it is possible to apply the existence result given by Theorem 6.1. Then, after proving a priori bound (independent of γ) for the solution of the truncated problem, we use a suitable Moser iteration argument [120] to prove that, if γ is small enough, this solution also satisfies the original problem (6.1).

Notations: If $A \subset \mathbb{R}^3$, we denote by $A^c = \mathbb{R}^3 \setminus A$. We use the notation $|u|_{L^q(A)}$ to indicate the $L^q(A)$-norm of a function $u : \mathbb{R}^3 \to \mathbb{R}$, and by $|u|_q$ its $L^q(\mathbb{R}^3)$-norm. We write $\mathcal{B}_r(x)$ to denote the ball centered at $x \in \mathbb{R}^3$ with radius $r > 0$, and, when $x = 0$, we put $\mathcal{B}_r = \mathcal{B}_r(0)$ and $\mathcal{B}_r^c := \mathcal{B}_r^c(0)$.

6.2 The modified problem

6.2.1 Work space stuff

We define $D^{s,p}(\mathbb{R}^3)$ as the closure of $C_c^\infty(\mathbb{R}^3)$ with respect to

$$[u]_{s,p}^p := \int_{\mathbb{R}^3}\int_{\mathbb{R}^3} \frac{|u(x)-u(y)|^p}{|x-y|^{3+sp}}dxdy.$$

Let us denote by $W^{s,p}(\mathbb{R}^3)$ the set of functions $u \in L^p(\mathbb{R}^3)$ such that $[u]_{s,p} < \infty$, endowed with the natural norm

$$\|u\|_{s,p}^p := [u]_{s,p}^p + |u|_p^p.$$

We have the following well-known embeddings, see the monograph by Molica Bisci, Rădulescu and Servadei [119].

Theorem 6.4 *Let $s \in (0,1)$ and $p \in [1,\infty)$ be such that $sp < 3$. Then there exists a constant $C_* := C_*(s,p) > 0$ such that, for any $u \in D^{s,p}(\mathbb{R}^3)$, we have*

$$|u|_{p_s^*}^p \leq C_*[u]_{s,p}^p.$$

Moreover, $W^{s,p}(\mathbb{R}^3)$ is continuously embedded in $L^q(\mathbb{R}^3)$ for any $q \in [p,p_s^*]$, and compactly in $L^q_{loc}(\mathbb{R}^3)$ for any $q \in [1,p_s^*)$.

We will often use the following vanishing-Lions type result (see Lemma 2.2 of Ambrosio and Isernia [4]).

Lemma 6.1 *Let $s \in (0,1)$ and $p \in (1,\infty)$ be such that $sp < 3$, and $r \in [p,p_s^*)$. If $\{u_n\}_{n\in\mathbb{N}}$ is a bounded sequence in $W^{s,p}(\mathbb{R}^3)$ and if*

$$\lim_{n\to\infty}\sup_{y\in\mathbb{R}^3}\int_{\mathcal{B}_R(y)} |u_n|^r dx = 0,$$

where $R > 0$, then $u_n \to 0$ in $L^\sigma(\mathbb{R}^3)$ for all $\sigma \in (p,p_s^*)$.

We also recall the following useful technical result (see Lemma 2.3 of Ambrosio and Isernia [4]).

Lemma 6.2 *Let $u \in W^{s,p}(\mathbb{R}^3)$ and let $\phi \in C_c^\infty(\mathbb{R}^3)$ be such that $0 \leq \phi \leq 1$, $\phi = 1$ in $\mathcal{B}_1$ and $\phi = 0$ in $\mathcal{B}_2^c$. Set $\phi_r(x) = \phi(x/r)$. Then,*

$$[u\phi_r - u]_{s,p} \to 0 \quad \text{and} \quad |u\phi_r - u|_p \to 0.$$

In order to study problem (P_ϵ), we use the change of variable $x \mapsto \epsilon x$ and we look for solutions to

$$\begin{cases} (a + b\,[u]_{s,p}^p)(-\Delta)_p^s u + V(\epsilon x)u^{p-1} = f(u) & \text{in } \mathbb{R}^3, \\ u \in W^{s,p}(\mathbb{R}^3), \quad u > 0 & \text{in } \mathbb{R}^3. \end{cases} \tag{$\hat{P}_\epsilon$}$$

Now, we introduce a penalized function. First of all, without loss of generality, we may assume that

$$0 \in \Lambda \text{ and } V(0) = V_0.$$

Let $K > 2p$ and $a_0 > 0$ be such that

$$f(a_0) = \frac{V_1}{K}a_0^{p-1} \tag{6.2}$$

and we define

$$\tilde{f}(t) := \begin{cases} f(t) & \text{if } t \leq a_0, \\ \dfrac{V_1}{K} t^{p-1} & \text{if } t > a_0, \end{cases}$$

and

$$g(x,t) := \begin{cases} \chi_\Lambda(x) f(t) + (1-\chi_\Lambda(x))\tilde{f}(t) & \text{if } t > 0, \\ 0 & \text{if } t \leq 0. \end{cases}$$

Then g satisfies the following properties:

(g_1) $\displaystyle\lim_{t\to 0^+} \frac{g(x,t)}{t^{2p-1}} = 0$ uniformly with respect to $x \in \mathbb{R}^3$,

(g_2) $g(x,t) \leq f(t)$ for all $x \in \mathbb{R}^3$, $t > 0$,

(g_3) (*i*) $0 < \vartheta G(x,t) \leq g(x,t)t$ for all $x \in \Lambda$ and $t > 0$,
(*ii*) $0 \leq pG(x,t) \leq g(x,t)t \leq \frac{V_1}{K} t^p$ for all $x \in \Lambda^c$ and $t > 0$,

(g_4) for each $x \in \Lambda$ the function $\dfrac{g(x,t)}{t^{2p-1}}$ is increasing in $(0,\infty)$, and for each $x \in \Lambda^c$ the function $\dfrac{g(x,t)}{t^{2p-1}}$ is increasing in $(0,a_0)$.

We remark that if u_ϵ is a solution of the problem

$$\begin{cases} (a + b\,[u]_{s,p}^p)(-\Delta)_p^s u + V(\epsilon x)u^{p-1} = g(\epsilon x, u) & \text{in } \mathbb{R}^3, \\ u \in W^{s,p}(\mathbb{R}^3), \quad u > 0 & \text{in } \mathbb{R}^3, \end{cases} \tag{6.3}$$

with $u_\epsilon(x) \leq a_0$ for all $x \in \Lambda_\epsilon^c$, where $\Lambda_\epsilon := \{x \in \mathbb{R}^3 : \epsilon x \in \Lambda\}$, then $g(\epsilon x, u_\epsilon) = f(u_\epsilon)$. Therefore, $v_\epsilon(x) = u_\epsilon(x/\epsilon)$ is a solution of problem (P_ϵ).

For any $\epsilon > 0$, we consider the following fractional space

$$\mathcal{H}_\epsilon := \left\{ u \in W^{s,p}(\mathbb{R}^3) : \int_{\mathbb{R}^3} V(\epsilon x)|u|^p \, dx < \infty \right\}$$

endowed with the norm

$$\|u\|_\epsilon^p := a[u]_{s,p}^p + \int_{\mathbb{R}^3} V(\epsilon x)|u|^p \, dx.$$

In order to study problem (6.3), we look for critical points of the C^1-functional $\mathcal{I}_\epsilon : \mathcal{H}_\epsilon \to \mathbb{R}$ defined by

$$\mathcal{I}_\epsilon(u) = \frac{1}{p}\|u\|_\epsilon^p + \frac{b}{2p}[u]_{s,p}^{2p} - \int_{\mathbb{R}^3} G(\epsilon x, u)\, dx.$$

For any $u, \varphi \in \mathcal{H}_\epsilon$ the differential of $\mathcal{I}_\epsilon$ is given by

$$\begin{aligned} \langle \mathcal{I}_\epsilon'(u), \varphi \rangle &= a \int_{\mathbb{R}^3}\int_{\mathbb{R}^3} \frac{|u(x)-u(y)|^{p-2}(u(x)-u(y))(\varphi(x)-\varphi(y))}{|x-y|^{3+sp}}\, dxdy \\ &+ b[u]_{s,p}^p \int_{\mathbb{R}^3}\int_{\mathbb{R}^3} \frac{|u(x)-u(y)|^{p-2}(u(x)-u(y))(\varphi(x)-\varphi(y))}{|x-y|^{3+sp}}\, dxdy \\ &+ \int_{\mathbb{R}^3} V(\epsilon x)|u|^{p-2}u\varphi\, dx - \int_{\mathbb{R}^3} g(\epsilon x, u)\varphi\, dx. \end{aligned}$$

We denote by $\mathcal{N}_\epsilon$ the Nehari manifold associated with $\mathcal{I}_\epsilon$, that is,

$$\mathcal{N} := \{u \in \mathcal{H}_\epsilon \setminus \{0\} \,:\, \langle I'_\epsilon(u), u\rangle = 0\},$$

and we define

$$c_\epsilon := \inf_{u\in\mathcal{N}} \mathcal{I}_\epsilon(u).$$

Next, we show that $\mathcal{I}_\epsilon$ has a mountain pass geometry in the sense described by Ambrosetti and Rabinowitz [1].

Lemma 6.3 *The functional $\mathcal{I}_\epsilon$ satisfies the following conditions:*

(i) *there exist $\alpha, \rho > 0$ such that $\mathcal{I}_\epsilon(u) \geq \alpha$ with $\|u\|_\epsilon = \rho$;*

(ii) *there exists $e \in \mathcal{H}_\epsilon$ with $\|e\|_\epsilon > \rho$ and $\mathcal{I}_\epsilon(e) < 0$.*

Proof. (i) Using assumptions (g_1), (g_2), (f_1) and (f_2), for any given $\zeta > 0$ there exists a positive constant C_ζ such that

$$|g(x,t)| \leq \zeta|t|^{p-1} + C_\zeta|t|^{\nu-1}.$$

Thus, using Sobolev embeddings we have

$$\mathcal{I}_\epsilon(u) = \frac{1}{p}\|u\|_\epsilon^p + \frac{b}{2p}[u]_{s,p}^{2p} - \int_{\mathbb{R}^3} G(\epsilon x, u)\,dx \geq \frac{1}{p}\|u\|_\epsilon^p - C\zeta\|u\|_\epsilon^p - \bar{C}_\zeta\|u\|_\epsilon^\nu.$$

Consequently, we can choose α, ρ such that $I_\epsilon(u) \geq \alpha$ and $\|u\|_\epsilon = \rho$.

(ii) Let $u \in C_c^\infty(\mathbb{R}^3)$ be such that $u \geq 0$, $u \not\equiv 0$ and $supp(u) \subset \Lambda_\epsilon$. In view of (g_3)–(i) and $\vartheta \in (2p, p_s^*)$, we can see that, for some constants $C_1, C_2 > 0$ and for any $t > 0$

$$\mathcal{I}_\epsilon(tu) \leq \frac{t^p}{p}\|u\|_\epsilon^p + \frac{bt^{2p}}{2p}[u]_{s,p}^{2p} - C_1 t^\vartheta \int_{\Lambda_\epsilon} u^\vartheta\,dx + C_2 \to -\infty \quad \text{as } t \to \infty,$$

which completes the proof. □

Invoking the mountain-pass theorem without (PS)-condition (see Willem [153]), we deduce the existence of a Palais-Smale sequence $\{u_n\}_{n\in\mathbb{N}} \subset \mathcal{H}_\epsilon$ such that

$$\mathcal{I}_\epsilon(u_n) = c_\epsilon + o(1) \text{ and } \mathcal{I}'_\epsilon(u_n) = o(1), \tag{6.4}$$

where

$$c_\epsilon := \inf_{\gamma\in\Gamma_\epsilon}\max_{t\in[0,1]} \mathcal{I}_\epsilon(\gamma(t)) \text{ and } \Gamma_\epsilon := \Big\{\gamma \in C^0([0,1],\mathcal{H}_\epsilon) : \gamma(0) = 0,\, \mathcal{I}_\epsilon(\gamma(1)) \leq 0\Big\}. \tag{6.5}$$

We use the following equivalent characterization of c_ϵ, which is more appropriate for our purpose:

$$c_\epsilon = \inf_{u\in\mathcal{H}_\epsilon\setminus\{0\}}\max_{t\geq 0} \mathcal{I}_\epsilon(tu).$$

Moreover, from the monotonicity of g, it is easy to check that for all $u \in \mathcal{H}_\epsilon \setminus \{0\}$ there exists a unique $t_0 = t_0(u) > 0$ such that

$$\mathcal{I}_\epsilon(t_0 u) = \max_{t\geq 0} \mathcal{I}_\epsilon(tu).$$

Lemma 6.4 *The functional $\mathcal{I}_\epsilon$ fulfills the Palais-Smale condition at any level $c \in \mathbb{R}$.*

Proof. Let $\{u_n\}_{n\in\mathbb{N}}$ be a sequence satisfying (6.4). Let us prove that $\{u_n\}_{n\in\mathbb{N}}$ is bounded in $\mathcal{H}_\epsilon$. Using (g_3), we obtain that

$$\begin{aligned}
c_\epsilon + o(1) &= \mathcal{I}_\epsilon(u_n) - \frac{1}{\vartheta}\langle \mathcal{I}'_\epsilon(u_n), u_n\rangle \\
&= \left(\frac{1}{p} - \frac{1}{\vartheta}\right)\|u_n\|_\epsilon^p + b\left(\frac{1}{2p} - \frac{1}{\vartheta}\right)[u_n]_{s,p}^{2p} + \frac{1}{\vartheta}\int_{\Lambda_\epsilon^c}[g(\epsilon x, u_n)u_n - \vartheta G(\epsilon x, u_n)]\,dx \\
&+\frac{1}{\vartheta}\int_{\Lambda_\epsilon}[g(\epsilon x, u_n)u_n - \vartheta G(\epsilon x, u_n)]\,dx \\
&\geq \left(\frac{1}{p} - \frac{1}{\vartheta}\right)\|u_n\|_\epsilon^p + \frac{1}{\vartheta}\int_{\Lambda_\epsilon^c}[g(\epsilon x, u_n)u_n - \vartheta G(\epsilon x, u_n)]\,dx \\
&\geq \left(\frac{1}{p} - \frac{1}{\vartheta}\right)\|u_n\|_\epsilon^p - \frac{(\vartheta-p)}{\vartheta p}\frac{1}{K}\int_{\Lambda_\epsilon^c} V(\epsilon x)|u_n|^p\,dx \\
&\geq \left(\frac{1}{p} - \frac{1}{\vartheta}\right)\left(1 - \frac{1}{K}\right)\|u_n\|_\epsilon^p.
\end{aligned}$$

Thanks to $\vartheta > 2p$ and $K > 2p > 3$, we deduce that $\{u_n\}_{n\in\mathbb{N}}$ is bounded in $\mathcal{H}_\epsilon$. From Theorem 6.4 we may assume that $u_n \rightharpoonup u$ in $\mathcal{H}_\epsilon$.

Now, we prove that for any $\eta > 0$ there exists $R = R(\eta) > 0$ such that

$$\limsup_{n\to\infty}\int_{\mathcal{B}_R^c}\left(a\int_{\mathbb{R}^3}\frac{|u_n(x)-u_n(y)|^p}{|x-y|^{3+sp}}\,dy + V(\epsilon x)|u_n|^p\right)dx < \eta. \tag{6.6}$$

For any $R > 0$, let $\psi_R \in C^\infty(\mathbb{R}^3)$ be such that $\psi_R = 0$ in $\mathcal{B}_R$, $\psi_R = 1$ in $\mathcal{B}_{2R}^c$, $0 \leq \psi_R \leq 1$, and $|\nabla\psi_R| \leq \frac{C}{R}$, where C is a constant independent of R. Since $\{\psi_R u_n\}_{n\in\mathbb{N}}$ is bounded in $\mathcal{H}_\epsilon$, it follows that $\langle \mathcal{I}'_\epsilon(u_n), \psi_R u_n\rangle = o_n(1)$, that is

$$\begin{aligned}
&(a + b\,[u_n]_{s,p}^p)\int_{\mathbb{R}^3}\int_{\mathbb{R}^3}\frac{|u_n(x)-u_n(y)|^p}{|x-y|^{3+sp}}\psi_R(x)\,dxdy + \int_{\mathbb{R}^3} V(\epsilon x)|u_n|^p\psi_R\,dx \\
&= o(1) + \int_{\mathbb{R}^3} g(\epsilon x, u_n)\psi_R u_n\,dx \\
&- (a + b\,[u_n]_{s,p}^p)\int_{\mathbb{R}^3}\int_{\mathbb{R}^3}\frac{|u_n(x)-u_n(y)|^{p-2}(u_n(x)-u_n(y))(\psi_R(x)-\psi_R(y))}{|x-y|^{3+sp}}u_n(y)\,dxdy.
\end{aligned}$$

Take $R > 0$ such that $\Lambda_\epsilon \subset \mathcal{B}_R$. By the definition of ψ_R and (g_3)–(ii) we get

$$\begin{aligned}
&a\int_{\mathbb{R}^3}\int_{\mathbb{R}^3}\frac{|u_n(x)-u_n(y)|^p}{|x-y|^{3+sp}}\psi_R\,dxdy + \left(1-\frac{1}{K}\right)\int_{\mathbb{R}^3} V(\epsilon x)|u_n|^p\psi_R\,dx \leq o(1)- \\
&(a + b\,[u_n]_{s,p}^p)\int_{\mathbb{R}^3}\int_{\mathbb{R}^3}\frac{|u_n(x)-u_n(y)|^{p-2}(u_n(x)-u_n(y))(\psi_R(x)-\psi_R(y))}{|x-y|^{3+sp}}u_n(y)\,dxdy.
\end{aligned} \tag{6.7}$$

Let us note that, from the boundedness of $\{u_n\}_{n\in\mathbb{N}}$ in $\mathcal{H}_\epsilon$, we can suppose that $a + b\,[u_n]_{s,p}^p \to \ell \in (0,\infty)$. Now, using the Hölder's inequality and the boundedness of $\{u_n\}_{n\in\mathbb{N}}$ in $\mathcal{H}_\epsilon$ we have

$$\begin{aligned}
&\left|\int_{\mathbb{R}^3}\int_{\mathbb{R}^3}\frac{|u_n(x)-u_n(y)|^{p-2}(u_n(x)-u_n(y))(\psi_R(x)-\psi_R(y))}{|x-y|^{3+sp}}u_n(y)\,dxdy\right| \\
&\quad\leq C\left(\int_{\mathbb{R}^3}\int_{\mathbb{R}^3}\frac{|\psi_R(x)-\psi_R(y)|^p}{|x-y|^{3+sp}}|u_n(y)|^p\,dxdy\right)^{\frac{1}{p}}.
\end{aligned} \tag{6.8}$$

On the other hand, by the definition of ψ_R, and using polar coordinates and the boundedness of $\{u_n\}_{n\in\mathbb{N}}$ in $\mathcal{H}_\epsilon$, we obtain

$$\begin{aligned}
&\int_{\mathbb{R}^3}\int_{\mathbb{R}^3}\frac{|\psi_R(x)-\psi_R(y)|^p}{|x-y|^{3+sp}}|u_n(x)|^p dxdy\\
&=\int_{\mathbb{R}^3}\int_{|y-x|>R}\frac{|\psi_R(x)-\psi_R(y)|^p}{|x-y|^{3+sp}}|u_n(x)|^p dxdy+\int_{\mathbb{R}^3}\int_{|y-x|\le R}\frac{|\psi_R(x)-\psi_R(y)|^p}{|x-y|^{3+sp}}|u_n(x)|^p dxdy\\
&\le C\int_{\mathbb{R}^3}|u_n(x)|^p\left(\int_{|y-x|>R}\frac{dy}{|x-y|^{3+sp}}\right)dx+\frac{C}{R^p}\int_{\mathbb{R}^3}|u_n(x)|^p\left(\int_{|y-x|\le R}\frac{dy}{|x-y|^{3+sp-p}}\right)dx\\
&\le C\int_{\mathbb{R}^3}|u_n(x)|^p\left(\int_{|z|>R}\frac{dz}{|z|^{3+sp}}\right)dx+\frac{C}{R^p}\int_{\mathbb{R}^3}|u_n(x)|^p\left(\int_{|z|\le R}\frac{dz}{|z|^{3+sp-p}}\right)dx\\
&\le C\int_{\mathbb{R}^3}|u_n(x)|^p dx\left(\int_R^\infty\frac{d\rho}{\rho^{sp+1}}\right)+\frac{C}{R^p}\int_{\mathbb{R}^3}|u_n(x)|^p dx\left(\int_0^R\frac{d\rho}{\rho^{sp-p+1}}\right)\\
&\le \frac{C}{R^{sp}}\int_{\mathbb{R}^3}|u_n(x)|^p dx+\frac{C}{R^p}R^{-sp+p}\int_{\mathbb{R}^3}|u_n(x)|^p dx\\
&\le \frac{C}{R^{sp}}\int_{\mathbb{R}^3}|u_n(x)|^p dx\le\frac{C}{R^{sp}}\to 0\quad\text{as } R\to\infty.
\end{aligned}$$

Gathering the last relation together with (6.7) and (6.8), we infer that (6.6) is satisfied. In particular, by Fatou's lemma we have

$$\int_{\mathcal{B}_R^c}\left(a\int_{\mathbb{R}^3}\frac{|u(x)-u(y)|^p}{|x-y|^{3+sp}}dy+V(\epsilon x)|u|^p\right)dx<\eta. \tag{6.9}$$

Moreover, from (6.6) and (6.9), we deduce that

$$u_n\to u \text{ in } L^\sigma(\mathbb{R}^3)\qquad\text{for any } \sigma\in[p,p_s^*). \tag{6.10}$$

Indeed, for any $\eta>0$ there exists $R=R(\eta)>0$ such that, for any n large enough, we have

$$\begin{aligned}
|u_n-u|_p^p&=|u_n-u|_{L^p(\mathcal{B}_R)}^p+|u_n-u|_{L^p(\mathcal{B}_R^c)}^p\\
&\le\eta+|u_n-u|_{L^p(\mathcal{B}_R^c)}^p\\
&\le\eta+\frac{1}{V_1}\int_{\mathcal{B}_R^c}V(\epsilon x)|u_n-u|^p\,dx\\
&\le\eta+C\int_{\mathcal{B}_R^c}\left(a\int_{\mathbb{R}^3}\frac{|(u_n(x)-u(x))-(u_n(y)-u(y))|^p}{|x-y|^{3+sp}}dy+V(\epsilon x)|u_n-u|^p\right)dx\\
&\le\kappa\eta,
\end{aligned}$$

that is $u_n\to u$ in $L^p(\mathbb{R}^3)$. Then, by interpolation, it follows that (6.10) is verified.

Now, from the boundedness of $\{u_n\}_{n\in\mathbb{N}}$ and the growth assumptions on g we get

$$\begin{aligned}
&\left|\int_{\mathbb{R}^3}(g(\epsilon x,u_n)u_n-g(\epsilon x,u)u)(u_n-u)\,dx\right|\\
&\le\left(|u_n|_p^{p-1}+|u|_p^{p-1}\right)|u_n-u|_p+C\left(|u_n|_\nu^{\nu-1}+|u|_\nu^{\nu-1}\right)|u_n-u|_\nu\\
&\le C|u_n-u|_p+C|u_n-u|_\nu
\end{aligned}$$

which together with (6.10) implies that

$$\lim_{n\to 0}\int_{\mathbb{R}^3}(g(\epsilon x,u_n)u_n-g(\epsilon x,u)u)(u_n-u)\,dx=0. \tag{6.11}$$

In what follows, we prove that $\|u_n - u\|_\epsilon \to 0$ as $n \to \infty$. Let $\varphi \in \mathcal{H}_\epsilon$ be fixed and denote by $B_\varphi : \mathcal{H}_\epsilon \to \mathbb{R}$ the linear functional on $\mathcal{H}_\epsilon$ defined as

$$B_\varphi(v) := \int_{\mathbb{R}^3}\int_{\mathbb{R}^3} \frac{|\varphi(x)-\varphi(y)|^{p-2}(\varphi(x)-\varphi(y))(v(x)-v(y))}{|x-y|^{3+sp}}\,dxdy$$

for all $v \in \mathcal{H}_\epsilon$. Moreover, by Hölder's inequality, it is clear that B_φ is continuous on $\mathcal{H}_\epsilon$.

It follows from $u_n \rightharpoonup u$ in $\mathcal{H}_\epsilon$ that

$$\lim_{n\to\infty}\left((a+b[u_n]_{s,p}^p)-(a+b[u]_{s,p}^p)\right)B_u(u_n-u)=0, \tag{6.12}$$

where we used the fact that $(a+b[u_n]_{s,p}^p)-(a+b[u]_{s,p}^p)$ is bounded in $\mathbb{R}$.

On the other hand, by $u_n \rightharpoonup u$ in $\mathcal{H}_\epsilon$, $\mathcal{I}'_\epsilon(u_n) \to 0$, and (6.10), we know that $\langle \mathcal{I}'_\epsilon(u_n) - \mathcal{I}'_\epsilon(u), u_n - u\rangle \to 0$ as $n \to \infty$. Then, by (6.11) and (6.12), we obtain

$$\begin{aligned}
o(1) &= \langle \mathcal{I}'_\epsilon(u_n) - \mathcal{I}'_\epsilon(u), u_n - u\rangle \\
&= (a+b[u_n]_{s,p}^p)B_{u_n}(u_n-u) - (a+b[u_n]_{s,p}^p)B_u(u_n-u) \\
&\quad + \left((a+b[u_n]_{s,p}^p)-(a+b[u]_{s,p}^p)\right)B_u(u_n-u) + \int_{\mathbb{R}^3} V(\epsilon x)(|u_n|^{p-2}u_n - |u|^{p-2}u)(u_n-u)\,dx \\
&\quad - \int_{\mathbb{R}^3} (g(\epsilon x,u_n)u_n - g(\epsilon x,u)u)(u_n-u)\,dx \\
&= (a+b[u_n]_{s,p}^p)(B_{u_n}(u_n-u)-B_u(u_n-u)) \\
&\quad + \int_{\mathbb{R}^3} V(\epsilon x)(|u_n|^{p-2}u_n - |u|^{p-2}u)(u_n-u)\,dx + o(1).
\end{aligned}$$

It follows that

$$\lim_{n\to\infty}\left((a+b[u_n]_{s,p}^p)(B_{u_n}(u_n-u)-B_u(u_n-u)) + \int_{\mathbb{R}^3} V(\epsilon x)(|u_n|^{p-2}u_n - |u|^{p-2}u)(u_n-u)\,dx\right)=0.$$

Since $(|z|^{p-2}z - |w|^{p-2}w)(z-w) \geq 0$ for all $z, w \in \mathbb{R}$, we can see that

$$(a+b[u_n]_{s,p}^p)(B_{u_n}(u_n-u)-B_u(u_n-u)) \geq 0,$$

and by (V_1) we also have

$$V(\epsilon x)(|u_n|^{p-2}u_n - |u|^{p-2}u)(u_n-u) \geq 0.$$

Therefore, we deduce that

$$\begin{aligned}
&\lim_{n\to\infty}(a+b[u_n]_{s,p}^p)(B_{u_n}(u_n-u)-B_u(u_n-u))=0, \\
&\lim_{n\to\infty}\int_{\mathbb{R}^3} V(\epsilon x)(|u_n|^{p-2}u_n - |u|^{p-2}u)(u_n-u)\,dx = 0.
\end{aligned} \tag{6.13}$$

In what follows, we shall apply Simon's inequalities, see Appendix B. We first suppose that $p \geq 2$. Then, by (6.13) and (B.1), we obtain

$$\begin{aligned}
[u_n-u]_{s,p}^p &= \int_{\mathbb{R}^3}\int_{\mathbb{R}^3} |u_n(x)-u_n(y)-u(x)+u(y)|^p|x-y|^{-(3+sp)}\,dxdy \\
&\leq c_p\int_{\mathbb{R}^3}\int_{\mathbb{R}^3}[|u_n(x)-u_n(y)|^{p-2}(u_n(x)-u_n(y)) - |u(x)-u(y)|^{p-2}(u(x)-u(y))] \\
&\quad \times (u_n(x)-u_n(y)-u(x)+u(y))|x-y|^{-(3+sp)}\,dxdy \\
&= c_p[B_{u_n}(u_n-u)-B_u(u_n-u)] = o(1).
\end{aligned}$$

In a similar way, by (V_1) and (6.13), we get

$$\int_{\mathbb{R}^3} V(\epsilon x)|u_n - u|^p \, dx \le c_p \int_{\mathbb{R}^3} V(\epsilon x)(|u_n|^{p-2}u_n - |u|^{p-2}u)(u_n - u) \, dx = o(1).$$

In conclusion, $\|u_n - u\|_\epsilon \to 0$ as $n \to \infty$.

Now, we consider the case when $1 < p < 2$. Since $u_n \rightharpoonup u$ in $\mathcal{H}_\epsilon$, there exists $\kappa > 0$ such that $\|u_n\|_\epsilon \le \kappa$ for all $n \in \mathbb{N}$. Hence, by (6.13), (B.1) and Hölder's inequality we can see that

$$\begin{aligned}
[u_n - u]_{s,p}^p &\le C_p \left(B_{u_n}(u_n - u) - B_u(u_n - u)\right)^{\frac{p}{2}} \left([u_n]_{s,p}^p + [u]_{s,p}^p\right)^{\frac{2-p}{2}} \\
&\le C_p \left(B_{u_n}(u_n - u) - B_u(u_n - u)\right)^{\frac{p}{2}} \left([u_n]_{s,p}^{\frac{p(2-p)}{2}} + [u]_{s,p}^{\frac{p(2-p)}{2}}\right) \\
&\le C_p' \left(B_{u_n}(u_n - u) - B_u(u_n - u)\right)^{\frac{p}{2}} = o(1),
\end{aligned}$$

where we used the following inequality

$$(a+b)^{\frac{2-p}{2}} \le a^{\frac{2-p}{2}} + b^{\frac{2-p}{2}} \quad \forall a, b \ge 0, \, 1 < p < 2.$$

In a similar way, we obtain that

$$\int_{\mathbb{R}^3} V(\epsilon x)|u_n - u|^p \, dx \le C_p'' \left(\int_{\mathbb{R}^3} V(\epsilon x)(|u_n|^{p-2}u_n - |u|^{p-2}u)(u_n - u) \, dx\right)^{\frac{p}{2}} = o(1).$$

Then we can infer that $\|u_n - u\|_\epsilon \to 0$ as $n \to \infty$. This ends the proof of lemma. □

As a byproduct of Lemma 6.3, Lemma 6.4 and mountain pass theorem [1], we deduce that for all $\epsilon > 0$ there exists $u_\epsilon \in \mathcal{H}_\epsilon$ such that $\mathcal{I}_\epsilon(u_\epsilon) = c_\epsilon$ and $\mathcal{I}_\epsilon'(u_\epsilon) = 0$, that is u_ϵ is a weak solution to (6.3).

Let $u^- := \min\{u, 0\}$. Using $\langle \mathcal{I}_\epsilon'(u), u^- \rangle = 0$, $g(x,t) = 0$ for $t \le 0$, and

$$|x - y|^{p-2}(x - y)(x^- - y^-) \ge |x^- - y^-|^p \quad \forall x, y \in \mathbb{R},$$

we can infer

$$\begin{aligned}
\|u^-\|_\epsilon^p &\le a \int_{\mathbb{R}^3} \int_{\mathbb{R}^3} \frac{|u(x) - u(y)|^{p-2}(u(x) - u(y))(u^-(x) - u^-(y))}{|x - y|^{N+sp}} \, dxdy \\
&+ \int_{\mathbb{R}^3} V(\epsilon x)|u|^{p-2}uu^- \, dx \\
&+ b[u]_{s,p}^p \int_{\mathbb{R}^3} \int_{\mathbb{R}^3} \frac{|u(x) - u(y)|^{p-2}(u(x) - u(y))(u^-(x) - u^-(y))}{|x - y|^{N+sp}} \, dxdy = 0,
\end{aligned}$$

which implies that $u^- = 0$, that is $u \ge 0$ in $\mathbb{R}^3$. By a Moser iteration argument [120], we obtain that $u \in L^\infty(\mathbb{R}^3) \cap C^0(\mathbb{R}^N)$ (see Lemma 6.8 below). By the maximum principle, we infer that $u > 0$ in $\mathbb{R}^3$. □

6.2.2 The limiting problem

Let us consider the following family of autonomous problems, where μ is a positive parameter,

$$\begin{cases} (a + b\,[u]_{s,p}^p)(-\Delta)_p^s u + \mu u^{p-1} = f(u) & \text{in } \mathbb{R}^3, \\ u \in W^{s,p}(\mathbb{R}^3), \quad u > 0 & \text{in } \mathbb{R}^3. \end{cases} \tag{6.14}$$

The energy functional associated with problem (6.14) is given by

$$\mathcal{E}_\mu(u) = \frac{1}{p}a[u]_{s,p}^p + \frac{b}{2p}[u]_{s,p}^{2p} + \mu|u|_p^p - \int_{\mathbb{R}^3} F(u)\,dx$$

which is well defined on the space $\mathcal{H}_\mu := W^{s,p}(\mathbb{R}^N)$ endowed with the norm

$$\|u\|_\mu^p := a[u]_{s,p}^p + \mu|u|_p^p.$$

Then $\mathcal{E}_\mu \in C^1(\mathcal{H}_\mu, \mathbb{R})$ and for any $u, \varphi \in \mathcal{H}_\mu$

$$\begin{aligned}\langle \mathcal{E}_\mu'(u), \varphi\rangle &= a\int_{\mathbb{R}^3}\int_{\mathbb{R}^3} \frac{|u(x)-u(y)|^{p-2}(u(x)-u(y))(\varphi(x)-\varphi(y))}{|x-y|^{3+sp}}\,dxdy \\ &+ b[u]_{s,p}^p \int_{\mathbb{R}^3}\int_{\mathbb{R}^3} \frac{|u(x)-u(y)|^{p-2}(u(x)-u(y))(\varphi(x)-\varphi(y))}{|x-y|^{3+sp}}\,dxdy \\ &+ \mu\int_{\mathbb{R}^3} |u|^{p-2}u\varphi\,dx - \int_{\mathbb{R}^3} f(u)\varphi\,dx.\end{aligned}$$

We denote by $\mathcal{N}_\mu$ the Nehari manifold associated with $\mathcal{E}_\mu$, that is,

$$\mathcal{N}_\mu := \Big\{u \in \mathcal{H}_\mu \setminus \{0\} : \langle \mathcal{E}_\mu'(u), u\rangle = 0\Big\},$$

and

$$d_\mu := \inf_{u\in\mathcal{N}_\mu} \mathcal{E}_\mu(u).$$

Arguing as in Lemma 6.3, we obtain that $\mathcal{E}_\mu$ has a mountain-pass geometry and

$$d_\mu = \inf_{\gamma\in\Gamma_\mu}\max_{t\in[0,1]} \mathcal{E}_\mu(\gamma(t)) = \inf_{u\in\mathcal{H}_\mu\setminus\{0\}}\max_{t\geq 0} \mathcal{E}_\mu(tu),$$

where

$$\Gamma_\mu := \Big\{\gamma \in C^0([0,1], \mathcal{H}_\mu) : \gamma(0) = 0,\ \mathcal{E}_\mu(\gamma(1)) < 0\Big\}.$$

Lemma 6.5 *Let $\{u_n\}_{n\in\mathbb{N}} \subset \mathcal{H}_\mu$ be a $(PS)_c$ sequence for $\mathcal{E}_\mu$ such that $u_n \rightharpoonup 0$. Then we have either*

(a) $u_n \to 0$ in $\mathcal{H}_\mu$, or

(b) there is a sequence $\{y_n\}_{n\in\mathbb{N}} \subset \mathbb{R}^3$ and constant $R, \beta > 0$ such that

$$\liminf_{n\to\infty}\int_{B_R(y_n)} |u_n|^p\,dx \geq \beta.$$

Proof. Assume that (b) is not true. Then, by Lemma 6.1 we deduce that

$$u_n \to 0 \quad \text{in } L^\sigma(\mathbb{R}^3) \text{ for all } \sigma \in (p, p_s^*). \tag{6.15}$$

From (6.15) and (f_1)–(f_2) we can infer that

$$\int_{\mathbb{R}^3} F(u_n)\,dx = \int_{\mathbb{R}^3} f(u_n)u_n\,dx = o(1) \quad \text{as } n \to \infty.$$

On the other hand, arguing as in Lemma 6.4, we know that $\{u_n\}_{n\in\mathbb{N}}$ is bounded in $\mathcal{H}_\mu$, and we may assume that $u_n \rightharpoonup u$ in $\mathcal{H}_\mu$. Taking into account that $\langle \mathcal{E}_\mu'(u_n), u_n\rangle = 0$, we get

$$\|u_n\|_\mu^p + b\,[u_n]_{s,p}^{2p} = \int_{\mathbb{R}^3} f(u_n)u_n\,dx = o(1),$$

which implies that $\|u_n\|_\mu \to 0$ as $n \to \infty$, that is (a) holds true. □

Now, we prove the following existence result for problem (6.14).

Theorem 6.5 *For all $\mu > 0$, problem (6.14) admits a positive ground state solution.*

Proof. Using the mountain-pass theorem without (PS)-condition, there exists a Palais-Smale sequence $\{u_n\}_{n\in\mathbb{N}} \subset \mathcal{H}_\mu$ for $\mathcal{E}_\mu$ at the level d_μ. Arguing as in Lemma 6.4, we know that $\{u_n\}_{n\in\mathbb{N}}$ is bounded in $\mathcal{H}_\mu$, so we may assume that

$$\begin{aligned} u_n &\rightharpoonup u \quad \text{in } \mathcal{H}_\mu, \\ u_n &\to u \quad \text{in } L^\sigma_{loc}(\mathbb{R}^3) \text{ for all } \sigma \in [1, p_s^*). \end{aligned}$$

Moreover, by Lemma 6.5, it follows that u is nontrivial. Now, for any $\varphi \in C_c^\infty(\mathbb{R}^3)$ we have

$$\begin{aligned} &a\int_{\mathbb{R}^3}\int_{\mathbb{R}^3} \frac{|u(x)-u(y)|^{p-2}(u(x)-u(y))(\varphi(x)-\varphi(y))}{|x-y|^{3+sp}}\,dxdy + \mu\int_{\mathbb{R}^3}|u|^{p-2}u\varphi\,dx + \\ &bB^p\int_{\mathbb{R}^3}\int_{\mathbb{R}^3} \frac{|u(x)-u(y)|^{p-2}(u(x)-u(y))(\varphi(x)-\varphi(y))}{|x-y|^{3+sp}}\,dxdy - \int_{\mathbb{R}^3} f(u)\varphi\,dx = 0, \end{aligned} \tag{6.16}$$

where $B^p := \lim_{n\to\infty}[u_n]^p_{s,p}$. By Fatou's lemma we have

$$[u]^p_{s,p} \le B^p. \tag{6.17}$$

Our aim is to prove that the equality holds in (6.17). Assume by contradiction that $[u]^p_{s,p} < B^p$. Taking $\varphi = u$ in (6.16) we have that $\langle \mathcal{E}'_\mu(u), u\rangle < 0$. From assumptions (f_1) and (f_2) we can see that $\langle \mathcal{E}'_\mu(t_1u), t_1u\rangle < 0$ for some $0 < t_1 \ll 1$. Thus, there exists $\tau \in (t_1, 1)$ such that $\langle \mathcal{E}'_\mu(\tau u), \tau u\rangle = 0$. Now, using $\tau \in (0,1)$, the fact that $t \mapsto \frac{1}{2p}f(t)t - F(t)$ is increasing for any $t \ge 0$, and Fatou's lemma we can infer that

$$\begin{aligned} d_\mu &\le \mathcal{E}_\mu(\tau u) - \frac{1}{2p}\langle \mathcal{E}'_\mu(\tau u), \tau u\rangle < \mathcal{E}_\mu(u) - \frac{1}{2p}\langle \mathcal{E}'_\mu(u), u\rangle \\ &\le \liminf_{n\to\infty}\left(\mathcal{E}_\mu(u_n) - \frac{1}{2p}\langle \mathcal{E}'_\mu(u_n), u_n\rangle\right) = d_\mu, \end{aligned} \tag{6.18}$$

which gives a contradiction. Hence $[u]^p_{s,p} = B^p$. Therefore, by (6.16), we deduce that $\mathcal{E}'_\mu(u) = 0$, that is $\mathcal{E}_\mu$ admits a nontrivial critical point $u \in \mathcal{H}_\mu$. By the maximum principle, we deduce that $u > 0$ in $\mathbb{R}^3$. Finally, proceeding as in (6.18) with $\tau = 1$, we conclude that u is a ground state solution of problem (6.14).

Let $u^- := \min\{u, 0\}$. Using $\langle \mathcal{E}'_\mu(u), u^-\rangle = 0$, $f(t) = 0$ for $t \le 0$, and

$$|x-y|^{p-2}(x-y)(x^- - y^-) \ge |x^- - y^-|^p \quad \forall x, y \in \mathbb{R},$$

we can infer

$$\begin{aligned} \|u^-\|^p_\mu &\le a\int_{\mathbb{R}^3}\int_{\mathbb{R}^3} \frac{|u(x)-u(y)|^{p-2}(u(x)-u(y))(u^-(x)-u^-(y))}{|x-y|^{N+sp}}\,dxdy \\ &+\mu\int_{\mathbb{R}^3}|u|^{p-2}uu^-\,dx + b[u]^p_{s,p}\int_{\mathbb{R}^3}\int_{\mathbb{R}^3} \frac{|u(x)-u(y)|^{p-2}(u(x)-u(y))(u^-(x)-u^-(y))}{|x-y|^{N+sp}}\,dxdy \\ &= \int_{\mathbb{R}^3} f(u)u^-\,dx = 0, \end{aligned}$$

which implies that $u^- = 0$, that is $u \ge 0$. By a Moser iteration argument [120], we obtain that $u \in L^\infty(\mathbb{R}^3)$ (see Lemma 6.8 below). From the maximum principle, we can infer that $u > 0$ in $\mathbb{R}^3$. □

Remark 6.2 *We conjecture that if* $s \in (0,1)$ *and* $p \in (1,\infty)$ *are such that* $sp < 3$, $f(u) = u^{q-1}$ *with* $q \in (p, p_s^*)$, *then it is possible to obtain an existence result for problem* (6.14) *if* $b > 0$ *is small enough. The idea is to apply the Struwe-Jeanjean monotonicity trick by considering the truncated functionals* $\mathcal{E}_{b,\lambda}^k : H_{rad}^s(\mathbb{R}^3) \to \mathbb{R}$, *with* $k \in \mathbb{N}$, $\lambda \in [\lambda_0, 1]$, *defined by*

$$\mathcal{E}_{b,\lambda}^k(u) = \frac{\|u\|_\mu^p}{p} + \frac{b}{2p}\chi\left(\frac{\|u\|_\mu^p}{k^p}\right)[u]_{s,p}^{2p} - \lambda \int_{\mathbb{R}^N} F(u)\,dx,$$

where χ *is a cut-off function with support in the ball* $\mathcal{B}_2$. *Then, once proved that there exists a sequence* $\{\lambda_j\}_{j\in\mathbb{N}} \subset [\lambda_0, 1]$, $\lambda_j \to 1$, *and* $\{u_j\}_{j\in\mathbb{N}} \subset H_{rad}^s(\mathbb{R}^3)$ *such that each* u_j *is a critical point of* $\mathcal{E}_{b,\lambda_j}^k$, *one has to show that for all* $k > 0$ *large, there exists* $b_0 > 0$ *such that, for all* $b \in (0, b_0)$, $\|u_j\|_\mu \leq k$ *for all* $j \in \mathbb{N}$. *After that, arguing as in Ambrosio [2], the existence result for problem* (6.14) *follows.*

We conclude this section by proving a very interesting relation between the levels c_ϵ and d_{V_0}.

Lemma 6.6 *We have* $\limsup_{\epsilon\to 0} c_\epsilon \leq d_{V_0}$.

Proof. Let ω be a positive ground state of problem (6.14) given by Theorem 6.5 with $\mu = V_0$. For any $\epsilon > 0$ let $\psi_\epsilon(x) := \psi(\epsilon x)$ where $\psi \in C_c^\infty(\mathbb{R}^3)$ is such that

$$\begin{cases} \psi(x) = 1 & \text{if } x \in \mathcal{B}_1, \\ \psi(x) = 0 & \text{if } x \in \mathcal{B}_2^c, \\ 0 \leq \psi \leq 1, \end{cases}$$

and consider the function $\omega_\epsilon(x) = \psi_\epsilon(x)\omega(x)$. For simplicity, let us assume that $supp(\psi) \subset \mathcal{B}_2 \subset \Lambda$.

By Lemma 6.2 and the dominated convergence theorem we can infer that, as $\epsilon \to 0$,

$$\omega_\epsilon \to \omega \in W^{s,p}(\mathbb{R}^3) \quad \text{and} \quad \mathcal{E}_{V_0}(\omega_\epsilon) \to \mathcal{E}_{V_0}(\omega) = d_{V_0}. \tag{6.19}$$

Now, for each $\epsilon > 0$ there exists $t_\epsilon > 0$ such that

$$\mathcal{I}_\epsilon(t_\epsilon\omega_\epsilon) = \max_{t\geq 0} \mathcal{I}_\epsilon(t\omega_\epsilon).$$

Hence, $\langle \mathcal{I}_\epsilon'(t_\epsilon\omega_\epsilon), \omega_\epsilon\rangle = 0$ and we have

$$t_\epsilon^p\|\omega_\epsilon\|_\epsilon^p + b\,t_\epsilon^{2p}[\omega_\epsilon]_{s,p}^{2p} = \int_{\mathbb{R}^3} f(t_\epsilon\omega_\epsilon)t_\epsilon\omega_\epsilon\,dx,$$

which implies that

$$\frac{a}{t_\epsilon^p}[\omega_\epsilon]_{s,p}^p + \frac{1}{t_\epsilon^p}\int_{\mathbb{R}^3} V(\epsilon x)|\omega_\epsilon|^p\,dx + b\,[\omega_\epsilon]_{s,p}^{2p} = \int_{\mathbb{R}^3} \frac{f(t_\epsilon\omega_\epsilon)}{(t_\epsilon\omega_\epsilon)^{2p-1}}\omega_\epsilon^{2p}\,dx. \tag{6.20}$$

By the growth assumptions on f it follows that $t_\epsilon \to t_0 > 0$. Our aim is to prove that $t_0 = 1$. Taking the limit as $\epsilon \to 0$ in (6.20) and using (6.19) we get

$$\frac{a}{t_0^p}[\omega]_{s,p}^p + \frac{1}{t_0^p}\int_{\mathbb{R}^3} V_0|\omega|^p\,dx + b\,[\omega]_{s,p}^{2p} = \int_{\mathbb{R}^3} \frac{f(t_0\omega)}{(t_0\omega)^{2p-1}}\omega^{2p}\,dx.$$

From the above relation, $\omega \in \mathcal{N}_{V_0}$ and (f_4) we deduce that $t_0 = 1$. On the other hand, we can note that

$$c_\epsilon \leq \max_{t\geq 0} \mathcal{I}_\epsilon(t\omega_\epsilon) = \mathcal{I}_\epsilon(t_\epsilon\omega_\epsilon) = \mathcal{E}_{V_0}(t_\epsilon\omega_\epsilon) + \frac{t_\epsilon^p}{p}\int_{\mathbb{R}^3}(V(\epsilon x) - V_0)\omega_\epsilon^p\,dx.$$

Hence, using (6.19), since $t_\epsilon \to 1$ and $V(\epsilon x)$ is bounded on the support of ω_ϵ, we conclude the proof. □

6.2.3 Proof of Theorem 6.1

For $\epsilon > 0$, let u_ϵ be the mountain pass solution of problem (6.3). For any $\epsilon_n \to 0^+$ we denote by

$$u_n := u_{\epsilon_n}, \quad \mathcal{I}_n := \mathcal{I}_{\epsilon_n}, \quad \mathcal{H}_n := \mathcal{H}_{\epsilon_n} \quad \text{and} \quad c_n := c_{\epsilon_n}.$$

Then, u_n satisfies

$$(a + b[u_n]^p_{s,p})(-\Delta)^s_p u_n + V(\epsilon x) u_n^{p-1} = g(\epsilon x, u_n) \quad \text{in } \mathbb{R}^3.$$

Lemma 6.7 *Let $\epsilon_n \to 0$ and $\{u_n\}_{n\in\mathbb{N}} \subset \mathcal{H}_n$ be such that $\mathcal{I}_n(u_n) = c_{\epsilon_n}$ and $\mathcal{I}'_n(u_n) = 0$. Then there exists $\{\tilde{y}_{\epsilon_n}\}_{n\in\mathbb{N}} \subset \mathbb{R}^3$ such that the translated sequence*

$$\tilde{u}_n(x) := u_n(x + \tilde{y}_{\epsilon_n})$$

has a subsequence which converges in $W^{s,p}(\mathbb{R}^3)$. Moreover, up to a subsequence, $\{y_{\epsilon_n}\}_{n\in\mathbb{N}} := \{\epsilon_n \tilde{y}_{\epsilon_n}\}_{n\in\mathbb{N}}$ is such that $y_{\epsilon_n} \to y_0$ for some $y_0 \in \Lambda$ such that $V(y_0) = V_0$.

Proof. Using $\langle \mathcal{I}'_n(u_n), u_n\rangle = 0$ and assumptions (g_1) and (g_2), we deduce that there is $\kappa > 0$ such that

$$\|u_n\|_{\epsilon_n} \geq \kappa > 0 \quad \text{for any } n \in \mathbb{N}.$$

Taking into account $\mathcal{I}_n(u_n) = c_n$, $\langle \mathcal{I}'_n(u_n), u_n\rangle = 0$ and Lemma 6.6, we can argue as in the proof of Lemma 2.3 to deduce that $\{u_n\}_{n\in\mathbb{N}}$ is bounded in $\mathcal{H}_n$.

Now, proceeding as in the proof of Lemma 6.5, we prove that there are a sequence $\{\tilde{y}_{\epsilon_n}\}_{n\in\mathbb{N}} \subset \mathbb{R}^3$ and constants $R, \beta > 0$ such that

$$\liminf_{n\to\infty} \int_{\mathcal{B}_R(\tilde{y}_{\epsilon_n})} |u_n|^p dx \geq \beta.$$

Hereafter, we denote by $\{\tilde{y}_n\}_{n\in\mathbb{N}}$ the sequence $\{\tilde{y}_{\epsilon_n}\}_{n\in\mathbb{N}}$. Set $\tilde{u}_n(x) := u_n(x + \tilde{y}_n)$. Then $\{\tilde{u}_n\}_{n\in\mathbb{N}}$ is bounded in $W^{s,p}(\mathbb{R}^3)$, and we may assume that

$$\tilde{u}_n \rightharpoonup \tilde{u} \quad \text{weakly in } W^{s,p}(\mathbb{R}^3). \tag{6.21}$$

Moreover, $\tilde{u} \neq 0$ in view of

$$\int_{\mathcal{B}_R} |\tilde{u}|^p dx \geq \beta. \tag{6.22}$$

Now, we set $y_n := \epsilon_n \tilde{y}_n$. Let us begin by proving that $\{y_n\}_{n\in\mathbb{N}}$ is bounded in $\mathbb{R}$. To this end, it is enough to show the following claim.

Claim 1. $\lim_{n\to\infty} \text{dist}(y_n, \overline{\Lambda}) = 0$.

Indeed, if the claim does not hold, there is $\delta > 0$ and a subsequence of $\{y_n\}_{n\in\mathbb{N}}$, still denoted by itself, such that

$$\text{dist}(y_n, \overline{\Lambda}) \geq \delta \quad \forall n \in \mathbb{N}.$$

Then we can find $r > 0$ such that $\mathcal{B}_r(y_n) \subset \Lambda^c$ for all $n \in \mathbb{N}$. Since $\tilde{u} \geq 0$ and $C^\infty_c(\mathbb{R}^3)$ is dense in $W^{s,p}(\mathbb{R}^3)$, we can find a sequence $\{\psi_j\}_{j\in\mathbb{N}} \subset C^\infty_c(\mathbb{R}^3)$ such that $\psi_j \geq 0$ and $\psi_j \to \tilde{u}$ in $W^{s,p}(\mathbb{R}^3)$. Fixed $j \in \mathbb{N}$ and using $\psi = \psi_j$ as test function in $\langle \mathcal{I}'_n(u_n), \psi\rangle = 0$ we get

$$\begin{aligned} &a \int_{\mathbb{R}^3}\int_{\mathbb{R}^3} \frac{|\tilde{u}_n(x) - \tilde{u}_n(y)|^{p-2}(\tilde{u}_n(x) - \tilde{u}_n(y))(\psi_j(x) - \psi_j(y))}{|x-y|^{3+sp}} dxdy \\ &+ b[\tilde{u}_n]^p_{s,p} \int_{\mathbb{R}^3}\int_{\mathbb{R}^3} \frac{|\tilde{u}_n(x) - \tilde{u}_n(y)|^{p-2}(\tilde{u}_n(x) - \tilde{u}_n(y))(\psi_j(x) - \psi_j(y))}{|x-y|^{3+sp}} dxdy \\ &+ \int_{\mathbb{R}^3} V(\tilde{\xi}_n)|\tilde{u}_n|^{p-2}\tilde{u}_n \psi_j \, dx = \int_{\mathbb{R}^3} g(\tilde{\xi}_n, \tilde{u}_n)\psi_j \, dx, \end{aligned} \tag{6.23}$$

where $\widetilde{\xi}_n := \epsilon_n x + \epsilon_n \tilde{y}_n$. Taking into account that $u_n \geq 0$, $\psi_j \geq 0$ and the definition of g, we can see that

$$\begin{aligned} \int_{\mathbb{R}^3} g(\widetilde{\xi}_n, \tilde{u}_n)\psi_j \, dx &= \int_{\mathcal{B}_{\frac{r}{\epsilon_n}}} g(\widetilde{\xi}_n, \tilde{u}_n)\psi_j \, dx + \int_{\mathcal{B}^c_{\frac{r}{\epsilon_n}}} g(\widetilde{\xi}_n, \tilde{u}_n)\psi_j \, dx \\ &\leq \int_{\mathcal{B}_{\frac{r}{\epsilon_n}}} \frac{V_1}{K}|\tilde{u}_n|^{p-2}\tilde{u}_n\psi_j \, dx + \int_{\mathcal{B}^c_{\frac{r}{\epsilon_n}}} f(\tilde{u_n})\psi_j \, dx. \end{aligned} \tag{6.24}$$

Gathering (6.23) and (6.24) we have

$$\begin{aligned} &a \int_{\mathbb{R}^3}\int_{\mathbb{R}^3} \frac{|\tilde{u}_n(x) - \tilde{u}_n(y)|^{p-2}(\tilde{u}_n(x) - \tilde{u}_n(y))(\psi_j(x) - \psi_j(y))}{|x-y|^{3+sp}} \, dxdy \\ &+ b[\tilde{u}_n]^p_{s,p} \int_{\mathbb{R}^3}\int_{\mathbb{R}^3} \frac{|\tilde{u}_n(x) - \tilde{u}_n(y)|^{p-2}(\tilde{u}_n(x) - \tilde{u}_n(y))(\psi_j(x) - \psi_j(y))}{|x-y|^{3+sp}} \, dxdy \\ &+ V_1\left(1 - \frac{1}{K}\right)\int_{\mathbb{R}^3} |\tilde{u}_n|^{p-2}\tilde{u}_n\psi_j \, dx \leq \int_{\mathcal{B}^c_{\frac{r}{\epsilon_n}}} f(\tilde{u}_n)\psi_j \, dx. \end{aligned} \tag{6.25}$$

From (6.21) and the facts that ψ_j as compact support in $\mathbb{R}^3$ and $\epsilon_n \to 0$ we can see that

$$\begin{aligned} &\int_{\mathbb{R}^3}\int_{\mathbb{R}^3} \frac{|\tilde{u}_n(x) - \tilde{u}_n(y)|^{p-2}(\tilde{u}_n(x) - \tilde{u}_n(y))(\psi_j(x) - \psi_j(y))}{|x-y|^{3+sp}} \, dxdy \\ &\quad \to \int_{\mathbb{R}^3}\int_{\mathbb{R}^3} \frac{|\tilde{u}(x) - \tilde{u}(y)|^{p-2}(\tilde{u}(x) - \tilde{u}(y))(\psi_j(x) - \psi_j(y))}{|x-y|^{3+sp}} \, dxdy \quad \text{as } n \to \infty \end{aligned}$$

and

$$\int_{\mathcal{B}^c_{\frac{r}{\epsilon_n}}} f(\tilde{u}_n)\psi_j \, dx \to 0 \quad \text{as } n \to \infty.$$

The above limits together with (6.25) and $[\tilde{u}_n]^p_{s,p} \to B^p$ imply that

$$\begin{aligned} &a \int_{\mathbb{R}^3}\int_{\mathbb{R}^3} \frac{|\tilde{u}(x) - \tilde{u}(y)|^{p-2}(\tilde{u}(x) - \tilde{u}(y))(\psi_j(x) - \psi_j(y))}{|x-y|^{3+sp}} \, dxdy \\ &+ bB^p \int_{\mathbb{R}^3}\int_{\mathbb{R}^3} \frac{|\tilde{u}(x) - \tilde{u}(y)|^{p-2}(\tilde{u}(x) - \tilde{u}(y))(\psi_j(x) - \psi_j(y))}{|x-y|^{3+sp}} \, dxdy \\ &+ V_1\left(1 - \frac{1}{K}\right)\int_{\mathbb{R}^3} |\tilde{u}|^{p-2}\tilde{u}\psi_j \, dx \leq 0. \end{aligned}$$

Taking the limit as $j \to \infty$ we have

$$a[\tilde{u}]^p_{s,p} + bB^p[\tilde{u}]^p_{s,p} + V_1\left(1 - \frac{1}{K}\right)|\tilde{u}|^p_p \leq 0.$$

This gives a contradiction in view of (6.22). Hence there exists a subsequence $\{y_n\}_{n\in\mathbb{N}}$ such that $y_n \to y_0 \in \overline{\Lambda}$.

Claim 2. $y_0 \in \Lambda$.

Using the definition of g and (6.23) we can see that

$$\begin{aligned} &a \int_{\mathbb{R}^3}\int_{\mathbb{R}^3} \frac{|\tilde{u}_n(x) - \tilde{u}_n(y)|^{p-2}(\tilde{u}_n(x) - \tilde{u}_n(y))(\psi_j(x) - \psi_j(y))}{|x-y|^{3+sp}} \, dxdy \\ &+ b[\tilde{u}_n]^p_{s,p} \int_{\mathbb{R}^3}\int_{\mathbb{R}^3} \frac{|\tilde{u}_n(x) - \tilde{u}_n(y)|^{p-2}(\tilde{u}_n(x) - \tilde{u}_n(y))(\psi_j(x) - \psi_j(y))}{|x-y|^{3+sp}} \, dxdy \\ &+ \int_{\mathbb{R}^3} V(\widetilde{\xi}_n)|\tilde{u}_n|^{p-2}\tilde{u}_n\psi_j \, dx \leq \int_{\mathbb{R}^3} f(\tilde{u}_n)\psi_j \, dx. \end{aligned}$$

Taking the limit as $n \to \infty$ we get

$$
\begin{aligned}
&a \int_{\mathbb{R}^3}\int_{\mathbb{R}^3} \frac{|\tilde{u}(x)-\tilde{u}(y)|^{p-2}(\tilde{u}(x)-\tilde{u}(y))(\psi_j(x)-\psi_j(y))}{|x-y|^{3+sp}}\,dxdy \\
&+ b[\tilde{u}]^p_{s,p} \int_{\mathbb{R}^3}\int_{\mathbb{R}^3} \frac{|\tilde{u}(x)-\tilde{u}(y)|^{p-2}(\tilde{u}(x)-\tilde{u}(y))(\psi_j(x)-\psi_j(y))}{|x-y|^{3+sp}}\,dxdy \\
&+ \int_{\mathbb{R}^3} V(y_0)|\tilde{u}|^{p-2}\tilde{u}\psi_j\,dx \le \int_{\mathbb{R}^3} f(\tilde{u})\psi_j\,dx.
\end{aligned}
$$

Passing to the limit as $j \to \infty$, we obtain

$$
a[\tilde{u}]^p_{s,p} + bB^p[\tilde{u}]^p_{s,p} + V(y_0)|\tilde{u}|^p_p \le \int_{\mathbb{R}^3} f(\tilde{u})\tilde{u}\,dx.
$$

Using Fatou's lemma we have $[\tilde{u}]^p_{s,p} = [u]^p_{s,p} \le B^p$, which combined with the above inequality yields

$$
a[\tilde{u}]^p_{s,p} + b[\tilde{u}]^{2p}_{s,p} + V(y_0)|\tilde{u}|^p_p \le \int_{\mathbb{R}^3} f(\tilde{u})\tilde{u}\,dx.
$$

Hence, there exists $\tau \in (0,1)$ such that $\tau\tilde{u} \in \mathcal{N}_{V(y_0)}$. Let $d_{V(y_0)}$ be the mountain pass level associated with $\mathcal{E}_{V(y_0)}$. By Lemma 6.6 we can see that

$$
d_{V(y_0)} \le \mathcal{E}_{V(y_0)}(\tau\tilde{u}) \le \liminf_{n\to\infty} \mathcal{I}_n(u_n) = \liminf_{n\to\infty} c_n \le d_{V_0}.
$$

Thereby, $d_{V(y_0)} \le d_{V_0}$, and this implies that $V(y_0) \le V_0 = V(0)$. This together with the definition of V_0 yields that $V(y_0) = V_0$. From assumption (V_2) we have that $y_0 \notin \partial\Lambda$, thus $y_0 \in \Lambda$.

Claim 3. $\tilde{u}_n \to \tilde{u}$ in $W^{s,p}(\mathbb{R}^3)$ as $n \to \infty$.
Consider the set $\widetilde{\Lambda}_n = \frac{\Lambda}{\epsilon_n} - \tilde{y}_n$ and define the functions

$$
\widetilde{\chi}^1_n(x) := \begin{cases} 1 & \text{if } x \in \widetilde{\Lambda}_n \\ 0 & \text{if } x \in \widetilde{\Lambda}^c_n \end{cases} \quad \text{and} \quad \widetilde{\chi}^2_n(x) := 1 - \widetilde{\chi}^1_n(x).
$$

Introduce the following functions:

$$
\begin{aligned}
h^1_n(x) &:= \left(\frac{1}{p}-\frac{1}{\vartheta}\right) V(\widetilde{\xi}_n)|\tilde{u}_n|^p \widetilde{\chi}^1_n(x), \quad h^1(x) := \left(\frac{1}{p}-\frac{1}{\vartheta}\right) V(y_0)|\tilde{u}|^p, \\
h^2_n(x) &:= \left[\left(\frac{1}{p}-\frac{1}{\vartheta}\right) V(\widetilde{\xi}_n)|\tilde{u}_n(x)|^p + \frac{1}{\vartheta} g(\widetilde{\xi}_n, \tilde{u}_n(x))\tilde{u}_n(x) - G(\widetilde{\xi}_n, \tilde{u}_n(x))\right] \widetilde{\chi}^2_n(x) \\
&\ge \left[\left(\frac{1}{p}-\frac{1}{\vartheta}\right) - \frac{1}{K}\right] V(\widetilde{\xi}_n)|\tilde{u}_n|^p \widetilde{\chi}^2_n(x), \\
h^3_n(x) &:= \left(\frac{1}{\vartheta} g(\widetilde{\xi}_n, \tilde{u}_n(x))\tilde{u}_n(x) - G(\widetilde{\xi}_n, \tilde{u}_n(x))\right) \widetilde{\chi}^1_n(x) = \left[\frac{1}{\vartheta} f(\tilde{u}_n(x))\tilde{u}_n(x) - F(\tilde{u}_n(x))\right] \widetilde{\chi}^1_n(x), \\
h^3(x) &:= \left[\frac{1}{\vartheta} f(\tilde{u}(x))\tilde{u}(x) - F(\tilde{u}(x))\right].
\end{aligned}
$$

In the light of (f_3), $K > 2p > \frac{\vartheta p}{\vartheta - p}$, and (g_3), we can see that the above functions are nonnegative.

By (6.21) and Claim 2, we can deduce that $\tilde{u}_n(x) \to \tilde{u}(x)$ a.e. $x \in \mathbb{R}^3$ and $\epsilon_n \tilde{y}_n \to y_0 \in \Lambda$, from which we can infer that $\widetilde{\chi}^1_n(x) \to 1$, $h^1_n(x) \to h^1(x)$, $h^2_n(x) \to 0$ and $h^3_n(x) \to h^3(x)$ a.e. $x \in \mathbb{R}^3$.

Now, using Lemma 6.6, the invariance of $\mathbb{R}^3$ by translation and Fatou's lemma we can deduce that

$$\begin{aligned}
d_{V_0} &\geq \limsup_{n\to\infty} c_n = \limsup_{n\to\infty} \left(\mathcal{I}_n(u_n) - \frac{1}{\vartheta}\langle \mathcal{I}_n'(u_n), u_n\rangle \right) \\
&= \limsup_{n\to\infty} \left[\left(\frac{1}{p} - \frac{1}{\vartheta}\right) a\,[\tilde{u}_n]_{s,p}^p + \left(\frac{1}{2p} - \frac{1}{\vartheta}\right) b\,[\tilde{u}_n]_{s,p}^{2p} + \int_{\mathbb{R}^3} (h_n^1(x) + h_n^2(x) + h_n^3(x))\,dx \right] \\
&\geq \liminf_{n\to\infty} \left[\left(\frac{1}{p} - \frac{1}{\vartheta}\right) a\,[\tilde{u}_n]_{s,p}^p + \left(\frac{1}{2p} - \frac{1}{\vartheta}\right) b\,[\tilde{u}_n]_{s,p}^{2p} + \int_{\mathbb{R}^3} (h_n^1(x) + h_n^2(x) + h_n^3(x))\,dx \right] \\
&\geq \left(\frac{1}{p} - \frac{1}{\vartheta}\right) a\,[\tilde{u}]_{s,p}^p + \left(\frac{1}{2p} - \frac{1}{\vartheta}\right) b\,[\tilde{u}]_{s,p}^{2p} + \int_{\mathbb{R}^3} (h^1(x) + h^3(x))\,dx \geq d_{V_0}.
\end{aligned}$$

Therefore,

$$\lim_{n\to\infty} [\tilde{u}_n]_{s,p}^p = [\tilde{u}]_{s,p}^p \tag{6.26}$$

and $h_n^1 \to h^1$, $h_n^2 \to 0$, and $h_n^3 \to h^3$ in $L^1(\mathbb{R}^3)$. Hence we can infer that

$$\lim_{n\to\infty} \int_{\mathbb{R}^3} V(\tilde{\xi}_n)|\tilde{u}_n|^p\,dx = \int_{\mathbb{R}^3} V(y_0)|\tilde{u}|^p\,dx$$

and thus

$$\lim_{n\to\infty} |\tilde{u}_n|_p^p = |\tilde{u}|_p^p. \tag{6.27}$$

Combining (6.26) with (6.27), and using the Brezis-Lieb lemma [14] we get

$$\|\tilde{u}_n - \tilde{u}\|_{s,p}^p \to 0 \quad \text{as } n \to \infty.$$

This completes the proof. □

Lemma 6.8 *Let $\{\tilde{u}_n\}_{n\in\mathbb{N}}$ be the sequence given in Lemma 6.7. Then, $\tilde{u}_n \in L^\infty(\mathbb{R}^3)$ and there exists $C > 0$ such that*

$$|\tilde{u}_n|_\infty \leq C \quad \forall n \in \mathbb{N}.$$

Moreover,

$$\tilde{u}_n(x) \to 0 \text{ as } |x| \to \infty \text{ uniformly in } n \in \mathbb{N}. \tag{6.28}$$

Proof. For any $L > 0$, let $\tilde{u}_{L,n} := \min\{\tilde{u}_n, L\}$. Let $\sigma > 1$ and define the function

$$\ell(\tilde{u}_n) := \ell_{L,\sigma}(\tilde{u}_n) = \tilde{u}_n\, \tilde{u}_{L,n}^{p(\sigma-1)} \in \mathcal{H}_\epsilon.$$

We note that ℓ is increasing, so for any $a, b \in \mathbb{R}$ it holds $(a-b)(\ell(a)-\ell(b)) \geq 0$. We introduce the functions

$$\mathcal{Q}(t) := \frac{|t|^p}{p} \quad \text{and} \quad \mathcal{L}(t) := \int_0^t (\ell'(\tau))^{\frac{1}{p}}\,d\tau.$$

Let us point out that

$$\mathcal{L}(\tilde{u}_n) \geq \frac{1}{\sigma}\tilde{u}_n \tilde{u}_{L,n}^{\sigma-1}.$$

Hence, from Theorem 6.4 and the above inequality we get

$$[\mathcal{L}(\tilde{u}_n)]_{s,p}^p \geq C_*^{-1}|\mathcal{L}(\tilde{u}_n)|_{p_s^*}^p \geq C_*^{-1}\frac{1}{\sigma^p}|\tilde{u}_n\tilde{u}_{L,n}^{\sigma-1}|_{p_s^*}^p. \tag{6.29}$$

On the other hand, for any $a, b \in \mathbb{R}$, it holds

$$\mathcal{Q}'(a-b)(\ell(a)-\ell(b)) \geq |\mathcal{L}(a)-\mathcal{L}(b)|^p.$$

Indeed, when $a > b$ we have

$$\begin{aligned}\mathcal{Q}'(a-b)(\ell(a)-\ell(b)) &= (a-b)^{p-1}(\ell(a)-\ell(b)) = (a-b)^{p-1}\int_b^a \ell'(\tau)\,d\tau \\ &= (a-b)^{p-1}\int_b^a (\mathcal{L}'(\tau))^p\,d\tau \geq \left(\int_b^a \mathcal{L}'(\tau)\,d\tau\right)^p = (\mathcal{L}(a)-\mathcal{L}(b))^p,\end{aligned}$$

where in the last inequality we used the Jensen inequality. A similar argument works when $a \leq b$. Therefore, we deduce that

$$\begin{aligned}&|\mathcal{L}(\tilde{u}_n)(x)-\mathcal{L}(\tilde{u}_n)(y)|^p \\ &\leq |\tilde{u}_n(x)-\tilde{u}_n(y)|^{p-2}(\tilde{u}_n(x)-\tilde{u}_n(y))(\tilde{u}_n(x)\tilde{u}_{L,n}^{p(\sigma-1)}(x)-\tilde{u}_n(y)\tilde{u}_{L,n}^{p(\sigma-1)}(y)).\end{aligned} \tag{6.30}$$

Using $\ell(\tilde{u}_n)$ as test function in (6.3) we get

$$\begin{aligned}&a\int_{\mathbb{R}^3}\int_{\mathbb{R}^3}\frac{|\tilde{u}_n(x)-\tilde{u}_n(y)|^{p-2}(\tilde{u}_n(x)-\tilde{u}_n(y))(\tilde{u}_n(x)\tilde{u}_{L,n}^{p(\sigma-1)}(x)-\tilde{u}_n(y)\tilde{u}_{L,n}^{p(\sigma-1)}(y))}{|x-y|^{3+sp}}\,dxdy \\ &+b[\tilde{u}_n]_{s,p}^p\int_{\mathbb{R}^3}\int_{\mathbb{R}^3}\frac{|\tilde{u}_n(x)-\tilde{u}_n(y)|^{p-2}(\tilde{u}_n(x)-\tilde{u}_n(y))(\tilde{u}_n(x)\tilde{u}_{L,n}^{p(\sigma-1)}(x)-\tilde{u}_n(y)\tilde{u}_{L,n}^{p(\sigma-1)}(y))}{|x-y|^{3+sp}}\,dxdy \\ &+\int_{\mathbb{R}^3}V(\widetilde{\xi}_n)|\tilde{u}_n|^p\tilde{u}_{L,n}^{p(\sigma-1)}\,dx = \int_{\mathbb{R}^3}g(\widetilde{\xi}_n,\tilde{u}_n)\tilde{u}_n\tilde{u}_{L,n}^{p(\sigma-1)}\,dx,\end{aligned} \tag{6.31}$$

where $\widetilde{\xi}_n := \epsilon_n x + \epsilon_n \tilde{y}_n$. Putting together (6.30) and (6.31) we get

$$\begin{aligned}&a[\mathcal{L}(\tilde{u}_n)]_{s,p}^p + \int_{\mathbb{R}^3}V(\widetilde{\xi}_n)|\tilde{u}_n|^p\tilde{u}_{L,n}^{p(\sigma-1)}\,dx \\ &\leq a\int_{\mathbb{R}^3}\int_{\mathbb{R}^3}\frac{|\tilde{u}_n(x)-\tilde{u}_n(y)|^{p-2}(\tilde{u}_n(x)-\tilde{u}_n(y))(\tilde{u}_n(x)\tilde{u}_{L,n}^{p(\sigma-1)}(x)-\tilde{u}_n(y)\tilde{u}_{L,n}^{p(\sigma-1)}(y))}{|x-y|^{3+sp}}\,dxdy \\ &+b[\tilde{u}_n]_{s,p}^p\int_{\mathbb{R}^3}\int_{\mathbb{R}^3}\frac{|\tilde{u}_n(x)-\tilde{u}_n(y)|^{p-2}(\tilde{u}_n(x)-\tilde{u}_n(y))(\tilde{u}_n(x)\tilde{u}_{L,n}^{p(\sigma-1)}(x)-\tilde{u}_n(y)\tilde{u}_{L,n}^{p(\sigma-1)}(y))}{|x-y|^{3+sp}}\,dxdy \\ &+\int_{\mathbb{R}^3}V(\widetilde{\xi}_n)|\tilde{u}_n|^p\tilde{u}_{L,n}^{p(\sigma-1)}\,dx = \int_{\mathbb{R}^3}g(\widetilde{\xi}_n,\tilde{u}_n)\tilde{u}_n\tilde{u}_{L,n}^{p(\sigma-1)}\,dx.\end{aligned} \tag{6.32}$$

From (6.29) and (6.32) we can infer that

$$\begin{aligned}|\tilde{u}_n\tilde{u}_{L,n}^{\sigma-1}|_{p_s^*}^p &\leq \sigma^p C_*[\mathcal{L}(\tilde{u}_n)]_{s,p}^p \leq \frac{\sigma^p C_*}{a}\int_{\mathbb{R}^3}g(\widetilde{\xi}_n,\tilde{u}_n)\tilde{u}_n\tilde{u}_{L,n}^{p(\sigma-1)}\,dx \\ &\leq \sigma^p C\int_{\mathbb{R}^3}g(\widetilde{\xi}_n,\tilde{u}_n)\tilde{u}_n\tilde{u}_{L,n}^{p(\sigma-1)}\,dx.\end{aligned} \tag{6.33}$$

Using the growth assumptions on g we have that for all $\zeta > 0$ there exists $C_\zeta > 0$ such that

$$|g(x,t)| \leq \zeta |t|^{p-1} + C_\zeta |t|^{p_s^*-1} \quad \text{for all } x \in \mathbb{R}^3 \text{ and } t \in \mathbb{R},$$

which together with (6.33) implies

$$|\tilde{u}_n \tilde{u}_{L,n}^{\sigma-1}|_{p_s^*}^p \leq C\sigma^p \left(\int_{\mathbb{R}^3} \zeta |\tilde{u}_n|^p \tilde{u}_{L,n}^{p(\sigma-1)} \, dx + \int_{\mathbb{R}^3} C_\zeta |\tilde{u}_n|^{p_s^*} \tilde{u}_{L,n}^{p(\sigma-1)} \, dx \right).$$

Choosing ζ sufficiently small we deduce that

$$|\tilde{u}_n \tilde{u}_{L,n}^{\sigma-1}|_{p_s^*}^p \leq C\sigma^p \int_{\mathbb{R}^3} |\tilde{u}_n|^{p_s^*} \tilde{u}_{L,n}^{p(\sigma-1)} \, dx. \tag{6.34}$$

Now, let $\sigma = \frac{p_s^*}{p}$ and fix $R > 0$. Using the fact that $0 \leq \tilde{u}_{L,n} \leq \tilde{u}_n$ and the Hölder's inequality, we obtain

$$\begin{aligned}
\int_{\mathbb{R}^3} \tilde{u}_n^{p_s^*} \tilde{u}_{L,n}^{p(\sigma-1)} \, dx &= \int_{\mathbb{R}^3} \tilde{u}_n^{p_s^*-p} (\tilde{u}_n \tilde{u}_{L,n}^{\frac{p_s^*-p}{p}})^p \, dx \\
&= \int_{\{\tilde{u}_n > R\}} \tilde{u}_n^{p_s^*-p} (\tilde{u}_n \tilde{u}_{L,n}^{\frac{p_s^*-p}{p}})^p \, dx + \int_{\{\tilde{u}_n < R\}} \tilde{u}_n^{p_s^*-p} (\tilde{u}_n \tilde{u}_{L,n}^{\frac{p_s^*-p}{p}})^p \, dx \\
&\leq \left(\int_{\{\tilde{u}_n > R\}} \tilde{u}_n^p \, dx \right)^{\frac{p_s^*-p}{p}} \left(\int_{\mathbb{R}^3} (\tilde{u}_n \tilde{u}_{L,n}^{\frac{p_s^*-p}{p}})^{p_s^*} \, dx \right)^{\frac{p}{p_s^*}} + R^{p_s^*-p} \int_{\{\tilde{u}_n < R\}} \tilde{u}_n^{p_s^*} \, dx.
\end{aligned}$$

Since $\{\tilde{u}_n\}_{n \in \mathbb{N}}$ strongly converges in $L^{p_s^*}(\mathbb{R}^3)$, we can see that for any R sufficiently large

$$\left(\int_{\{\tilde{u}_n > R\}} \tilde{u}_n^p \, dx \right)^{\frac{p_s^*-p}{p}} \leq \frac{1}{2C\sigma^p},$$

and thus we deduce

$$\int_{\mathbb{R}^3} \tilde{u}_n^{p_s^*} \tilde{u}_{L,n}^{p(\sigma-1)} \, dx \leq \frac{1}{2C\sigma^p} \left(\int_{\mathbb{R}^3} (\tilde{u}_n \tilde{u}_{L,n}^{\frac{p_s^*-p}{p}})^{p_s^*} \, dx \right)^{\frac{p}{p_s^*}} + R^{p_s^*-p} \int_{\mathbb{R}^3} \tilde{u}_n^{p_s^*} \, dx. \tag{6.35}$$

Putting together (6.34) and (6.35) we have

$$|\tilde{u}_n \tilde{u}_{L,n}^{\sigma-1}|_{p_s^*}^p \leq C\sigma^p R^{p_s^*-p} \int_{\mathbb{R}^3} \tilde{u}_n^{p_s^*} \, dx < \infty,$$

and letting $L \to \infty$ we deduce that $\tilde{u}_n \in L^{\frac{(p_s^*)^2}{p}}(\mathbb{R}^3)$.

Now, taking the limit as $L \to \infty$ in (6.34) we have

$$|\tilde{u}_n|_{\sigma p_s^*}^{\sigma p} \leq C\sigma^p \int_{\mathbb{R}^3} \tilde{u}_n^{p_s^*+p(\sigma-1)} \, dx$$

which implies

$$\left(\int_{\mathbb{R}^3} \tilde{u}_n^{\sigma p_s^*} \, dx \right)^{\frac{1}{p_s^*(\sigma-1)}} \leq (C\sigma)^{\frac{1}{\sigma-1}} \left(\int_{\mathbb{R}^3} \tilde{u}_n^{p_s^*+p(\sigma-1)} \, dx \right)^{\frac{1}{p(\sigma-1)}}.$$

For $m \geq 1$ we define σ_{m+1} inductively so that $p_s^* + p(\sigma_{m+1} - 1) = p_s^* \sigma_m$ and $\sigma_1 = \frac{p_s^*}{p}$. Then we have

$$\left(\int_{\mathbb{R}^3} \tilde{u}_n^{p_s^* \sigma_{m+1}} \, dx \right)^{\frac{1}{p_s^*(\sigma_{m+1}-1)}} \leq (C\sigma_{m+1})^{\frac{1}{\sigma_{m+1}-1}} \left(\int_{\mathbb{R}^3} \tilde{u}_n^{p_s^* \sigma_m} \, dx \right)^{\frac{1}{p_s^*(\sigma_m - 1)}}.$$

Set

$$\mathcal{D}_m := \left(\int_{\mathbb{R}^3} \tilde{u}_n^{p_s^* \sigma_m} \, dx \right)^{\frac{1}{p_s^*(\sigma_m - 1)}}.$$

Using an iteration argument, we can find $C_0 > 0$ independent of m such that

$$\mathcal{D}_{m+1} \leq \prod_{k=1}^{m} (C\sigma_{k+1})^{\frac{1}{\sigma_{k+1}-1}} \mathcal{D}_1 \leq C_0 \mathcal{D}_1.$$

Taking the limit as $m \to \infty$ we get $|\tilde{u}_n|_\infty \leq C$ for all $n \in \mathbb{N}$.

Now, we note that $\tilde{u}_n$ is a solution to

$$(-\Delta)_p^s \tilde{u}_n = [g(\epsilon_n x + \epsilon_n \tilde{y}_n, \tilde{u}_n) - V(\epsilon_n x + \epsilon_n \tilde{y}_n) \tilde{u}_n^{p-1}](a + b[\tilde{u}_n]_{s,p}^p)^{-1} =: h_n \quad \text{in } \mathbb{R}^3.$$

Moreover, $h_n \in L^\infty(\mathbb{R}^3)$ and $|h_n|_\infty \leq C$ for all $n \in \mathbb{N}$. Indeed, this last inequality is a consequence of the growth assumptions on g, $|\tilde{u}_n|_\infty \leq C$ and $a \leq a + b[\tilde{u}_n]_{s,p}^p \leq C$ for all $n \in \mathbb{N}$. Then, using Corollary 5.5 in [75] we can deduce that $\tilde{u}_n \in C^{0,\alpha}(\mathbb{R}^3)$ for some $\alpha > 0$. From this fact and (6.21) we infer that (6.28) holds true. □

Corollary 6.1 *There exists $n_0 \in \mathbb{N}$ such that*

$$u_n(x) < a_0 \quad \forall n \geq n_0 \text{ and } \forall x \in \Lambda_{\epsilon_n}^c.$$

Hence, u_n is a solution of problem $(\hat{P}_{\epsilon_n})$.

Proof. By Lemma 6.7 we can find $\{\tilde{y}_n\}_{n\in\mathbb{N}} \subset \mathbb{R}^3$ such that $\tilde{u}_n = u_n(\cdot + \tilde{y}_n) \to \tilde{u}$ in $W^{s,p}(\mathbb{R}^3)$ and $y_n = \epsilon_n \tilde{y}_n \to y_0$ for some $y_0 \in \Lambda$ such that $V(y_0) = V_0$.

Now, if we choose $r > 0$ such that $\mathcal{B}_r(y_0) \subset \mathcal{B}_{2r}(y_0) \subset \Lambda$, then $\mathcal{B}_{\frac{r}{\epsilon_n}}\left(\frac{y_0}{\epsilon_n}\right) \subset \Lambda_{\epsilon_n}$. Hence, there exists $n_1 \in \mathbb{N}$ such that for any $y \in \mathcal{B}_{\frac{r}{\epsilon_n}}(\tilde{y}_n)$ we have

$$\left| y - \frac{y_0}{\epsilon_n} \right| \leq |y - \tilde{y}_n| + \left| \tilde{y}_n - \frac{y_0}{\epsilon_n} \right| < \frac{2r}{\epsilon_n} \quad \forall n \geq n_1,$$

and consequently

$$\Lambda_{\epsilon_n}^c \subset \mathcal{B}_{\frac{r}{\epsilon_n}}^c(\tilde{y}_n) \quad \text{for any } n \geq n_1.$$

In the light of Lemma 6.8, there exists $R > 0$ such that

$$\tilde{u}_n(x) < a_0 \quad \text{for } |x| \geq R \text{ and } \forall n \in \mathbb{N},$$

which yields

$$u_n(x) = \tilde{u}_n(x - \tilde{y}_n) < a_0 \quad \text{for } x \in \mathcal{B}_R^c(\tilde{y}_n) \text{ and } \forall n \in \mathbb{N}.$$

On the other hand, there exists $n_2 \in \mathbb{N}$ such that

$$\mathcal{B}_{\frac{r}{\epsilon_n}}^c(\tilde{y}_n) \subset \mathcal{B}_R^c(\tilde{y}_n) \quad \forall n \geq n_2.$$

Hence, choosing $n_0 = \max\{n_1, n_2\}$, we can infer that

$$\Lambda^c_{\epsilon_n} \subset \mathcal{B}^c_{\frac{r}{\epsilon_n}}(\tilde{y}_n) \subset \mathcal{B}^c_R(\tilde{y}_n) \quad \forall n \geq n_0,$$

and then

$$u_n(x) < a_0 \quad \forall x \in \Lambda^c_{\epsilon_n} \text{ and } \forall n \geq n_0.$$

The proof is now complete. □

Proof of Theorem 6.1 Let u_ϵ be a nonnegative solution of problem (6.3). Then, there exists $\epsilon_0 > 0$ such that

$$u_\epsilon(x) < a_0 \quad \forall x \in \Lambda^c_\epsilon \text{ and } \forall \epsilon \in (0, \epsilon_0),$$

that is u_ϵ is a solution of problem $(\hat{P}_\epsilon)$ for $\epsilon \in (0, \epsilon_0)$. Consider $v_\epsilon(x) := u_\epsilon(x/\epsilon)$ for any $\epsilon \in (0, \epsilon_0)$ and note that v_ϵ is a solution to (P_ϵ). Let η_ϵ be a global maximum point of v_ϵ. It is easy to see that there exists $\tau_0 > 0$ such that $v_\epsilon(\eta_\epsilon) \geq \tau_0$ for any $\epsilon > 0$. Set $z_\epsilon := \frac{\eta_\epsilon}{\epsilon} - \tilde{y}_\epsilon$. Then z_ϵ is a global maximum point of $\tilde{u}_\epsilon(x) = u_\epsilon(x + \tilde{y}_\epsilon)$ and $\tilde{u}_\epsilon(z_\epsilon) \geq \tau_0$ for any $\epsilon > 0$. We claim that

$$\lim_{\epsilon \to 0} V(\eta_\epsilon) = V_0.$$

If the above limit is not true, then there exist $\epsilon_n \to 0$ and $\delta > 0$ such that

$$V(\eta_{\epsilon_n}) \geq V_0 + \delta \quad \forall n \in \mathbb{N}. \tag{6.36}$$

By Lemma 6.8, we know that $\tilde{u}_{\epsilon_n}(x) \to 0$ as $|x| \to \infty$ uniformly in $n \in \mathbb{N}$. Therefore, $\{z_{\epsilon_n}\}_{n \in \mathbb{N}}$ is bounded. Moreover, for some subsequence, there is $y_0 \in \Lambda$ such that $\epsilon_n \tilde{y}_{\epsilon_n} \to y_0$ and $V(y_0) = V_0$. Therefore, $\eta_{\epsilon_n} = \epsilon_n z_{\epsilon_n} + \epsilon_n \tilde{y}_{\epsilon_n} \to y_0$ which combined with the continuity of V yields $V(\eta_{\epsilon_n}) \to V_0$. This contradicts (6.36). Accordingly, $V(\eta_\epsilon) \to V_0$ as $\epsilon \to 0$.

Next, we provide a decay estimate for v_ϵ. Using (6.28) and (g_1), there exists $R_1 > 0$ such that

$$g(\epsilon x, \tilde{u}_\epsilon(x)) \leq \frac{V_1}{2} \tilde{u}_\epsilon(x)^{p-1} \quad \forall x \in \mathcal{B}^c_{R_1}.$$

Therefore,

$$\begin{aligned}
(-\Delta)^s_p \tilde{u}_\epsilon + \frac{V_1}{2(a + bA_1^p)} \tilde{u}_\epsilon^{p-1} &\leq (-\Delta)^s_p \tilde{u}_\epsilon + \frac{V_1}{2(a + b[\tilde{u}_\epsilon]^p_{s,p})} \tilde{u}_\epsilon^{p-1} \\
&= \frac{1}{a + b[\tilde{u}_\epsilon]^p_{s,p}} \left[g(\epsilon x + \epsilon \tilde{y}_\epsilon, \tilde{u}_\epsilon) - \left(V(\epsilon x + \epsilon \tilde{y}_\epsilon) - \frac{V_1}{2} \right) \tilde{u}_\epsilon^{p-1} \right] \\
&\leq \frac{1}{a + b[\tilde{u}_\epsilon]^p_{s,p}} \left[g(\epsilon x + \epsilon \tilde{y}_\epsilon, \tilde{u}_\epsilon) - \frac{V_1}{2} \tilde{u}_\epsilon^{p-1} \right] \leq 0 \quad \text{in } \mathcal{B}^c_{R_1},
\end{aligned} \tag{6.37}$$

where $A_1 > 0$ is such that $a + b[u_\epsilon]^p_{s,p} \leq a + bA_1^p$, for any $\epsilon \in (0, \epsilon_0)$. Applying Lemma 7.1 in [42], we can find a continuous positive function w and a positive constant C such that

$$0 < w(x) \leq \frac{C}{1 + |x|^{3+sp}} \tag{6.38}$$

and

$$(-\Delta)^s_p w + \frac{V_1}{2(a + bA_1^p)} w^{p-1} \geq 0 \quad \text{in } \mathcal{B}^c_{R_2}, \tag{6.39}$$

for some $R_2 > 0$. Thanks to the continuity of $\tilde{u}_\epsilon$ and w, there exists $C_1 > 0$ such that

$$\psi_\epsilon := \tilde{u}_\epsilon - C_1 w \leq 0 \quad \text{for } |x| = R_3,$$

where $R_3 := \max\{R_1, R_2\}$.

Taking $\phi = \max\{\psi_\epsilon, 0\} \in W_0^{s,p}(\mathcal{B}_R^c)$ as test function in (6.37) and using (6.39) with $\tilde{w} = C_1 w$, we can deduce that

$$\begin{aligned} 0 &\geq \int_{\mathbb{R}^3}\int_{\mathbb{R}^3} \frac{|\tilde{u}_\epsilon(x) - \tilde{u}_\epsilon(y)|^{p-2}(\tilde{u}_\epsilon(x) - \tilde{u}_\epsilon(y))(\phi(x) - \phi(y))}{|x-y|^{3+sp}} dxdy + \frac{V_1}{2(a + bA_1^p)} \int_{\mathbb{R}^3} \tilde{u}_\epsilon^{p-1}\phi\, dx \\ &\geq \int_{\mathbb{R}^3}\int_{\mathbb{R}^3} \frac{\mathcal{G}_\epsilon(x,y)}{|x-y|^{3+sp}}(\phi(x) - \phi(y))\, dxdy + \frac{V_1}{2(a + bA_1^p)} \int_{\mathbb{R}^3} [\tilde{u}_\epsilon^{p-1} - \tilde{w}^{p-1}]\phi\, dx, \end{aligned} \tag{6.40}$$

where

$$\mathcal{G}_\epsilon(x,y) := |\tilde{u}_\epsilon(x) - \tilde{u}_\epsilon(y)|^{p-2}(\tilde{u}_\epsilon(x) - \tilde{u}_\epsilon(y)) - |\tilde{w}(x) - \tilde{w}(y)|^{p-2}(\tilde{w}(x) - \tilde{w}(y)).$$

Therefore, if we prove that

$$\int_{\mathbb{R}^3}\int_{\mathbb{R}^3} \frac{\mathcal{G}_\epsilon(x,y)}{|x-y|^{N+sp}}(\phi(x) - \phi(y))dxdy \geq 0, \tag{6.41}$$

it follows from (6.40) that

$$0 \geq \frac{V_1}{2(a + bA_1^p)} \int_{\{\tilde{u}_\epsilon \geq \tilde{w}\}} [\tilde{u}_\epsilon^{p-1} - \tilde{w}^{p-1}](\tilde{u}_\epsilon - \tilde{w})dx \geq 0$$

which yields that

$$\{x \in \mathbb{R}^3 : |x| \geq R_3 \text{ and } \tilde{u}_\epsilon(x) \geq \tilde{w}\} = \emptyset.$$

To achieve our purpose, we first note that for all $c, d \in \mathbb{R}$ it holds

$$|d|^{p-2}d - |c|^{p-2}c = (p-1)(d-c)\int_0^1 |c + t(d-c)|^{p-2}dt.$$

Taking $d = \tilde{u}_\epsilon(x) - \tilde{u}_\epsilon(y)$ and $c = \tilde{w}(x) - \tilde{w}(y)$ we can see that

$$|d|^{p-2}d - |c|^{p-2}c = (p-1)(d-c)I(x,y),$$

where $I(x,y) \geq 0$ stands for the integral. Now, recalling that

$$(x-y)(x^+ - y^+) \geq |x^+ - y^+|^2 \text{ for all } x, y \in \mathbb{R},$$

we have

$$\begin{aligned} (d-c)(\phi(x) - \phi(y)) &= [(\tilde{u}_\epsilon - \tilde{w})(x) - (\tilde{u}_\epsilon - \tilde{w})(y)][(\tilde{u}_\epsilon - \tilde{w})^+(x) - (\tilde{u}_\epsilon - \tilde{w})^+(y)] \\ &\geq |(\tilde{u}_\epsilon - \tilde{w})^+(x) - (\tilde{u}_\epsilon - \tilde{w})^+(y)|^2, \end{aligned}$$

which gives $(|d|^{p-2}d - |c|^{p-2}c)(\phi(x) - \phi(y)) \geq 0$, that is (6.41) holds true.

Therefore, $\psi_\epsilon \leq 0$ in $\mathcal{B}_{R_3}^c$, which implies that $\tilde{u}_\epsilon \leq C_1 w$ in $\mathcal{B}_{R_3}^c$, that is $\tilde{u}_\epsilon(x) \leq C(1 + |x|^{3+sp})^{-1}$ in $\mathcal{B}_{R_3}^c$. Consequently,

$$\begin{aligned} v_\epsilon(x) = u_\epsilon\left(\frac{x}{\epsilon}\right) &= \tilde{u}_\epsilon\left(\frac{x}{\epsilon} - \tilde{y}_\epsilon\right) \leq C_1 w\left(\frac{x}{\epsilon} - \tilde{y}_\epsilon\right) \\ &\leq \frac{C}{1 + |\frac{x}{\epsilon} - \tilde{y}_\epsilon|^{3+sp}} = \frac{C\epsilon^{3+sp}}{\epsilon^{3+sp} + |x - \epsilon\tilde{y}_\epsilon|^{3+sp}} \\ &\leq \frac{C\epsilon^{3+sp}}{\epsilon^{3+sp} + |x - \eta_\epsilon|^{3+sp}} \end{aligned}$$

and this ends the proof of Theorem 6.1. □

6.3 Multiplicity of solutions

In this section we deal with the multiplicity of positive solutions to problem (P_ϵ). To achieve this result, we need some auxiliary properties.

6.3.1 The generalized Nehari method

Let us denote by

$$\mathcal{N}_\epsilon := \{u \in \mathcal{H}_\epsilon : \langle \mathcal{I}'_\epsilon(u), u\rangle = 0\}$$

the Nehari manifold associated with problem (6.3), and define

$$\mathcal{H}_\epsilon^+ := \{u \in \mathcal{H}_\epsilon : |supp(u^+) \cap \Lambda_\epsilon| > 0\} \subset \mathcal{H}_\epsilon.$$

Let $\mathbb{S}_\epsilon$ be the unit sphere of $\mathcal{H}_\epsilon$ and set $\mathbb{S}_\epsilon^+ := \mathbb{S}_\epsilon \cap \mathcal{H}_\epsilon^+$. By the definition of $\mathbb{S}_\epsilon^+$ and using the fact that $\mathcal{H}_\epsilon^+$ is open in $\mathcal{H}_\epsilon$, it follows that $\mathbb{S}_\epsilon^+$ is an incomplete C^1-manifold of codimension 1, modeled on $\mathcal{H}_\epsilon$ and contained in the open $\mathcal{H}_\epsilon^+$. Thus, $\mathcal{H}_\epsilon = T_u\mathbb{S}_\epsilon^+ \oplus \mathbb{R}u$ for each $u \in \mathbb{S}_\epsilon^+$, where

$$T_u\mathbb{S}_\epsilon^+ := \left\{v \in \mathcal{H}_\epsilon : B_u(v) + \int_{\mathbb{R}^3} V(\epsilon x)|u|^{p-2}uv\,dx = 0\right\}$$

and

$$B_u(v) = \int_{\mathbb{R}^3}\int_{\mathbb{R}^3} \frac{|u(x)-u(y)|^{p-2}(u(x)-u(y))(v(x)-v(y))}{|x-y|^{3+sp}}\,dxdy.$$

The next result is fundamental to overcome the non-differentiability of the Nehari manifold $\mathcal{N}$ and the incompleteness of $\mathbb{S}_\epsilon^+$.

Lemma 6.9 *Assume that (V_1)-(V_2') and (f_1)-(f_4) hold. Then, the following properties hold.*

> *(i) For each $u \in \mathcal{H}_\epsilon^+$, let $h_u : \mathbb{R}^+ \to \mathbb{R}$ be defined by $h_u(t) := \mathcal{I}_\epsilon(tu)$. Then there exists a unique $t_u > 0$ such that $h'_u(t) > 0$ for all $t \in (0, t_u)$ and $h'_u(t) < 0$ for all $t \in (t_u, \infty)$.*
>
> *(ii) There exists $\tau > 0$ independent of u such that $t_u \geq \tau$ for any $u \in \mathbb{S}_\epsilon^+$. Moreover, for each compact set $\mathbb{K} \subset \mathbb{S}_\epsilon^+$ there is a constant $C_\mathbb{K} > 0$ such that $t_u \leq C_\mathbb{K}$ for any $u \in \mathbb{K}$.*
>
> *(iii) The map $\hat{m}_\epsilon : \mathcal{H}_\epsilon^+ \to \mathcal{N}$ given by $\hat{m}_\epsilon(u) = t_u u$ is continuous and $m_\epsilon := \hat{m}_\epsilon|_{\mathbb{S}_\epsilon^+}$ is a homeomorphism between $\mathbb{S}_\epsilon^+$ and $\mathcal{N}$. Moreover, $m_\epsilon^{-1}(u) = \frac{u}{\|u\|_\epsilon}$.*
>
> *(iv) If there is a sequence $\{u_n\}_{n\in\mathbb{N}} \subset \mathbb{S}_\epsilon^+$ such that $\operatorname{dist}(u_n, \partial\mathbb{S}_\epsilon^+) \to 0$, then $\|m_\epsilon(u_n)\|_\epsilon \to \infty$ and $\mathcal{I}_\epsilon(m_\epsilon(u_n)) \to \infty$.*

Proof. (*i*) We know that $h_u \in C^1(\mathbb{R}^+, \mathbb{R})$, and by Lemma 6.3, we have that $h_u(0) = 0$, $h_u(t) > 0$ for $t > 0$ small enough and $h_u(t) < 0$ for $t > 0$ sufficiently large. Then there exists a global maximum point $t_u > 0$ for h_u such that $h'_u(t_u) = 0$, that is $t_u u \in \mathcal{N}$.

Now, we aim to prove the uniqueness of a such t_u. Assume by contradiction that there exist $t_1 > t_2 > 0$ such that $h'_u(t_1) = h'_u(t_2) = 0$, or equivalently

$$t_1^{p-1}\|u\|_\epsilon^p + bt_1^{2p-1}[u]_{s,p}^{2p} = \int_{\mathbb{R}^3} g(\epsilon x, t_1 u)u\,dx, \tag{6.42}$$

$$t_2^{p-1}\|u\|_\epsilon^p + bt_2^{2p-1}[u]_{s,p}^{2p} = \int_{\mathbb{R}^3} g(\epsilon x, t_2 u)u\,dx. \tag{6.43}$$

Dividing both members of relation (6.42) by t_1^{2p-1} we get

$$\frac{\|u\|_\epsilon^p}{t_1^p} + b[u]_{s,p}^{2p} = \int_{\mathbb{R}^3} \frac{g(\epsilon x, t_1 u)}{(t_1 u)^{2p-1}} u^{2p} dx,$$

and similarly, dividing both members of (6.43) by t_2^{2p-1}, we obtain

$$\frac{\|u\|_\epsilon^p}{t_2^p} + b[u]_{s,p}^{2p} = \int_{\mathbb{R}^3} \frac{g(\epsilon x, t_2 u)}{(t_2 u)^{2p-1}} u^{2p} dx.$$

Subtracting the above identities, and taking into account the definition of g we can see that

$$\begin{aligned}
\left(\frac{1}{t_1^p} - \frac{1}{t_2^p}\right) \|u\|_\epsilon^p &= \int_{\mathbb{R}^3} \left[\frac{g(\epsilon x, t_1 u)}{(t_1 u)^{2p-1}} - \frac{g(\epsilon x, t_2 u)}{(t_2 u)^{2p-1}}\right] u^{2p} dx \\
&\geq \int_{\Lambda_\epsilon^c \cap \{t_2 u > a_0\}} \left[\frac{g(\epsilon x, t_1 u)}{(t_1 u)^{2p-1}} - \frac{g(\epsilon x, t_2 u)}{(t_2 u)^{2p-1}}\right] u^{2p} dx \\
&+ \int_{\Lambda_\epsilon^c \cap \{t_2 u \leq a_0 < t_1 u\}} \left[\frac{g(\epsilon x, t_1 u)}{(t_1 u)^{2p-1}} - \frac{g(\epsilon x, t_2 u)}{(t_2 u)^{2p-1}}\right] u^{2p} dx \\
&+ \int_{\Lambda_\epsilon^c \cap \{t_1 u < a_0\}} \left[\frac{g(\epsilon x, t_1 u)}{(t_1 u)^{2p-1}} - \frac{g(\epsilon x, t_2 u)}{(t_2 u)^{2p-1}}\right] u^{2p} dx \\
&\geq \frac{1}{K}\left(\frac{1}{t_1^p} - \frac{1}{t_2^p}\right) \int_{\Lambda_\epsilon^c \cap \{t_2 u > a_0\}} V_0 u^p dx \\
&+ \int_{\Lambda_\epsilon^c \cap \{t_2 u \leq a_0 < t_1 u\}} \left[\frac{V_0}{K} \frac{1}{(t_1 u)^p} - \frac{f(t_2 u)}{(t_2 u)^{2p-1}}\right] u^{2p} dx.
\end{aligned}$$

Accordingly, in view of $t_1 > t_2$, we have

$$\begin{aligned}
\|u\|_\epsilon^p &\leq \frac{1}{K} \int_{\Lambda_\epsilon^c \cap \{t_2 u > a_0\}} V_0 u^p dx + \frac{t_1^p t_2^p}{t_2^p - t_1^p} \int_{\Lambda_\epsilon^c \cap \{t_2 u \leq a_0 < t_1 u\}} \left[\frac{V_0}{K} \frac{1}{(t_1 u)^p} - \frac{f(t_2 u)}{(t_2 u)^{2p-1}}\right] u^{2p} dx \\
&= \frac{1}{K} \int_{\Lambda_\epsilon^c \cap \{t_2 u > a_0\}} V_0 u^p dx - \frac{t_2^p}{t_1^p - t_2^p} \int_{\Lambda_\epsilon^c \cap \{t_2 u \leq a_0 < t_1 u\}} \frac{V_0}{K} u^p dx \\
&+ \frac{t_1^p}{t_1^p - t_2^p} \int_{\Lambda_\epsilon^c \cap \{t_2 u \leq a_0 < t_1 u\}} \frac{f(t_2 u)}{(t_2 u)^{p-1}} u^p dx \leq \frac{1}{K} \int_{\Lambda_\epsilon^c} V_0 u^p dx \leq \frac{1}{K} \|u\|_\epsilon^p.
\end{aligned}$$

Since $u \neq 0$ and $K > 1$, we get a contradiction.

(*ii*) Let $u \in \mathbb{S}_\epsilon^+$. By (*i*) there exists $t_u > 0$ such that $h_u'(t_u) = 0$, or equivalently

$$t_u^{p-1} \leq t_u^{p-1} \|u\|_\epsilon^p + b t_u^{2p-1} [u]_{s,p}^{2p} = \int_{\mathbb{R}^3} g(\epsilon x, t_u u)\, u\, dx.$$

From (g_1)–(g_2) and Theorem 6.4, for all $\xi > 0$, we obtain

$$t_u^{p-1} \leq \int_{\mathbb{R}^3} g(\epsilon x, t_u u)\, u\, dx \leq \xi t_u^{p-1} C_1 + C_\xi t_u^{\nu-1} C_2,$$

and choosing ξ sufficiently small, we can find $\tau > 0$, independent of u, such that $t_u \geq \tau$.

Now, let $\mathbb{K} \subset \mathbb{S}_\epsilon^+$ be a compact set, and assume by contradiction that there exists a sequence $\{u_n\}_{n \in \mathbb{N}} \subset \mathbb{K}$ such that $t_n := t_{u_n} \to \infty$. Therefore, there exists $u \in \mathbb{K}$ such that $u_n \to u$ in $\mathcal{H}_\epsilon$. From the proof of (*ii*) in Lemma 6.3 we get

$$\mathcal{I}_\epsilon(t_n u_n) \to -\infty. \tag{6.44}$$

On the other hand, fixed $v \in \mathcal{N}$, by $\langle \mathcal{I}'_\epsilon(v), v\rangle = 0$ and (g_3), we can infer

$$\begin{aligned}
\mathcal{I}_\epsilon(v) &= \mathcal{I}_\epsilon(v) - \frac{1}{\vartheta}\langle \mathcal{I}'_\epsilon(v), v\rangle \\
&\geq \left(\frac{\vartheta - p}{p\vartheta}\right)\|v\|_\epsilon^p + \left(\frac{\vartheta - 2p}{2p\vartheta}\right) b[v]_{s,p}^{2p} + \frac{1}{\vartheta}\int_{\Lambda_\epsilon^c} [g(\epsilon x, v)v - \vartheta G(\epsilon x, v)]\, dx \\
&\geq \left(\frac{\vartheta - p}{p\vartheta}\right)\|v\|_\epsilon^p - \left(\frac{\vartheta - p}{p\vartheta}\right)\frac{1}{K}\int_{\Lambda_\epsilon^c} V(\epsilon x) v^p dx \geq \left(1 - \frac{1}{K}\right)\left(\frac{\vartheta - p}{p\vartheta}\right)\|v\|_\epsilon^p.
\end{aligned}$$

Taking into account that $\{t_{u_n} u_n\}_{n\in\mathbb{N}} \subset \mathcal{N}$, $K > 2p > 3$, $\vartheta > 2p$, from the above inequality and (6.44), we obtain a contradiction.

(iii) Firstly, we note that $\hat{m}_\epsilon$, m_ϵ and m_ϵ^{-1} are well defined. Indeed, by *(i)*, for each $u \in \mathcal{H}_\epsilon^+$ there exists a unique $m_\epsilon(u) \in \mathcal{N}$. On the other hand, if $u \in \mathcal{N}$ then $u \in \mathcal{H}_\epsilon^+$. Otherwise, if $u \notin \mathcal{H}_\epsilon^+$, we get

$$|supp(u^+) \cap \Lambda_\epsilon| = 0,$$

which together with $(g3)$-(ii) yields

$$\begin{aligned}
\|u\|_\epsilon^p &\leq \int_{\mathbb{R}^3} g(\epsilon x, u)\, u\, dx = \int_{\Lambda_\epsilon^c} g(\epsilon x, u)\, u\, dx + \int_{\Lambda_\epsilon} g(\epsilon x, u)\, u\, dx \\
&= \int_{\Lambda_\epsilon^c} g(\epsilon x, u^+)\, u^+\, dx \leq \frac{1}{K}\int_{\Lambda_\epsilon^c} V(\epsilon x)|u|^p dx \leq \frac{1}{K}\|u\|_\epsilon^p
\end{aligned} \tag{6.45}$$

and this leads to a contradiction because $K > 1$. Consequently, $m_\epsilon^{-1}(u) = \frac{u}{\|u\|_\epsilon} \in \mathbb{S}_\epsilon^+$, m_ϵ^{-1} is well defined and continuous. From $u \in \mathbb{S}_\epsilon^+$, we can see that

$$m_\epsilon^{-1}(m_\epsilon(u)) = m_\epsilon^{-1}(t_u u) = \frac{t_u u}{\|t_u u\|_\epsilon} = \frac{u}{\|u\|_\epsilon} = u$$

which implies that m_ϵ is a bijection.

Next, we prove that $\hat{m}_\epsilon$ is a continuous function. Let $\{u_n\}_{n\in\mathbb{N}} \subset \mathcal{H}_\epsilon^+$ and $u \in \mathcal{H}_\epsilon^+$ such that $u_n \to u$ in $\mathcal{H}_\epsilon$. Since $\hat{m}(tu) = \hat{m}(u)$ for all $t > 0$, we may assume that $\|u_n\|_\epsilon = \|u\|_\epsilon = 1$ for all $n \in \mathbb{N}$. By *(ii)* there exists $t_0 > 0$ such that $t_n := t_{u_n} \to t_0$. Since $t_n u_n \in \mathcal{N}$, we have

$$t_n^p\|u_n\|_\epsilon^p + b t_n^{2p}[u_n]_{s,p}^{2p} = \int_{\mathbb{R}^3} g(\epsilon x, t_n u_n)\, t_n u_n\, dx.$$

Letting $n \to \infty$, we obtain

$$t_0^p\|u\|_\epsilon^p + b t_0^{2p}[u]_{s,p}^{2p} = \int_{\mathbb{R}^3} g(\epsilon x, t_0 u)\, t_0 u\, dx,$$

which implies that $t_0 u \in \mathcal{N}$. By *(i)*, we deduce that $t_u = t_0$, and this shows that $\hat{m}_\epsilon(u_n) \to \hat{m}_\epsilon(u)$ in $\mathcal{H}_\epsilon^+$. Therefore, $\hat{m}_\epsilon$ and m_ϵ are continuous functions.

(iv) Let $\{u_n\}_{n\in\mathbb{N}} \subset \mathbb{S}_\epsilon^+$ be such that $\text{dist}(u_n, \partial\mathbb{S}_\epsilon^+) \to 0$. Observing that for each $r \in [p, p_s^*]$ and $n \in \mathbb{N}$ it holds

$$|u_n^+|_{L^r(\Lambda_\epsilon)} \leq \inf_{v\in\partial\mathbb{S}_\epsilon^+} |u_n - v|_{L^r(\Lambda_\epsilon)} \leq C_r \inf_{v\in\partial\mathbb{S}_\epsilon^+} \|u_n - v\|_\epsilon,$$

then by (g_1), (g_2), (g_3)–(ii), we obtain

$$\begin{aligned}\int_{\mathbb{R}^3} G(\epsilon x, tu_n)\,dx &= \int_{\Lambda_\epsilon^c} G(\epsilon x, tu_n)\,dx + \int_{\Lambda_\epsilon} G(\epsilon x, tu_n)\,dx \\ &\leq \frac{t^p}{K}\int_{\Lambda_\epsilon^c} V(\epsilon x)|u_n|^p dx + \int_{\Lambda_\epsilon} F(tu_n)\,dx \\ &\leq \frac{t^p}{K}\|u_n\|_\epsilon^p + C_1 t^{2p}\int_{\Lambda_\epsilon}(u_n^+)^{2p}dx + C_2 t^\nu \int_{\Lambda_\epsilon}(u_n^+)^\nu dx \\ &\leq \frac{t^p}{K} + C_1' t^{2p}\mathrm{dist}(\mathrm{u_n}, \partial\mathbb{S}_\epsilon^+)^{2\mathrm{p}} + C_2' t^\nu \mathrm{dist}(\mathrm{u_n}, \partial\mathbb{S}_\epsilon^+)^\nu\end{aligned}$$

from which we find

$$\limsup_{n\to\infty}\int_{\mathbb{R}^3} G(\epsilon x, tu_n)\,dx \leq \frac{t^p}{K} \quad \forall t>0. \tag{6.46}$$

Taking in mind the definition of $m_\epsilon(u_n)$ and using (6.46), we have

$$\begin{aligned}\liminf_{n\to\infty}\mathcal{I}_\epsilon(m_\epsilon(u_n)) &\geq \liminf_{n\to\infty}\mathcal{I}_\epsilon(tu_n) \\ &\geq \liminf_{n\to\infty}\left[\frac{t^p}{p}\|u_n\|_\epsilon^p + \frac{bt^{2p}}{2p}[u_n]_{s,p}^{2p}\right] - \frac{t^p}{K} \geq \left(\frac{K-p}{pK}\right)t^p.\end{aligned}$$

This implies that for all $t>0$

$$\liminf_{n\to\infty}\left\{\frac{1}{p}\|m_\epsilon(u_n)\|_\epsilon^p + \frac{b}{2p}[m_\epsilon(u_n)]_{s,p}^{2p}\right\} \geq \liminf_{n\to\infty}\mathcal{I}_\epsilon(m_\epsilon(u_n)) \geq \left(\frac{K-p}{pK}\right)t^p,$$

and recalling that $K>2p>p$, from the arbitrariness of t, we get $\mathcal{I}_\epsilon(m_\epsilon(u_n))=\infty$ and $\|m_\epsilon(u_n)\|_\epsilon \to \infty$ as $n\to\infty$. □

Let us define the maps

$$\hat{\psi}_\epsilon : \mathcal{H}_\epsilon^+ \to \mathbb{R} \quad \text{and} \quad \psi_\epsilon : \mathbb{S}_\epsilon^+ \to \mathbb{R},$$

by $\hat{\psi}_\epsilon(u) := \mathcal{I}_\epsilon(\hat{m}_\epsilon(u))$ and $\psi_\epsilon := \hat{\psi}_\epsilon|_{\mathbb{S}_\epsilon^+}$.

The next result is a consequence of Lemma 6.9 and Corollary 2.3 of Szulkin and Weth [144].

Proposition 6.1 *Assume that* (V_1)–(V_2') *and* (f_1)–(f_4) *are satisfied. Then the following properties hold.*

(a) $\hat{\psi}_\epsilon \in C^1(\mathcal{H}_\epsilon^+, \mathbb{R})$ *and*

$$\langle \hat{\psi}_\epsilon'(u), v\rangle = \frac{\|\hat{m}_\epsilon(u)\|_\epsilon}{\|u\|_\epsilon}\langle \mathcal{I}_\epsilon'(\hat{m}_\epsilon(u)), v\rangle \quad \forall u\in\mathcal{H}_\epsilon^+, \forall v\in\mathcal{H}_\epsilon.$$

(b) $\psi_\epsilon \in C^1(\mathbb{S}_\epsilon^+, \mathbb{R})$ *and*

$$\langle \psi_\epsilon'(u), v\rangle = \|m_\epsilon(u)\|_\epsilon\langle \mathcal{I}_\epsilon'(m_\epsilon(u)), v\rangle, \quad \forall v\in T_u\mathbb{S}_\epsilon^+.$$

(c) *If* $\{u_n\}_{n\in\mathbb{N}}$ *is a* $(PS)_d$ *sequence for* ψ_ϵ, *then* $\{m_\epsilon(u_n)\}_{n\in\mathbb{N}}$ *is a* $(PS)_d$ *sequence for* $\mathcal{I}_\epsilon$. *If* $\{u_n\}_{n\in\mathbb{N}}\subset\mathcal{N}$ *is a bounded* $(PS)_d$ *sequence for* $\mathcal{I}_\epsilon$, *then* $\{m_\epsilon^{-1}(u_n)\}_{n\in\mathbb{N}}$ *is a* $(PS)_d$ *sequence for* ψ_ϵ.

(d) u *is a critical point of* ψ_ϵ *if and only if* $m_\epsilon(u)$ *is a critical point for* $\mathcal{I}_\epsilon$. *Moreover, the corresponding critical values coincide and*

$$\inf_{u\in\mathbb{S}_\epsilon^+}\psi_\epsilon(u) = \inf_{u\in\mathcal{N}}\mathcal{I}_\epsilon(u).$$

Remark 6.3 *As in Szulkin and Weth [144], we can see that the infimum of $\mathcal{I}_\epsilon$ over $\mathcal{N}$ has the following minimax characterization:*

$$c_\epsilon = \inf_{u\in\mathcal{N}} \mathcal{I}_\epsilon(u) = \inf_{u\in\mathcal{H}_\epsilon^+} \max_{t>0} \mathcal{I}_\epsilon(tu) = \inf_{u\in\mathbb{S}_\epsilon^+} \max_{t>0} \mathcal{I}_\epsilon(tu).$$

Now, we prove the following result.

Corollary 6.2 *Let $d \in \mathbb{R}$. Then ψ_ϵ satisfies the $(PS)_d$ condition on $\mathbb{S}_\epsilon^+$.*

Proof. Let $\{u_n\}_{n\in\mathbb{N}} \subset \mathbb{S}_\epsilon^+$ be a (PS) sequence for ψ_ϵ at the level d. Then

$$\psi_\epsilon(u_n) \to d \quad \text{and} \quad \psi_\epsilon'(u_n) \to 0 \text{ in } (T_{u_n}\mathbb{S}_\epsilon^+)'.$$

By Proposition 6.1-(c) it follows that $\{m_\epsilon(u_n)\}_{n\in\mathbb{N}}$ is a $(PS)_d$ sequence for $\mathcal{I}_\epsilon$ in $\mathcal{H}_\epsilon$. Then, by Lemma 6.4, we can see that $\mathcal{I}_\epsilon$ fulfills the $(PS)_d$ condition in $\mathcal{H}_\epsilon$, so there exists $u \in \mathbb{S}_\epsilon^+$ such that, up to a subsequence,

$$m_\epsilon(u_n) \to m_\epsilon(u) \text{ in } \mathcal{H}_\epsilon.$$

Applying Lemma 6.9-(iii), we conclude that $u_n \to u$ in $\mathbb{S}_\epsilon^+$. □

Now, we deal with the autonomous problem (6.14). We denote by $\mathcal{H}_\mu^+$ the open subset of $\mathcal{H}_\mu$ defined as

$$\mathcal{H}_\mu^+ := \{u \in \mathcal{H}_\mu : |supp(u^+)| > 0\},$$

and $\mathbb{S}_\mu^+ := \mathbb{S}_\mu \cap \mathcal{H}_\mu^+$, where $\mathbb{S}_\mu$ is the unit sphere of $\mathcal{H}_\mu$. We note that $\mathbb{S}_\mu^+$ is an incomplete C^1-manifold of codimension 1 modeled on $\mathcal{H}_\mu$ and contained in $\mathcal{H}_\mu^+$. Thus, $\mathcal{H}_\mu = T_u\mathbb{S}_\mu^+ \oplus \mathbb{R}u$ for each $u \in \mathbb{S}_\mu^+$, where $T_u\mathbb{S}_\mu^+ := \big\{u \in \mathcal{H}_\mu : B_u(v) + \mu\int_{\mathbb{R}^3} |u|^{p-2}uv\,dx = 0\big\}$.

Arguing as before, we can see that the following results hold.

Lemma 6.10 *Assume that (f_1)–(f_4) hold. Then the following properties hold.*

- (i) *For each $u \in \mathcal{H}_\mu^+$, let $h : \mathbb{R}^+ \to \mathbb{R}$ be defined by $h_u(t) := \mathcal{E}_\mu(tu)$. Then there exists a unique $t_u > 0$ such that $h_u'(t) > 0$ for all $t \in (0, t_u)$ and $h_u'(t) < 0$ for all $t \in (t_u, \infty)$.*
- (ii) *There exists $\tau > 0$ independent of u such that $t_u \geq \tau$ for any $u \in \mathbb{S}_\mu^+$. Moreover, for each compact set $\mathbb{K} \subset \mathbb{S}_\mu^+$ there is a constant $C_\mathbb{K} > 0$ such that $t_u \leq C_\mathbb{K}$ for any $u \in \mathbb{K}$.*
- (iii) *The map $\hat{m}_\mu : \mathcal{H}_\mu^+ \to \mathcal{N}_\mu$ given by $\hat{m}_\mu(u) = t_u u$ is continuous and $m_\mu := \hat{m}_\mu|_{\mathbb{S}_\mu^+}$ is a homeomorphism between $\mathbb{S}_\mu^+$ and $\mathcal{N}_\mu$. Moreover, $m_\mu^{-1}(u) = \frac{u}{\|u\|_\mu}$.*
- (iv) *If there is a sequence $\{u_n\}_{n\in\mathbb{N}} \subset \mathbb{S}_\mu^+$ such that $\operatorname{dist}(u_n, \partial\mathbb{S}_\mu^+) \to 0$, then $\|m_\mu(u_n)\|_\mu \to \infty$ and $\mathcal{E}_\mu(m_\mu(u_n)) \to \infty$.*

Let us define the maps

$$\hat{\psi}_\mu : \mathcal{H}_\mu^+ \to \mathbb{R} \quad \text{and} \quad \psi_\mu : \mathbb{S}_\mu^+ \to \mathbb{R},$$

by $\hat{\psi}_\mu(u) := \mathcal{E}_\mu(\hat{m}_0(u))$ and $\psi_\mu := \hat{\psi}_\mu|_{\mathbb{S}_\mu^+}$.

Proposition 6.2 *Assume that (f_1)–(f_4) are satisfied. Then,*

(a) *$\hat{\psi}_\mu \in C^1(\mathcal{H}_\mu^+, \mathbb{R})$ and*

$$\langle \hat{\psi}_\mu'(u), v\rangle = \frac{\|\hat{m}_\mu(u)\|_\mu}{\|u\|_\mu} \langle \mathcal{E}_\mu'(\hat{m}_\mu(u)), v\rangle \quad \forall u \in \mathcal{H}_\mu^+, \forall v \in \mathcal{H}_\mu.$$

(*b*) $\psi_\mu \in C^1(\mathbb{S}^+_\mu, \mathbb{R})$ *and*

$$\langle \psi'_\mu(u), v\rangle = \|m_\mu(u)\|_\mu \langle \mathcal{E}'_\mu(m_\mu(u)), v\rangle, \quad \forall v \in T_u\mathbb{S}^+_\mu.$$

(*c*) *If* $\{u_n\}_{n\in\mathbb{N}}$ *is a* $(PS)_d$ *sequence for* ψ_μ, *then* $\{m_\mu(u_n)\}_{n\in\mathbb{N}}$ *is a* $(PS)_d$ *sequence for* $\mathcal{E}_\mu$. *If* $\{u_n\}_{n\in\mathbb{N}} \subset \mathcal{N}_\mu$ *is a bounded* $(PS)_d$ *sequence for* $\mathcal{E}_\mu$, *then* $\{m_\mu^{-1}(u_n)\}_{n\in\mathbb{N}}$ *is a* $(PS)_d$ *sequence for* ψ_μ.

(*d*) u *is a critical point of* ψ_μ *if and only if* $m_\mu(u)$ *is a critical point for* $\mathcal{E}_\mu$. *Moreover, the corresponding critical values coincide and*

$$\inf_{u\in\mathbb{S}^+_\mu} \psi_\mu(u) = \inf_{u\in\mathcal{N}_\mu} \mathcal{E}_\mu(u).$$

Remark 6.4 *As in Szulkin and Weth [144], we can see that the infimum of* $\mathcal{E}_\mu$ *over* $\mathcal{N}_\mu$ *has the following minimax characterization:*

$$d_\mu = \inf_{u\in\mathcal{N}_\mu} \mathcal{E}_\mu(u) = \inf_{u\in\mathcal{H}^+_\mu} \max_{t>0} \mathcal{E}_\mu(tu) = \inf_{u\in\mathbb{S}^+_\mu} \max_{t>0} \mathcal{E}_\mu(tu).$$

Next, we give a compactness result for the autonomous problem which we will use later.

Lemma 6.11 *Let* $\{u_n\}_{n\in\mathbb{N}} \subset \mathcal{N}_\mu$ *be a sequence such that* $\mathcal{E}_\mu(u_n) \to d_\mu$. *Then,* $\{u_n\}_{n\in\mathbb{N}}$ *has a convergent subsequence in* $\mathcal{H}_\mu$.

Proof. Since $\{u_n\}_{n\in\mathbb{N}} \subset \mathcal{N}_\mu$ and $\mathcal{E}_\mu(u_n) \to d_\mu$, we can apply Lemma 6.10-(*iii*), Proposition 6.2-(*d*) and the definition of d_μ to infer that

$$v_n := m^{-1}(u_n) = \frac{u_n}{\|u_n\|_\mu} \in \mathbb{S}^+_\mu \quad \forall n \in \mathbb{N}$$

and

$$\psi_\mu(v_n) = \mathcal{E}_\mu(u_n) \to d_\mu = \inf_{v\in\mathbb{S}^+_\mu} \psi_\mu(v).$$

Let us introduce the map $\mathcal{F} : \overline{\mathbb{S}}^+_\mu \to \mathbb{R} \cup \{\infty\}$ defined by setting

$$\mathcal{F}(u) := \begin{cases} \psi_\mu(u) & \text{if } u \in \mathbb{S}^+_\mu, \\ \infty & \text{if } u \in \partial\mathbb{S}^+_\mu. \end{cases}$$

We note that

(i) $(\overline{\mathbb{S}}^+_\mu, \delta_\mu)$, where $\delta_\mu(u, v) := \|u - v\|_\mu$, is a complete metric space;

(ii) $\mathcal{F} \in C(\overline{\mathbb{S}}^+_\mu, \mathbb{R} \cup \{\infty\})$, by Lemma 6.10-(*iv*);

(iii) $\mathcal{F}$ is bounded below, by Proposition 6.2-(*d*).

Hence, invoking the Ekeland variational principle to $\mathcal{F}$, we can find $\{\hat{v}_n\}_{n\in\mathbb{N}} \subset \mathbb{S}^+_\mu$ such that $\{\hat{v}_n\}_{n\in\mathbb{N}}$ is a $(PS)_{d_\mu}$ sequence for ψ_μ on $\mathbb{S}^+_\mu$ and $\|\hat{v}_n - v_n\|_\mu = o_n(1)$. Then, using Proposition 6.2, Theorem 6.5 and arguing as in the proof of Corollary 6.2, we obtain the conclusion. □

Remark 6.5 *By Lemma 6.6,* (V_1) *and* (V'_2), *we obtain that* $\lim_{\epsilon\to 0} c_\epsilon = d_{V_0}$.

6.3.2 The barycenter map

In this subsection, our main purpose is to apply the Ljusternik-Schnirelmann category theory to prove a multiplicity result for problem (6.3). We begin with some technical results.

Lemma 6.12 *Let $\epsilon_n \to 0$ and $\{u_n\}_{n\in\mathbb{N}} \subset \mathcal{N}_{\epsilon_n}$ be such that $\mathcal{I}_{\epsilon_n}(u_n) \to d_{V_0}$. Then there exists $\{\tilde{y}_n\}_{n\in\mathbb{N}} \subset \mathbb{R}^3$ such that the translated sequence*

$$\tilde{u}_n(x) := u_n(x + \tilde{y}_n)$$

has a subsequence which converges in $\mathcal{H}_{V_0}$. Moreover, up to a subsequence, the sequence $\{y_n\}_{n\in\mathbb{N}} := \{\epsilon_n\tilde{y}_n\}_{n\in\mathbb{N}}$ is such that $y_n \to y_0 \in M$.

Proof. Since $\langle \mathcal{I}'_{\epsilon_n}(u_n), u_n\rangle = 0$ and $\mathcal{I}_{\epsilon_n}(u_n) \to d_{V_0}$, it is easy to see that $\{u_n\}_{n\in\mathbb{N}}$ is bounded. Let us observe that $\|u_n\|_{\epsilon_n} \not\to 0$ since $d_{V_0} > 0$. Therefore, arguing as in Lemma 6.5, we can find a sequence $\{\tilde{y}_n\}_{n\in\mathbb{N}} \subset \mathbb{R}^3$ and constants $R, \alpha > 0$ such that

$$\liminf_{n\to\infty} \int_{B_R(\tilde{y}_n)} |u_n|^p dx \geq \alpha.$$

Set $\tilde{u}_n(x) := u_n(x + \tilde{y}_n)$. Then, $\{\tilde{u}_n\}_{n\in\mathbb{N}}$ is bounded in $\mathcal{H}_{V_0}$ and we can assume that

$$\tilde{u}_n \rightharpoonup \tilde{u} \text{ weakly in } \mathcal{H}_{V_0},$$

for some $\tilde{u} \neq 0$. Let $\{t_n\}_{n\in\mathbb{N}} \subset (0, +\infty)$ be such that $\tilde{v}_n := t_n\tilde{u}_n \in \mathcal{N}_{V_0}$ (see Lemma 6.10-(i)), and set $y_n := \epsilon_n\tilde{y}_n$. Then, from $u_n \in \mathcal{N}_{\epsilon_n}$ and (g_2), we can see that

$$\begin{aligned} d_{V_0} \leq \mathcal{E}_{V_0}(\tilde{v}_n) &\leq \frac{a}{p}[\tilde{v}_n]^p_{s,p} + \frac{1}{p}\int_{\mathbb{R}^3} V(\epsilon_n x + y_n)|\tilde{v}_n|^p\,dx + \frac{b}{2p}[\tilde{v}_n]^{2p}_{s,p} - \int_{\mathbb{R}^3} F(\tilde{v}_n)\,dx \\ &\leq \frac{at_n^p}{p}[u_n]^p_{s,p} + \frac{t_n^p}{p}\int_{\mathbb{R}^3} V(\epsilon_n x)|u_n|^p\,dx + \frac{b}{2p}[u_n]^{2p}_{s,p} - \int_{\mathbb{R}^3} G(\epsilon_n x, t_n u_n)\,dx \\ &= \mathcal{I}_{\epsilon_n}(t_n u_n) \leq \mathcal{I}_{\epsilon_n}(u_n) = d_{V_0} + o(1), \end{aligned} \tag{6.47}$$

which gives

$$\mathcal{E}_{V_0}(\tilde{v}_n) \to d_{V_0} \text{ and } \{\tilde{v}_n\}_{n\in\mathbb{N}} \subset \mathcal{N}_{V_0}. \tag{6.48}$$

In particular, relation (6.48) implies that $\{\tilde{v}_n\}_{n\in\mathbb{N}}$ is bounded in $\mathcal{H}_{V_0}$, so we may assume that $\tilde{v}_n \rightharpoonup \tilde{v}$. Obviously, $\{t_n\}_{n\in\mathbb{N}}$ is bounded and we have $t_n \to t_0 \geq 0$. If $t_0 = 0$, from the boundedness of $\{\tilde{u}_n\}_{n\in\mathbb{N}}$, we get $\|\tilde{v}_n\|_{V_0} = t_n\|\tilde{u}_n\|_{V_0} \to 0$, that is $\mathcal{E}_{V_0}(\tilde{v}_n) \to 0$ in contrast with the fact $d_{V_0} > 0$. Then, $t_0 > 0$. From the uniqueness of the weak limit we have $\tilde{v} = t_0\tilde{u}$ and $\tilde{u} \neq 0$. By Lemma 6.11, we deduce that

$$\tilde{v}_n \to \tilde{v} \text{ in } \mathcal{H}_{V_0}, \tag{6.49}$$

which implies that $\tilde{u}_n = \dfrac{\tilde{v}_n}{t_n} \to \dfrac{\tilde{v}}{t_0} = \tilde{u}$ in $\mathcal{H}_{V_0}$, and

$$\mathcal{E}_{V_0}(\tilde{v}) = d_{V_0} \text{ and } \langle \mathcal{E}'_{V_0}(\tilde{v}), \tilde{v}\rangle = 0.$$

Next, we show that $\{y_n\}_{n\in\mathbb{N}}$ has a subsequence such that $y_n \to y_0 \in M$. Assume by contradiction that $\{y_n\}_{n\in\mathbb{N}}$ is not bounded, that is there exists a subsequence, still denoted by $\{y_n\}_{n\in\mathbb{N}}$, such that $|y_n| \to +\infty$. Take $R > 0$ such that $\Lambda \subset \mathcal{B}_R$. We may suppose that $|y_n| > 2R$ for n large enough, so, for any $x \in \mathcal{B}_{R/\epsilon_n}$, we get $|\epsilon_n x + y_n| \geq |y_n| - |\epsilon_n x| > R$. Then, we deduce that

$$\begin{aligned} \|\tilde{u}_n\|^p_{V_0} &\leq \|\tilde{u}_n\|^p_{V_0} + b[\tilde{u}_n]^{2p}_{s,p} = \int_{\mathbb{R}^3} g(\epsilon_n x + y_n, \tilde{u}_n)\tilde{u}_n\,dx \\ &\leq \int_{\mathcal{B}_{R/\epsilon_n}} \tilde{f}(\tilde{u}_n)\tilde{u}_n\,dx + \int_{\mathcal{B}^c_{R/\epsilon_n}} f(\tilde{u}_n)\tilde{u}_n\,dx. \end{aligned}$$

Since $\tilde{u}_n \to \tilde{u}$ in $\mathcal{H}_{V_0}$, from the dominated convergence theorem we can see that

$$\int_{\mathcal{B}^c_{R/\epsilon_n}} f(\tilde{u}_n)\tilde{u}_n\,dx = o(1).$$

Recalling that $\tilde{f}(\tilde{u}_n)\tilde{u}_n \le \frac{V_0}{K}|\tilde{u}_n|^p$, we get

$$\|\tilde{u}_n\|^p_{V_0} \le \frac{1}{K}\int_{\mathcal{B}_{R/\epsilon_n}} V_0|\tilde{u}_n|^p\,dx + o(1),$$

which yields

$$\left(1-\frac{1}{K}\right)\|\tilde{u}_n\|^p_{V_0} \le o(1).$$

Then we obtain a contradiction thanks to $\tilde{u}_n \to \tilde{u} \neq 0$. Thus, $\{y_n\}_{n\in\mathbb{N}}$ is bounded and, up to a subsequence, we may assume that $y_n \to y_0$. If $y_0 \notin \overline{\Lambda}$, then there exists $r > 0$ such that $y_n \in \mathcal{B}_{r/2}(y_0) \subset \overline{\Lambda}^c$ for any n large enough. Reasoning as before, we get a contradiction. Hence, $y \in \overline{\Lambda}$.

Next, we prove that $V(y_0) = V_0$. Assume by contradiction that $V(y_0) > V_0$. Taking into account (6.49), Fatou's lemma and the invariance of $\mathbb{R}^3$ by translations, we have

$$\begin{aligned} d_{V_0} &= \mathcal{E}_{V_0}(\tilde{v}) < \liminf_{n\to\infty}\Big[\frac{1}{p}[\tilde{v}_n]^p_{s,p} + \frac{1}{p}\int_{\mathbb{R}^3} V(\epsilon_n x + y_n)|\tilde{v}_n|^p\,dx + \frac{b}{2p}[\tilde{v}_n]^{2p}_{s,p} - \int_{\mathbb{R}^3} F(\tilde{v}_n)\,dx\Big] \\ &\le \liminf_{n\to\infty} \mathcal{I}_{\epsilon_n}(t_n u_n) \le \liminf_{n\to\infty} \mathcal{I}_{\epsilon_n}(u_n) = d_{V_0} \end{aligned}$$

which gives a contradiction. □

Now, we aim to relate the number of positive solutions of problem (6.3) to the topology of the set M. For this reason, we take $\delta > 0$ such that

$$M_\delta = \{x \in \mathbb{R}^3 : \operatorname{dist}(x, M) \le \delta\} \subset \Lambda,$$

and we consider $\eta \in C^\infty_c(\mathbb{R}_+, [0,1])$ such that $\eta(t) = 1$ if $0 \le t \le \frac{\delta}{2}$ and $\eta(t) = 0$ if $t \ge \delta$. For any $y \in M$, we define

$$\Psi_{\epsilon,y}(x) := \eta(|\epsilon x - y|)w\left(\frac{\epsilon x - y}{\epsilon}\right)$$

where $w \in \mathcal{H}_{V_0}$ is a positive ground state solution to the autonomous problem (6.14) (whose existence is guaranteed by Theorem 6.5). Let $t_\epsilon > 0$ be the unique number such that

$$\max_{t\ge 0} \mathcal{I}_\epsilon(t\Psi_{\epsilon,y}) = \mathcal{I}_\epsilon(t_\epsilon \Psi_{\epsilon,y}).$$

Finally, we consider the map $\Phi_\epsilon : M \to \mathcal{N}$ defined by setting

$$\Phi_\epsilon(y) := t_\epsilon \Psi_{\epsilon,y}.$$

Lemma 6.13 *The functional* Φ_ϵ *satisfies*

$$\lim_{\epsilon\to 0} \mathcal{I}_\epsilon(\Phi_\epsilon(y)) = d_{V_0} \textit{ uniformly in } y \in M.$$

Proof. Assume by contradiction that there exist $\delta_0 > 0$, $\{y_n\}_{n\in\mathbb{N}} \subset M$ and $\epsilon_n \to 0$ such that

$$|\mathcal{I}_{\epsilon_n}(\Phi_{\epsilon_n}(y_n)) - d_{V_0}| \ge \delta_0. \tag{6.50}$$

Let us observe that, by using the change of variable $z = \dfrac{\epsilon_n x - y_n}{\epsilon_n}$, if $z \in \mathcal{B}_{\frac{\delta}{\epsilon_n}}$, it follows that $\epsilon_n z \in \mathcal{B}_\delta$ and then $\epsilon_n z + y_n \in \mathcal{B}_\delta(y_n) \subset M_\delta \subset \Lambda$. Then, recalling that $G = F$ in Λ and $\eta(t) = 0$ for $t \geq \delta$, we have

$$\begin{aligned}
\mathcal{I}_{\epsilon_n}(\Phi_{\epsilon_n}(y_n)) &= \frac{t^p_{\epsilon_n}}{p}\|\Psi_{\epsilon_n,y_n}\|^p_{\epsilon_n} + b\frac{t^{2p}_{\epsilon_n}}{2p}[\Psi_{\epsilon_n,y_n}]^{2p}_{s,p} - \int_{\mathbb{R}^3} G(\epsilon_n x, t_{\epsilon_n}\Psi_{\epsilon_n,y_n})\, dx \\
&= \frac{t^p_{\epsilon_n}}{p}\left([\eta(|\epsilon_n z|)w]^p_{s,p} + \int_{\mathbb{R}^3} V(\epsilon_n z + y_n)(\eta(|\epsilon_n z|)w(z))^p\, dz\right) \\
&+ b\frac{t^{2p}_{\epsilon_n}}{2p}[\eta(|\epsilon_n z|)w]^{2p}_{s,p} - \int_{\mathbb{R}^3} F(t_{\epsilon_n}\eta(|\epsilon_n z|)w(z))\, dz.
\end{aligned}$$

Now, we verify that the sequence $\{t_{\epsilon_n}\}_{n\in\mathbb{N}}$ satisfies $t_{\epsilon_n} \to 1$ as $\epsilon_n \to 0$. By the definition of t_{ϵ_n}, it follows that $\langle \mathcal{I}'_{\epsilon_n}(\Phi_{\epsilon_n}(y_n)), \Phi_{\epsilon_n}(y_n)\rangle = 0$, which gives

$$\frac{1}{t^p_{\epsilon_n}}\|\Psi_{\epsilon_n,y_n}\|^p_{\epsilon_n} + b[\Psi_{\epsilon_n,y_n}]^{2p}_{s,p} = \int_{\mathbb{R}^3}\left[\frac{f(t_{\epsilon_n}\eta(|\epsilon_n z|)w(z))}{(t_{\epsilon_n}\eta(|\epsilon_n z|)w(z))^{2p-1}}\right](\eta(|\epsilon_n z|)w(z))^{2p}\, dz, \tag{6.51}$$

where we used the fact that $g = f$ on Λ. Since $\eta = 1$ in $\mathcal{B}_{\frac{\delta}{2}} \subset \mathcal{B}_{\frac{\delta}{\epsilon_n}}$ for all n large enough, from (6.51) it follows that

$$\frac{1}{t^p_{\epsilon_n}}\|\Psi_{\epsilon_n,y_n}\|^p_{\epsilon_n} + b[\Psi_{\epsilon_n,y_n}]^{2p}_{s,p} \geq \int_{\mathcal{B}_{\frac{\delta}{2}}}\left[\frac{f(t_{\epsilon_n}w(z))}{(t_{\epsilon_n}w(z))^{2p-1}}\right]|w(z)|^{2p}\, dz.$$

Since w is continuous, we can find a vector $\hat{z} \in \mathbb{R}^3$ such that

$$w(\hat{z}) = \min_{z\in\bar{\mathcal{B}}_{\frac{\delta}{2}}} w(z) > 0.$$

Then, by (f_4), we deduce that

$$\frac{1}{t^p_{\epsilon_n}}\|\Psi_{\epsilon_n,y_n}\|^p_{\epsilon_n} + b[\Psi_{\epsilon_n,y_n}]^{2p}_{s,p} \geq \left[\frac{f(t_{\epsilon_n}w(\hat{z}))}{(t_{\epsilon_n}w(\hat{z}))^{2p-1}}\right]|w(\hat{z})|^{2p}|\mathcal{B}_{\frac{\delta}{2}}|. \tag{6.52}$$

Now, assume by contradiction that $t_{\epsilon_n} \to \infty$. Let us observe that Lemma 6.2 yields

$$\|\Psi_{\epsilon_n,y_n}\|_{\epsilon_n} \to \|w\|_{V_0} \in (0,\infty). \tag{6.53}$$

From $t_{\epsilon_n} \to \infty$ and (6.53), it follows that

$$\frac{1}{t^p_{\epsilon_n}}\|\Psi_{\epsilon_n,y_n}\|^p_{\epsilon_n} + b[\Psi_{\epsilon_n,y_n}]^{2p}_{s,p} \to b[w]^{2p}_{s,p}. \tag{6.54}$$

On the other hand, by (f_3), we have

$$\lim_{n\to\infty}\frac{f(t_{\epsilon_n}w(\hat{z}))}{(t_{\epsilon_n}w(\hat{z}))^{2p-1}} = \infty. \tag{6.55}$$

Putting together (6.52), (6.54) and (6.55) we get a contradiction. Therefore, $\{t_{\epsilon_n}\}_{n\in\mathbb{N}}$ is bounded and, up to a subsequence, we may assume that $t_{\epsilon_n} \to t_0$ for some $t_0 \geq 0$. Indeed, from (6.51), (6.53), (f_1)–(f_2) we can see that $t_0 > 0$. Hence, letting $n \to \infty$ in (6.51), we deduce from (6.53) and the dominated convergence theorem that

$$\frac{1}{t^p_0}\|w\|^p_{V_0} + b[w]^{2p}_{s,p} = \int_{\mathbb{R}^3}\frac{f(t_0 w)}{(t_0 w)^{2p-1}}\, w^{2p}\, dx. \tag{6.56}$$

Since $w \in \mathcal{N}_{V_0}$, we can see that

$$\|w\|_{V_0}^p + b[w]_{s,p}^{2p} = \int_{\mathbb{R}^3} f(w)w\,dx. \tag{6.57}$$

In the light of (6.56), (6.57) and (f_4), we deduce that $t_0 = 1$. Accordingly, taking the limit as $n \to \infty$ in (6.51), we obtain

$$\lim_{n\to\infty} \mathcal{I}_{\epsilon_n}(\Phi_{\epsilon_n,y_n}) = \mathcal{E}_{V_0}(w) = d_{V_0},$$

which contradicts (6.50). The proof is now complete. □

At this point, we are in the position to define the barycenter map. For any $\delta > 0$ given by Lemma 6.13, we take $\rho = \rho(\delta) > 0$ such that $M_\delta \subset \mathcal{B}_\rho$, and we consider $\Upsilon : \mathbb{R}^3 \to \mathbb{R}^3$ given by

$$\Upsilon(x) := \begin{cases} x & \text{if } |x| < \rho, \\ \frac{\rho x}{|x|} & \text{if } |x| \geq \rho. \end{cases}$$

We define the barycenter map $\beta_\epsilon : \mathcal{N} \to \mathbb{R}^3$ as follows

$$\beta_\epsilon(u) := \frac{\displaystyle\int_{\mathbb{R}^3} \Upsilon(\epsilon x)|u(x)|^p\,dx}{\displaystyle\int_{\mathbb{R}^3} |u(x)|^p\,dx}.$$

Arguing as in Lemma 3.14 of Ambrosio and Isernia [4] we can prove the following result.

Lemma 6.14 *The function β_ϵ satisfies the following limit*

$$\lim_{\epsilon\to 0} \beta_\epsilon(\Phi_\epsilon(y)) = y \text{ uniformly in } y \in M.$$

Now, we introduce the following subset of $\mathcal{N}$:

$$\widetilde{\mathcal{N}}_\epsilon := \{u \in \mathcal{N} : \mathcal{I}_\epsilon(u) \leq d_{V_0} + h_1(\epsilon)\},$$

where $h_1(\epsilon) := \sup_{y\in M} |\mathcal{I}_\epsilon(\Phi_\epsilon(y)) - d_{V_0}| \to 0$ as $\epsilon \to 0$ by Lemma 6.13. By the definition of $h_1(\epsilon)$, it follows that, for all $y \in M$ and $\epsilon > 0$, $\Phi_\epsilon(y) \in \widetilde{\mathcal{N}}_\epsilon$ and $\widetilde{\mathcal{N}}_\epsilon \neq \emptyset$. Moreover, as in Lemma 3.15 in [4], we can see that the following lemma holds true.

Lemma 6.15 *For any $\delta > 0$ there holds*

$$\lim_{\epsilon\to 0} \sup_{u\in\widetilde{\mathcal{N}}_\epsilon} \operatorname{dist}(\beta_\epsilon(u), M_\delta) = 0.$$

Before proving the multiplicity result for the modified problem (6.3), we recall the following useful abstract result whose proof can be found in Benci and Cerami [9].

Lemma 6.16 *Let I, I_1 and I_2 be closed sets with $I_1 \subset I_2$, and let $\pi : I \to I_2$ and $\psi : I_1 \to I$ be two continuous maps such that $\pi \circ \psi$ is homotopically equivalent to the embedding $j : I_1 \to I_2$. Then $cat_I(I) \geq cat_{I_2}(I_1)$.*

Theorem 6.6 *Assume that (V_1)–(V_2') and (f_1)–(f_4) hold. Then, given $\delta > 0$ there exists $\bar{\epsilon}_\delta > 0$ such that, for any $\epsilon \in (0, \bar{\epsilon}_\delta)$, problem (6.3) has at least $cat_{M_\delta}(M)$ positive solutions.*

Proof. For any $\epsilon > 0$, we consider the map $\alpha_\epsilon : M \to \mathbb{S}_\epsilon^+$ defined as $\alpha_\epsilon(y) = m_\epsilon^{-1}(\Phi_\epsilon(y))$. By Lemma 6.13, we have

$$\lim_{\epsilon\to 0} \psi_\epsilon(\alpha_\epsilon(y)) = \lim_{\epsilon\to 0} \mathcal{I}_\epsilon(\Phi_\epsilon(y)) = d_{V_0} \text{ uniformly in } y \in M. \tag{6.58}$$

Set

$$\widetilde{\mathcal{S}}_\epsilon^+ := \{w \in \mathbb{S}_\epsilon^+ : \psi_\epsilon(w) \le d_{V_0} + h_1(\epsilon)\},$$

where $h_1(\epsilon) := \sup_{y\in M} |\psi_\epsilon(\alpha_\epsilon(y)) - d_{V_0}| \to 0$ as $\epsilon \to 0$ by (6.58). Since $\psi_\epsilon(\alpha_\epsilon(y)) \in \widetilde{\mathcal{S}}_\epsilon^+$ we deduce that $\widetilde{\mathcal{S}}_\epsilon^+ \neq \emptyset$ for all $\epsilon > 0$. From Lemma 6.13, Lemma 6.9-(*iii*), Lemma 6.15 and Lemma 6.14, we can find $\bar{\epsilon} = \bar{\epsilon}_\delta > 0$ such that the following diagram is well defined for any $\epsilon \in (0, \bar{\epsilon})$.

$$M \stackrel{\Phi_\epsilon}{\to} \Phi_\epsilon(M) \stackrel{m_\epsilon^{-1}}{\to} \alpha_\epsilon(M) \stackrel{m_\epsilon}{\to} \Phi_\epsilon(M) \stackrel{\beta_\epsilon}{\to} M_\delta.$$

In view of Lemma 6.14, and decreasing $\bar{\epsilon}$ if necessary, we can see that $\beta_\epsilon(\Phi_\epsilon(y)) = y + \theta(\epsilon, y)$ for all $y \in M$, for some function $\theta(\epsilon, y)$ such that $|\theta(\epsilon, y)| < \frac{\delta}{2}$ uniformly in $y \in M$ and for all $\epsilon \in (0, \bar{\epsilon})$. Then, we can see that $H(t, y) := y + (1 - t)\theta(\epsilon, y)$ with $(t, y) \in [0, 1] \times M$, is a homotopy between $\beta_\epsilon \circ \Phi_\epsilon = (\beta_\epsilon \circ m_\epsilon) \circ (m_\epsilon^{-1} \circ \Phi_\epsilon)$ and the inclusion map $id : M \to M_\delta$. This fact together with Lemma 6.16 implies that

$$cat_{\alpha_\epsilon(M)}\alpha_\epsilon(M) \ge cat_{M_\delta}(M). \tag{6.59}$$

Therefore, by Corollary 6.2 and Corollary 28 in [144], with $c = c_\epsilon \le d_{V_0} + h_1(\epsilon) = d$ and $K = \alpha_\epsilon(M)$, we can see that Ψ_ϵ has at least $cat_{\alpha_\epsilon(M)}\alpha_\epsilon(M)$ critical points on $\widetilde{\mathcal{S}}_\epsilon^+$. Taking into account Proposition 6.1-(*d*) and (6.59), we can infer that $\mathcal{I}_\epsilon$ admits at least $cat_{M_\delta}(M)$ critical points in $\widetilde{\mathcal{N}}_\epsilon$. □

6.3.3 Proof of Theorem 6.2

Take $\delta > 0$ such that $M_\delta \subset \Lambda$. We begin by proving that there exists $\tilde{\epsilon}_\delta > 0$ such that for any $\epsilon \in (0, \tilde{\epsilon}_\delta)$ and any solution $u_\epsilon \in \widetilde{\mathcal{N}}_\epsilon$ of (6.3), it holds

$$|u_\epsilon|_{L^\infty(\Lambda_\epsilon^c)} < a_0. \tag{6.60}$$

Suppose by contradiction that for some subsequence $\{\epsilon_n\}_{n\in\mathbb{N}}$ such that $\epsilon_n \to 0$, we can find $u_{\epsilon_n} \in \widetilde{\mathcal{N}}_{\epsilon_n}$ such that $\mathcal{I}'_{\epsilon_n}(u_{\epsilon_n}) = 0$ and

$$|u_{\epsilon_n}|_{L^\infty(\Lambda_{\epsilon_n}^c)} \ge a_0. \tag{6.61}$$

Since $\mathcal{I}_{\epsilon_n}(u_{\epsilon_n}) \le d_{V_0} + h_1(\epsilon_n)$ and $h_1(\epsilon_n) \to 0$, we can proceed as in the first part of the proof of Lemma 6.7, to deduce that $\mathcal{I}_{\epsilon_n}(u_{\epsilon_n}) \to d_{V_0}$. Then, by Lemma 6.12, we can find $\{\tilde{y}_n\}_{n\in\mathbb{N}} \subset \mathbb{R}^3$ such that $\tilde{u}_n = u_{\epsilon_n}(\cdot + \tilde{y}_n) \to \tilde{u}$ in $W^{s,p}(\mathbb{R}^3)$ and $\epsilon_n \tilde{y}_n \to y_0 \in M$.

Now, if we choose $r > 0$ such that $\mathcal{B}_r(y_0) \subset \mathcal{B}_{2r}(y_0) \subset \Lambda$, we can see that $\mathcal{B}_{\frac{r}{\epsilon_n}}(\frac{y_0}{\epsilon_n}) \subset \Lambda_{\epsilon_n}$. In particular, for any $y \in \mathcal{B}_{\frac{r}{\epsilon_n}}(\tilde{y}_n)$ it holds

$$\left|y - \frac{y_0}{\epsilon_n}\right| \le |y - \tilde{y}_n| + \left|\tilde{y}_n - \frac{y_0}{\epsilon_n}\right| < \frac{1}{\epsilon_n}(r + o(1)) < \frac{2r}{\epsilon_n} \text{ for } n \text{ sufficiently large.}$$

Therefore, for any n big enough we have $\Lambda_{\epsilon_n}^c \subset \mathcal{B}^c_{\frac{r}{\epsilon_n}}(\tilde{y}_n)$. On the other hand, using (6.28), we know that

$$\tilde{u}_n(x) \to 0 \text{ as } |x| \to \infty$$

uniformly in $n \in \mathbb{N}$. Hence, there exists $R > 0$ such that

$$\tilde{u}_n(x) < a_0 \text{ for all } |x| \geq R, n \in \mathbb{N}.$$

Consequently, $u_{\epsilon_n}(x) < a_0$ for any $x \in \mathcal{B}^c_R(\tilde{y}_n)$ and $n \in \mathbb{N}$.

Since there exists $\nu \in \mathbb{N}$ such that for any $n \geq \nu$, we have

$$\Lambda^c_{\epsilon_n} \subset \mathcal{B}^c_{\frac{r}{\epsilon_n}}(\tilde{y}_n) \subset \mathcal{B}^c_R(\tilde{y}_n),$$

we deduce that $u_{\epsilon_n}(x) < a_0$ for any $x \in \Lambda^c_{\epsilon_n}$ and $n \geq \nu$, which is in contrast with relation (6.61).

Let $\bar{\epsilon}_\delta > 0$ given by Theorem 6.6 and we fix $\epsilon \in (0, \epsilon_\delta)$, where $\epsilon_\delta := \min\{\tilde{\epsilon}_\delta, \bar{\epsilon}_\delta\}$. In view of Theorem 6.6, we know that the problem (6.3) admits at least $cat_{M_\delta}(M)$ nontrivial solutions. Let us denote by u_ϵ one of these solutions. Since $u_\epsilon \in \widetilde{\mathcal{N}}_\epsilon$ satisfies (6.60), by the definition of g it follows that u_ϵ is a solution of $(\hat{P}_\epsilon)$. Then $\hat{u}(x) := u(x/\epsilon)$ is a solution to (P_ϵ), and we can conclude that (P_ϵ) has at least $cat_{M_\delta}(M)$ solutions.

Finally, we study the behavior of the maximum points of solutions to problem $(\hat{P}_\epsilon)$. Take $\epsilon_n \to 0^+$ and consider a sequence $\{u_n\}_{n\in\mathbb{N}} \subset \mathcal{H}_{\epsilon_n}$ of solutions to $(\hat{P}_\epsilon)$. Let us observe that (g_1) implies that we can find $\gamma > 0$ such that

$$g(\epsilon x, t)t \leq \frac{V_0}{K} t^p \text{ for any } x \in \mathbb{R}^3, t \leq \gamma. \tag{6.62}$$

Arguing as before, we can find $R > 0$ such that

$$|u_n|_{L^\infty(B^c_R(\tilde{y}_n))} < \gamma. \tag{6.63}$$

Moreover, up to extract a subsequence, we may assume that

$$|u_n|_{L^\infty(B_R(\tilde{y}_n))} \geq \gamma. \tag{6.64}$$

Indeed, if (6.64) does not hold, in view of (6.63) we can see that $|u_n|_\infty < \gamma$. Then, thanks to $\langle \mathcal{I}'_{\epsilon_n}(u_n), u_n \rangle = 0$ and (6.62), we get

$$\|u_n\|^p_{\epsilon_n} \leq \int_{\mathbb{R}^3} g(\epsilon_n x, u_n) u_n \, dx \leq \frac{V_0}{K} \int_{\mathbb{R}^3} |u_n|^p \, dx$$

which yields $\|u_n\|_{\epsilon_n} = 0$, and this is an absurd. Accordingly, relation (6.64) holds true.

Taking into account (6.63) and (6.64) we can deduce that the maximum points $p_n \in \mathbb{R}^3$ of u_n belong to $\mathcal{B}_R(\tilde{y}_n)$. Therefore, $p_n = \tilde{y}_n + q_n$ for some $q_n \in \mathcal{B}_R$. Hence, $\eta_n := \epsilon_n \tilde{y}_n + \epsilon_n q_n$ is the maximum point of $\hat{u}_n(x) = u_n(x/\epsilon_n)$. Since $|q_n| < R$ for any $n \in \mathbb{N}$ and $\epsilon_n \tilde{y}_n \to y_0 \in M$ (in view of Lemma 6.12), by the continuity of V, we can infer that

$$\lim_{n\to\infty} V(\eta_{\epsilon_n}) = V(y_0) = V_0,$$

which ends the proof of theorem. □

6.4 Critical and supercritical fractional Kirchhoff problems

This section is devoted to the existence of positive solutions to

$$\begin{cases} \left(\epsilon^{sp} a + \epsilon^{2sp-3} b \, [u]^p_{s,p}\right) (-\Delta)^s_p u + V(x) u^{p-1} = u^{q-1} + \gamma u^{r-1} & \text{in } \mathbb{R}^3, \\ u \in W^{s,p}(\mathbb{R}^3), \quad u > 0 & \text{in } \mathbb{R}^3. \end{cases}$$

After rescaling, we study the following Kirchhoff problem

$$\begin{cases} \left(a + b\,[u]_{s,p}^{p}\right)(-\Delta)_p^s u + V(\epsilon x)u^{p-1} = u^{q-1} + \gamma u^{r-1} & \text{in } \mathbb{R}^3, \\ u \in W^{s,p}(\mathbb{R}^3), \quad u > 0 & \text{in } \mathbb{R}^3, \end{cases} \tag{6.65}$$

where $\gamma > 0$ and the powers q and r are such that $2p < q < p_s^* \leq r$.

In what follows, we truncate the nonlinearity $\phi(u) := u^{q-1} + \gamma u^{r-1}$ in a suitable way. Let $K > 0$ be a real number, whose value will be fixed later, and we set

$$\phi_\gamma(t) := \begin{cases} 0 & \text{if } t < 0, \\ t^{q-1} + \gamma t^{r-1} & \text{if } 0 \leq t < K, \\ (1 + \gamma K^{r-q})t^{q-1} & \text{if } t \geq K. \end{cases}$$

Then ϕ_γ satisfies the following properties:

(ϕ_1) $\displaystyle\lim_{t\to 0} \frac{\phi_\gamma(t)}{t^{2p-1}} = 0$;

(ϕ_2) $\displaystyle\lim_{t\to \infty} \frac{\phi_\gamma(t)}{t^{\nu-1}} = 0$ for some $\nu \in (q, p_s^*)$;

(ϕ_3) $0 < q\Phi_\gamma(t) \leq t\phi_\gamma(t)$ for all $t > 0$, where $\Phi_\gamma(t) = \int_0^t \phi_\gamma(\tau)\,d\tau$;

(ϕ_4) $t \mapsto \dfrac{\phi_\gamma(t)}{t^{2p-1}}$ is increasing in $(0, \infty)$.

Moreover

$$\phi_\gamma(t) \leq (1 + \gamma K^{r-q})t^{q-1} \text{ for all } t \geq 0. \tag{6.66}$$

Therefore, we consider the following truncated problem

$$\begin{cases} \left(a + b\,[u]_{s,p}^{p}\right)(-\Delta)_p^s u + V(\epsilon x)u^{p-1} = \phi_\gamma(u) & \text{in } \mathbb{R}^3, \\ u \in W^{s,p}(\mathbb{R}^3), \quad u > 0 & \text{in } \mathbb{R}^3. \end{cases} \tag{6.67}$$

We remark that weak solutions of (6.67) are critical points of the energy functional $\mathcal{I}_{\epsilon,\gamma} : \mathcal{H}_\epsilon \to \mathbb{R}$ defined by

$$\mathcal{I}_{\epsilon,\gamma}(u) = \frac{1}{p}\|u\|_\epsilon^p + \frac{b}{2p}[u]_{s,p}^{2p} - \int_{\mathbb{R}^3} \Phi_\gamma(u)\,dx.$$

We also consider the autonomous functional

$$\mathcal{E}_{V_0,\gamma}(u) = \frac{1}{p}\|u\|_{V_0}^p + \frac{b}{2p}[u]_{s,p}^{2p} - \int_{\mathbb{R}^3} \Phi_\gamma(u)\,dx.$$

Using Theorem 6.1, we know that for any $\gamma \geq 0$ there exists $\bar{\epsilon}(\gamma) > 0$ such that, for any $\epsilon \in (0, \bar{\epsilon}(\gamma))$, problem (6.67) admits a positive solution $u_{\epsilon,\gamma}$.

Now, we prove that it is possible to estimate the $\mathcal{H}_\epsilon$-norm of these solutions uniformly with respect to γ.

Lemma 6.17 *There exists $\bar{C} > 0$ such that $\|u_{\epsilon,\gamma}\|_\epsilon \leq \bar{C}$ for any $\epsilon > 0$ sufficiently small and uniformly in γ.*

Proof. A simple inspection of the proof of Theorem 6.1 shows that any solution $u_{\epsilon,\gamma}$ of problem (6.67) satisfies the following inequality

$$\mathcal{I}_{\epsilon,\gamma}(u_{\epsilon,\gamma}) \leq d_{V_0,\gamma} + h_\gamma(\epsilon),$$

where $d_{V_0,\gamma}$ is the mountain pass level related to the functional $\mathcal{E}_{V_0,\gamma}$, and $h_\gamma(\epsilon) \to 0$ as $\epsilon \to 0$. Then, decreasing $\bar{\epsilon}(\gamma)$ if necessary, we may suppose that

$$\mathcal{I}_{\epsilon,\gamma}(u_{\epsilon,\gamma}) \leq d_{V_0,\gamma} + 1 \quad \text{for any } \epsilon \in (0, \bar{\epsilon}(\gamma)). \tag{6.68}$$

Using the fact that $d_{V_0,\gamma} \leq d_{V_0,0}$ for any $\gamma \geq 0$, we deduce that

$$\mathcal{I}_{\epsilon,\gamma}(u_{\epsilon,\gamma}) \leq d_{V_0,0} + 1 \quad \text{for any } \epsilon \in (0, \bar{\epsilon}(\gamma)). \tag{6.69}$$

From (ϕ_3) and $q > 2p$, we infer that

$$\begin{aligned}
\mathcal{I}_{\epsilon,\gamma}(u_{\epsilon,\gamma}) &= \mathcal{I}_{\epsilon,\gamma}(u_{\epsilon,\gamma}) - \frac{1}{q}\langle \mathcal{I}'_{\epsilon,\gamma}(u_{\epsilon,\gamma}), u_{\epsilon,\gamma}\rangle \\
&= \left(\frac{1}{p} - \frac{1}{q}\right) \|u_{\epsilon,\gamma}\|_\epsilon^p + \left(\frac{1}{2p} - \frac{1}{q}\right) [u_{\epsilon,\gamma}]_{s,p}^{2p} + \int_{\mathbb{R}^3} \frac{1}{q}\phi_\gamma(u_{\epsilon,\gamma})u_{\epsilon,\gamma} - \Phi_\gamma(u_{\epsilon,\gamma})\, dx \\
&\geq \left(\frac{1}{p} - \frac{1}{q}\right) \|u_{\epsilon,\gamma}\|_\epsilon^p.
\end{aligned} \tag{6.70}$$

Putting together relations (6.69) and (6.70), we have

$$\|u_{\epsilon,\gamma}\|_\epsilon \leq \left[\left(\frac{pq}{q-p}\right)(d_{V_0,0} + 1)\right]^{\frac{1}{p}} \quad \text{for any } \epsilon \in (0, \bar{\epsilon}(\gamma)).$$

The proof is now complete. □

Next, we claim that $u_{\epsilon,\gamma}$ is a solution of the original problem (6.65). To do this, we will show that we can find $K_0 > 0$ such that for any $K \geq K_0$, there exists $\gamma_0 = \gamma_0(K) > 0$ such that

$$|u_{\epsilon,\gamma}|_\infty \leq K \quad \text{for all } \gamma \in [0, \gamma_0]. \tag{6.71}$$

In order to achieve our purpose, we use the Moser iteration technique [120]. For simplicity, we set $u := u_{\epsilon,\gamma}$. For any $L > 0$, we define $u_L := \min\{u, L\} \geq 0$, where $\sigma > 1$ will be chosen later, and let $w_L = u u_L^{\sigma-1}$. Taking $u_L^{p(\sigma-1)}u$ in (6.67), we can see that

$$\begin{aligned}
&a\int_{\mathbb{R}^3}\int_{\mathbb{R}^3} \frac{|u(x)-u(y)|^{p-2}(u(x)-u(y))(u(x)u_L^{p(\sigma-1)}(x) - u(y)u_L^{p(\sigma-1)}(y))}{|x-y|^{3+sp}}\, dxdy \\
&+ b[u]_{s,p}^p \int_{\mathbb{R}^3}\int_{\mathbb{R}^3} \frac{|u(x)-u(y)|^{p-2}(u(x)-u(y))(u(x)u_L^{p(\sigma-1)}(x) - u(y)u_L^{p(\sigma-1)}(y))}{|x-y|^{3+sp}}\, dxdy \\
&= \int_{\mathbb{R}^3} \phi_\gamma(u)u_L^{p(\sigma-1)}u\, dx - \int_{\mathbb{R}^3} V(\epsilon x)|u|^p u_L^{p(\sigma-1)}\, dx.
\end{aligned} \tag{6.72}$$

Using (6.29) and (6.30) with $\tilde{u}_n$ and $\tilde{u}_{L,n}$ replaced by u and u_L respectively, we can note that

$$\begin{aligned}
&a\frac{C_*^{-1}}{\sigma^p}|w_L|_{p_s^*}^p \leq a[\mathcal{L}(u)]_{s,p}^p \\
&\leq a\int_{\mathbb{R}^3}\int_{\mathbb{R}^3} \frac{|u(x)-u(y)|^{p-2}(u(x)-u(y))(u(x)u_L^{p(\sigma-1)}(x) - u(y)u_L^{p(\sigma-1)}(y))}{|x-y|^{3+sp}}\, dxdy \\
&+ b[u]_{s,p}^p \int_{\mathbb{R}^3}\int_{\mathbb{R}^3} \frac{|u(x)-u(y)|^{p-2}(u(x)-u(y))(u(x)u_L^{p(\sigma-1)}(x) - u(y)u_L^{p(\sigma-1)}(y))}{|x-y|^{3+sp}}\, dxdy.
\end{aligned} \tag{6.73}$$

On the other hand, by (6.66) and (V_1), we obtain

$$\int_{\mathbb{R}^3} \phi_\gamma(u) u_L^{p(\sigma-1)} u\, dx - \int_{\mathbb{R}^3} V(\epsilon x)|u|^p u_L^{p(\sigma-1)}\, dx \le (1+\gamma K^{r-q}) \int_{\mathbb{R}^3} u^q u_L^{p(\sigma-1)}\, dx. \quad (6.74)$$

Putting together (6.72), (6.73) and (6.74) we get

$$|w_L|_{p_s^*}^p \le \frac{\sigma^p}{a} C_*(1+\gamma K^{r-q}) \int_{\mathbb{R}^3} u^q u_L^{p(\sigma-1)}\, dx. \quad (6.75)$$

Now, by Hölder's inequality, we have

$$\int_{\mathbb{R}^3} u^q u_L^{p(\sigma-1)}\, dx \le |u|_{p_s^*}^{q-p} |w_L|^p_{\frac{pp_s^*}{p_s^*-(q-p)}}$$

which together with (6.75) yields

$$|w_L|_{p_s^*}^p \le \frac{\sigma^p}{a} C_*(1+\gamma K^{r-q}) |u|_{p_s^*}^{q-p} |w_L|^p_{\frac{pp_s^*}{p_s^*-(q-p)}}. \quad (6.76)$$

Set

$$C_{\gamma,K} := \frac{C_*}{a}(1+\gamma K^{r-q}) \quad \text{and} \quad \alpha^* := \alpha^*(s,p,q) = \frac{pp_s^*}{p_s^*-(q-p)}$$

so that (6.76) becomes

$$|w_L|_{p_s^*}^p \le \sigma^p C_{\gamma,K} |u|_{p_s^*}^{q-p} |w_L|_{\alpha^*}^p. \quad (6.77)$$

On the other hand, by Theorem 6.4 and Lemma 6.17 we know that

$$|u|_{p_s^*}^p \le C_* \|u\|_\epsilon^p \le C_* \bar{C}^p,$$

which together with (6.77) gives

$$|w_L|_{p_s^*}^p \le \sigma^p C_{\gamma,K} M_1 |w_L|_{\alpha^*}^p, \quad (6.78)$$

where $M_1 := (C_* \bar{C}^p)^{\frac{q-p}{p}}$.

Now, we observe that if $u^\sigma \in L^{\alpha^*}(\mathbb{R}^3)$, by the definition of w_L, $u_L \le u$, and (6.78), it follows that

$$|w_L|_{p_s^*}^p \le \sigma^p C_{\gamma,K} M_1 |u|_{\sigma\alpha^*}^{p\sigma} < \infty. \quad (6.79)$$

Passing to the limit as $L \to +\infty$ in (6.79) and using Fatou's lemma we have

$$|u|_{p_s^*\sigma} \le (C_{\gamma,K} M_1)^{\frac{1}{p\sigma}} \sigma^{\frac{1}{\sigma}} |u|_{\sigma\alpha^*} \quad (6.80)$$

whenever $u^{\sigma\alpha^*} \in L^1(\mathbb{R}^3)$.

Now, we set $\sigma := \frac{p_s^*}{\alpha^*} > 1$, and we observe that, being $u \in L^{p_s^*}(\mathbb{R}^3)$, the above inequality holds for this choice of σ. Then, using the fact that $\sigma^2\alpha^* = p_s^*\sigma$, it follows that (6.80) holds with σ replaced by σ^2. Therefore, we can see that

$$|u|_{p_s^*\sigma^2} \le (C_{\gamma,K} M_1)^{\frac{1}{p\sigma^2}} \sigma^{\frac{2}{\sigma^2}} |u|_{\sigma^2\alpha^*} \le (C_{\gamma,K} M_1)^{\frac{1}{p}(\frac{1}{\sigma}+\frac{1}{\sigma^2})} \sigma^{\frac{1}{\sigma}+\frac{2}{\sigma^2}} |u|_{\sigma\alpha^*}.$$

Iterating this process and recalling that $\sigma\alpha^* := p_s^*$, we can infer that for every $m \in \mathbb{N}$

$$|u|_{p_s^*\sigma^m} \le (C_{\gamma,K} M_1)^{\sum_{j=1}^m \frac{1}{p\sigma^j}} \sigma^{\sum_{j=1}^m j\sigma^{-j}} |u|_{p_s^*}. \quad (6.81)$$

Taking the limit in (6.81) as $m \to +\infty$ and using Lemma 6.17, we get

$$|u|_\infty \le (C_{\gamma,K} M_1)^{\delta_1} \sigma^{\delta_2} M_2, \tag{6.82}$$

where $M_2 := C_*^{\frac{1}{p}} \bar{C}$ and

$$\delta_1 := \frac{1}{p} \sum_{j=1}^{\infty} \frac{1}{\sigma^j} < \infty \quad \text{and} \quad \delta_2 := \sum_{j=1}^{\infty} \frac{j}{\sigma^j} < \infty.$$

Next, we find some suitable values of K and γ such that the following inequality holds

$$(C_{\gamma,K} M_1)^{\delta_1} \sigma^{\delta_2} M_2 \le K,$$

or equivalently

$$1 + \gamma K^{r-q} \le (K M_1^{-1})^{\frac{1}{\delta_1}} \left(\frac{C_* M_1}{a} \right)^{-1} \sigma^{-\frac{\delta_2}{\delta_1}}.$$

Take $K > 0$ such that

$$(K M_1^{-1})^{\frac{1}{\delta_1}} \left(\frac{C_* M_1}{a} \right)^{-1} \sigma^{-\frac{\delta_2}{\delta_1}} - 1 > 0,$$

and fix $\gamma_0 > 0$ such that

$$\gamma \le \gamma_0 \le \left[(K M_1^{-1})^{\frac{1}{\delta_1}} \left(\frac{C_* M_1}{a} \right)^{-1} \sigma^{-\frac{\delta_2}{\delta_1}} - 1 \right] \frac{1}{K^{r-q}}.$$

Therefore, thanks to (6.82), we can infer that

$$|u|_\infty \le K \text{ for all } \gamma \in [0, \gamma_0],$$

that is $u = u_{\epsilon,\gamma}$ is a solution of (6.65). This ends the proof of Theorem 6.3. □

Finally, we point out that, by assuming $q > 2p$ (since we aim to use Theorem 6.1), the combined effect of concave-convex type growth $1 < q < 2p$, $r \ge p_s^*$ has been excluded. Anyway, when $\epsilon = 1$, $s \in (0,1)$ and $p \in (1,\infty)$ are such that $sp < 3$, the potential V is constant, and $p < q < p_s^* \le r$, we suspect that it is possible to obtain an existence result to (6.65) with $b > 0$ sufficiently small. Indeed, one can truncate the nonlinearity $\phi(u)$ as before, and taking into account Remark 6.2, we can deduce an existence result for problem (6.67), provided that $b > 0$ is small enough. Combining this fact with a Moser iteration argument, the desired existence result for problem (6.65) follows.

6.5 Historical comments

When $a = \epsilon = 1$, $b = 0$ and $p = 2$, equation (P_ϵ) becomes a fractional Schrödinger equation of the type

$$(-\Delta)^s u + V(x)u = f(x,u) \quad \text{in } \mathbb{R}^3 \tag{6.83}$$

introduced by Laskin [90] in the study of fractional quantum mechanics. Equation (6.83) has been widely considered in these last years, and several existence and multiplicity results to (6.83) have been established by applying suitable techniques and assuming different conditions on the potential V and on the nonlinearity f. On the other hand, when $s = \epsilon = 1$ and $p = 2$, problem (P_ϵ) reduces to a classical Kirchhoff equation of the type

$$-\left(a + b\int_{\mathbb{R}^3} |\nabla u|^2\, dx\right)\Delta u + V(x)u = f(x,u) \quad \text{in } \mathbb{R}^3, \tag{6.84}$$

which is related to the stationary analogue of the Kirchhoff equation

$$\rho u_{tt} - \left(\frac{p_0}{h} + \frac{E}{2L}\int_0^L |u_x|^2 dx\right)u_{xx} = 0 \tag{6.85}$$

proposed by Kirchhoff [88] as an extension of the classical D'Alembert's wave equation for describing the transversal oscillations of a stretched string. The parameters appearing in problem (6.85) have the following meaning: L is the length of the string, h is the area of the cross-section, E is the young modulus (elastic modulus) of the material, ρ is the mass density, and p_0 is the initial tension. We notice that nonlocal boundary value problems like (6.84) model several physical and biological systems where u describes a process which depends on the average of itself, as for example, the population density.

Concerning perturbed Kirchhoff problems, He and Zou [70] proved a multiplicity result for the following Kirchhoff equation

$$-\left(a\varepsilon^2 + b\varepsilon\int_{\mathbb{R}^3} |\nabla u|^2 dx\right)\Delta u + V(x)u = g(u) \quad \text{in } \mathbb{R}^3, \tag{6.86}$$

provided that $\epsilon > 0$ is sufficiently small, under the following condition on V introduced by Rabinowitz [133]:

$$V_\infty = \liminf_{|x|\to\infty} V(x) > \inf_{x\in\mathbb{R}^3} V(x), \text{ where } V_\infty \leq \infty,$$

and g is a subcritical nonlinearity. We refer to Figueiredo and Santos Junior [54] applied the generalized Nehari manifold method and Ljusternik-Schnirelmann theory to deduce a multiplicity result for a class of Kirchhoff equations requiring that the potential V fulfills (V_1)–(V_2). Later, He, Li and Peng [72] obtained the existence and multiplicity of solutions to (6.86), when (V_1)–(V_2) are in force, and $g(u) = f(u) + u^5$, where $f \in C^1$ is a subcritical nonlinearity which does not satisfy the Ambrosetti-Rabinowitz condition [1].

In the nonlocal framework, Fiscella and Valdinoci [57] proposed a stationary fractional Kirchhoff variational model in a bounded domain $\Omega \subset \mathbb{R}^N$ with homogeneous Dirichlet boundary conditions and involving a critical nonlinearity

$$\begin{cases} M\left(\int_{\mathbb{R}^N} |(-\Delta)^{\frac{s}{2}}u|^2 dx\right)(-\Delta)^s u = \lambda f(x,u) + |u|^{\Phi-2}u & \text{in } \Omega, \\ u = 0 & \text{in } \mathbb{R}^N \setminus \Omega, \end{cases} \tag{6.87}$$

where $M : \mathbb{R}^+ \to \mathbb{R}^+$ is an increasing continuous positive Kirchhoff function whose typical example is given by $M(t) = a + bt$, with $a > 0$ and $b \geq 0$, f is a superlinear function with subcritical growth at infinity, and $\lambda > 0$ is a parameter. This model takes care of the nonlocal aspect of the tension arising from nonlocal measurements of the fractional length of the string.

On the other hand, a great attention has been devoted to the study of fractional Kirchhoff problems involving the fractional p-Laplace operator $(-\Delta)_p^s$. For instance, Pucci, Xiang and Zhang [131] obtained a multiplicity result for a nonhomogeneous fractional Kirchhoff-Schrödinger equation assuming that the potential V satisfies a Bartsch-Wang type condition. Fiscella and Pucci [56] dealt with stationary fractional Kirchhoff p-Laplacian equations involving a Hardy potential and different critical nonlinearities.

6.6 Final remarks, perspectives, and open problems

This chapter deals with the study of concentration properties of solutions for the fractional p-Kirchhoff equation in the three-dimensional case. *Why the concentration analysis in this case is important and innovative?* There are at least three reasons for this study, as follows.

(i) *Problem complexity*: the analysis of this class of Kirchhoff equations tackles a triply nonlocal problem (fractional + Kirchhoff + potential well) on the whole space.

(ii) *Methodological innovation*: the study provides a blueprint for asymptotic analysis via the limit problem strategy and uniform estimates.

(iii) *Theoretical impact*: the study highlights new theoretical ideas based on potential-induced concentration, distinct from critical exponent bubbling.

The content of this chapter is based on the results established in [6]. This study was extended by Liu, Rădulescu and Yuan [108] in the context of fractional Kirchhoff equations with discontinuous reaction and by Sun, Fu and Liang [143] in the two-dimensional case (fractional p-Kirchhoff problems with critical exponential growth).

The main perspectives generated by the analysis developed in this chapter are the following:

(i) *Physical perspective:* The Kirchhoff term introduces a nonlocal dependence of the elasticity on the total strain energy $[u]_{s,p}^p$, making the equation relevant to problems arising in may applied fields, such as fractional elasticity theory, nonlocal continuum mechanics, phase transition in heterogeneous media with long-range interactions, or pattern formation where the medium's properties depend on the overall deformation.

(ii) *Analytical perspective:* This study shows how competing effects (Kirchhoff nonlocality, fractional diffusion, critical growth) interact. By summarizing, their effects are the following:

– the Kirchhoff term can prevent concentration phenomena in certain regimes;

– the existence of positive solutions depends on a threshold in the parameter γ associated with the critical/supercritical reaction term;

– compactness recovery is possible below certain energy levels linked to the best fractional Sobolev constant.

Some open problems inspired by the model studied in this chapter are the following.

Open Problem 1. (i) Study what happens as $b \to 0$, which corresponds to the case of degenerate fractional p-Kirchhoff equations.

(ii) Study the case of singular Kirchhoff nonlocal terms of the form $M(t) = a + bt^{-\alpha}$, where $\alpha > 0$).

The next research direction is concerned with the analysis of sign-changing solutions.

Open Problem 2. This chapter deals with positive solutions. What about the existence of nodal solutions and their concentration properties? Does the nonlocal Kirchhoff term affect the number of sign-changing solutions?

Consider the following fractional p-Kirchhoff equation with Hardy potential:

$$\begin{cases} \left(\epsilon^{sp}a + \epsilon^{2sp-3}b\,|x|^{-sp}\,[u]_{s,p}^p\right)(-\Delta)_p^s u + V(x)u^{p-1} = f(u) & \text{in } \mathbb{R}^3, \\ u \in W^{s,p}(\mathbb{R}^3), \quad u > 0 & \text{in } \mathbb{R}^3. \end{cases} \tag{6.88}$$

Open Problem 3. Study the concentration properties of solutions to problem (6.88) in relationship with the interplay between Kirchhoff term, Hardy singularity, and critical nonlinearity.

Another interesting direction is about related problems in relationship with the fractional order s and the fractional p-Laplace operator.

Open Problem 4. Study the asymptotic analysis as $s \to 0^+$ and $s \to 1^-$. In the second case, does the solution converge to a solution of the classical p-Kirchhoff equation?

The anisotropic case associated with variable exponents complicates the functional setting.

Open Problem 5. Study problem (P_ϵ) in the case of the fractional $p(x)$-Laplace operator (if the constant p is replaced by a variable function $p(x)$).

Finally, we propose the following new research direction concerning problem (P_ϵ) in the framework of fractional double-phase operators.

Open Problem 6. Replace $(-\Delta)_p^s$ with the nonhomogeneous fractional operator $(-\Delta)_p^s + (-\Delta)_q^s$, where $1 < p < q$.

6.7 Glossary

Fractional p-Kirchhoff Equation: This is a prototypical equation for studying doubly nonlocal nonlinear phenomena. It stands at the intersection of nonlocal calculus of variations, nonlinear elliptic PDEs, and mathematical modeling of complex materials.

Morse Theory: This is a refined method at the interplay between the topology of a manifold and the critical points of a function defined on it. In PDEs, it is used to study the multiplicity, stability, and bifurcation of solutions to variational problems.

Ljusternik–Schnirelmann Theory: It is a powerful method to generate critical points, which complements the Morse theory by providing existence and multiplicity without needing nondegeneracy or detailed local analysis.

Moser Iteration: It is a fundamental tool to prove boundedness and L^∞ estimates for solutions of elliptic equations.

7

Magnetic Kirchhoff equations with critical growth

Why magnetic Kirchhoff equations? Because moving from Kirchhoff to magnetic Kirchhoff equations shifts the focus from classical elasticity to quantum mechanics, introduces the deep physical principle of gauge invariance, and creates a much more challenging and interesting set of mathematical problems centered on new types of functional spaces and concentration phenomena. Introducing the magnetic field in the study of Kirchhoff equations creates a host of new analytical difficulties, such as:

(i) *Loss of compactness*: the standard embedding property is no longer compact in the magnetic case and new techniques like the magnetic concentration-compactness principle are needed to overcome this.

(ii) *Concentration properties*: the magnetic field can influence the precise location and shape of the concentration of solutions.

(iii) *Interaction of nonlocalities*: the problem features a fascinating interplay between the Kirchhoff nonlocality and the magnetic nonlocality.

In this chapter, we study the following nonlinear magnetic Kirchhoff equation with critical growth

$$\begin{cases} -\left(a\epsilon^2 + b\epsilon\,[u]^2_{A/\epsilon}\right)\Delta_{A/\epsilon}u + V(x)u = f(|u|^2)u + |u|^4u \quad \text{in } \mathbb{R}^3, \\ u \in H^1(\mathbb{R}^3, \mathbb{C}), \end{cases}$$

where $\epsilon > 0$ is a parameter, $a, b > 0$ are constants, $V : \mathbb{R}^3 \to \mathbb{R}$ and $A : \mathbb{R}^3 \to \mathbb{R}^3$ are continuous potentials, and $f : \mathbb{R} \to \mathbb{R}$ is a nonlinear term with subcritical growth. Under a local assumption on the potential V, combining variational methods, penalization techniques and the Ljusternik-Schnirelmann theory, we establish multiplicity and concentration properties of solutions to the above problem for ε small. A feature of the analysis developed in this chapter is that the function f is assumed to be only continuous, which allows to consider larger classes of nonlinearities in the reaction.

The study developed in what follows allows to handle the simultaneous effects of:

(i) *nonlocal Kirchhoff coefficient* affecting the principal part;

(ii) *magnetic field* breaking Euclidean symmetry;

(iii) *critical exponent* causing lack of compactness;

(iv) *competing nonlinearities*: the interaction between the subcritical term and the critical nonlinearity creates a delicate balance in the energy estimates;

(v) *magnetic concentration-compactness*: adapting Lions' concentration-compactness principle to the magnetic setting is a key tools, as the magnetic field affects how mass can concentrate or vanish at infinity.

DOI: 10.1201/9781003797159-7

A hierarchy of the basic variational tools used in this chapter is now described:

(a) *mountain pass theorem*, as the primary tool to prove the existence of nontrivial solutions;

(b) *Palais-Smale condition*, as the central analytical challenge (the critical growth directly threatens the compactness).

All adapted to the intricate nonlocal structure of the Kirchhoff problem.

7.1 Statement of the problem

We are concerned with multiplicity properties and concentration phenomena for the following nonlinear magnetic Kirchhoff equation with critical growth

$$\begin{cases} -\left(a\epsilon^2 + b\epsilon\,[u]_{A/\epsilon}^2\right)\Delta_{A/\epsilon}u + V(x)u = f(|u|^2)u + |u|^4u \quad \text{in } \mathbb{R}^3, \\ u \in H^1(\mathbb{R}^3, \mathbb{C}), \end{cases} \tag{7.1}$$

where $\epsilon > 0$ is a parameter, $a, b > 0$ are constants, $V : \mathbb{R}^3 \to \mathbb{R}$ is a continuous function, the magnetic potential $A : \mathbb{R}^3 \to \mathbb{R}^3$ is Hölder's continuous with exponent $\alpha \in (0, 1]$, and $-\Delta_A u$ is the magnetic Laplace operator with the following form

$$-\Delta_A u := \left(\frac{1}{i}\nabla - A(x)\right)^2 u = -\Delta u - \frac{2}{i}A(x)\cdot\nabla u + |A(x)|^2 u - \frac{1}{i}u\,\mathrm{div}(A(x)),$$

while the definition of $[u]_A^2$ will be given in the next section.

The main purpose in this chapter is to investigate the qualitative properties of solutions to problem (7.1) by combining a local assumption on V and adapting the penalization method and the Ljusternik-Schnirelmann category theory.

We make the following assumptions on the potential V:

(V_1) There exists $V_0 > 0$ such that $V(x) \geq V_0$ for all $x \in \mathbb{R}^3$;

(V_2) There exists a bounded open set $\Lambda \subset \mathbb{R}^3$ such that

$$V_0 = \min_{x\in\Lambda} V(x) < \min_{x\in\partial\Lambda} V(x).$$

Observe that

$$M := \{x \in \Lambda : V(x) = V_0\} \neq \emptyset.$$

Assume that $f \in C(\mathbb{R}, \mathbb{R})$ is a nonlinearity satisfying:

(f_1) $f(t) = 0$ if $t \leq 0$, and $\lim_{t\to 0^+} \frac{f(t)}{t} = 0$;

(f_2) There exist $\sigma, q \in (4, 6)$ and $\mu > 0$ such that

$$f(t) \geq \mu t^{\frac{\sigma-2}{2}} \quad \forall t > 0, \quad \text{and} \quad \lim_{t\to+\infty} \frac{f(t)}{t^{\frac{q-2}{2}}} = 0;$$

(f_3) There is a positive constant $4 < \theta < 6$ such that

$$0 < \frac{\theta}{2}F(t) \leq tf(t),\ \forall t > 0, \text{ where } F(t) = \int_0^t f(s)ds;$$

(f_4) The mapping $t \mapsto \frac{f(t)}{t}$ is strictly increasing in $(0, \infty)$.

The main result in this chapter establishes the following multiplicity and concentration property of solutions.

Theorem 7.1 *Assume that V satisfies (V_1), (V_2) and f satisfies (f_1)-(f_4). Then, for any $\delta > 0$ such that*

$$M_\delta := \{x \in \mathbb{R}^3 : \text{dist}\,(x, M) < \delta\} \subset \Lambda,$$

there exists $\varepsilon_\delta > 0$ such that, for any $0 < \varepsilon < \varepsilon_\delta$, problem (7.1) has at least $cat_{M_\delta}(M)$ nontrivial solutions. Moreover, for every sequence $\{\varepsilon_n\}$ such that $\varepsilon_n \to 0^+$ as $n \to +\infty$, if we denote by u_{ε_n} one of these solutions of problem (7.1) for $\varepsilon = \varepsilon_n$ and $\eta_{\varepsilon_n} \in \mathbb{R}^3$ is the global maximum point of $|u_{\varepsilon_n}|$, then

$$\lim_{\varepsilon_n \to 0^+} V(\eta_{\varepsilon_n}) = V_0.$$

Due to the presence of the magnetic field $A(x)$, problem (7.1) cannot be changed into a pure real-valued problem, hence we must deal directly with a complex-valued problem. This fact creates several new difficulties in employing the methods to deal with our problem. On the other hand, due to the presence of the nonlocal term, it is possible that the weak limit of a bounded Palais-Smale sequence of the Kirchhoff equation is not a solution of this problem. Moreover, since the problem we deal with has critical growth, we need more refined estimates to overcome the lack of compactness.

7.2 Abstract setting

For $u : \mathbb{R}^3 \to \mathbb{C}$, let us denote by

$$\nabla_A u := \Big(\frac{\nabla}{i} - A\Big)u.$$

Consider the function spaces

$$D_A^1(\mathbb{R}^3, \mathbb{C}) := \{u \in L^6(\mathbb{R}^3, \mathbb{C}) : |\nabla_A u| \in L^2(\mathbb{R}^3, \mathbb{R})\},$$

and

$$H_A^1(\mathbb{R}^3, \mathbb{C}) := \{u \in D_A^1(\mathbb{R}^3, \mathbb{C}) : u \in L^2(\mathbb{R}^3, \mathbb{C})\}.$$

Then $D_A^1(\mathbb{R}^3, \mathbb{C})$ and $H_A^1(\mathbb{R}^3, \mathbb{C})$ are Hilbert spaces endowed with the scalar products

$$\langle u, v\rangle_D := \text{Re} \int_{\mathbb{R}^3} \nabla_A u \overline{\nabla_A v} dx, \quad \text{for any } u, v \in D_A^1(\mathbb{R}^3, \mathbb{C}),$$

$$\langle u, v\rangle_H := \text{Re} \int_{\mathbb{R}^3} \Big(\nabla_A u \overline{\nabla_A v} + u\overline{v}\Big) dx, \quad \text{for any } u, v \in H_A^1(\mathbb{R}^3, \mathbb{C}),$$

where Re and the bar denote the real part of a complex number and the complex conjugation, respectively. Let $\|u\|_D$ and $\|u\|_A$ denote the norms induced by inner products $\langle u, v\rangle_D$ and $\langle u, v\rangle_A$, and $\langle u, u\rangle_D = [u]_A^2$.

On $H_A^1(\mathbb{R}^3, \mathbb{C})$, we will frequently use the following diamagnetic inequality of Lieb and Loss [98, Theorem 7.21]

$$|\nabla_A u(x)| \geq |\nabla |u(x)||. \tag{7.2}$$

Moreover, making a simple change of variables, since $\Delta_{A_\varepsilon} = \epsilon^2 \Delta_{A/\epsilon}$ and $[u]_{A_\varepsilon}^2 = \frac{1}{\epsilon}[u]_{A/\epsilon}^2$, we observe that problem (7.1) is equivalent to

$$-\left(a + b\,[u]_{A_\varepsilon}^2\right)\Delta_{A_\varepsilon} u + V_\varepsilon(x)u = f(|u|^2)u + |u|^4 u \quad \text{in } \mathbb{R}^3, \tag{7.3}$$

where $A_\varepsilon(x) = A(\varepsilon x)$ and $V_\varepsilon(x) = V(\varepsilon x)$.

Let D_ε and H_ε be the Hilbert spaces obtained as the closure of $C_c^\infty(\mathbb{R}^N, \mathbb{C})$ with respect to the scalar products

$$\langle u, v\rangle_{D_\epsilon} := \mathrm{Re} \int_{\mathbb{R}^3} a \nabla_{A_\varepsilon} u \overline{\nabla_{A_\varepsilon} v} dx$$

and

$$\langle u, v\rangle_\epsilon := \mathrm{Re} \int_{\mathbb{R}^3} \left(a \nabla_{A_\varepsilon} u \overline{\nabla_{A_\varepsilon} v} + V_\varepsilon(x) u \overline{v}\right) dx$$

respectively, where $a > 0$. We denote by $\|\cdot\|_{D_\varepsilon}$ and $\|\cdot\|_\varepsilon$ the norms induced by inner products $\langle \cdot, \cdot\rangle_{D_\epsilon}$ and $\langle \cdot, \cdot\rangle_\epsilon$.

The diamagnetic inequality (7.2) implies that, if $u \in H_{A_\varepsilon}^1(\mathbb{R}^3, \mathbb{C})$, then $|u| \in H^1(\mathbb{R}^3, \mathbb{R})$ and $\|u\| \leq C\|u\|_\varepsilon$. Therefore, the embedding $H_\varepsilon \hookrightarrow L^r(\mathbb{R}^3, \mathbb{C})$ is continuous for $2 \leq r \leq 6$ and the embedding $H_\varepsilon \hookrightarrow L_{\mathrm{loc}}^r(\mathbb{R}^3, \mathbb{C})$ is compact for $1 \leq r < 6$.

For compact supported functions in $H^1(\mathbb{R}^3, \mathbb{R})$, we have the following property.

Lemma 7.1 *If $u \in H^1(\mathbb{R}^3, \mathbb{R})$ and u has compact support, then $\omega := e^{iA(0)\cdot x} u \in H_\varepsilon$.*

Proof. Assume that $\mathrm{supp}(u) \subset B_R(0)$. Since V is continuous, we have

$$\int_{\mathbb{R}^3} V_\varepsilon(x)|\omega|^2 dx = \int_{B_R(0)} V_\varepsilon(x)|\omega|^2 dx \leq C\|u\|_2^2 < +\infty.$$

Moreover, since V and A are continuous, we obtain

$$\begin{aligned}
\int_{\mathbb{R}^3} |\nabla_{A_\varepsilon}\omega|^2 dx &= \int_{\mathbb{R}^3} |\nabla \omega|^2 dx + \int_{\mathbb{R}^3} |A_\varepsilon(x)|^2 |\omega|^2 dx + 2\mathrm{Re} \int_{\mathbb{R}^3} i A_\varepsilon(x) \overline{\omega} \nabla \omega dx \\
&\leq 2 \int_{\mathbb{R}^3} |\nabla \omega|^2 dx + 2 \int_{\mathbb{R}^3} |A_\varepsilon(x)|^2 |\omega|^2 dx \\
&\leq C \left[\int_{\mathbb{R}^3} |\nabla u|^2 dx + \int_{\mathbb{R}^3} |u|^2 dx\right] < +\infty
\end{aligned}$$

and we conclude the proof. □

7.3 The modified problem

To study problem (7.1), or equivalently, equation (7.3), we modify in a suitable way the nonlinearity f so that, for $\varepsilon > 0$ small enough, the solutions of such modified problem are also solutions of the original one. More precisely, choosing $K > 2$, there exists a unique number $a_0 > 0$ verifying $f(a_0) + a_0^2 = V_0/K$ by (f_4), where V_0 is given in (V_1). Consider the function

$$\tilde{f}(t) := \begin{cases} f(t) + (t^+)^2, & t \le a_0, \\ V_0/K, & t > a_0, \end{cases}$$

and introduce the penalized nonlinearity $g : \mathbb{R}^3 \times \mathbb{R} \to \mathbb{R}$ by setting

$$g(x,t) := \chi_\Lambda(x)(f(t) + (t^+)^2) + (1 - \chi_\Lambda(x))\tilde{f}(t), \tag{7.4}$$

where χ_Λ is the characteristic function on Λ and $G(x,t) := \displaystyle\int_0^t g(x,s)ds$.

In view of (f_1)–(f_4), we observe that g is a Carathéodory function satisfying the following properties:

(g_1) $g(x,t) = 0$ for each $t \le 0$;

(g_2) $\lim\limits_{t\to 0^+} \frac{g(x,t)}{t} = 0$ uniformly in $x \in \mathbb{R}^3$;

(g_3) $g(x,t) \le f(t) + t^2$ for all $t \ge 0$ and any $x \in \mathbb{R}^3$;

(g_4) $0 < \theta G(x,t) \le 2g(x,t)t$ for each $x \in \Lambda$ and $t > 0$;

(g_5) $0 < G(x,t) \le g(x,t)t \le V_0 t/K$, for each $x \in \Lambda^c$, $t > 0$;

(g_6) for each $x \in \Lambda$, the function $t \mapsto \frac{g(x,t)}{t}$ is strictly increasing in $t \in (0,+\infty)$ and for each $x \in \Lambda^c$, the function $t \mapsto \frac{g(x,t)}{t}$ is strictly increasing in $(0, a_0)$.

Next, we consider the modified problem

$$-\left(a + b\,[u]^2_{A_\varepsilon}\right)\Delta_{A_\varepsilon} u + V_\varepsilon(x)u = g(\varepsilon x, |u|^2)u \quad \text{in } \mathbb{R}^3. \tag{7.5}$$

Note that, if u is a nontrivial solution of problem (7.5) with

$$|u(x)|^2 \le a_0 \quad \text{for all } x \in \Lambda_\varepsilon^c, \quad \Lambda_\varepsilon := \{x \in \mathbb{R}^3 : \varepsilon x \in \Lambda\},$$

then u is a nontrivial solution of problem (7.3).

The energy functional associated to problem (7.5) is

$$J_\varepsilon(u) := \frac{a}{2}[u]^2_{A_\varepsilon} + \frac{1}{2}\int_{\mathbb{R}^3} V_\varepsilon(x)|u|^2 dx + \frac{b}{4}[u]^4_{A_\varepsilon} - \frac{1}{2}\int_{\mathbb{R}^3} G(\varepsilon x, |u|^2)dx$$

for all $u \in H_\varepsilon$. Then $J_\varepsilon \in C^1(H_\varepsilon, \mathbb{R})$ and its critical points are weak solutions of the modified problem (7.5).

We denote by $\mathcal{N}_\varepsilon$ the Nehari manifold of J_ε, that is,

$$\mathcal{N}_\varepsilon := \{u \in H_\varepsilon \backslash \{0\} : J'_\varepsilon(u)[u] = 0\},$$

and define the number c_ε by

$$c_\varepsilon = \inf_{u \in \mathcal{N}_\varepsilon} J_\varepsilon(u).$$

Let H_ε^+ be an open subset H_ε given by

$$H_\varepsilon^+ = \{u \in H_\varepsilon : |\text{supp}(u) \cap \Lambda_\varepsilon| > 0\},$$

and $S_\varepsilon^+ = S_\varepsilon \cap H_\varepsilon^+$, where S_ε is the unit sphere of H_ε. Note that S_ε^+ is a non-complete $C^{1,1}$-manifold of codimension 1, modeled on H_ε and contained in H_ε^+. Therefore, $H_\varepsilon = T_u S_\varepsilon^+ \bigoplus \mathbb{R}u$ for each $u \in T_u S_\varepsilon^+$, where $T_u S_\varepsilon^+ = \{v \in H_\varepsilon : \langle u, v\rangle_\epsilon = 0\}$.

Now we show that the functional J_ε satisfies the mountain pass geometry.

Lemma 7.2 *For any fixed $\varepsilon > 0$, the functional J_ε satisfies the following properties:*

(i) *there exist $\beta, r > 0$ such that $J_\varepsilon(u) \geq \beta$ if $\|u\|_\varepsilon = r$;*

(ii) *there exists $e \in H_\varepsilon$ with $\|e\|_\varepsilon > r$ such that $J_\varepsilon(e) < 0$.*

Proof. (i) By (g_2), (g_3) and (f_2), for any $\zeta > 0$ small, there exists $C_\zeta > 0$ such that

$$G(\varepsilon x, |u|^2) \leq \zeta |u|^4 + C_\zeta |u|^6 \quad \text{for all } x \in \mathbb{R}^3.$$

By the Sobolev embedding it follows that

$$\begin{aligned} J_\varepsilon(u) &\geq \frac{a}{2}[u]^2_{A_\varepsilon} + \frac{1}{2}\int_{\mathbb{R}^3} V_\varepsilon(x)|u|^2 dx + \frac{b}{4}[u]^4_{A_\varepsilon} - \frac{\zeta}{2}\int_{\mathbb{R}^3} |u|^4 dx - \frac{C_\zeta}{2}\int_{\mathbb{R}^3} |u|^6 dx \\ &\geq \frac{1}{2}\|u_n\|^2_\varepsilon - C_1\zeta\|u_n\|^4_\varepsilon - C_2 C_\zeta \|u_n\|^6_\varepsilon. \end{aligned}$$

Hence we can choose some $\beta, r > 0$ such that $J_\varepsilon(u) \geq \beta$ if $\|u\|_\varepsilon = r$ small.

(ii) For each $u \in H_\varepsilon \setminus \{0\}$ with $\text{supp}(u) \subset \Lambda_\varepsilon$. By the definition of g and (f_3), we have

$$\begin{aligned} J_\varepsilon(tu) &= \frac{t^2}{2}\|u\|^2_\varepsilon + \frac{bt^4}{4}[u]^4_{A_\varepsilon} - \frac{1}{2}\int_{\Lambda_\varepsilon} F(t^2|u|^2)dx - \frac{t^6}{6}\int_{\Lambda_\varepsilon} |u|^6 dx, \\ &\leq \frac{t^2}{2}\|u\|^2_\varepsilon + \frac{bt^4}{4}[u]^4_{A_\varepsilon} - \frac{t^6}{6}\int_{\Lambda_\varepsilon} |u|^6 dx, \end{aligned}$$

hence $J_\varepsilon(tu) \to -\infty$ as $t \to +\infty$ and the conclusion follows. □

Since f is only continuous, the next results are important because they allow to overcome the non-differentiability of $\mathcal{N}_\varepsilon$ and the incompleteness of S^+_ε.

Lemma 7.3 *Assume that (V_1)–(V_2) and (f_1)–(f_4) are satisfied, then the following properties hold:*

(*A*1) *For any $u \in H^+_\varepsilon$, let $g_u : \mathbb{R}^+ \to \mathbb{R}$ be given by $g_u(t) = J_\varepsilon(tu)$. Then there exists a unique $t_u > 0$ such that $g'_u(t) > 0$ in $(0, t_u)$ and $g'_u(t) < 0$ in (t_u, ∞).*

(*A*2) *There exists $\tau > 0$ independent on u such that $t_u \geq \tau$ for all $u \in S^+_\varepsilon$. Moreover, for each compact $\mathcal{W} \subset S^+_\varepsilon$ there is $C_\mathcal{W}$ such that $t_u \leq C_\mathcal{W}$, for all $u \in \mathcal{W}$.*

(*A*3) *The map $\widehat{m}_\varepsilon : H^+_\varepsilon \to \mathcal{N}_\varepsilon$ given by $\widehat{m}_\varepsilon(u) = t_u u$ is continuous and $m_\varepsilon = \widehat{m}_\varepsilon|_{S^+_\varepsilon}$ is a homeomorphism between S^+_ε and $\mathcal{N}_\varepsilon$. Moreover, $m^{-1}_\varepsilon(u) = \frac{u}{\|u\|_\varepsilon}$.*

(*A*4) *If there is a sequence $\{u_n\} \subset S^+_\varepsilon$ such that $\text{dist}\,(u_n, \partial S^+_\varepsilon) \to 0$, then $\|m_\varepsilon(u_n)\|_\varepsilon \to \infty$ and $J_\varepsilon(m_\varepsilon(u_n)) \to \infty$.*

Proof. (*A*1) Arguing as in the proof of Lemma 7.2, we have $g_u(0) = 0$, $g_u(t) > 0$ for $t > 0$ small and $g_u(t) < 0$ for $t > 0$ large. Therefore, $\max_{t \geq 0} g_u(t)$ is achieved at a global maximum point $t = t_u$ verifying $g'_u(t_u) = 0$ and $t_u u \in \mathcal{N}_\varepsilon$.

Now, we show that t_u is unique. Arguing by contradiction, suppose that there exist $t_1 > t_2 > 0$ such that $g'_u(t_1) = g'_u(t_2) = 0$. Then, for $i = 1, 2$,

$$t_i a[u]^2_{A_\varepsilon} + t_i \int_{\mathbb{R}^3} V_\varepsilon(x)|u|^2 dx + t_i^3 b[u]^4_{A_\varepsilon} = \int_{\mathbb{R}^3} g(\varepsilon x, t_i^2|u|^2) t_i |u|^2 dx.$$

Hence,

$$\frac{a[u]^2_{A_\varepsilon} + \int_{\mathbb{R}^3} V_\varepsilon(x)|u|^2 dx}{t_i^2} + b[u]^4_{A_\varepsilon} = \int_{\mathbb{R}^3} \frac{g(\varepsilon x, t_i^2|u|^2)|u|^2}{t_i^2} dx,$$

which implies that

$$\Big(\frac{1}{t_1^2}-\frac{1}{t_2^2}\Big)\Big(a[u]_{A_\varepsilon}^2+\int_{\mathbb{R}^3}V_\varepsilon(x)|u|^2dx\Big)=\int_{\mathbb{R}^3}\Big(\frac{g(\varepsilon x,t_1^2|u|^2)}{t_1^2|u|^2}-\frac{g(\varepsilon x,t_2^2|u|^2)}{t_2^2|u|^2}\Big)|u|^4dx$$
$$\geq\int_{\Lambda_\varepsilon^c\cap\{t_2^2|u|^2\leq a_0\leq t_1^2|u|^2\}}\Big(\frac{g(\varepsilon x,t_1^2|u|^2)}{t_1^2|u|^2}-\frac{g(\varepsilon x,t_2^2|u|^2)}{t_2^2|u|^2}\Big)|u|^4dx$$
$$+\int_{\Lambda_\varepsilon^c\cap\{a_0\leq t_2^2|u|^2\}}\Big(\frac{g(\varepsilon x,t_1^2|u|^2)}{t_1^2|u|^2}-\frac{g(\varepsilon x,t_2^2|u|^2)}{t_2^2|u|^2}\Big)|u|^4dx$$
$$=\int_{\Lambda_\varepsilon^c\cap\{t_2^2|u|^2\leq a_0\leq t_1^2|u|^2\}}\Big(\frac{V_0}{K}\frac{1}{t_1^2|u|^2}-\frac{f(t_2^2|u|^2)+t_2^4|u|^4}{t_2^2|u|^2}\Big)|u|^4dx$$
$$+\frac{1}{K}\left(\frac{1}{t_1^2}-\frac{1}{t_2^2}\right)\int_{\Lambda_\varepsilon^c\cap\{a_0\leq t_2^2|u|^2\}}V_0|u|^2dx.$$

Since $t_1>t_2>0$, we have

$$\Big(a[u]_{A_\varepsilon}^2+\int_{\mathbb{R}^3}V_\varepsilon(x)|u|^2dx\Big)$$
$$\leq\frac{t_1^2t_2^2}{t_2^2-t_1^2}\int_{\Lambda_\varepsilon^c\cap\{t_2^2|u|^2\leq a_0\leq t_1^2|u|^2\}}\Big(\frac{V_0}{K}\frac{1}{t_1^2|u|^2}-\frac{f(t_2^2|u|^2)+t_2^4|u|^4}{t_2^2|u|^2}\Big)|u|^4dx$$
$$+\frac{1}{K}\int_{\Lambda_\varepsilon^c\cap\{a_0\leq t_2^2|u|^2\}}V_0|u|^2dx\leq\frac{1}{K}\int_{\Lambda_\varepsilon^c}V_0|u|^2dx\leq\frac{1}{K}\|u\|_\varepsilon^2,$$

which is a contradiction. Therefore, $\max_{t\geq 0}g_u(t)$ is achieved at a unique $t=t_u$ so that $g_u'(t)=0$ and $t_uu\in\mathcal{N}_\varepsilon$.

(*A*2) For $\forall\, u\in S_\varepsilon^+$, by (*A*1), there exists a unique $t_u>0$ such that

$$t_u+t_u^3b[u]_{A_\varepsilon}^4=\int_{\mathbb{R}^3}g(\varepsilon x,t_u^2|u|^2)t_u|u|^2dx.$$

From (g_2), the Sobolev embeddings and $q>4$, we get

$$t_u\leq\zeta t_u^3\int_{\mathbb{R}^3}|u|^4dx+C_\zeta t_u^{q-1}\int_{\mathbb{R}^3}|u|^qdx+t_u^5\int_{\mathbb{R}^3}|u|^6dx\leq C_1\zeta t_u^3+C_2C_\zeta t_u^{q-1}+C_3t_u^5,$$

which implies that $t_u\geq\tau$ for some $\tau>0$. If $\mathcal{W}\subset S_\varepsilon^+$ is compact, suppose by contradiction that there is $\{u_n\}\subset\mathcal{W}$ with $t_n:=t_{u_n}\to\infty$. Since the set $\mathcal{W}$ is compact, there exists a $u\in\mathcal{W}$ such that $u_n\to u$ in H_ε. Moreover, using the proof of Lemma 7.2(ii), we have that $J_\varepsilon(t_nu_n)\to-\infty$.

On the other hand, let $v_n:=t_nu_n\in\mathcal{N}_\varepsilon$, from the definition of g and (g_4), (g_5) and $4<\theta<6$, it yields that

$$J_\varepsilon(v_n)=J_\varepsilon(v_n)-\frac{1}{\theta}J_\varepsilon'(v_n)[v_n]\geq\Big(\frac{1}{2}-\frac{1}{\theta}\Big)\|v_n\|_\varepsilon^2+\left(\frac{1}{4}-\frac{1}{\theta}\right)b[v_n]_{A_\varepsilon}^4$$
$$+\int_{\Lambda_\varepsilon^c}\Big(\frac{1}{\theta}g(\varepsilon x,|v_n|^2)|v_n|^2-\frac{1}{2}G(\varepsilon x,|v_n|^2)\Big)dx$$
$$\geq\Big(\frac{1}{2}-\frac{1}{\theta}\Big)\Big(\|v_n\|_\varepsilon^2-\frac{1}{K}\int_{\mathbb{R}^3}V(\varepsilon x)|v_n|^2dx\Big)\geq\Big(\frac{1}{2}-\frac{1}{\theta}\Big)\left(1-\frac{1}{K}\right)\|v_n\|_\varepsilon^2.$$

Thus, substituting $v_n:=t_nu_n$ and $\|v_n\|_\varepsilon=t_n$, we obtain

$$0<\Big(\frac{1}{2}-\frac{1}{\theta}\Big)\left(1-\frac{1}{K}\right)\leq\frac{J_\varepsilon(v_n)}{t_n^2}\leq 0$$

as $n \to \infty$, which yields a contradiction. This proves $(A2)$.

$(A3)$ We first observe that $\widehat{m}_\varepsilon$, m_ε and m_ε^{-1} are well defined. Indeed, by $(A2)$, for each $u \in H_\varepsilon^+$, there is a unique $\widehat{m}_\varepsilon(u) \in \mathcal{N}_\varepsilon$. On the other hand, if $u \in \mathcal{N}_\varepsilon$, then $u \in H_\varepsilon^+$. Otherwise, we have $|\mathrm{supp}(u) \cap \Lambda_\varepsilon| = 0$ and by (g_5), it follows that

$$\begin{aligned}\|u\|_\varepsilon^2 + b[u]_{A_\varepsilon}^4 &= \int_{\mathbb{R}^3} g(\varepsilon x, |u|^2)|u|^2 dx = \int_{\Lambda_\varepsilon^c} g(\varepsilon x, |u|^2)|u|^2 dx \\ &\le \frac{1}{K}\int_{\mathbb{R}^3} V(\varepsilon x)|u|^2 dx \le \frac{1}{K}\|u\|_\varepsilon^2,\end{aligned}$$

which is impossible since $K > 2$ and $u \neq 0$. Therefore, $m_\varepsilon^{-1}(u) = \frac{u}{\|u\|_\varepsilon} \in S_\varepsilon^+$ is well defined and continuous. From

$$m_\varepsilon^{-1}\big(m_\varepsilon(u)\big) = m_\varepsilon^{-1}(t_u u) = \frac{t_u u}{t_u\|u\|_\varepsilon} = u, \ \ \forall\, u \in S_\varepsilon^+,$$

we conclude that m_ε is a bijection. We now prove that $\widehat{m}_\varepsilon : H_\varepsilon^+ \to \mathcal{N}_\varepsilon$ is continuous. Let $\{u_n\} \subset H_\varepsilon^+$ and $u \in H_\varepsilon^+$ be such that $u_n \to u$ in H_ε. By $(A2)$, there is a $t_0 > 0$ such that $t_n := t_{u_n} \to t_0$. Using $t_n u_n \in \mathcal{N}_\varepsilon$, we obtain

$$t_n^2 a[u_n]_{A_\varepsilon}^2 + t_n^2\int_{\mathbb{R}^3} V_\varepsilon(x)|u_n|^2 dx + t_n^4 b[u_n]_{A_\varepsilon}^4 = \int_{\mathbb{R}^3} g(\varepsilon x, t_n^2|u_n|^2)t_n^2|u_n|^2 dx, \ \ \forall\, n \in \mathbb{N},$$

and passing to the limit as $n \to \infty$ in the last inequality, we obtain

$$t_0^2 a[u]_{A_\varepsilon}^2 + t_0^2\int_{\mathbb{R}^3} V_\varepsilon(x)|u|^2 dx + t_0^4 b[u]_{A_\varepsilon}^4 = \int_{\mathbb{R}^3} g(\varepsilon x, t_0^2|u|^2)t_0^2|u|^2 dx.$$

We obtain that $t_0 u \in \mathcal{N}_\varepsilon$ and $t_u = t_0$. This proves $\widehat{m}_\varepsilon(u_n) \to \widehat{m}_\varepsilon(u)$ in H_ε^+. Thus, $\widehat{m}_\varepsilon$ and m_ε are continuous functions and $(A3)$ is proved.

$(A4)$ Let $\{u_n\} \subset S_\varepsilon^+$ be a subsequence such that $\mathrm{dist}\,(u_n, \partial S_\varepsilon^+) \to 0$, then for each $v \in \partial S_\varepsilon^+$ and $n \in N$, we have $|u_n| = |u_n - v|$ a.e. in Λ_ε. Therefore, by (V_1), (V_2) and the Sobolev embedding, for any $t \in [2, 6]$, there exists a constant $C_t > 0$ such that

$$\begin{aligned}\|u_n\|_{L^t(\Lambda_\varepsilon)} &\le \inf_{v\in\partial S_\varepsilon^+}\|u_n - v\|_{L^t(\Lambda_\varepsilon)} \le C_t\Big(\inf_{v\in\partial S_\varepsilon^+}\int_{\Lambda_\varepsilon}(|\nabla_{A_\varepsilon} u_n - v|^2 + V_\varepsilon(x)|u_n - v|^2)dx\Big)^{\frac{1}{2}} \\ &\le C_t\,\mathrm{dist}\,(u_n, \partial S_\varepsilon^+)\end{aligned}$$

for all $n \in N$. By (g_2), (g_3) and (g_5), for each $t > 0$, we have

$$\begin{aligned}\int_{\mathbb{R}^N} G(\varepsilon x, t^2|u_n|^2)dx &\le \int_{\Lambda_\varepsilon}\Big(F(t^2|u_n|^2) + \frac{t^6|u_n|^6}{6}\Big)dx + \frac{t^2}{K}\int_{\Lambda_\varepsilon^c} V(\varepsilon x)|u_n|^2 dx \\ &\le C_1 t^4\int_{\Lambda_\varepsilon}|u_n|^4 dx + C_2 t^q\int_{\Lambda_\varepsilon}|u_n|^q dx + \frac{t^6}{6}\int_{\Lambda_\varepsilon}|u_n|^6 dx + \frac{t^2}{K}\|u_n\|_\varepsilon^2 \\ &\le C_3 t^4\mathrm{dist}\,(u_n, \partial S_\varepsilon^+)^4 + C_4 t^q\mathrm{dist}\,(u_n, \partial S_\varepsilon^+)^q + C_5 t^6\mathrm{dist}\,(u_n, \partial S_\varepsilon^+)^6 + \frac{t^2}{K}.\end{aligned}$$

Therefore,

$$\limsup_{n\to\infty}\int_{\mathbb{R}^3} G(\varepsilon x, t^2|u_n|^2)dx \le \frac{t^2}{K}, \ \forall\, t > 0.$$

On the other hand, from the definition of m_ε and the last inequality, for all $t > 0$, we have

$$\liminf_{n\to\infty} J_\varepsilon(m_\varepsilon(u_n)) \ge \liminf_{n\to\infty} J_\varepsilon(tu_n) \ge \liminf_{n\to\infty}\frac{t^2}{2}\|u_n\|_\varepsilon^2 - \frac{t^2}{K} = \frac{K-2}{2K}t^2,$$

which implies

$$\liminf_{n\to\infty}\left\{\frac{1}{2}\|m_\varepsilon(u_n)\|_\varepsilon^2+\frac{b}{4}[m_\varepsilon(u_n)]_{A_\varepsilon}^4\right\}\geq\liminf_{n\to\infty}J_\varepsilon(m_\varepsilon(u_n))\geq\frac{K-2}{2K}t^2,\ \forall t>0.$$

From the arbitrary of $t>0$, it is easy to see that $\|m_\varepsilon(u_n)\|_\varepsilon\to\infty$ and $J_\varepsilon(m_\varepsilon(u_n))\to\infty$ as $n\to\infty$. We conclude the proof of Lemma 7.3. □

Now we define the function

$$\widehat{\Psi}_\varepsilon : H_\varepsilon^+\to\mathbb{R},$$

by $\widehat{\Psi}_\varepsilon(u)=J_\varepsilon(\widehat{m}_\varepsilon(u))$ and we denote $\Psi_\varepsilon:=(\widehat{\Psi}_\varepsilon)|_{S_\varepsilon^+}$.

From Lemma 7.3, arguing as in Corollary 10 of Szulkin and Weth [144], we obtain the following property.

Lemma 7.4 *Assume that (V_1)–(V_2) and (f_1)–(f_4) are satisfied, then*

$(B1)$ $\widehat{\Psi}_\varepsilon\in C^1(H_\varepsilon^+,\mathbb{R})$ and

$$\widehat{\Psi}_\varepsilon'(u)v=\frac{\|\widehat{m}_\varepsilon(u)\|_\epsilon}{\|u\|_\epsilon}J_\varepsilon'(\widehat{m}_\varepsilon(u))[v],\ \ \forall\, u\in H_\varepsilon^+ \text{ and}\,\forall\, v\in H_\varepsilon.$$

$(B2)$ $\Psi_\varepsilon\in C^1(S_\varepsilon^+,\mathbb{R})$ and

$$\Psi_\varepsilon'(u)v=\|m_\varepsilon(u)\|_\epsilon J_\varepsilon'(\widehat{m}_\varepsilon(u))[v],\ \ \forall\, v\in T_uS_\varepsilon^+.$$

$(B3)$ If $\{u_n\}$ is a $(PS)_c$ sequence of Ψ_ε, then $\{m_\varepsilon(u_n)\}$ is a $(PS)_c$ sequence of J_ε. If $\{u_n\}\subset\mathcal{N}_\varepsilon$ is a bounded $(PS)_c$ sequence of J_ε, then $\{m_\varepsilon^{-1}(u_n)\}$ is a $(PS)_c$ sequence of Ψ_ε.

$(B4)$ u is a critical point of Ψ_ε if and only if $m_\varepsilon(u)$ is a critical point of J_ε. Moreover, the corresponding critical values coincide and

$$\inf_{S_\varepsilon^+}\Psi_\varepsilon.=\inf_{\mathcal{N}_\varepsilon}J_\varepsilon.$$

As in Szulkin and Weth [144], we have the following variational characterization of the infimum of J_ε over $\mathcal{N}_\varepsilon$:

$$c_\varepsilon=\inf_{u\in\mathcal{N}_\varepsilon}J_\varepsilon(u)=\inf_{u\in H_\varepsilon^+}\sup_{t>0}J_\varepsilon(tu)=\inf_{u\in S_\varepsilon^+}\sup_{t>0}J_\varepsilon(tu). \tag{7.6}$$

The following result is important to prove the $(PS)_c$ condition for the functional J_ε.

Lemma 7.5 *Let $c>0$ and $\{u_n\}$ is a $(PS)_c$ sequence for J_ϵ, then $\{u_n\}$ is bounded in H_ϵ.*

Proof. Assume that $\{u_n\}\subset H_\epsilon$ is a $(PS)_c$ sequence for J_ϵ, that is, $J_\epsilon(u_n)\to c$ and $J_\epsilon'(u_n)\to 0$. By (g_4), (g_5) and $4<\theta<6$, we have

$$\begin{aligned}
c+o(1)+o(1)\|u_n\|_\varepsilon &= J_\varepsilon(u_n)-\frac{1}{\theta}J_\varepsilon'(u_n)[u_n]=\Big(\frac{1}{2}-\frac{1}{\theta}\Big)\|u_n\|_\varepsilon^2+\Big(\frac{1}{4}-\frac{1}{\theta}\Big)b[u_n]_{A_\varepsilon}^4\\
&\quad+\int_{\mathbb{R}^3}\Big(\frac{1}{\theta}g(\varepsilon x,|u_n|^2)|u_n|^2-\frac{1}{2}G(\varepsilon x,|u_n|^2)\Big)dx\\
&\geq\Big(\frac{1}{2}-\frac{1}{\theta}\Big)\|u_n\|_\varepsilon^2+\int_{\Lambda_\varepsilon^c}\Big(\frac{1}{\theta}g(\varepsilon x,|u_n|^2)|u_n|^2-\frac{1}{2}G(\varepsilon x,|u_n|^2)\Big)dx\\
&\geq\Big(\frac{1}{2}-\frac{1}{\theta}\Big)\|u_n\|_\varepsilon^2-\frac{1}{2}\int_{\Lambda_\varepsilon^c}G(\varepsilon x,|u_n|^2)dx\\
&\geq\Big(\frac{1}{2}-\frac{1}{\theta}\Big)\|u_n\|_\varepsilon^2-\frac{1}{2K}\int_{\mathbb{R}^3}V(\varepsilon x)|u_n|^2dx\geq\Big(\frac{1}{2}-\frac{1}{\theta}-\frac{1}{2K}\Big)\|u_n\|_\varepsilon^2.
\end{aligned}$$

Since $K > 2$, from the last inequality, we obtain that $\{u_n\}$ is bounded in H_ε. □

Define

$$c_0 = \frac{abS^3}{4} + \frac{S^6}{24}\Big((b^2 + 4aS^{-3})^{3/2} + b^3\Big), \tag{7.7}$$

where S is the best constant for the Sobolev embedding $D^{1,2}(\mathbb{R}^3, \mathbb{R}) \hookrightarrow L^6(\mathbb{R}^3, \mathbb{R})$.

The following lemma provides a range of levels in which the functional J_ϵ verifies the Palais-Smale condition.

Lemma 7.6 *The functional J_ϵ satisfies the $(PS)_c$ condition for any $c \in (0, c_0)$.*

Proof. Let $(u_n) \subset H_\epsilon$ be a $(PS)_c$ for J_ϵ. By Lemma 7.5, (u_n) is bounded in H_ϵ. Thus, up to a subsequence, $u_n \rightharpoonup u$ in H_ϵ and $u_n \to u$ in $L^r_{\text{loc}}(\mathbb{R}^3, \mathbb{C})$ for all $1 \le r < 6$ as $n \to +\infty$.

Step 1. For the fixed $\epsilon > 0$, let $R > 0$ be such that $\Lambda_\epsilon \subset B_{R/2}(0)$. We show that for any given $\zeta > 0$, for R large enough,

$$\limsup_{n\to\infty} \int_{B^c_R(0)} (a|\nabla_{A_\epsilon} u_n|^2 + V_\epsilon(x)|u_n|^2)dx \le \zeta. \tag{7.8}$$

Let $\phi_R \in C^\infty(\mathbb{R}^3, \mathbb{R})$ be a cut-off function such that

$$\phi_R = 0 \quad x \in B_{R/2}(0), \quad \phi_R = 1 \quad x \in B^c_R(0), \quad 0 \le \phi_R \le 1, \quad \text{and} \quad |\nabla \phi_R| \le C/R$$

where $C > 0$ is a constant independent of R. Since the sequence $(\phi_R u_n)$ is bounded in H_ϵ, we have

$$J'_\epsilon(u_n)[\phi_R u_n] = o(1).$$

Therefore,

$$a \operatorname{Re} \int_{\mathbb{R}^3} \nabla_{A_\epsilon} u_n \overline{\nabla_{A_\epsilon}(\phi_R u_n)}dx + \int_{\mathbb{R}^3} V_\epsilon(x)|u_n|^2 \phi_R dx + b[u_n]^2_{A_\epsilon} \operatorname{Re} \int_{\mathbb{R}^3} \nabla_{A_\epsilon} u_n \overline{\nabla_{A_\epsilon}(\phi_R u_n)}dx$$
$$= \int_{\mathbb{R}^3} g(\epsilon x, |u_n|^2)|u_n|^2 \phi_R dx + o(1).$$

Since $\overline{\nabla_{A_\epsilon}(\phi_R u_n)} = i\overline{u_n}\nabla\phi_R + \phi_R \overline{\nabla_{A_\epsilon} u_n}$, using (g_5), we have

$$\begin{aligned}
&\int_{\mathbb{R}^3} (a|\nabla_{A_\epsilon} u_n|^2 + V_\epsilon(x)|u_n|^2)\phi_R dx \\
&\le \int_{\mathbb{R}^3} g(\epsilon x, |u_n|^2)|u_n|^2 \phi_R dx - \left(a + b[u_n]^2_{A_\epsilon}\right) \operatorname{Re} \int_{\mathbb{R}^3} i\overline{u_n} \nabla_{A_\epsilon} u_n \nabla \phi_R dx + o(1) \\
&\le \frac{1}{K} \int_{\mathbb{R}^3} V_\epsilon(x)|u_n|^2 \phi_R dx + C\left| \operatorname{Re} \int_{\mathbb{R}^3} i\overline{u_n} \nabla_{A_\epsilon} u_n \nabla \phi_R dx \right| + o(1).
\end{aligned}$$

By the definition of ϕ_R, the Hölder's inequality and the boundedness of (u_n) in H_ε, we obtain

$$\left(1 - \frac{1}{K}\right) \int_{\mathbb{R}^3} (a|\nabla_{A_\varepsilon} u_n|^2 + V_\varepsilon(x)|u_n|^2)\phi_R dx \le \frac{C}{R}\|u_n\|_2 \|\nabla_{A_\varepsilon} u_n\|_2 + o(1) \le \frac{C_1}{R} + o(1)$$

and so relation (7.8) holds.

Step 2. For any $R > 0$, the following limit holds:

$$\limsup_{n\to\infty} \int_{B_R(0)} (a|\nabla_{A_\epsilon} u_n|^2 + V_\epsilon(x)|u_n|^2)dx = \int_{B_R(0)} (a|\nabla_{A_\epsilon} u|^2 + V_\epsilon(x)|u|^2)dx. \tag{7.9}$$

Let $\phi_\rho \in C^\infty(\mathbb{R}^3, \mathbb{R})$ be a cut-off function such that

$$\phi_\rho = 1 \quad x \in B_\rho(0), \quad \phi_\rho = 0 \quad x \in B_{2\rho}^c(0), \quad 0 \le \phi_\rho \le 1, \quad \text{and} \quad |\nabla \phi_\rho| \le C/\rho$$

where $C > 0$ is a constant independent of ρ. Let

$$P_n(x) = M(u_n)|\nabla_{A_\varepsilon} u_n - \nabla_{A_\varepsilon} u|^2 + V_\epsilon(x)|u_n - u|^2$$

where $M(u_n) = a + b\int_{\mathbb{R}^3} |\nabla_{A_\varepsilon} u_n|^2 dx$. For the fixed $R > 0$, choosing $\rho > R > 0$, we have

$$\begin{aligned}\int_{B_R} P_n(x)dx &\le \int_{\mathbb{R}^3} P_n(x)\phi_\rho(x)dx \\ &= M(u_n)\int_{\mathbb{R}^3} |\nabla_{A_\varepsilon} u_n - \nabla_{A_\varepsilon} u|^2\phi_\rho(x)dx + \int_{\mathbb{R}^3} V_\epsilon(x)|u_n - u|^2\phi_\rho(x)dx \\ &= J_{n,\rho}^1 - J_{n,\rho}^2 + J_{n,\rho}^3 + J_{n,\rho}^4, \end{aligned} \tag{7.10}$$

where

$$J_{n,\rho}^1 = M(u_n)\int_{\mathbb{R}^3} |\nabla_{A_\epsilon} u_n|^2\phi_\rho(x)dx + \int_{\mathbb{R}^3} V_\epsilon(x)|u_n|^2\phi_\rho(x)dx - \int_{\mathbb{R}^3} g(\epsilon x, |u_n|^2)|u_n|^2\phi_\rho dx,$$

$$J_{n,\rho}^2 = M(u_n)\,\mathrm{Re}\int_{\mathbb{R}^3} \nabla_{A_\epsilon} u_n \overline{\nabla_{A_\epsilon} u}\phi_\rho(x)dx + \mathrm{Re}\int_{\mathbb{R}^3} V_\epsilon(x)u_n\overline{u}\phi_\rho(x)dx - \mathrm{Re}\int_{\mathbb{R}^3} g(\epsilon x, |u_n|^2)u_n\overline{u}\phi_\rho(x)dx,$$

$$J_{n,\rho}^3 = -M(u_n)\,\mathrm{Re}\int_{\mathbb{R}^3} (\nabla_{A_\epsilon} u_n - \nabla_{A_\epsilon} u)\overline{\nabla_{A_\epsilon} u}\phi_\rho(x)dx - \mathrm{Re}\int_{\mathbb{R}^3} V_\epsilon(x)(u_n - u)\overline{u}\phi_\rho(x)dx,$$

and

$$J_{n,\rho}^4 = \mathrm{Re}\int_{\mathbb{R}^3} g(\epsilon x, |u_n|^2)u_n\overline{(u_n - u)}\phi_\rho(x)dx.$$

We obtain

$$J_{n,\rho}^1 = J_\epsilon'(u_n)[\phi_\rho u_n] - M(u_n)\,\mathrm{Re}\int_{\mathbb{R}^3} i\overline{u_n}\nabla_{A_\epsilon} u_n \nabla\phi_\rho dx$$

and

$$J_{n,\rho}^2 = J_\epsilon'(u_n)[\phi_\rho u] - M(u_n)\,\mathrm{Re}\int_{\mathbb{R}^3} i\overline{u}\nabla_{A_\epsilon} u_n \nabla\phi_\rho dx.$$

Then

$$\lim_{\rho\to\infty}\limsup_{n\to\infty}\left|J_{n,\rho}^1\right| = 0, \qquad \lim_{\rho\to\infty}\limsup_{n\to\infty}\left|J_{n,\rho}^2\right| = 0.$$

On the other hand, since the sequence $\{u_n\}$ is bounded in H_ϵ, we assume that $\int_{\mathbb{R}^3} |\nabla_{A_\varepsilon} u_n|^2 dx \to l^2$. Then

$$\begin{aligned} J_{n,\rho}^3 =& -(a + bl^2)\,\mathrm{Re}\int_{\mathbb{R}^3} (\nabla_{A_\epsilon} u_n - \nabla_{A_\epsilon} u)\overline{\nabla_{A_\epsilon}(u\phi_\rho(x))}dx - \mathrm{Re}\int_{\mathbb{R}^3} V_\epsilon(x)(u_n - u)\overline{(u\phi_\rho(x))}dx \\ &+ (a + bl^2)\,\mathrm{Re}\int_{\mathbb{R}^3} (\nabla_{A_\epsilon} u_n - \nabla_{A_\epsilon} u)i\overline{u}\nabla\phi_\rho dx + o(1) \\ =& -(a + bl^2)\langle u_n - u, u\phi_\rho(x)\rangle_\epsilon + (a + bl^2)\,\mathrm{Re}\int_{\mathbb{R}^3} (\nabla_{A_\epsilon} u_n - \nabla_{A_\epsilon} u)i\overline{u}\nabla\phi_\rho dx + o(1), \end{aligned}$$

hence

$$\lim_{\rho\to\infty}\limsup_{n\to\infty}\left|J_{n,\rho}^3\right| = 0.$$

Now we prove that

$$\lim_{\rho\to\infty}\limsup_{n\to\infty}\left|J_{n,\rho}^4\right| = 0. \tag{7.11}$$

First we show

$$\lim_{n\to\infty}\int_{\Lambda_\varepsilon}|u_n|^6dx=\int_{\Lambda_\varepsilon}|u|^6dx. \tag{7.12}$$

Using the boundedness of (u_n) in D_ϵ and the diamagnetic inequality (7.2), we may assume that

$$|\nabla|u_n||^2\rightharpoonup\mu \quad\text{and}\quad |u_n|^6\rightharpoonup\nu \tag{7.13}$$

in the sense of measures. By the concentration-compactness principle, we can find an at most countable index I, sequences $(x_i)\subset\mathbb{R}^3,(\mu_i),(\nu_i)\subset(0,\infty)$ such that

$$\begin{aligned}&\mu\geq|\nabla|u||^2+\sum_{i\in I}\mu_i\delta_{x_i},\\&\nu=|u|^6+\sum_{i\in I}\nu_i\delta_{x_i}\quad\text{and}\quad S\nu_i^{1/3}\leq\mu_i\end{aligned} \tag{7.14}$$

for any $i\in I$, where δ_{x_i} is the Dirac mass at the point x_i.

Let us show that $(x_i)_{i\in I}\cap\Lambda_\varepsilon=\emptyset$. Assume, by contradiction, that $x_i\in\Lambda_\varepsilon$ for some $i\in I$. For any $\rho>0$, we define $\psi_\rho(x)=\psi(\frac{x-x_i}{\rho})$ where $\psi\in C_0^\infty(\mathbb{R}^3,[0,1])$ is such that $\psi=1$ in B_1, $\psi=0$ in $\mathbb{R}^3\setminus B_2$ and $\|\nabla\psi\|_{L^\infty(\mathbb{R}^3,\mathbb{R})}\leq 2$. We suppose that $\rho>0$ is such that $\operatorname{supp}(\psi_\rho)\subset\Lambda_\varepsilon$. Since $\{\psi_\rho u_n\}$ is bounded in H_ε, we can see that $J_\varepsilon'(u_n)[\psi_\rho u_n]=o_n(1)$, that is,

$$\begin{aligned}&\left(a+b[u_n]_{A_\varepsilon}^2\right)\int_{\mathbb{R}^3}|\nabla_{A_\varepsilon}u_n|^2\psi_\rho dx+\left(a+b[u_n]_{A_\varepsilon}^2\right)\operatorname{Re}\int_{\mathbb{R}^3}i\overline{u_n}\nabla_{A_\varepsilon}u_n\nabla\psi_\rho dx+\int_{\mathbb{R}^3}V_\epsilon(x)|u_n|^2\psi_\rho dx\\&=\int_{\mathbb{R}^3}g(\varepsilon x,|u_n|^2)|u_n|^2\psi_\rho dx+o(1)=\int_{\mathbb{R}^N}f(|u_n|^2)|u_n|^2\psi_\rho dx+\int_{\mathbb{R}^N}|u_n|^6\psi_\rho dx+o(1).\end{aligned}$$

Using the diamagnetic inequality (7.2) again, it follows that

$$\begin{aligned}&\left(a+b\int_{\mathbb{R}^3}|\nabla|u_n||^2\psi_\rho dx\right)\int_{\mathbb{R}^3}|\nabla|u_n||^2\psi_\rho dx+\left(a+b[u_n]_{A_\varepsilon}^2\right)\operatorname{Re}\int_{\mathbb{R}^3}i\overline{u_n}\nabla_{A_\varepsilon}u_n\nabla\psi_\rho dx\\&\leq\int_{\mathbb{R}^N}f(|u_n|^2)|u_n|^2\psi_\rho dx+\int_{\mathbb{R}^N}|u_n|^6\psi_\rho dx+o(1).\end{aligned} \tag{7.15}$$

Due to the fact that f has the subcritical growth and ψ_ρ has the compact support, we have that

$$\lim_{\rho\to0}\lim_{n\to\infty}\int_{\mathbb{R}^3}f(|u_n|^2)|u_n|^2\psi_\rho dx=\lim_{\rho\to0}\int_{\mathbb{R}^3}f(|u|^2)|u|^2\psi_\rho dx=0. \tag{7.16}$$

Now, we show that

$$\lim_{\rho\to0}\limsup_{n\to\infty}\left|\int_{\mathbb{R}^3}i\overline{u_n}\nabla_{A_\varepsilon}u_n\nabla\psi_\rho dx\right|=0. \tag{7.17}$$

Because of the boundedness of $\{u_n\}$ in H_ε, using the Hölder's inequality, the strong convergence of $(|u_n|)$ in $L^2_{\mathrm{loc}}(\mathbb{R}^3,\mathbb{R})$, $|u|\in L^6(\mathbb{R}^3,\mathbb{R})$, $|\nabla\psi_\rho|\leq C\rho^{-1}$ and $|B_{2\rho}(x_i)|\sim\rho^3$, we have that

$$\begin{aligned}0&\leq\lim_{\rho\to0}\limsup_{n\to\infty}\left|\int_{\mathbb{R}^3}i\overline{u_n}\nabla_{A_\varepsilon}u_n\nabla\psi_\rho dx\right|\leq\lim_{\rho\to0}\limsup_{n\to\infty}\int_{\mathbb{R}^3}|\overline{u_n}\nabla\psi_\rho||\nabla_{A_\varepsilon}u_n|dx\\&\leq\lim_{\rho\to0}\lim_{n\to\infty}\left(\int_{B_{2\rho}(x_i)}|\overline{u_n}\nabla\psi_\rho|^2dx\right)^{1/2}[u_n]_{A_\varepsilon}\leq C\lim_{\rho\to0}\left(\int_{B_{2\rho}(x_i)}|u|^2dx\right)^{1/2}=0\end{aligned}$$

which shows that relation (7.17) holds.

Then, taking into account (7.13), (7.15), (7.16) and (7.17), we can conclude that $\nu_i \geq a\mu_i + b\mu_i^2$. Together with the inequality $S\nu_i^{1/3} \leq \mu_i$ in (7.14), we have

$$\mu_i \geq \frac{S^3}{2}(b + \sqrt{b^2 + 4aS^{-3}}). \tag{7.18}$$

Now, from (f_3), (g_4) and (g_5), we have

$$\begin{aligned}
c &= J_\varepsilon(u_n) - \frac{1}{4}J'_\varepsilon(u_n)[u_n] + o(1) \\
&= \frac{1}{4}\|u_n\|_\varepsilon^2 + \int_{\mathbb{R}^3} \Big(\frac{1}{4}g(\varepsilon x, |u_n|^2)|u_n|^2 - \frac{1}{2}G(\varepsilon x, |u_n|^2)\Big)dx + o(1) \\
&> \frac{1}{4}\|u_n\|_\varepsilon^2 + \int_{\Lambda_\varepsilon^c} \Big(\frac{1}{4}g(\varepsilon x, |u_n|^2)|u_n|^2 - \frac{1}{2}G(\varepsilon x, |u_n|^2)\Big)dx + \frac{1}{12}\int_{\Lambda_\varepsilon} |u_n|^6 dx + o(1) \\
&\geq \frac{1}{4}\Big(\int_{\Lambda_\varepsilon} a\psi_\rho|\nabla|u_n||^2 dx + \int_{\Lambda_\varepsilon^c} V_\varepsilon(x)|u_n|^2\Big) - \frac{1}{2}\int_{\Lambda_\varepsilon^c} G(\varepsilon x, |u_n|^2)dx + \frac{1}{12}\int_{\Lambda_\varepsilon} |u_n|^6 dx + o(1) \\
&\geq \frac{a}{4}\int_{\Lambda_\varepsilon} \psi_\rho|\nabla|u_n||^2 dx + \Big(\frac{1}{4} - \frac{1}{2K}\Big)\int_{\Lambda_\varepsilon^c} V_\varepsilon(x)|u_n|^2 dx + \frac{1}{12}\int_{\Lambda_\varepsilon} \psi_\rho|u_n|^6 dx + o(1) \\
&\geq \frac{a}{4}\int_{\Lambda_\varepsilon} \psi_\rho|\nabla|u_n||^2 dx + \frac{1}{12}\int_{\Lambda_\varepsilon} \psi_\rho|u_n|^6 dx + o(1).
\end{aligned}$$

From the above arguments, (7.14) and (7.18), we have

$$\begin{aligned}
c \;&\geq \frac{1}{4}\sum_{\{i\in I: x_i\in\Lambda_\varepsilon\}} a\psi_\rho(x_i)\mu_i + \frac{1}{12}\sum_{\{i\in I: x_i\in\Lambda_\varepsilon\}} \psi_\rho(x_i)\nu_i \geq \frac{a}{4}\mu_i + \frac{1}{12}\nu_i \\
&\geq \frac{aS}{4}\cdot\frac{bS^2 + \sqrt{b^2S^4 + 4aS}}{2} + \frac{1}{12}\Big(\frac{bS^2 + \sqrt{b^2S^4 + 4aS}}{2}\Big)^3 = c_0,
\end{aligned}$$

which gives a contradiction. This means that relation (7.12) holds.

Next, we observe that

$$\Big|J_{n,\rho}^4\Big| \leq \int_{(\mathbb{R}^3\setminus\Lambda_\varepsilon)\cap B_{2\rho}(0)} \Big|g(\epsilon x, |u_n|^2)u_n\overline{(u_n - u)}\Big|dx + \int_{\Lambda_\varepsilon\cap B_{2\rho}(0)} \Big|g(\epsilon x, |u_n|^2)u_n\overline{(u_n - u)}\Big|dx.$$

By the Sobolev compact embedding $H_\varepsilon \hookrightarrow L^r_{\text{loc}}(\mathbb{R}^3, \mathbb{C})$ for $1 \leq r < 6$, and (g_5), we have

$$\int_{(\mathbb{R}^3\setminus\Lambda_\varepsilon)\cap B_{2\rho}(0)} \Big|g(\epsilon x, |u_n|^2)u_n\overline{(u_n - u)}\Big|dx \to 0, \text{ as } n \to \infty.$$

Moreover, using the Sobolev compact embedding $H_\varepsilon \hookrightarrow L^r_{\text{loc}}(\mathbb{R}^3, \mathbb{C})$ for $1 \leq r < 6$, again, and (7.12), we have,

$$\int_{\Lambda_\varepsilon\cap B_{2\rho}(0)} \Big|g(\epsilon x, |u_n|^2)u_n\overline{(u_n - u)}\Big|dx \to 0, \text{ as } n \to \infty.$$

Thus, (7.11) holds. Moreover, by (7.10), it follows that

$$0 \leq \limsup_{n\to\infty} \int_{B_R} P_n(x)dx \leq \limsup_{n\to\infty} \Big(|J_{n,\rho}^1| + |J_{n,\rho}^2| + |J_{n,\rho}^3| + |J_{n,\rho}^4|\Big) = 0.$$

Then,

$$\limsup_{n\to\infty} \int_{B_R} P_n(x)dx = 0.$$

Thus, relation (7.9) holds.

Step 3. From (7.8) and (7.9), we have

$$\begin{aligned}
\|u\|_\varepsilon^2 &\le \liminf_{n\to\infty}\|u_n\|_\varepsilon^2 \le \limsup_{n\to\infty}\|u_n\|_\varepsilon^2 \\
&\le \limsup_{n\to\infty}\Big\{\int_{B_R(0)}(a|\nabla_{A_\epsilon}u_n|^2+V_\epsilon(x)|u_n|^2)dx+\int_{B_R^c(0)}(a|\nabla_{A_\epsilon}u_n|^2+V_\epsilon(x)|u_n|^2)dx\Big\} \\
&\le \int_{B_R(0)}(a|\nabla_{A_\epsilon}u|^2+V_\epsilon(x)|u|^2)dx+\zeta.
\end{aligned}$$

Passing to the limit as $\zeta\to 0$ we have $R\to\infty$, which implies that

$$\|u\|_\varepsilon^2 \le \liminf_{n\to\infty}\|u_n\|_\varepsilon^2 \le \limsup_{n\to\infty}\|u_n\|_\varepsilon^2 \le \|u\|_\varepsilon^2.$$

Then $u_n\to u$ in H_ϵ and we complete the proof. □

Since f is only assumed to be continuous, the following result is required for the multiplicity property in the next section.

Corollary 7.1 *The functional Ψ_ε satisfies the $(PS)_c$ condition on S_ε^+ at any level $c\in(0,c_0)$.*

Proof. Let $\{u_n\}\subset S_\varepsilon^+$ be a $(PS)_c$ sequence for Ψ_ε. Then $\Psi_\varepsilon(u_n)\to c$ and $\|\Psi_\varepsilon'(u_n)\|_*\to 0$, where $\|\cdot\|_*$ is the norm in the dual space $(T_{u_n}S_\varepsilon^+)^*$. By Lemma 7.4($B3$), we know that $\{m_\varepsilon(u_n)\}$ is a $(PS)_c$ sequence for J_ε in H_ε. From Lemma 7.6, we know that there exists a $u\in S_\varepsilon^+$ such that, up to a subsequence, $m_\varepsilon(u_n)\to m_\varepsilon(u)$ in H_ε. By Lemma 7.3($A3$), we obtain

$$u_n\to u \text{ in } S_\varepsilon^+,$$

and the proof is complete. □

7.4 Multiple solutions of the modified problem

7.4.1 The autonomous problem

For our purpose, we need to study the following limit problem

$$-\Big(a+b\int_{\mathbb{R}^3}|\nabla u|^2dx\Big)\Delta u+V_0u=f(|u|^2)u+|u|^4u,\quad u:\mathbb{R}^3\to\mathbb{R},\tag{7.19}$$

whose associated C^1-functional, defined in $H^1(\mathbb{R}^3,\mathbb{R})$, is

$$I_{V_0}(u):=\frac12\int_{\mathbb{R}^3}(a|\nabla u|^2+V_0u^2)dx+\frac b4\Big(\int_{\mathbb{R}^3}|\nabla u|^2dx\Big)^2-\frac12\int_{\mathbb{R}^3}F(u^2)dx-\frac16\int_{\mathbb{R}^3}(u^+)^6dx.$$

Let

$$\mathcal{N}_0:=\{u\in H^1(\mathbb{R}^3,\mathbb{R})\setminus\{0\}:I_{V_0}'(u)[u]=0\}$$

and

$$c_{V_0}:=\inf_{u\in\mathcal{N}_0}I_{V_0}(u).$$

By (f_1) and (f_4), for each $u \in H^1(\mathbb{R}^3, \mathbb{R})\backslash\{0\}$, there is a unique $t(u) > 0$ such that

$$I_{V_0}(t(u)u) = \max_{t\geq 0} I_{V_0}(tu) \quad \text{and} \quad t(u)u \in \mathcal{N}_{V_0}.$$

Then, using the assumptions on f, arguing as in Willem [153, Lemma 4.1 and Theorem 4.2] we have that

$$0 < c_{V_0} = \inf_{u\in H^1(\mathbb{R}^3,\mathbb{R})\backslash\{0\}} \max_{t\geq 0} I_{V_0}(tu).$$

Now, we estimate the ground state energy c_{V_0} and show that $c_{V_0} \in (0, c_0)$.

Lemma 7.7 *Assume that (f_1)–(f_4) hold, then $0 < c_{V_0} < c_0$.*

Proof. For $\delta > 0$, the function $\omega_\delta : \mathbb{R}^3 \to \mathbb{R}$ defined by

$$\omega_\delta(x) = 3^{1/4}\frac{\delta^{1/4}}{(\delta + |x|^2)^{1/2}}$$

is a family of functions on which S is attained. Let $\phi \in C_0^\infty(\mathbb{R}^3, [0,1])$ be a cut-off function with

$$\phi = 1 \quad x \in B_{\rho/2}(0), \quad \phi = 0 \quad x \in B_\rho^c(0),$$

then we define the test function by $v_\delta = \omega_\delta/\|\omega_\delta\|_6$, where $u_\delta = \phi v_\delta$. Since

$$I_{V_0}(tv_\delta) := \frac{t^2}{2}\int_{\mathbb{R}^3}(a|\nabla v_\delta|^2 + V_0 v_\delta^2)dx + \frac{bt^4}{4}\Big(\int_{\mathbb{R}^3}|\nabla v_\delta|^2 dx\Big)^2 - \frac{1}{2}\int_{\mathbb{R}^3}F(t^2v_\delta^2)dx - \frac{t^6}{6}\int_{\mathbb{R}^3}v_\delta^6 dx,$$

under the conditions (f_1)–(f_4), we obtain that for small $\delta > 0$, we have $\max_{t>0} I_{V_0}(tv_\delta) < c_0$. Thus, since $0 < c_{V_0} < c_0$, we conclude the proof. □

Let $H_0 := H^1(\mathbb{R}^3, \mathbb{R})$ and define by H_0^+ the open set of H_0 given by

$$H_0^+ = \{u \in H_0 : |\text{supp}(u^+)| > 0\},$$

and $S_0^+ = S_0 \cap H_0^+$, where S_0 be the unit sphere of H_0.

Then S_0^+ is a non-complete $C^{1,1}$-manifold of codimension 1, modeled on H_0 and contained in H_0^+. Therefore, $H_0 = T_u S_0^+ \bigoplus \mathbb{R}u$ for each $u \in T_u S_0^+$, where $T_u S_0^+ = \{v \in H_0 : \langle u, v\rangle_0 = 0\}$.

Arguing as in the proof of Lemma 7.3, we have the following property.

Lemma 7.8 *Let V_0 be given in (V_1) and suppose that (f_1)–(f_4) are satisfied, then the following properties hold:*

- *(a1) For any $u \in H_0^+$, let $g_u : \mathbb{R}^+ \to \mathbb{R}$ be given by $g_u(t) = I_{V_0}(tu)$. Then there exists a unique $t_u > 0$ such that $g_u'(t) > 0$ in $(0, t_u)$ and $g_u'(t) < 0$ in (t_u, ∞).*
- *(a2) There is a $\tau > 0$ independent on u such that $t_u > \tau$ for all $u \in S_0^+$. Moreover, for each compact $\mathcal{W} \subset S_0^+$ there is $C_\mathcal{W}$ such that $t_u \leq C_\mathcal{W}$, for all $u \in \mathcal{W}$.*
- *(a3) The map $\widehat{m} : H_0^+ \to \mathcal{N}_0$ given by $\widehat{m}(u) = t_u u$ is continuous and $m_0 = \widehat{m}_0|_{S_0^+}$ is a homeomorphism between S_0^+ and $\mathcal{N}_0$. Moreover, $m^{-1}(u) = \frac{u}{\|u\|_0}$.*
- *(a4) If there is a sequence $\{u_n\} \subset S_0^+$ such that $dist(u_n, \partial S_0^+) \to 0$, then $\|m(u_n)\|_0 \to \infty$ and $I_{V_0}(m(u_n)) \to \infty$.*

We consider the functional defined by

$$\widehat{\Psi}_0(u) = I_{V_0}(\widehat{m}(u)) \quad \text{and} \quad \Psi_0 := \widehat{\Psi}_0|_{S_0}.$$

Arguing as in Willem [153]Proposition 9 and Corollary 10, we have the following property.

Lemma 7.9 *Let V_0 be given in (V_1) and suppose that (f_1)–(f_4) are satisfied, then*

(*b*1) $\widehat{\Psi}_0 \in C^1(H_0^+, \mathbb{R})$ *and*

$$\widehat{\Psi}_0'(u)v = \frac{\|\widehat{m}(u)\|_0}{\|u\|_0} I_{V_0}'(\widehat{m}(u))[v], \ \ \forall\, u \in H_0^+ \text{ and } \forall v \in H_0.$$

(*b*2) $\Psi_0 \in C^1(S_0^+, \mathbb{R})$ *and*

$$\Psi_0'(u)v = \|m(u)\|_0 I_{V_0}'(\widehat{m}(u))[v], \ \ \forall v \in T_u S_0^+.$$

(*b*3) *If $\{u_n\}$ is a $(PS)_c$ sequence of Ψ_0, then $\{m(u_n)\}$ is a $(PS)_c$ sequence of I_{V_0}. If $\{u_n\} \subset \mathcal{N}_0$ is a bounded $(PS)_c$ sequence of I_{V_0}, then $\{m^{-1}(u_n)\}$ is a $(PS)_c$ sequence of Ψ_0.*

(*b*4) *u is a critical point of Ψ_0 if and only if $m(u)$ is a critical point of I_{V_0}. Moreover, the corresponding critical values coincide and $\inf_{S_0^+} \Psi_0 = \inf_{\mathcal{N}_0} I_{V_0}$.*

We have the following variational characterization of the infimum of I_{V_0} over $\mathcal{N}_0$:

$$c_{V_0} = \inf_{u \in \mathcal{N}_0} I_{V_0}(u) = \inf_{u \in H_0^+} \sup_{t>0} I_{V_0}(tu) = \inf_{u \in S_0^+} \sup_{t>0} I_{V_0}(tu). \tag{7.20}$$

The next result is useful in later arguments.

Lemma 7.10 *Let $\{u_n\} \subset H_0$ be a $(PS)_{c_{V_0}}$ sequence for I_{V_0}, then problem (7.19) has a positive ground state solution.*

Proof. By (f_3), it follows that

$$c_{V_0} + o(1) + o(1)\|u_n\| = I_{V_0}(u_n) - \frac{1}{4}\langle I_{V_0}'(u_n), u_n\rangle \geq \frac{1}{4}\|u_n\|^2.$$

Thus, (u_n) is bounded in $H^1(\mathbb{R}^3, \mathbb{R})$. We can assume that there exists $u \in H^1(\mathbb{R}^3, \mathbb{R})$ such that $u_n \rightharpoonup u$ in $H^1(\mathbb{R}^3, \mathbb{R})$, $u_n \to u$ in $L^2_{\text{loc}}(\mathbb{R}^3, \mathbb{R})$ and $u_n \to u$ a.e. in $x \in \mathbb{R}^3$.

Now we show that (u_n) is non-vanishing. Otherwise, assume that (u_n) is vanishing, then Lions' concentration-compactness principle implies that $u_n \to 0$ in $L^r(\mathbb{R}^3, \mathbb{R})$ for any $2 < r < 6$. By the compact embedding and (f_1)–(f_2), we have $\int_{\mathbb{R}^3} F(u_n^2)dx \to 0$ and $\int_{\mathbb{R}^3} f(u_n^2)u_n^2 dx \to 0$ as $n \to \infty$. Moreover, using the former limits, it follows that

$$c_{V_0} = \frac{1}{2}\int_{\mathbb{R}^3} (a|\nabla u_n|^2 + V_0 u_n^2)dx + \frac{b}{4}\Big(\int_{\mathbb{R}^3} |\nabla u_n|^2 dx\Big)^2 - \frac{1}{6}\int_{\mathbb{R}^3} (u_n^+)^6 dx + o(1) \tag{7.21}$$

and

$$\int_{\mathbb{R}^3} (a|\nabla u_n|^2 + V_0 u_n^2)dx + b\Big(\int_{\mathbb{R}^3} |\nabla u_n|^2 dx\Big)^2 = \int_{\mathbb{R}^3} (u_n^+)^6 dx + o(1). \tag{7.22}$$

Suppose that

$$\lim_{n\to\infty} \int_{\mathbb{R}^3} (u_n^+)^6 dx = l \geq 0.$$

If $l = 0$, then relations (7.21) and (7.22) imply that $c_{V_0} = 0$, which contradicts with $c_{V_0} > 0$. Then $l > 0$, using the definition of S, we have

$$a\int_{\mathbb{R}^3} |\nabla u_n|^2 dx + b\Big(\int_{\mathbb{R}^3} |\nabla u_n|^2 dx\Big)^2 \leq \int_{\mathbb{R}^3} (u_n^+)^6 dx + o(1) \leq S^{-3}\Big(\int_{\mathbb{R}^3} |\nabla u_n|^2 dx\Big)^3 + o(1). \tag{7.23}$$

If $\int_{\mathbb{R}^3} |\nabla u_n|^2 dx \to 0$, then

$$\int_{\mathbb{R}^3} (u_n^+)^6 dx \leq S^{-3}\Big(\int_{\mathbb{R}^3} |\nabla u_n|^2 dx\Big)^3 \to 0,$$

which contradicts with $l > 0$. Thus $\int_{\mathbb{R}^3} |\nabla u_n|^2 dx \not\to 0$. By (7.23), we have

$$\int_{\mathbb{R}^3} |\nabla u_n|^2 dx \geq \frac{S^3}{2}(b + \sqrt{b^2 + 4aS^{-3}}) + o(1). \tag{7.24}$$

Using (7.23) again, it follows that

$$\Big(\int_{\mathbb{R}^3} |\nabla u_n|^2 dx\Big)^2 \geq aS^3 + \frac{bS^6}{2}(b + \sqrt{b^2 + 4aS^{-3}}) + o(1). \tag{7.25}$$

From (7.24) and (7.25), we can obtain

$$\begin{aligned} c_{V_0} &= \frac{1}{3}\int_{\mathbb{R}^3} (a|\nabla u_n|^2 + V_0 u_n^2) dx + \frac{b}{12}\Big(\int_{\mathbb{R}^3} |\nabla u_n|^2 dx\Big)^2 + o(1) \\ &\geq \frac{a}{3}\int_{\mathbb{R}^3} |\nabla u_n|^2 dx + \frac{b}{12}\Big(\int_{\mathbb{R}^3} |\nabla u_n|^2 dx\Big)^2 + o(1) \geq c_0 + o(1) \end{aligned}$$

which contradicts with $c_{V_0} < c_0$. Thus, $\{u_n\}$ is non-vanishing, that is, there exist $R, \eta > 0$, and $\{\tilde{y}_n\} \subset \mathbb{R}^3$ such that

$$\lim_{n\to\infty} \int_{B_R(\tilde{y}_n)} u_n^2 dx \geq \eta. \tag{7.26}$$

By the invariance of the translation of I_{V_0}, we may assume that $\{\tilde{y}_n\}$ is bounded, so $u \neq 0$.

Now we claim that $I'_{V_0}(u) = 0$. Assume that there exists $\iota \geq 0$ such that $\int_{\mathbb{R}^3} |\nabla u_n|^2 dx \to \iota^2$. Since $I'_{V_0}(u_n) \to 0$, then u is a solution of the following equation

$$-(a + b\iota^2)\Delta u + V_0 u = f(|u|^2)u + (u^+)^5, \quad u \in H^1(\mathbb{R}^3, \mathbb{R}). \tag{7.27}$$

It suffices to show that $\iota^2 = \int_{\mathbb{R}^3} |\nabla u|^2 dx$. From the weakly lower semi-continuous of the norm, it is easy to obtain that $\iota^2 \geq \int_{\mathbb{R}^3} |\nabla u|^2 dx$. Then, by (7.27), we have $\langle I'_{V_0}(u), u\rangle \leq 0$. By (f_4), there exists $t \leq 1$ such that $tu \in \mathcal{N}_0$. By (7.27) again, it follows that

$$\begin{aligned} c_{V_0} \leq& I_{V_0}(tu) - \frac{1}{4}I'_{V_0}(tu)[tu] \\ =& \frac{1}{4}\int_{\mathbb{R}^3} (a|\nabla tu|^2 + V_0(tu)^2) dx + \int_{\mathbb{R}^3} \Big(\frac{1}{4}f(|tu|^2)|tu|^2 - \frac{1}{2}F(|tu|^2)\Big) dx + \frac{1}{12}\int_{\mathbb{R}^3} (tu^+)^6 dx \\ \leq& \frac{1}{4}\int_{\mathbb{R}^3} (a|\nabla u|^2 + V_0 u^2) dx + \int_{\mathbb{R}^3} \Big(\frac{1}{4}f(|u|^2)|u|^2 - \frac{1}{2}F(|u|^2)\Big) dx + \frac{1}{12}\int_{\mathbb{R}^3} (u^+)^6 dx \\ \leq& \frac{1}{4}\int_{\mathbb{R}^3} (a|\nabla u_n|^2 + V_0 u_n^2) dx + \int_{\mathbb{R}^3} \Big(\frac{1}{4}f(|u_n|^2)|u_n|^2 - \frac{1}{2}F(|u_n|^2)\Big) dx + \frac{1}{12}\int_{\mathbb{R}^3} (u_n^+)^6 dx + o(1) \\ =& I_{V_0}(u_n) - \frac{1}{4}I'_{V_0}(u_n)[u_n] + o(1) = c_{V_0} + o(1). \end{aligned}$$

Thus, $t = 1$ and

$$\begin{aligned} &\lim_{n\to\infty} \frac{1}{4}\int_{\mathbb{R}^3} (a|\nabla u_n|^2 + V_0 u_n^2) dx + \int_{\mathbb{R}^3} \Big(\frac{1}{4}f(|u_n|^2)|u_n|^2 - \frac{1}{2}F(|u_n|^2)\Big) dx + \frac{1}{12}\int_{\mathbb{R}^3} (u_n^+)^6 dx \\ &= \frac{1}{4}\int_{\mathbb{R}^3} (a|\nabla u|^2 + V_0 u^2) dx + \int_{\mathbb{R}^3} \Big(\frac{1}{4}f(|u|^2)|u|^2 - \frac{1}{2}F(|u|^2)\Big) dx + \frac{1}{12}\int_{\mathbb{R}^3} (u^+)^6 dx. \end{aligned}$$

Then $\iota^2 = \int_{\mathbb{R}^3} |\nabla u|^2 dx$ and $u_n \to \omega$ in $H^1(\mathbb{R}^3, \mathbb{R})$. Therefore, $I'_{V_0}(u) = 0$ and $I_{V_0}(u) = c_{V_0}$. Then u is a ground state of problem. From the assumption of f, $u \geq 0$. Moreover, using the standard argument, we obtain that $u(x) > 0$ for $x \in \mathbb{R}^N$. The proof is complete. □

Arguing as in the proof of Lemma 2.3 in Zhang and Zou [161], there exists a positive radial ground state solution of problem (7.19), which implies that this solution decays exponentially at infinity with its gradient; moreover, this ground state solution is of class $C^2(\mathbb{R}^3, \mathbb{R}) \cap L^\infty(\mathbb{R}^3, \mathbb{R})$.

Lemma 7.11 *Let* $\{u_n\} \subset \mathcal{N}_0$ *be such that* $I_0(u_n) \to c_{V_0}$. *Then* $\{u_n\}$ *has a convergent subsequence in* H_0.

Proof. Since $\{u_n\} \subset \mathcal{N}_0$, from Lemma 7.8($a3$), Lemma 7.9($b4$) and the definition of c_{V_0}, we have

$$v_n = m^{-1}(u_n) = \frac{u_n}{\|u_n\|_0} \in S_0^+, \quad \forall n \in \mathbb{N},$$

and

$$\Psi_0(v_n) = I_0(u_n) \to c_{V_0} = \inf_{u \in S_0^+} \Psi_0(u).$$

Although S_0^+ is not a complete C^1 manifold, we still can apply the Ekeland variational principle to the functional $\mathcal{E}_0 : H \to \mathbb{R} \cup \{\infty\}$ defined by $\mathcal{E}_0(u) := \widehat{\Psi}_0(u)$ if $u \in S_0^+$ and $\mathcal{E}_0(u) := \infty$ if $u \in \partial S_0^+$, where $H = \overline{S_0^+}$ is the complete metric space equipped with the metric $d(u, v) := \|u - v\|_0$. In fact, by Lemma 7.8(a4), $\mathcal{E}_0 \in C(H, \mathbb{R} \cup \{\infty\})$, and from Lemma 7.9(b4), $\mathcal{E}_0$ is bounded below. Therefore, there exists a sequence $\{\tilde{v}_n\} \subset S_0^+$ such that $\{\tilde{v}_n\}$ is a $(PS)_{c_{V_0}}$ sequence for Ψ_0 on S_0^+ and

$$\|\tilde{v}_n - v_n\|_0 = o(1).$$

Arguing as in Lemma 7.10 and Lemma 7.9, we conclude the proof. □

Now, we establish the relationship between c_ϵ and c_{V_0}.

Lemma 7.12 *The numbers* c_ϵ *and* c_{V_0} *satisfy the following inequality*

$$\lim_{\epsilon \to 0} c_\epsilon = c_{V_0} < c_0.$$

Proof. Let $\eta \in C_c^\infty(\mathbb{R}^3, [0, 1])$ be a cut-off function such that $\eta = 1$ in $B_{\rho/2}$ and $\mathrm{supp}(\eta) = B_\rho \subset \Lambda$ for some $\rho > 0$. Let us define $\omega_\epsilon(x) := \eta_\epsilon(x)\omega(x)e^{iA(0)\cdot x}$, where $\eta_\epsilon(x) = \eta(\epsilon x)$ for $\epsilon > 0$, ω is a positive and radial ground state solution of problem (7.19). We observe that $|\omega_\epsilon| = \eta_\epsilon \omega$ and $\omega_\epsilon \in H_\varepsilon$ in view of Lemma 7.1. Arguing as in the proof of Lemma 4.6 in Ji and Rădulescu [83], we obtain

$$\lim_{\epsilon \to 0} \|\omega_\epsilon\|_\epsilon^2 = \|\omega\|_{V_0}^2, \tag{7.28}$$

$$\lim_{\epsilon \to 0} [\omega_\epsilon]_{A_\varepsilon}^2 = \int_{\mathbb{R}^3} |\nabla \omega|^2 dx, \tag{7.29}$$

and

$$\lim_{\epsilon \to 0} \int_{\mathbb{R}^3} |\nabla \omega_\epsilon|^6 dx = \int_{\mathbb{R}^3} |\nabla \omega|^6 dx. \tag{7.30}$$

Let $t_\epsilon > 0$ be the unique number such that

$$J_\epsilon(t_\epsilon \omega_\epsilon) = \max_{t \geq 0} J_\epsilon(t\omega_\epsilon).$$

Then t_ϵ satisfies

$$t_\epsilon^2\Big(a[\omega_\epsilon]_{A_\varepsilon}^2+\int_{\mathbb{R}^3}V_\varepsilon(x)|\omega_\epsilon|^2dx\Big)+bt_\epsilon^4[\omega_\epsilon]_{A_\varepsilon}^4=\int_{\mathbb{R}^3}g(\epsilon x,t_\epsilon^2|\omega_\epsilon|^2)t_\epsilon^2|\omega_\epsilon|^2dx$$
$$=\int_{\mathbb{R}^3}f(t_\epsilon^2|\omega_\epsilon|^2)t_\epsilon^2|\omega_\epsilon|^2dx+\int_{\mathbb{R}^3}t_\epsilon^6|\omega_\epsilon|^6dx,$$

where we use $\mathrm{supp}(\eta)\subset\Lambda$ and the definition of $g(x,t)$. Moreover, combining the facts that $\eta=1$ in $B_{\rho/2}$, u is a positive continuous function and hypothesis (f_4), we have

$$\begin{aligned}\frac{1}{t_\epsilon^2}\Big(a[\omega_\epsilon]_{A_\varepsilon}^2+\int_{\mathbb{R}^3}V_\varepsilon(x)|\omega_\epsilon|^2dx\Big)+b[\omega_\epsilon]_{A_\varepsilon}^4&=\frac{1}{t_\epsilon^2}\int_{\mathbb{R}^3}f(t_\epsilon^2|\omega_\epsilon|^2)|\omega_\epsilon|^2dx+\int_{\mathbb{R}^3}t_\epsilon^2|\omega_\epsilon|^6dx\\&\geq\frac{1}{t_\epsilon^2}\int_{\mathbb{R}^3}f(t_\epsilon^2\eta^2(|\epsilon x|)\omega^2(x))\eta^2(|\epsilon x|)\omega^2(x)dz\\&\geq\frac{1}{t_\epsilon^2}\int_{B_{\rho/(2\epsilon)}(0)}f(t_\epsilon^2\omega^2(z))\omega^2(z)dz\\&\geq\frac{1}{t_\epsilon^2}\int_{B_{\rho/2}(0)}f(t_\epsilon^2\omega^2(z))\omega^2(z)dz\\&\geq\frac{f(t_\epsilon^2\gamma^2)}{t_\epsilon^2}\int_{B_{\rho/2}(0)}\omega^2(z)dz\end{aligned}$$

for all $0<\epsilon<1$ and where $\gamma=\min\{\omega(z):|z|\leq\rho/2\}$.

If $t_\epsilon\to+\infty$ as $\epsilon\to0$, by (f_4), we deduce that $[\omega_\epsilon]_{A_\varepsilon}^4\to+\infty$ which contradicts (7.29). Therefore, up to a subsequence, we may assume that $t_\epsilon\to t_0\geq0$ as $\epsilon\to0$.

If $t_\epsilon\to0$, using the fact that f is increasing, the Lebesgue dominated convergence theorem and relation (7.30), we obtain

$$a[\omega_\epsilon]_{A_\varepsilon}^2+\int_{\mathbb{R}^3}V_\varepsilon(x)|\omega_\epsilon|^2dx+bt_\epsilon^2[\omega_\epsilon]_{A_\varepsilon}^4=\int_{\mathbb{R}^3}f(t_\epsilon^2|\omega_\epsilon|^2)|\omega_\epsilon|^2dx+\int_{\mathbb{R}^3}t_\epsilon^4|\omega_\epsilon|^6dx\to0,\ \text{as }\epsilon\to0,$$

which contradicts (7.28). Thus, we have $t_0>0$ and

$$t_0^2\int_{\mathbb{R}^3}(a|\nabla\omega|^2+V_0\omega^2)dx+bt_0^4\Big(\int_{\mathbb{R}^3}|\nabla\omega|^2dx\Big)^2=\int_{\mathbb{R}^2}f(t_0^2\omega^2)t_0^2\omega^2dx+\int_{\mathbb{R}^3}t_0^6|\omega|^6dx,$$

so that $t_0\omega\in\mathcal{N}_{V_0}$. Since $\omega\in\mathcal{N}_{V_0}$, we obtain that $t_0=1$ and so, using the Lebesgue dominated convergence theorem, we get

$$\lim_{\epsilon\to0}\int_{\mathbb{R}^3}F(|t_\epsilon\omega_\epsilon|^2)dx=\int_{\mathbb{R}^2}F(\omega^2)dx.$$

Hence

$$\lim_{\epsilon\to0}J_\epsilon(t_\epsilon\omega_\epsilon)=I_{V_0}(u)=c_{V_0}.$$

Since $c_\epsilon\leq\max_{t\geq0}J_\epsilon(t\omega_\epsilon)=J_\epsilon(t_\epsilon\omega_\epsilon)$, we can conclude that $\limsup_{\epsilon\to0}c_\epsilon\leq c_{V_0}$. Moreover, by (7.6), (7.20) and $I_{V_0}(|u|)\leq J_\epsilon(u)$ for any $u\in H_\varepsilon$, we have $c_{V_0}\leq c_\epsilon$. Then $c_{V_0}\leq\liminf_{\epsilon\to0}c_\epsilon$. Combining with the previous arguments, we conclude that $\lim_{\epsilon\to0}c_\epsilon=c_{V_0}<c_0$. The proof is now complete. □

Remark 7.1 *From Lemma 7.8 and Lemma 7.6, we see that for $\epsilon>0$ small, problem* (7.5) *has a ground state solution u_ϵ such that $J_\varepsilon(u_\epsilon)=c_\epsilon$ and $J_\varepsilon'(u_\epsilon)=0$.*

7.4.2 Auxiliary asymptotic properties

In what follows, we prove a multiplicity result for the modified problem (7.5) using the Ljusternik-Schnirelmann category theory. We first provide some useful preliminaries.

Let $\delta > 0$ be such that $M_\delta \subset \Lambda$, $\omega \in H^1(\mathbb{R}^3, \mathbb{R})$ be a positive ground state solution of the limit problem (7.19), and $\eta \in C^\infty(\mathbb{R}^+, [0,1])$ be a nonincreasing cut-off function defined in $[0, +\infty)$ such that $\eta(t) = 1$ if $0 \leq t \leq \delta/2$ and $\eta(t) = 0$ if $t \geq \delta$.

For any $y \in M$, let us introduce the function

$$\Psi_{\varepsilon,y}(x) := \eta(|\varepsilon x - y|)\omega\Big(\frac{\varepsilon x - y}{\varepsilon}\Big) \exp\Big(i\tau_y\Big(\frac{\varepsilon x - y}{\varepsilon}\Big)\Big),$$

where

$$\tau_y(x) := \sum_i^3 A_i(y)x_i.$$

Let $t_\varepsilon > 0$ be the unique positive number such that

$$\max_{t \geq 0} J_\varepsilon(t\Psi_{\varepsilon,y}) = J_\varepsilon(t_\varepsilon \Psi_{\varepsilon,y}).$$

Note that $t_\varepsilon \Psi_{\varepsilon,y} \in \mathcal{N}_\varepsilon$.

Let us define $\Phi_\varepsilon : M \to \mathcal{N}_\varepsilon$ as

$$\Phi_\varepsilon(y) := t_\varepsilon \Psi_{\varepsilon,y}.$$

By construction, $\Phi_\varepsilon(y)$ has compact support for any $y \in M$.

Moreover, arguing as in Lemma 7.8, the energy of the above functions has the following behavior as $\varepsilon \to 0^+$.

Lemma 7.13 *The limit*

$$\lim_{\varepsilon \to 0^+} J_\varepsilon(\Phi_\varepsilon(y)) = c_{V_0}$$

holds uniformly in $y \in M$.

Now we define the barycenter map.

Let $\rho > 0$ be such that $M_\delta \subset B_\rho$ and consider $\Upsilon : \mathbb{R}^3 \to \mathbb{R}^3$ defined by setting

$$\Upsilon(x) := \begin{cases} x, & \text{if } |x| < \rho, \\ \rho x/|x|, & \text{if } |x| \geq \rho. \end{cases}$$

The barycenter map $\beta_\varepsilon : \mathcal{N}_\varepsilon \to \mathbb{R}^3$ is defined by

$$\beta_\varepsilon(u) := \frac{1}{\|u\|_4^4} \int_{\mathbb{R}^3} \Upsilon(\varepsilon x)|u(x)|^4 dx.$$

We have the following asymptotic property.

Lemma 7.14 *The limit*

$$\lim_{\varepsilon \to 0^+} \beta_\varepsilon(\Phi_\varepsilon(y)) = y$$

holds uniformly in $y \in M$.

Proof. Assume by contradiction that there exist $\kappa > 0$, $\{y_n\} \subset M$ and $\varepsilon_n \to 0$ such that

$$|\beta_{\varepsilon_n}(\Phi_{\varepsilon_n}(y_n)) - y_n| \geq \kappa. \tag{7.31}$$

Using the change of variable $z = (\varepsilon_n x - y_n)/\varepsilon_n$, we can see that

$$\beta_{\varepsilon_n}(\Phi_{\varepsilon_n}(y_n)) = y_n + \frac{\displaystyle\int_{\mathbb{R}^3} (\Upsilon(\varepsilon_n z + y_n) - y_n)\eta^4(|\varepsilon_n z|)\omega^4(z)dz}{\displaystyle\int_{\mathbb{R}^3} \eta^4(|\varepsilon_n z|)\omega^4(z)dz}.$$

Taking into account that $\{y_n\} \subset M \subset M_\delta \subset B_\rho$ and applying the Lebesgue dominated convergence theorem, we obtain

$$|\beta_{\varepsilon_n}(\Phi_{\varepsilon_n}(y_n)) - y_n| = o(1),$$

which contradicts relation (7.31). The proof is now complete. □

Now, we prove a useful compactness property.

Proposition 7.1 *Let $\varepsilon_n \to 0^+$ and $(u_n) \subset \mathcal{N}_{\varepsilon_n}$ be such that $J_{\varepsilon_n}(u_n) \to c_{V_0}$. Then there exists $(\tilde{y}_n) \subset \mathbb{R}^N$ such that the sequence $\{|v_n|\} \subset H^1(\mathbb{R}^3, \mathbb{R})$, where $v_n(x) := u_n(x + \tilde{y}_n)$, has a convergent subsequence in $H^1(\mathbb{R}^3, \mathbb{R})$. Moreover, up to a subsequence, $y_n := \varepsilon_n \tilde{y}_n \to y \in M$ as $n \to +\infty$.*

Proof. Since $J'_{\varepsilon_n}(u_n)[u_n] = 0$ and $J_{\varepsilon_n}(u_n) \to c_{V_0}$, arguing as in the proof of Lemma 7.5, we have that there exists $C > 0$ such that $\|u_n\|_{\varepsilon_n} \leq C$ for all $n \in \mathbb{N}$.

Arguing as in the proof of Lemma 7.10 and recalling that $c_{V_0} > 0$, we have that there exist a sequence $\{\tilde{y}_n\} \subset \mathbb{R}^3$ and constants R, $\beta > 0$ such that

$$\liminf_{n\to\infty} \int_{B_R(\tilde{y}_n)} |u_n|^2 dx \geq \beta. \tag{7.32}$$

Now, let us consider the sequence $\{|v_n|\} \subset H^1(\mathbb{R}^3, \mathbb{R})$, where $v_n(x) := u_n(x + \tilde{y}_n)$. By the diamagnetic inequality (7.2), we get that $\{|v_n|\}$ is bounded in $H^1(\mathbb{R}^3, \mathbb{R})$, and using (7.32), we may assume that $|v_n| \rightharpoonup v$ in $H^1(\mathbb{R}^3, \mathbb{R})$ for some $v \neq 0$.

Let now $t_n > 0$ be such that $\tilde{v}_n := t_n|v_n| \in \mathcal{N}_{V_0}$, and set $y_n := \varepsilon_n \tilde{y}_n$.

Using the diamagnetic inequality (7.2) again, we have

$$c_{V_0} \leq I_0(\tilde{v}_n) \leq \max_{t\geq 0} J_{\varepsilon_n}(tu_n) = J_{\varepsilon_n}(u_n) = c_{V_0} + o(1),$$

which yields $I_0(\tilde{v}_n) \to c_{V_0}$ as $n \to +\infty$.

Since the sequences $\{|v_n|\}$ and $\{\tilde{v}_n\}$ are bounded in $H^1(\mathbb{R}^3, \mathbb{R})$ and $|v_n| \not\to 0$ in $H^1(\mathbb{R}^3, \mathbb{R})$, then $\{t_n\}$ is also bounded and so, up to a subsequence, we may assume that $t_n \to t_0 \geq 0$.

We claim that $t_0 > 0$. Indeed, if $t_0 = 0$, then, since $\{|v_n|\}$ is bounded, we have $\tilde{v}_n \to 0$ in $H^1(\mathbb{R}^3, \mathbb{R})$, that is, $I_0(\tilde{v}_n) \to 0$, which contradicts $c_{V_0} > 0$.

Thus, up to a subsequence, we may assume that $\tilde{v}_n \rightharpoonup \tilde{v} := t_0 v \neq 0$ in $H^1(\mathbb{R}^3, \mathbb{R})$, and, by Lemma 7.11, we can deduce that $\tilde{v}_n \to \tilde{v}$ in $H^1(\mathbb{R}^3, \mathbb{R})$, which gives $|v_n| \to v$ in $H^1(\mathbb{R}^3, \mathbb{R})$.

Now we show the final part, namely that $\{y_n\}$ has a subsequence such that $y_n \to y \in M$. Assume by contradiction that $\{y_n\}$ is not bounded and so, up to a subsequence, $|y_n| \to +\infty$ as $n \to +\infty$. Choose $R > 0$ such that $\Lambda \subset B_R(0)$. Then for n large enough, we have $|y_n| > 2R$, and, for any $x \in B_{R/\varepsilon_n}(0)$,

$$|\varepsilon_n x + y_n| \geq |y_n| - \varepsilon_n|x| > R.$$

Since $u_n \in \mathcal{N}_{\varepsilon_n}$, using (V_1) and the diamagnetic inequality (7.2), we get that

$$\begin{aligned}
&\int_{\mathbb{R}^3} (a|\nabla|v_n||^2 + V_0|v_n|^2)dx \leq a[v_n]^2_{A_\varepsilon} + \int_{\mathbb{R}^3} V(\varepsilon_n x + y_n)|v_n|^2 dx + b[v_n]^4_{A_\varepsilon} \\
&\leq \int_{\mathbb{R}^3} g(\varepsilon_n x + y_n, |v_n|^2)|v_n|^2 dx \leq \int_{B_{R/\varepsilon_n}(0)} \tilde{f}(|v_n|^2)|v_n|^2 dx \\
&+ \int_{B^c_{R/\varepsilon_n}(0)} f(|v_n|^2)|v_n|^2 dx + \int_{B^c_{R/\varepsilon_n}(0)} |v_n|^6 dx. \qquad (7.33)
\end{aligned}$$

Since $|v_n| \to v$ in $H^1(\mathbb{R}^3, \mathbb{R})$ and $\tilde{f}(t) \leq V_0/K$, we can see that (7.33) yields

$$\min\left\{1, V_0\left(1 - \frac{1}{K}\right)\right\} \int_{\mathbb{R}^3} (|\nabla|v_n||^2 + |v_n|^2)dx = o(1),$$

that is $|v_n| \to 0$ in $H^1(\mathbb{R}^3, \mathbb{R})$, which contradicts to $v \not\equiv 0$.

Therefore, we may assume that $y_n \to y_0 \in \mathbb{R}^3$. Assume by contradiction that $y_0 \notin \overline{\Lambda}$. Then there exists $r > 0$ such that for every n large enough we have that $|y_n - y_0| < r$ and $B_{2r}(y_0) \subset \overline{\Lambda}^c$. Then, if $x \in B_{r/\varepsilon_n}(0)$, we have that $|\varepsilon_n x + y_n - y_0| < 2r$ so that $\varepsilon_n x + y_n \in \overline{\Lambda}^c$ and so, arguing as before, we reach a contradiction. Thus, $y_0 \in \overline{\Lambda}$.

To prove that $V(y_0) = V_0$, we suppose by contradiction that $V(y_0) > V_0$. Using the Fatou's lemma, the change of variable $z = x + \tilde{y}_n$ and $\max_{t\geq 0} J_{\varepsilon_n}(tu_n) = J_{\varepsilon_n}(u_n)$, we obtain

$$\begin{aligned}
c_{V_0} = I_0(\tilde{v}) &< \frac{1}{2}\int_{\mathbb{R}^3} (a|\nabla\tilde{v}|^2 + V(y_0)|\tilde{v}|^2)dx + \frac{b}{4}\Big(\int_{\mathbb{R}^3} |\nabla\tilde{v}|^2 dx\Big)^2 - \frac{1}{2}\int_{\mathbb{R}^3} F(|\tilde{v}|^2)dx \\
&\leq \liminf_{n\to\infty} \Big(\frac{1}{2}\int_{\mathbb{R}^3} (a|\nabla\tilde{v}_n|^2 + V(\varepsilon_n x + y_n)|\tilde{v}_n|^2)dx + \frac{b}{4}\Big(\int_{\mathbb{R}^3} |\nabla\tilde{v}_n|^2 dx\Big)^2 - \frac{1}{2}\int_{\mathbb{R}^3} F(|\tilde{v}_n|^2)dx\Big) \\
&= \liminf_{n\to\infty} \Big(\frac{t_n^2}{2}\int_{\mathbb{R}^3} (a|\nabla|u_n||^2 + V(\varepsilon_n z)|u_n|^2)dx + \frac{t_n^4 b}{4}\Big(\int_{\mathbb{R}^3} |\nabla u_n|^2 dx\Big)^2 - \frac{1}{2}\int_{\mathbb{R}^3} F(|t_n u_n|^2)dx\Big) \\
&\leq \liminf_{n\to\infty} J_{\varepsilon_n}(t_n u_n) \leq \liminf_{n\to\infty} J_{\varepsilon_n}(u_n) = c_{V_0}
\end{aligned}$$

which is impossible and the proof is complete. □

Let now

$$\tilde{\mathcal{N}}_\varepsilon := \{u \in \mathcal{N}_\varepsilon : J_\varepsilon(u) \leq c_{V_0} + h(\varepsilon)\},$$

where $h : \mathbb{R}^+ \to \mathbb{R}^+$, $h(\varepsilon) \to 0$ as $\varepsilon \to 0^+$.

Fixing $y \in M$, by Lemma 7.13, $|J_\varepsilon(\Phi_\varepsilon(y)) - c_{V_0}| \to 0$ as $\varepsilon \to 0^+$, we get that $\tilde{\mathcal{N}}_\varepsilon \neq \emptyset$ for any $\varepsilon > 0$ small enough.

We have the following relation between $\tilde{\mathcal{N}}_\varepsilon$ and the barycenter map.

Lemma 7.15 *We have*

$$\lim_{\varepsilon\to 0^+} \sup_{u\in\tilde{\mathcal{N}}_\varepsilon} \operatorname{dist}(\beta_\varepsilon(u), M_\delta) = 0.$$

Proof. Let $\varepsilon_n \to 0^+$ as $n \to +\infty$. For any $n \in \mathbb{N}$, there exists $u_n \in \tilde{\mathcal{N}}_{\varepsilon_n}$ such that

$$\sup_{u\in\tilde{\mathcal{N}}_{\varepsilon_n}} \inf_{y\in M_\delta} |\beta_{\varepsilon_n}(u) - y| = \inf_{y\in M_\delta} |\beta_{\varepsilon_n}(u_n) - y| + o(1).$$

Therefore, it is enough to prove that there exists $\{y_n\} \subset M_\delta$ such that

$$\lim_{n\to\infty} |\beta_{\varepsilon_n}(u_n) - y_n| = 0.$$

By the diamagnetic inequality (7.2), we can see that $I_{V_0}(t|u_n|) \leq J_{\varepsilon_n}(tu_n)$ for any $t \geq 0$. Therefore, recalling that $\{u_n\} \subset \tilde{\mathcal{N}}_{\varepsilon_n} \subset \mathcal{N}_{\varepsilon_n}$, we can deduce that

$$c_{V_0} \leq \max_{t\geq 0} I_{V_0}(t|u_n|) \leq \max_{t\geq 0} J_{\varepsilon_n}(tu_n) = J_{\varepsilon_n}(u_n) \leq c_{V_0} + h(\varepsilon_n) \qquad (7.34)$$

which implies that $J_{\varepsilon_n}(u_n) \to c_{V_0}$ as $n \to +\infty$. Then Proposition 7.1 implies that there exists $\{\tilde{y}_n\} \subset \mathbb{R}^3$ such that $y_n = \varepsilon_n \tilde{y}_n \in M_\delta$ for n large enough.

Thus, making the change of variable $z = x - \tilde{y}_n$, we get

$$\beta_{\varepsilon_n}(u_n) = y_n + \frac{\int_{\mathbb{R}^3}(\Upsilon(\varepsilon_n z + y_n) - y_n)|u_n(z+\tilde{y}_n)|^4 dz}{\int_{\mathbb{R}^3}|u_n(z+\tilde{y}_n)|^4 dz}.$$

Since, up to a subsequence, $|u_n|(\cdot + \tilde{y}_n)$ converges strongly in $H^1(\mathbb{R}^3, \mathbb{R})$ and $\varepsilon_n z + y_n \to y \in M$ for any $z \in \mathbb{R}^3$, we conclude. □

7.4.3 Multiplicity of solutions for the modified problem

The following result establishes a relation between the topology of M and the number of solutions of the modified problem (7.5).

Theorem 7.2 *For any $\delta > 0$ such that $M_\delta \subset \Lambda$, there exists $\tilde{\varepsilon}_\delta > 0$ such that, for any $\varepsilon \in (0, \tilde{\varepsilon}_\delta)$, problem (7.5) has at least $\mathrm{cat}_{M_\delta}(M)$ nontrivial solutions.*

Proof. For any $\epsilon > 0$, we define the function $\pi_\epsilon : M \to S_\varepsilon^+$ by

$$\pi_\epsilon(y) = m_\varepsilon^{-1}(\Phi_\epsilon(y)), \ \forall y \in M.$$

By Lemma 7.13 and Lemma 7.4($B4$), we obtain

$$\lim_{\epsilon \to 0} \Psi_\epsilon(\pi_\epsilon(y)) = \lim_{\epsilon \to 0} J_\epsilon(\Phi_\epsilon(y)) = c_{V_0}, \text{ uniformly in } y \in M.$$

Hence, there is a number $\hat{\epsilon} > 0$ such that the set $\tilde{S}_\varepsilon^+ := \{u \in S_\varepsilon^+ : \Psi_\varepsilon(u) \le c_{V_0} + h(\varepsilon)\}$ is nonempty, for all $\epsilon \in (0, \hat{\epsilon})$, since $\pi_\epsilon(M) \subset \tilde{S}_\varepsilon^+$. Here h is given in the definition of $\tilde{\mathcal{N}}_\varepsilon$.

Given $\delta > 0$, by Lemma 7.13, Lemma 7.3($A3$), Lemma 7.14, and Lemma 7.15, we can find $\tilde{\varepsilon}_\delta > 0$ such that for any $\varepsilon \in (0, \tilde{\varepsilon}_\delta)$, the following diagram

$$M \xrightarrow{\Phi_\varepsilon} \Phi_\varepsilon(M) \xrightarrow{m_\varepsilon^{-1}} \pi_\epsilon(M) \xrightarrow{m_\varepsilon} \Phi_\varepsilon(M) \xrightarrow{\beta_\varepsilon} M_\delta$$

is well defined and continuous. From Lemma 7.14, we can choose a function $\Theta(\epsilon, z)$ with $|\Theta(\epsilon, z)| < \frac{\delta}{2}$ uniformly in $z \in M$, for all $\epsilon \in (0, \hat{\epsilon})$ such that $\beta_\varepsilon(\Phi_\varepsilon(z)) = z + \Theta(\epsilon, z)$ for all $z \in M$. Define $H(t, z) = z + (1-t)\Theta(\epsilon, z)$. Then $H : [0,1] \times M \to M_\delta$ is continuous. Clearly, $H(0, z) = \beta_\varepsilon(\Phi_\varepsilon(z))$, $H(1, z) = z$ for all $z \in M$. That is, $H(t, z)$ is a homotopy between $\beta_\varepsilon \circ \Phi_\varepsilon = (\beta_\varepsilon \circ m_\varepsilon) \circ \pi_\epsilon$ and the embedding $\iota : M \to M_\delta$. Thus, this fact implies that

$$\mathrm{cat}_{\pi_\epsilon(M)}(\pi_\epsilon(M)) \ge \mathrm{cat}_{M_\delta}(M). \tag{7.35}$$

By Corollary 7.1 and the abstract category theorem [144], Ψ_ε has at least $\mathrm{cat}_{\pi_\epsilon(M)}(\pi_\epsilon(M))$ critical points on S_ε^+. Therefore, from Lemma 7.4($B4$) and (7.35), we have that J_ε has at least $\mathrm{cat}_{M_\delta}(M)$ critical points in $\tilde{\mathcal{N}}_\varepsilon$ which implies that problem (7.5) has at least $\mathrm{cat}_{M_\delta}(M)$ solutions. □

7.5 Proof of Theorem 7.1

The idea is to show that the solutions u_ε obtained in Theorem 7.2 satisfy

$$|u_\varepsilon(x)|^2 \le a_0 \text{ for } x \in \Lambda_\varepsilon^c$$

for ε small. The key ingredient is the following result.

Lemma 7.16 *Let $\varepsilon_n \to 0^+$ and $u_n \in \tilde{\mathcal{N}}_{\varepsilon_n}$ be a solution of problem (7.5) for $\varepsilon = \varepsilon_n$. Then $J_{\varepsilon_n}(u_n) \to c_{V_0}$. Moreover, there exists $\{\tilde{y}_n\} \subset \mathbb{R}^N$ such that, if $v_n(x) := u_n(x + \tilde{y}_n)$, we have that $\{|v_n|\}$ is bounded in $L^\infty(\mathbb{R}^N, \mathbb{R})$ and*

$$\lim_{|x|\to+\infty} |v_n(x)| = 0 \quad \text{uniformly in } n \in \mathbb{N}.$$

We use the Moser iteration method to prove the above lemma. For our problem (7.5), there is more one nonlocal term and the nonlinear term has the critical growth for problem. The proof follows as in Ji and Rădulescu [83, Lemma 5.1], just after making some appropriate changes.

Proof of Theorem 7.1. Let $\delta > 0$ be such that $M_\delta \subset \Lambda$. We want to show that there exists $\hat{\varepsilon}_\delta > 0$ such that for any $\varepsilon \in (0, \hat{\varepsilon}_\delta)$ and any $u_\varepsilon \in \tilde{\mathcal{N}}_\varepsilon$ solution of problem (7.5), it holds

$$\|u_\varepsilon\|^2_{L^\infty(\Lambda^c_\varepsilon)} \le a_0. \tag{7.36}$$

We argue by contradiction and assume that there is a sequence $\varepsilon_n \to 0$ such that for every n there exists $u_n \in \tilde{\mathcal{N}}_{\varepsilon_n}$ which satisfies $J'_{\varepsilon_n}(u_n) = 0$ and

$$\|u_n\|^2_{L^\infty(\Lambda^c_{\varepsilon_n})} > a_0. \tag{7.37}$$

Arguing as in Lemma 7.16, we have that $J_{\varepsilon_n}(u_n) \to c_{V_0}$, and therefore we can use Proposition 7.1 to obtain a sequence $\{\tilde{y}_n\} \subset \mathbb{R}^3$ such that $y_n := \varepsilon_n \tilde{y}_n \to y_0$ for some $y_0 \in M$. Then, we can find $r > 0$, such that $B_r(y_n) \subset \Lambda$, and so $B_{r/\varepsilon_n}(\tilde{y}_n) \subset \Lambda_{\varepsilon_n}$ for all n large enough.

Using Lemma 7.16, there exists $R > 0$ such that $|v_n|^2 \le a_0$ in $B^c_R(0)$ and n large enough, where $v_n = u_n(\cdot + \tilde{y}_n)$. Hence $|u_n|^2 \le a_0$ in $B^c_R(\tilde{y}_n)$ and n large enough. Moreover, if n is so large that $r/\varepsilon_n > R$, then $\Lambda^c_{\varepsilon_n} \subset B^c_{r/\varepsilon_n}(\tilde{y}_n) \subset B^c_R(\tilde{y}_n)$, which gives $|u_n|^2 \le a$ for any $x \in \Lambda^c_{\varepsilon_n}$. This contradicts relation (7.37) and proves the claim.

Let now $\varepsilon_\delta := \min\{\hat{\varepsilon}_\delta, \tilde{\varepsilon}_\delta\}$, where $\tilde{\varepsilon}_\delta > 0$ is given by Theorem 7.2. Then we have at least $\mathrm{cat}_{M_\delta}(M)$ nontrivial solutions for problem (7.5). If $u_\varepsilon \in \tilde{\mathcal{N}}_\varepsilon$ is one of these solutions, then, by (7.36) and the definition of g, we conclude that u_ε is also a solution to problem (7.3).

Finally, we study the behavior of the maximum points of $|\hat{u}_\varepsilon|$, where $\hat{u}_\varepsilon(x) := u_\varepsilon(x/\varepsilon)$ is a solution to problem (7.1), as $\varepsilon \to 0^+$.

Take $\varepsilon_n \to 0^+$ and the sequence $\{u_n\}$ where each u_n is a solution of (7.5) for $\varepsilon = \varepsilon_n$. From the definition of g, there exists $\gamma \in (0, a_0)$ such that

$$g(\varepsilon x, t^2)t^2 \le \frac{V_0}{K} t^2, \quad \text{for all } x \in \mathbb{R}^N,\ |t| \le \gamma.$$

Arguing as above we can take $R > 0$ such that, for n large enough,

$$\|u_n\|_{L^\infty(B^c_R(\tilde{y}_n))} < \gamma. \tag{7.38}$$

Up to a subsequence, we may also assume that for n large enough

$$\|u_n\|_{L^\infty(B_R(\tilde{y}_n))} \ge \gamma. \tag{7.39}$$

Indeed, if (7.39) does not hold, up to a subsequence, if necessary, we have $\|u_n\|_\infty < \gamma$. Thus, since $J'_{\varepsilon_n}(u_{\varepsilon_n}) = 0$, using (g_5) and the diamagnetic inequality (7.2) that

$$\begin{aligned}\int_{\mathbb{R}^3} (a|\nabla|u_n||^2 + V_0|u_n|^2)dx + b\Big(\int_{\mathbb{R}^3} |\nabla|u_n||^2 dx\Big)^2 &\le \int_{\mathbb{R}^3} g(\varepsilon_n x, |u_n|^2)|u_n|^2 dx \\ &\le \frac{V_0}{K}\int_{\mathbb{R}^3} |u_n|^2 dx.\end{aligned}$$

Since $K > 2$, we obtain $\|u_n\| = 0$, which is a contradiction.

Taking into account (7.38) and (7.39), we can infer that the global maximum points p_n of $|u_{\varepsilon_n}|$ belongs to $B_R(\tilde{y}_n)$, that is $p_n = q_n + \tilde{y}_n$ for some $q_n \in B_R$. Recalling that the associated solution of problem (7.1) is $\hat{u}_n(x) = u_n(x/\varepsilon_n)$, we can see that a maximum point η_{ε_n} of $|\hat{u}_n|$ is $\eta_{\varepsilon_n} = \varepsilon_n\tilde{y}_n + \varepsilon_n q_n$. Since $q_n \in B_R$, $\varepsilon_n\tilde{y}_n \to y_0$ and $V(y_0) = V_0$, the continuity of V allows to conclude that

$$\lim_{n\to\infty} V(\eta_{\varepsilon_n}) = V_0.$$

The proof is now complete. □

7.6 Historical comments

For problem (7.1), there is a vast literature concerning the existence and concentration of solutions for the case without magnetic potential, that is, if $A \equiv 0$. The first result in this direction was given by Floer and Weinstein in [58], where the case $N = 1$ and $f = i_{\mathbb{R}}$ is considered. In [43], del Pino and Felmer studied the following problem

$$\begin{cases} -\varepsilon^2\Delta u + V(x)u = f(u) \text{ in } \Omega, \\ u = 0 \text{ on } \partial\Omega, \ u > 0 \text{ in } \Omega, \end{cases}$$

where Ω is a possibly unbounded domain in $\mathbb{R}^N$, $N \geq 3$. By introducing a penalization method, they found solutions of the above problem that concentrate around the local minimum of V. More precisely, the authors assumed that there exists a bounded open set $\Lambda \subset \Omega$ such that

$$\inf_{x\in\Lambda} V(x) < \min_{x\in\partial\Lambda} V(x), \tag{7.40}$$

and the nonlinearity f satisfies the subcritical growth condition. In [70], He and Zou considered the following fractional Schrödinger equation

$$\epsilon^{2s}(-\Delta)^s u + V(x)u = f(u) + u^{2_s^*-1} \text{ in } \mathbb{R}^N,$$

where V satisfies the local assumption (7.40), and f is subcritical. By using the Nehari manifold method and the Ljusternik-Schnirelmann category theory, the authors obtained the multiplicity of positive solutions. We notice that f is only continuous in [70], hence the Nehari manifold is only a topological manifold. Thus, the critical point theory in C^1 manifold cannot be applied in this framework. To overcome this difficulty, He and Zou [70] applied the method developed by Szulkin and Weth [144]. There are also many results about the existence, multiplicity and concentration of solutions for Kirchhoff equations, that is, provided that $A = 0$ and $a, b > 0$. In [127], Perera and Zhang studied the existence of solutions for Kirchhoff equation by using the Yang index and critical groups. Later on, He and Zou [70] studied the existence, multiplicity and concentration properties of positive solutions for problem (7.1) without critical term and magnetic field by using the Nehari manifold method, the penalization technique and the Ljusternik-Schnirelmann category theory. We notice that the nonlinear term f is a C^1 function in this chapter, which allows to apply the critical point theory in C^1 manifolds. Next, He and Zou [71] applied the method introduced by Szulkin and Weth [144] and studied multiplicity and concentration behavior of positive solutions for a fractional Kirchhoff equation where the nonlinear term f is only continuous.

The first result involving the magnetic field was obtained by Esteban and Lions [49]. They used the concentration-compactness principle and minimization arguments to obtain solutions for $\varepsilon > 0$ fixed and $N = 2, 3$. In particular, related with the content of this chapter, we also refer to Ji and Rădulescu [83] who used Nehari manifold analysis, penalization techniques and the Ljusternik-Schnirelmann category theory to study multiplicity and concentration results for a magnetic Schrödinger equation in which the subcritical nonlinearity $f \in C(\mathbb{R}, \mathbb{R})$. Multiplicity and concentrating phenomena for magnetic Kirchhoff equations with subcritical growth and continuous reaction were studied by Ji and Rădulescu [85], while the supercritical case was studied by Gao and Tan [61].

7.7 Final remarks, perspectives, and open problems

This chapter deals with the study of concentration properties of solutions for the magnetic Kirchhoff equation with critical growth in the three-dimensional case.

Why magnetic Kirchhoff equations with critical growth?

Because this is a triply nontrivial problem, by combining three basic concepts, as basic features of equations of this type:

(a) *Magnetic*: it breaks translation invariance;

(b) *Kirchhoff*: it couples local and nonlocal behavior;

(c) *Critical growth*: it creates concentration phenomena.

The most promising immediate perspectives associated with the analysis developed in this chapter seem to be the following direction:

(i) *Fractional magnetic Kirchhoff problems*: the analysis of the same equation after replacing the magnetic Laplace operator $(-\Delta)_A u$ with the fractional magnetic Laplacian $(-\Delta)^s_A u$. In this case, study concentration properties for $s \in (0, 1)$ with critical growth $2^*_s = 2N/(N - 2s)$.

(ii) *Variable order magnetic Kirchhoff equations*: the study of problems driven by the magnetic $p(x)$-Laplacian defined as $-\text{div}\,(|\nabla_A u|^{p(x)-2}\nabla_A u)$.

(iii) *Double-phase problems* driven by the operator

$$-\text{div}\,(|\nabla_A u|^{p-2}\nabla_A u + a(x)|\nabla_A u|^{q-2}\nabla_A u),$$

where $a \geq 0$ is a suitable modulating potential.

(iv) *Higher-order magnetic Kirchhoff equations*: the analysis of a related Kirchhoff problem driven by the biharmonic magnetic operator $\Delta^2_A u$.

The content of this chapter is based on the results established in [84]. This study was complemented by Deng and Luo [44] in the context of magnetic Kirchhoff equations with competing potentials and by Liu and Qian [105] for fractional Choquard-Kirchhoff equations with magnetic field.

Some open problems generated by the model studied in this chapter are the following.

Open Problem 1. Study the concentration of solutions to the magnetic Kirchhoff equation with Ginzburg-Landau nonlinearity

$$\begin{cases} -\left(a\epsilon^2 + b\epsilon\,[u]^2_{A/\epsilon}\right)\Delta_{A/\epsilon}u + V(x)u = \dfrac{1}{\epsilon^2}\,u(1-|u|^2) & \text{in } \mathbb{R}^3, \\ u \in H^1(\mathbb{R}^3, \mathbb{C}). \end{cases}$$

Another interesting direction is concerned with general nonlinearities which do not fulfill the Ambrosetti-Rabinowitz condition (f_3).

Open Problem 2. Study concentration properties of solutions to problem (7.1) with Berestycki-Lions nonlinearity.

The following open problem is concerned with the critical case with Hardy nonlinearity. Interesting issues in this case are the magnetic Hardy inequalities and the concentration-compactness with magnetic Hardy potentials.

Open Problem 3. Study the multiplicity and concentration of solutions to the following magnetic Kirchhoff equation with Hardy singular potential

$$\begin{cases} -\left(a\epsilon^2 + b\epsilon\,[u]^2_{A/\epsilon}\right)\Delta_{A/\epsilon}u + V(x)u = f(|u|^2)u + \dfrac{|u|^{2^*-2}u}{|x|^\alpha} & \text{in } \mathbb{R}^N, \\ u \in H^1(\mathbb{R}^N, \mathbb{C}). \end{cases}$$

The next problem is concerned with the study of the magnetic Kirchhoff equation for other unbounded domains.

Open Problem 4. Study concentration properties of solutions to problem (7.1) in the case of exterior domains or cylindrical domains. How does concentration occur without boundary constraints?

What about the case of domains with corners or cracks? Does concentration still occur at specific geometric points?

Finally, we propose a thorough analysis in the framework of evolution equations with magnetic potential. In this case, the nonlocal term complicates stability analysis.

Open Problem 5. Study orbital stability of concentrating solutions for time-dependent magnetic Kirchhoff equations and describe the gradient flow structure with magnetic field. Establish blow-up properties versus concentration phenomena in the time-dependent setting.

7.8 Glossary

Magnetic Laplace Operator: This is a fundamental operator in mathematical physics that generalizes the classical Laplacian to include the effects of magnetic fields.

Magnetic Field in Kirchhoff Equations: It creates a rich mathematical structure with deep physical interpretations, mainly in the context of quantum mechanics, continuum mechanics, and Ginzburg-Landau superconductivity theory.

Magnetic Kirchhoff Equation: It is a complex-valued, nonlocal elliptic equation that combines quantum mechanics (magnetic Laplacian) with continuum mechanics (Kirchhoff nonlocality).

Diamagnetic Inequality: This is a fundamental result that relates the gradient of a function's magnitude to its magnetic gradient. The inequality shows that the magnetic gradient norm dominates the ordinary gradient norm of the magnitude.

Magnetic Sobolev Space: This function space generalizes standard Sobolev spaces to accommodate the magnetic gradient ∇_A. Current relevant research directions include magnetic Strichartz estimates, magnetic Bourgain spaces, and magnetic modulation spaces.

8

Fractional Kirchhoff equations with discontinuous reaction

Why Kirchhoff equations with discontinuous reaction? This study is important since it shifts the problem from the smooth world of standard critical point theory to the more complex terrain of nonsmooth analysis, specifically for applications where the response is not continuous or the forcing term is not a function but a multivalued map. The key feature of this analysis is that it combines three major analytical complexities (nonlocality, nondifferentiability and potential-driven localization) into a single challenging problem. This triple interaction implies the following challenge layers:

(i) The operator is nonlocal in two ways, making compactness hard to obtain.

(ii) The discontinuous reaction implies that the associated energy is not differentiable (only locally Lipschitz), so standard critical point theory fails.

(iii) In order to establish concentration of solutions and to prove localization, one must show decay outside a prescribed set and rule out energy leaking to infinity.

This chapter is concerned with the analysis of concentration properties of solutions to the following fractional Kirchhoff equation with discontinuous nonlinearity

$$\begin{cases} \left(\epsilon^{2\alpha}a+\epsilon^{4\alpha-3}b\displaystyle\int_{\mathbb{R}^3}|(-\Delta)^{\frac{\alpha}{2}}u|^2\mathrm{d}x\right)(-\Delta)^\alpha u+V(x)u=H(u-\beta)f(u) & \text{in } \mathbb{R}^3,\\ u\in H^\alpha(\mathbb{R}^3),\quad u>0 & \text{in } \mathbb{R}^3, \end{cases}$$

where $\epsilon,\beta>0$ are small parameters, $\alpha\in(\frac{3}{4},1)$ and a,b are positive constants, $(-\Delta)^\alpha$ is the fractional Laplacian operator, H is the Heaviside function, V is a positive continuous potential, and f is a superlinear continuous function with subcritical growth. The main result in this chapter deals with the existence and multiplicity of solutions, as well as with concentration properties of solutions for small values of two positive parameters. The main feature is that the reaction term is allowed to be discontinuous with respect to the unknown function, which naturally arises in the applied sciences. For instance, equations of this type can model various physical processes in continuum mechanics, phase transition, and materials science where long-range interactions (fractional diffusion), nonlocal elasticity (Kirchhoff term), and discontinuous responses (stick-slip friction) are present.

The problem simultaneously combines three layers of nonlocality, which are given by the fractional Laplace operator, the Kirchhoff term, and a nonsmooth nonlinearity that introduces a multivalued structure. The analysis often involves comparing the original problem with a limit or auxiliary local problem (where ϵ is effectively set to 1 or the Kirchhoff term is simplified) to gain information about the concentration profile. This allows to provide a unified treatment for a complex model featuring multiple nonlocal effects and nonsmoothness.

DOI: 10.1201/9781003797159-8

8.1 Statement of the problem

Consider the following fractional Kirchhoff equation:

$$\begin{cases} \left(\epsilon^{2\alpha}a + \epsilon^{4\alpha-3}b\displaystyle\int_{\mathbb{R}^3}|(-\Delta)^{\frac{\alpha}{2}}u|^2\mathrm{d}x\right)(-\Delta)^\alpha u + V(x)u = H(u-\beta)f(u) & \text{in } \mathbb{R}^3, \\ u \in H^\alpha(\mathbb{R}^3), \quad u > 0 & \text{in } \mathbb{R}^3, \end{cases} \tag{K}$$

where $\alpha \in (0,1)$ and a, b are positive constants, $\epsilon, \beta > 0$ are positive parameters, H is the Heaviside function given by

$$H(t) := \begin{cases} 1, & \text{if } t > 0, \\ 0, & \text{if } t \le 0. \end{cases}$$

The operator $(-\Delta)^\alpha$ is the fractional Laplacian defined as $\mathcal{F}^{-1}(|\xi|^{2\alpha}\mathcal{F}(u))$, where $\mathcal{F}$ denotes the Fourier transform on $\mathbb{R}^3$. The potential $V : \mathbb{R}^3 \to \mathbb{R}$ is a continuous function satisfying the following conditions introduced by Rabinowitz in [133]:

(V_0) there exist $V_0, V_\infty > 0$ such that

$$V_0 := \inf_{x\in\mathbb{R}^3} V(x) < \liminf_{|y|\to\infty} V(y) = V_\infty,$$

and $f : \mathbb{R} \to \mathbb{R}$ is a continuous function fulfilling the following hypotheses:

(f_1) $f(t) = 0$ for all $t < 0$ and $f(t) = o(t^3)$ as $t \to 0^+$.

(f_2) There exists $4 < p < 2^*_\alpha - 1$ such that

$$\lim_{t\to\infty} \frac{f(t)}{t^p} = 0,$$

where $\alpha \in (\frac{3}{4}, 1)$, $2^*_\alpha = \frac{6}{3-2\alpha}$ is the fractional critical exponent.

(f_3) The function $t \to \frac{f(t)}{t^3}$ is increasing in $(0, \infty)$.

(f_4) $f(t) \ge \gamma t^\sigma$ for all $t > 0$ with some $\gamma > 0$ and $\sigma \in (3, p-1)$.

Conditions (f_1)–(f_3) imply that

$$F(t) \ge 0, \quad 4F(t) \le f(t)t, \quad \forall t \in \mathbb{R}, \tag{8.1}$$

where $F(t) = \int_0^t f(s)ds$.

In order to study problem (K), we use the change of variable $x \mapsto \epsilon x$ and we look for solutions to

$$\begin{cases} \left(a + b\displaystyle\int_{\mathbb{R}^3}|(-\Delta)^{\frac{\alpha}{2}}u|^2\mathrm{d}x\right)(-\Delta)^\alpha u + V(\epsilon x)u = H(u-\beta)f(u) \quad \text{in } \mathbb{R}^3, \\ \qquad u \in H^\alpha(\mathbb{R}^3), \quad u > 0 \quad \text{in } \mathbb{R}^3. \end{cases} \tag{K$_\epsilon$}$$

Despite the challenges posed by the discontinuous reaction term (the Heaviside function), the main result in this chapter not only guarantees the existence of positive solutions, but it also describes their shape by establishing that they concentrate as the parameters become small.

Theorem 8.1 *Assume that conditions* (V_0) *and* $(f_1)-(f_4)$ *hold. Then, there exist* $\epsilon^*, \beta^* > 0$ *such that problem* (K_ϵ) *has a positive solution* $u_{\epsilon,\beta}$ *for all* $\epsilon \in (0, \epsilon^*)$ *and* $\beta \in (0, \beta^*)$. *Moreover, there exists a maximum point* $x_{\epsilon,\beta} \in \mathbb{R}^3$ *of* $u_{\epsilon,\beta}$ *such that*

$$\lim_{(\epsilon,\beta)\to(0,0)} V(\epsilon x_{\epsilon,\beta}) = V_0.$$

For such a $x_{\epsilon,\beta}$, *the function* $v_\epsilon(x) \equiv u_\epsilon(x + x_{\epsilon,\beta})$ *converges to a positive ground state solution of the autonomous Kirchhoff equation*

$$\left(a + b\int_{\mathbb{R}^3} |(-\Delta)^{\frac{\alpha}{2}} u|^2 \mathrm{d}x\right)(-\Delta)^\alpha u + V_0 u = f(u), \quad u \in H^\alpha(\mathbb{R}^3).$$

We observe that the associated energy functional is only locally Lipschitz due to the effect of the Heaviside function, so that we are not able to use variational methods for C^1-functionals. Next, we establish a relationship between the mountain pass level and infimum of the energy functional on the Nehari manifold. Then, we use the Fatou lemma to prove the existence of positive ground state solutions for the corresponding limit equation. However, the Nehari manifold method does not work for locally Lipschitz continuous functionals, and so some new technique need to be developed to obtain fine estimates to the mountain pass levels. Finally, with the presence of the Kirchhoff term, the main obstacle arises in getting the compactness of the corresponding locally Lipschitz continuous energy functional. Precisely, this does not hold in general, namely, it is not clear that for any $\varphi \in C_0^\infty(\mathbb{R}^N)$,

$$\int_{\mathbb{R}^3} |(-\Delta)^{\frac{\alpha}{2}} u_n|^2 \mathrm{d}x \int_{\mathbb{R}^N} (-\Delta)^{\frac{\alpha}{2}} u_n (-\Delta)^{\frac{\alpha}{2}} \varphi \mathrm{d}x \to \int_{\mathbb{R}^3} |(-\Delta)^{\frac{\alpha}{2}} u|^2 \mathrm{d}x \int_{\mathbb{R}^N} (-\Delta)^{\frac{\alpha}{2}} u (-\Delta)^{\frac{\alpha}{2}} \varphi \mathrm{d}x,$$

where $\{u_n\}_{n\in\mathbb{N}}$ is a (PS)-sequence of the energy functional satisfying $u_n \rightharpoonup u$ in $H^\alpha(\mathbb{R}^3)$. Then, it is not certain that weak limits are critical points of energy functional. For this reason, it is necessary to give a specific profile decomposition of the Kirchhoff term $(\int_{\mathbb{R}^3} |(-\Delta)^{\frac{\alpha}{2}} u_n|^2 \mathrm{d}x)^2$ which enables us to establish refined energy estimate to the mountain pass level. Finally, the above information together with the mountain pass geometry behaviors of energy functional helps us to obtain the compactness.

8.2 Variational setting

In what follows, we outline the variational framework for problem (K) and recall some preliminary results. For details, we refer to the monograph by Molica Bisci, Rădulescu and Servadei [119]. For any $\alpha \in (0, 1)$, the fractional Sobolev space $H^\alpha(\mathbb{R}^3)$ is defined by

$$H^\alpha(\mathbb{R}^3) := \left\{ u \in L^2(\mathbb{R}^3) : \frac{|u(x) - u(y)|}{|x - y|^{\frac{3+2\alpha}{2}}} \in L^2(\mathbb{R}^3 \times \mathbb{R}^3) \right\}.$$

We have

$$\int_{\mathbb{R}^3}\int_{\mathbb{R}^3} \frac{|u(x) - u(y)|^2}{|x - y|^{3+2\alpha}} \mathrm{d}x\mathrm{d}y = C_\alpha^{-1} \int_{\mathbb{R}^3} |(-\Delta)^{\frac{\alpha}{2}} u|^2 \mathrm{d}x,$$

where

$$C_\alpha = \frac{1}{2}\left(\int_{\mathbb{R}^3} \frac{1-\cos\zeta_1}{|\zeta|^{3+2\alpha}} d\zeta\right)^{-1}.$$

We endow the space $H^\alpha(\mathbb{R}^3)$ with the norm

$$\|u\|_{H^\alpha(\mathbb{R}^3)} := \left(\int_{\mathbb{R}^3} |u|^2 \mathrm{d}x + \int_{\mathbb{R}^3} |(-\Delta)^{\frac{\alpha}{2}} u|^2 \mathrm{d}x\right)^{1/2}.$$

The function space $H^\alpha(\mathbb{R}^3)$ is also the completion of $C_0^\infty(\mathbb{R}^3)$ with $\|\cdot\|_{H^\alpha(\mathbb{R}^3)}$ and it is continuously embedded into $L^q(\mathbb{R}^3)$ for $q \in [2, 2_\alpha^*]$. The homogeneous space $D^{\alpha,2}(\mathbb{R}^3)$ is

$$D^{\alpha,2}(\mathbb{R}^3) := \left\{u \in L^{2_\alpha^*}(\mathbb{R}^3) : \frac{|u(x)-u(y)|}{|x-y|^{\frac{3+2\alpha}{2}}} \in L^2(\mathbb{R}^3 \times \mathbb{R}^3)\right\},$$

and it is also the completion of $C_0^\infty(\mathbb{R}^3)$ with respect to the norm

$$\|u\|_{D^{\alpha,2}} := \left(\int_{\mathbb{R}^3} |(-\Delta)^{\frac{\alpha}{2}} u|^2 \mathrm{d}x\right)^{1/2}.$$

Let

$$E := \left\{u \in H^\alpha(\mathbb{R}^3) : \int_{\mathbb{R}^3} V(x)u^2 \mathrm{d}x < \infty\right\}$$

be the Hilbert space equipped with the inner product

$$\langle u, v\rangle_E := a\int_{\mathbb{R}^3} (-\Delta)^{\frac{\alpha}{2}} u (-\Delta)^{\frac{\alpha}{2}} v \,\mathrm{d}x + \int_{\mathbb{R}^3} V(x)uv \,\mathrm{d}x,$$

and the corresponding induced norm

$$\|u\| := \left(\int_{\mathbb{R}^3} a|(-\Delta)^{\frac{\alpha}{2}} u|^2 \mathrm{d}x + \int_{\mathbb{R}^3} V(x)u^2 \mathrm{d}x\right)^{1/2}.$$

We now recall some definitions and basic results on the critical point theory of locally Lipschitz continuous functionals as developed by Chang [23], Clarke [37], and Motreanu and Rădulescu [121].

Let E be a real Banach space. A functional $I : E \to \mathbb{R}$ is locally Lipschitz continuous, $I \in Lip_{loc}(E, \mathbb{R})$ for short, if given $u \in E$ there is an open neighborhood $V := V_u \subset E$ and some constant $K = K_V > 0$ such that

$$|I(v_2) - I(v_1)| \leq K\|v_2 - v_1\|, \quad v_i \in V,\ i = 1, 2.$$

The directional derivative of I at u in the direction of $v \in E$ is defined by

$$I^0(u; v) = \limsup_{h\to 0\ \lambda\downarrow 0} \frac{I(u+h+\lambda v) - I(u+h)}{\lambda}.$$

So $I^0(u;\cdot)$ is continuous, convex and its subdifferential at $z \in E$ is given by

$$\partial I^0(u; z) = \{\mu \in E^* :\ I^0(u; v) \geq I^0(u; z) + \langle \mu, v - z\rangle,\ v \in E\},$$

where $\langle\cdot,\cdot\rangle$ is the duality pairing between E^* and E. The generalized gradient of I at u is the set

$$\partial I(u) = \{\mu \in E^* : \langle \mu, v \rangle \leq I^0(u; v), v \in E\}.$$

Since $I^0(u;0) = 0$, $\partial I(u)$ is the subdifferential of $I^0(u;0)$. A few definitions and properties will be recalled below. $\partial I(u) \subset E^*$ is convex, non-empty and weak*-compact,

$$\lambda I(u) = \min\{\|\mu\|_{E^*} : \mu \in \partial I(u)\},$$

and

$$\partial I(u) = \{I'(u)\}, \quad \text{if } I \in C^1(E, \mathbb{R}).$$

A critical point of I is an element $u_0 \in E$ such that $0 \in \partial I(u_0)$ and a critical value of I is a real number c such that $I(u_0) = c$ for some critical point $u_0 \in E$.

By a solution for (K_ϵ), we understand a function $u \in W_{loc}^{2\alpha,\frac{p+1}{p}}(\mathbb{R}^3) \cap H^\alpha(\mathbb{R}^3)$ verifying

$$\begin{cases} \left(a + b \int_{\mathbb{R}^3} |(-\Delta)^{\frac{\alpha}{2}} u|^2 \mathrm{d}x\right)(-\Delta)^\alpha u + V(\epsilon x) u \in [\underline{f}_H(u(x)), \overline{f}_H(u(x))] \text{ a.e. in } \mathbb{R}^3 \\ u \in H^\alpha(\mathbb{R}^3), \ u > 0 \text{ in } \mathbb{R}^3 \end{cases} \tag{$\overline{K}$}$$

where $f_H(t) = H(t-\beta)f(t)$, $\overline{f}_H(t) = \limsup_{\delta \to 0^+} \overline{f}_H(t+\delta)$ and $\underline{f}_H(t) = \liminf_{\delta \to 0^+} f_H(t-\delta)$.

Now we recall the following mountain pass theorem for locally Lipschitz functionals, which was established in Rădulescu [134].

Theorem 8.2 *Let $I \in Lip_{loc}(E, \mathbb{R})$ with $I(0) = 0$ and satisfying:*

(i) There exist $r > 0$ and $\rho > 0$ such that $I(u) \geq \rho$ for $\|u\| = r$, $u \in E$.

(ii) There is $e \in E \setminus B_r(0)$ with $I(e) < 0$.

Set

$$c := \inf_{\gamma \in \Gamma} \max_{s \in [0,1]} I(\gamma(s)) > 0,$$

where

$$\Gamma := \{\gamma \in C([0,1], E) \ : \ \gamma(0) = 0, I(\gamma(1)) < 0\}.$$

Then $c \geq \rho$ and there is a sequence $\{u_n\} \subset E$ verifying $I(u_n) \to c$ and $I'(u_n) \to 0$ in E'.

For the next two auxiliary properties, we refer to Chang [23] and Clarke [37].

Lemma 8.1 *Let $\{u_n\} \subset E$ and $\{\rho_n\} \subset E^*$ with $\rho_n \in \partial I(u_n)$. If $u_n \to u$ in E and $\rho_n \rightharpoonup \rho$ in E^*, then $\rho_0 \in \partial I(u)$.*

Lemma 8.2 *Let $R > 0$ and $\Psi(u) = \int_{\mathbb{R}^3} G(u)\mathrm{d}x$ and $\Psi_R(v) = \int_{B_R(0)} G(v)\mathrm{d}x$, where $G(t) = \int_0^t g(s)ds$. Then, $\Psi \in Lip_{loc}(L^{p+1}(\mathbb{R}^3), \mathbb{R})$, $\Psi_R \in Lip_{loc}(L^{p+1}(B_R(0)), \mathbb{R})$, $\partial\Psi(u) \in L^{\frac{p+1}{p}}(\mathbb{R}^3)$ and $\partial\Psi_R(u) \in L^{\frac{p+1}{p}}(B_R(0))$. Moreover, if $\rho \in \partial\Psi(u)$ and $\zeta \in \partial\Psi_R(v)$, then*

$$\rho(x) \in [\underline{g}(u(x)), \overline{g}(u(x))] \quad \text{a. e. in } \mathbb{R}^3$$

and

$$\zeta(x) \in [\underline{g}(v(x)), \overline{g}(v(x))] \quad \text{a. e. in } B_R(0).$$

The following Lions-type lemma in fractional Sobolev spaces is due to Secchi [135].

Lemma 8.3 *Assume that* $\{u_n\}_{n\in\mathbb{N}}$ *is bounded in* $H^\alpha(\mathbb{R}^3)$ *and*

$$\lim_{n\to\infty}\sup_{y\in\mathbb{R}^3}\int_{B_r(y)}|u_n|^2\mathrm{d}x=0,$$

for some $r>0$. *Then* $u_n\to 0$ *in* $L^s(\mathbb{R}^3)$ *for all* $s\in(2,2^*_\alpha)$.

8.3 Existence of solutions to problem (K$_\epsilon$) if $\epsilon = 1$

The energy functional associated with problem (K$_\epsilon$) if $\epsilon=1$ is $I_\beta : E\to\mathbb{R}$ defined as

$$I_\beta(u)=\frac{1}{2}\|u\|^2+\frac{b}{4}\left(\int_{\mathbb{R}^3}|(-\Delta)^{\frac{\alpha}{2}}u|^2\mathrm{d}x\right)^2-\int_{\mathbb{R}^3}F_H(u)\mathrm{d}x,\quad u\in E,$$

with $F_H(u)=\int_0^u f_H(s)ds$. We remark that $I_\beta\in Lip_{loc}(H^\alpha(\mathbb{R}^3),\mathbb{R})$.

We now verify the functional I_β satisfies the mountain pass geometry.

Lemma 8.4 *The following properties hold.*

(i) There exist $r>0$ *and* $\rho>0$ *such that* $I_\beta(u)\geq\rho$ *for* $\|u\|=r$, $u\in E$.

(ii) There is $v\in E\setminus B_r(0)$ *with* $I_\beta(v)<0$.

Proof. (i) Observe from the definition of H that for any $\beta>0$, there exists $C_\beta>0$ such that

$$f_H(t)\leq C_\beta t^p,\quad \text{for all } t\in\mathbb{R}. \tag{8.1}$$

So from Sobolev's imbedding inequality, we infer that

$$I_\beta(u)\geq\frac{1}{2}\|u\|^2-\frac{C_\beta}{p+1}\int_{\mathbb{R}^3}|u|^{p+1}\mathrm{d}x\geq\frac{1}{2}\|u\|^2-C\|u\|^{p+1}.$$

Hence there exist $r,\rho>0$, independent of β, such that for $\|u\|=r$, $I_\beta(u)\geq\rho>0$.

(ii) Take $e\in C_0^\infty(\mathbb{R}^3)\setminus\{0\}$ with $e>0$ and $\Theta:=|\{x:\,e(x)>\beta\}|>0$, then by (f_4) one has

$$\begin{aligned}I_\beta(te)&=\frac{t^2}{2}\|e\|^2+\frac{bt^4}{4}\left(\int_{\mathbb{R}^3}|(-\Delta)^{\frac{\alpha}{2}}e|^2\mathrm{d}x\right)^2-\int_{\mathbb{R}^3}F_H(te)\mathrm{d}x\\&\leq\frac{t^2}{2}\|e\|^2+\frac{bt^4}{4}\left(\int_{\mathbb{R}^3}|(-\Delta)^{\frac{\alpha}{2}}e|^2\mathrm{d}x\right)^2-\int_{\{te(x)>\beta\}}F(te)-F(\beta)\mathrm{d}x\\&\leq\frac{t^2}{2}\|e\|^2+\frac{bt^4}{4}\left(\int_{\mathbb{R}^3}|(-\Delta)^{\frac{\alpha}{2}}e|^2\mathrm{d}x\right)^2-\frac{\gamma t^{\sigma+1}}{\sigma+1}\int_\Theta|e|^{\sigma+1}\mathrm{d}x+F(\beta)\cdot|supp\{e(x)\}|,\end{aligned}$$

which implies that

$$I_\beta(te)\to-\infty,\quad\text{as}\quad t\to+\infty. \tag{8.2}$$

So, take $t_0>0$ large enough, then $v=t_0e$ satisfies $v\in E\setminus B_r(0)$ with $I_\beta(v)<0$. □

Combining Lemma 8.4 with Theorem 8.2, there exists a sequence $\{u_n\}\subset E$ satisfying

$$I_\beta(u_n)\to c\quad\text{and}\quad\lambda I_\beta(u_n)\to 0, \tag{8.3}$$

where c is the mountain pass level of the functional I_β.

Lemma 8.5 *The sequence $\{u_n\}$ is bounded in E.*

Proof. Set

$$A(u) := \frac{1}{2}\|u\|^2 + \frac{b}{4}\left(\int_{\mathbb{R}^3} |(-\Delta)^{\frac{\alpha}{2}} u|^2 \mathrm{d}x\right)^2, \quad B(u) := \int_{\mathbb{R}^3} F_H(u)\mathrm{d}x.$$

Let $\{w_n\} \subset H^\alpha(\mathbb{R}^3)^*$ be such that $\lambda I_\beta(u_n) = \|\omega_n\|_{E^*}$ and $\omega_n = A'(u_n) - \rho_n$, where $\{\rho_n\} \subset \partial B(u_n)$. Then

$$\|u_n\|^2 + b\left(\int_{\mathbb{R}^3} |(-\Delta)^{\frac{\alpha}{2}} u_n|^2 \mathrm{d}x\right)^2 = \langle \omega_n + \rho_n, u_n\rangle. \tag{8.4}$$

Since $\rho_n \in [\underline{f}_H(u_n(x)), \overline{f}_H(u_n(x))]$ a.e. in $\mathbb{R}^3$, then by (8.1) we have

$$\langle \rho_n, u_n\rangle \geq \int_{\mathbb{R}^3} \underline{f}_H(u_n)u_n \mathrm{d}x \geq \int_{\{u_n>\beta\}} f(u_n)u_n \mathrm{d}x \geq 4\int_{\{u_n>\beta\}} F(u_n)\mathrm{d}x.$$

It then follows from (8.4) that

$$\begin{aligned} c_\beta + o(1) = I_\beta(u_n) - \frac{1}{4}\langle \omega_n, u_n\rangle &= \frac{1}{4}\|u_n\|^2 + \frac{1}{4}\langle \rho_n, u_n\rangle - \int_{\{u_n>\beta\}} F_H(u_n)\mathrm{d}x \\ &\geq \frac{1}{4}\|u_n\|^2, \end{aligned} \tag{8.5}$$

which concludes the proof. □

Recalling (8.3) and Lemma 8.5, there exists a subsequence of $\{u_n\}$ (still denoted by $\{u_n\}$) such that

$$u_n \rightharpoonup u_0 \quad \text{in } E, \qquad u_n \to u_0 \quad \text{in } L^s_{loc}(\mathbb{R}^3), \;\; s \in (2, 2^*_\alpha). \tag{8.6}$$

Lemma 8.6 *Let $\{\rho_n\} \subset \partial B(u_n)$ with $\rho_n \rightharpoonup \rho_0$ in $L^{\frac{p+1}{p}}(\mathbb{R}^3)$. Then*

$$\rho_0(x) \in [\underline{f}_H(u(x)), \overline{f}_H(u(x))] \quad a.e \ in\ \mathbb{R}^3.$$

Proof. Since $u_n \rightharpoonup u$ in E, we have $u_n \to u$ in $L^{p+1}_{loc}(\mathbb{R}^3)$. Hence, for any $\varphi \in C_0^\infty(\mathbb{R}^3)$, we obtain

$$\int_{\mathbb{R}^3} u_n\varphi \mathrm{d}x \to \int_{\mathbb{R}^3} u\varphi \mathrm{d}x, \quad \int_{\mathbb{R}^3} \rho_n\varphi \mathrm{d}x \to \int_{\mathbb{R}^3} \rho_0\varphi \mathrm{d}x.$$

Recalling Lemma 8.1, we obtain

$$\rho_0(x) \in [\underline{f}_H(u(x)), \overline{f}_H(u(x))] \quad a.e. \text{ in } \mathbb{R}^3.$$

The proof is now complete. □

Lemma 8.7 *Let $\{u_n\} \subset E$ be the sequence defined in (8.3). Then there exist a sequence $\{y_n\} \subset \mathbb{R}^3$ and constants $R, \tau > 0$ such that*

$$\liminf_{n\to\infty} \int_{B_R(y_n)} u_n^2 \mathrm{d}x \geq \tau > 0. \tag{8.7}$$

Proof. Suppose by contradiction that (8.7) does not hold. Then it follows from Lemma 8.3 that $u_n \to 0$ in $L^s(\mathbb{R}^3)$ for $s \in (2, 2_\alpha^*)$. By (8.1) and (8.4), there exists $\{\rho_n\} \subset \partial B(u_n)$ with $\rho_n \in [\underline{f}_H(u_n(x)), \overline{f}_H(u_n(x))]$ a.e in $\mathbb{R}^3$ such that

$$\|u_n\|^2 \leq \int_{\mathbb{R}^3} \rho_n u_n \mathrm{d}x \leq \int_{\mathbb{R}^3} \overline{f}_H(u_n) u_n \mathrm{d}x \leq \int_{\{u_n \geq \beta\}} f(u_n) u_n \mathrm{d}x \to 0. \tag{8.8}$$

This implies that $u_n \to 0$ in E. Furthermore, $I_\beta(u_n) \to 0$ as $n \to \infty$, which contradicts $c_\beta > 0$. The proof is complete. □

Let c_∞ be the mountain pass level associated with the functional $I_\infty : H^\alpha(\mathbb{R}^3) \to \mathbb{R}$ defined by

$$I_\infty(u) = \frac{1}{2} \int_{\mathbb{R}^3} (a|(-\Delta)^{\frac{\alpha}{2}} u|^2 + V_\infty u^2) \mathrm{d}x + \frac{b}{4} \left(\int_{\mathbb{R}^3} |(-\Delta)^{\frac{\alpha}{2}} u|^2 \mathrm{d}x \right)^2 - \int_{\mathbb{R}^3} F(u) \mathrm{d}x. \tag{8.9}$$

Lemma 8.8 *Assume that there exists $\beta_1 > 0$ small such that $(1+\beta^2)c_\beta < c_\infty$ for fixed $\beta \in (0, \beta_1)$, then $u_0 \neq 0$.*

Proof. Suppose on the contrary that $u_0 = 0$. By Sobolev embeddings together with Lemma 8.7, we obtain $\{y_n\}$ is unbounded. Thus, up to subsequence, $|y_n| \to +\infty$.

Set $v_n(x) := u_n(x + y_n)$. We observe that $\{v_n\}$ is bounded in $H^\alpha(\mathbb{R}^3)$. Recalling Lemma 8.7, up to subsequence, there exists $v \in E \setminus \{0\}$ such that for $s \in [1, 2_\alpha^*)$

$$v_n \rightharpoonup v \quad \text{in} \quad E, \qquad v_n \to v \quad \text{in} \quad L^s_{loc}(\mathbb{R}^3). \tag{8.10}$$

For any $R > 0$, let $\varphi_R \in C_0^\infty(\mathbb{R}^3)$ be such that $\varphi_R(x) = 1$ in B_R and $\varphi_R(x) = 0$ in $\mathbb{R}^3 \setminus \{B_{2R}\}$, with $0 \leq \varphi_R \leq 1$ and $|\nabla \varphi_R| \leq \frac{C}{R}$, where C is a constant independent of R. Since the sequence $\{(\varphi_R v_n)(\cdot - y_n)\}$ is bounded in E, we obtain

$$\begin{aligned}(a + b\|u_n\|^2_{D^{2,\alpha}}) \int_{\mathbb{R}^3} (-\Delta)^{\frac{\alpha}{2}} u_n (-\Delta)^{\frac{\alpha}{2}} (\varphi_R v_n)(\cdot - y_n) \mathrm{d}x + \int_{\mathbb{R}^3} V(x) u_n (\varphi_R v_n)(\cdot - y_n) \mathrm{d}x \\ = \int_{\mathbb{R}^3} \rho_n (\varphi_R v_n)(\cdot - y_n) \mathrm{d}x + o(1), \quad v \in E.\end{aligned} \tag{8.11}$$

Using the fact that $\rho_n \in [\underline{f}_H(u_n(x)), \overline{f}_H(u_n(x))]$ a.e in $\mathbb{R}^3$, we have

$$\int_{\mathbb{R}^3} \rho_n (\varphi_R v_n)(\cdot - y_n) \mathrm{d}x \leq \int_{\mathbb{R}^3} \overline{f}_H(v_n) v_n \varphi_R \mathrm{d}x \leq \int_{\mathbb{R}^3} f(v_n) v_n \varphi_R \mathrm{d}x. \tag{8.12}$$

It follows from (8.10) that

$$\int_{\mathbb{R}^3} f(v_n) v_n \varphi_R \mathrm{d}x = \int_{\mathbb{R}^3} f(v) v \varphi_R \mathrm{d}x + o(1). \tag{8.13}$$

Observe that

$$\begin{aligned}\int_{\mathbb{R}^3} (-\Delta)^{\frac{\alpha}{2}} u_n (-\Delta)^{\frac{\alpha}{2}} (\varphi_R v_n)(\cdot - y_n) \mathrm{d}x = C_\alpha \int_{\mathbb{R}^3} \int_{\mathbb{R}^3} \frac{[v_n(x) - v_n(y)]^2 \varphi_R(x)}{|x-y|^{3+2\alpha}} \mathrm{d}x\mathrm{d}y \\ + C_\alpha \int_{\mathbb{R}^3} \int_{\mathbb{R}^3} \frac{[v_n(x) - v_n(y)][\varphi_R(x) - \varphi_R(y)]}{|x-y|^{3+2\alpha}} v_n(y) \mathrm{d}x\mathrm{d}y.\end{aligned} \tag{8.14}$$

Using the Hölder's inequality, the boundedness of $\{u_n\}$ in E and $|\nabla\varphi_R| \leq \frac{C}{R}$, we have

$$\begin{aligned}
&\int_{\mathbb{R}^3}\int_{\mathbb{R}^3} \frac{[v_n(x)-v_n(y)][\varphi_R(x)-\varphi_R(y)]}{|x-y|^{3+2\alpha}} v_n(y)\mathrm{d}x\mathrm{d}y \\
&\leq \|v_n\|^2_{D^{2,\alpha}} \left(\int_{\mathbb{R}^3}\int_{\mathbb{R}^3} \frac{[\varphi_R(x)-\varphi_R(y)]^2}{|x-y|^{3+2\alpha}} v_n^2(y)\mathrm{d}x\mathrm{d}y \right)^{1/2} \\
&\leq \|v_n\|^2_{D^{2,\alpha}} \left(\frac{C}{R^2} \int_{\mathbb{R}^3}\int_{|x-y|\leq 4R} \frac{1}{|x-y|^{1+2\alpha}} v_n^2(y)\mathrm{d}x\mathrm{d}y \right)^{1/2} \\
&\leq \frac{C}{R}\|v_n\|^2_{D^{2,\alpha}} \left(\int_{\mathbb{R}^3} v_n^2(y)\mathrm{d}y \int_0^{4R} \frac{1}{r^{2\alpha-1}} dr \right)^{1/2} \leq \frac{C}{R^\alpha},
\end{aligned} \tag{8.15}$$

where we have used polar coordinate transformation in the third inequality. Taking into account relations (8.11)–(8.15), by Fatou's lemma we infer that

$$\begin{aligned}
&(a+b\|v\|^2_{D^{2,\alpha}})C_\alpha \int_{\mathbb{R}^3}\int_{\mathbb{R}^3} \frac{[v(x)-v(y)]^2\varphi_R(x)}{|x-y|^{3+2\alpha}}\mathrm{d}x\mathrm{d}y + \int_{\mathbb{R}^3} V_\infty v^2\varphi_R \mathrm{d}x \\
&\leq \int_{\mathbb{R}^3} f(v)v\varphi_R\mathrm{d}x + \frac{C}{R^\alpha}.
\end{aligned} \tag{8.16}$$

Let $R \to +\infty$ in (8.16), then

$$(a+b\|v\|^2_{D^{2,\alpha}})\int_{\mathbb{R}^3} |(-\Delta)^{\frac{\alpha}{2}} v|^2\mathrm{d}x + \int_{\mathbb{R}^3} V_\infty v^2 \mathrm{d}x \leq \int_{\mathbb{R}^3} f(v)v\mathrm{d}x. \tag{8.17}$$

Since $v \neq 0$, there exists $t \in (0,1)$ such that $tv \in \mathcal{N}$, where $\mathcal{N}$ is the Nehari manifold associated with I_∞ given by

$$\mathcal{N} := \{u \in E\setminus\{0\} :\ I'_\infty(u)u = 0\}.$$

As a consequence, using (f_3) and Fatou's lemma, we have

$$\begin{aligned}
I_\infty(tv) - \frac{1}{4}I'_\infty(tv)tv &= \frac{t^2}{4}\int_{\mathbb{R}^3} [(|(-\Delta)^{\frac{\alpha}{2}} v|^2 + V_\infty v^2]\mathrm{d}x + \int_{\mathbb{R}^3} \left[\frac{1}{4}f(tv)tv - F(tv)\right]\mathrm{d}x \\
&\leq \liminf_{n\to\infty} \left(\frac{1}{4}\|u_n\|^2 + \int_{\mathbb{R}^3} \left(\frac{1}{4}f(u_n)u_n - F(u_n)\right)\mathrm{d}x \right).
\end{aligned} \tag{8.18}$$

Recalling the definitions of $\underline{f}_H$ and $\overline{f}_H$, we deduce from (f_1) that there exists $\beta_1 > 0$ such that $f(t) \leq V_0 t^3$ for $t \in (0, \beta_1)$ and then by (8.5) and $c_\beta < c_\infty$, we have for large n

$$\begin{aligned}
\int_{\mathbb{R}^3} f(u_n)u_n\mathrm{d}x &= \int_{\{u_n\leq\beta\}} f(u_n)u_n\mathrm{d}x + \int_{\{u_n>\beta\}} f(u_n)u_n\mathrm{d}x \\
&\leq \beta^2\|u_n\|^2 + \int_{\mathbb{R}^3} \underline{f}_H(u_n)u_n\mathrm{d}x \leq 4\beta^2(c_\beta + o(1)) + \int_{\mathbb{R}^3} \underline{f}_H(u_n)u_n\mathrm{d}x
\end{aligned} \tag{8.19}$$

for any fixed $\beta \in (0,\beta_1)$. Putting (8.19) into (8.18), setting $c_\infty = \inf_{u\in\mathcal{N}} I(u)$ and using (8.18), we obtain

$$\begin{aligned}
c_\infty &\leq I_\infty(tv) - \frac{1}{4}I'_\infty(tv)tv \\
&\leq \liminf_{n\to\infty} \left(\frac{1}{4}\|u_n\|^2 + \beta^2(c_\beta + o(1)) + \int_{\mathbb{R}^3} \left(\frac{1}{4}\underline{f}_H(u_n)u_n - F_H(u_n)\right)\mathrm{d}x \right) \\
&\leq \liminf_{n\to\infty} \left(\frac{1}{4}\|u_n\|^2 + \beta^2(c_\beta + o(1)) + \int_{\mathbb{R}^3} \left(\frac{1}{4}\rho_n u_n - F_H(u_n)\right)\mathrm{d}x \right) \\
&\leq \liminf_{n\to\infty} \left(I_\beta(u_n) - \frac{1}{4}\langle A'(u_n) - \rho_n, u_n\rangle + \beta^2(c_\beta + o(1)) \right) = (1+\beta^2)c_\beta,
\end{aligned}$$

which is a contradiction. Therefore, $u_0 \geq 0$ and $u_0 \neq 0$. □

The next result establishes the existence of mountain pass solutions to problem (K_ϵ) with $\epsilon = 1$, that is, there exists $u_0 \in E$ satisfying

$$I_\beta(u_0) = c_\beta \qquad \text{and} \qquad 0 \in \partial I_\beta(u_0),$$

where c_β is the mountain pass level associated with I_β.

Lemma 8.9 *Assume that there exists $\beta_1 > 0$ such that $(1+\beta^2)c_\beta < c_\infty$ for fixed $\beta \in (0, \beta_1)$. Then problem (K_ϵ) with $\epsilon = 1$ has a nontrivial solution $u_0 \in E$, and the set $\Lambda := \{x \in \mathbb{R}^3 : u_0(x) = \beta\}$ has null measure. Moreover,*

$$I_\beta(u_0) = \max_{t \geq 0} I_\beta(tu_0).$$

Proof. In view of Lemma 8.8, u_0 given in (8.6) is nonzero. Hence, there exists $\mathcal{B} > 0$ such that $\|u_n\|^2_{D^{\alpha,2}} \to \mathcal{B}$ as $n \to \infty$, where $\{u_n\}$ has been defined in (8.3). Indeed, we need to prove that $u_0 \in W^{2\alpha,\frac{p+1}{p}}_{loc}(\mathbb{R}^3) \cap H^\alpha(\mathbb{R}^3)$ and u_0 solves

$$\left(a + b\int_{\mathbb{R}^3} |(-\Delta)^{\frac{\alpha}{2}}u|^2 \mathrm{d}x\right)(-\Delta)^\alpha u + V(x)u \in [\underline{f}_H(u(x)), \overline{f}_H(u(x))] \quad a.e. \text{ in } \mathbb{R}^3.$$

Since $\{u_n\} \subset E$ is a $(\mathrm{PS})_{c_\beta}$ sequence, there exists $\{\rho_n\} \subset \partial B(u_n)$ such that

$$\begin{aligned}
&\int_{\mathbb{R}^3} (a(-\Delta)^{\frac{\alpha}{2}}u_n(-\Delta)^{\frac{\alpha}{2}}v + V(x)u_n v)\mathrm{d}x \\
&+ b\Big(\int_{\mathbb{R}^3} |(-\Delta)^{\frac{\alpha}{2}}u_n|^2\mathrm{d}x\Big)^2 \int_{\mathbb{R}^3}(-\Delta)^{\frac{\alpha}{2}}u_n(-\Delta)^{\frac{\alpha}{2}}v\mathrm{d}x = \int_{\mathbb{R}^3} \rho_n v\mathrm{d}x, \quad v \in E,
\end{aligned} \tag{8.20}$$

with $\rho_n \in [\underline{f}_H(u_n(x)), \overline{f}_H(u_n(x))]$ a.e. in $\mathbb{R}^3$. It then follows from the boundedness of $\{u_n\}$, the definition of H, (f_1) and (f_2) that the sequence $\{\rho_n\}$ is bounded in $L^{\frac{p+1}{p}}(\mathbb{R}^3)$. Thus, up to subsequence, there exists $\rho_0 \in L^{\frac{p+1}{p}}(\mathbb{R}^3)$ such that

$$\rho_n \rightharpoonup \rho_0 \qquad \text{in} \quad L^{\frac{p+1}{p}}(\mathbb{R}^3), \tag{8.21}$$

which implies by (8.6) that for any $v \in E$

$$\int_{\mathbb{R}^3}\big[a(-\Delta)^{\frac{\alpha}{2}}u_0(-\Delta)^{\frac{\alpha}{2}}v + V(x)u_0v\big]\mathrm{d}x + b\mathcal{B}\int_{\mathbb{R}^3}(-\Delta)^{\frac{\alpha}{2}}u_0(-\Delta)^{\frac{\alpha}{2}}v\mathrm{d}x = \int_{\mathbb{R}^3}\rho_0 v\mathrm{d}x. \tag{8.22}$$

Recalling Lemma 8.6, one has $\rho_0 \in [\underline{f}_H(u_0(x)), \overline{f}_H(u_0(x))]$ a.e in $\mathbb{R}^3$. From elliptic regularity theory, we deduce that $u \in W^{2\alpha,\frac{p+1}{p}}(\mathbb{R}^3)$. As a consequence, from (8.22), we immediately obtain that u_0 is a nonnegative weak solution of the following equation

$$\begin{cases} (a + b\mathcal{B})(-\Delta)^\alpha u + V(x)u = H(u-\beta)f(u) & \text{a. e. in } \mathbb{R}^3, \\ u \in H^\alpha(\mathbb{R}^3), \end{cases} \tag{8.23}$$

whose energy functional is

$$J_\beta(u) = \frac{a + b\mathcal{B}}{2}\int_{\mathbb{R}^3}|(-\Delta)^{\frac{\alpha}{2}}u|^2\mathrm{d}x + \frac{1}{2}\int_{\mathbb{R}^3}V(x)u^2\mathrm{d}x - \int_{\mathbb{R}^3}F_H(u)\mathrm{d}x.$$

By using the Stampacchia theorem, we deduce that $\Lambda = \{x \in \mathbb{R}^3 : u_0(x) = \beta\}$ has null measure for β small enough. Using $v = u_0^- := \min\{u_0, 0\}$ in (8.22) and the fact that $\rho_0 \in [\underline{f}_H(u_0(x)), \overline{f}_H(u_0(x))]$ a.e. in $\mathbb{R}^3$, we have

$$(a + b\mathcal{B}) \int_{\mathbb{R}^3} ((-\Delta)^\alpha u_0 u_0^- + V(x)|u_0^-|^2)\mathrm{d}x = \int_{\mathbb{R}^3} \rho_0 u_0^- \,\mathrm{d}x = 0.$$

So, by the maximum principle, $u_0 \geq 0$. Moreover, by the iteration method, we obtain that $u_0 \in L^\infty(\mathbb{R}^3)$. Using Proposition 2.9 in Silvestre [137] and $2\alpha > 1$, we have $u_0 \in C^{1,s}(\mathbb{R}^3)$ for $s \in (0, 2\alpha - 1)$. Using the maximum principle of Silvestre [137], we conclude that $u > 0$ in $\mathbb{R}^3$.

Note that one of the following cases must occur.

Case 1. $\mathcal{B} = \|u_0\|^2_{D^{\alpha,2}}$. Thus, $u_n \to u_0$ in $D^{\alpha,2}(\mathbb{R}^3)$ as $n \to \infty$. By relation (8.23), u_0 is a nontrivial weak solution of problem (K_1). Moreover, by the Hölder's inequality and the fractional Gagliardo–Nirenberg–Sobolev inequality, we have

$$\begin{aligned}\left|\int_{\mathbb{R}^3} (\rho_n u_n - \rho_0 u_0)\mathrm{d}x\right| &= \left|\int_{\mathbb{R}^3} \rho_n (u_n - u_0)\mathrm{d}x\right| + \left|\int_{\mathbb{R}^3} (\rho_n - \rho_0) u_0 \mathrm{d}x\right| \\ &\leq \|\rho_n\|_{\frac{p+1}{p}} \|u_n - u_0\|^{p+1}_{p+1} + o(1) \\ &\leq C\|\rho_n\|_{\frac{p+1}{p}} \|u_n - u_0\|^{\frac{3(p-1)}{2\alpha}}_{D^{\alpha,2}} \|u_n - u_0\|_2^{p+1-\frac{3(p-1)}{2\alpha}} + o(1) \\ &= o(1).\end{aligned}$$

As a consequence, letting $v = u_n$ in (8.20) and $v = u_0$ in (8.22), we have

$$\|u_n - u_0\|^2 = \int_{\mathbb{R}^3} (\rho_n u_n - \rho_0 u_0)\mathrm{d}x + o(1) = o(1),$$

hence $u_n \to u_0$ in E as $n \to \infty$.

Case 2. $\|u_0\|^2_{D^{\alpha,2}} < \mathcal{B}$. We claim that there exists $t^* \in (0,1)$ such that

$$\max_{t \geq 0} I_\beta(tu_0) = I_\beta(t^* u_0). \tag{8.24}$$

Since $h(t) := I_\beta(tu_0)$ is locally Lipschitz, then h is differentiable almost everywhere. A direct computation shows that there exist $\delta, t_0 > 0$ such that

$$h(t) > 0, \quad \text{for all } t \in (0,\delta) \qquad \text{and} \qquad h(t) < 0 \quad \text{for all } t \geq t_0, \tag{8.25}$$

which implies that there exists $t^* > 0$ such that relation (8.24) holds. Let $I \subset \mathbb{R}$ denote the set of the points where h' does not exist, hence $|I| = 0$. The claim is proved by showing the following assertion:

(a) $h'(t) > 0$ for any $t \in (0, t^*) \cap I^c$ and $h'(t) < 0$ for any $t \in (t^*, +\infty) \cap I^c$, where $t^* \in (0,1)$.

Using the chain rule for locally Lipschitz functions, there exists $w \in \partial I_\beta(tu_0)$ such that $h'(t) = \langle w, u_0\rangle$. Thus, there exists $\rho_0 \in \partial B(tu_0)$ satisfying

$$h'(t) = t\left(\int_{\mathbb{R}^3} a|(-\Delta)^{\frac{\alpha}{2}} u_0|^2 \mathrm{d}x + \int_{\mathbb{R}^3} V(x) u_0^2 \mathrm{d}x\right) + t^3 b\|u_0\|^4_{D^{\alpha,2}} - \int_{\mathbb{R}^3} \rho_0 u_0 \mathrm{d}x.$$

It follows from $0 \in \partial J_\beta(u_0)$, $|\Lambda| = 0$ and relation (8.22) that

$$(a + b\mathcal{B}) \int_{\mathbb{R}^3} |(-\Delta)^{\frac{\alpha}{2}} u_0|^2 \mathrm{d}x + \int_{\mathbb{R}^3} V(x) u_0^2 \mathrm{d}x = \int_{\mathbb{R}^3} f_H(u_0) u_0 \mathrm{d}x. \tag{8.26}$$

Based on the above facts, we obtain

$$h'(t) = t\Big(\int_{\mathbb{R}^3} f_H(u_0)u_0\mathrm{d}x - b\mathcal{B}\|u_0\|^2_{D^{\alpha,2}}\Big) + t^3 b\|u_0\|^4_{D^{\alpha,2}} - \int_{\mathbb{R}^3}\rho_0 u_0\mathrm{d}x.$$

This relation together with $|\Lambda| = 0$ and the fact that $\rho_0 \in [\underline{f}_H(tu_0(x)), \overline{f}_H(tu_0(x))]$ a.e in $\mathbb{R}^3$ implies that

$$\begin{aligned}
h'(t) &\le t\Big(\int_{\mathbb{R}^3} f_H(u_0)u_0\mathrm{d}x - \int_{\mathbb{R}^3}\underline{f}_H(tu_0)u_0\mathrm{d}x\Big) + t^3 b\|u_0\|^4_{D^{\alpha,2}} - b\mathcal{B}t\|u_0\|^2_{D^{\alpha,2}}\\
&= t^3\Big(\int_{\mathbb{R}^3}\frac{\underline{f}_H(u_0)}{t^2}u_0\mathrm{d}x - \int_{\mathbb{R}^3}\frac{\underline{f}_H(tu_0)}{t^3}u_0\mathrm{d}x + b\|u_0\|^4_{D^{\alpha,2}} - \frac{b\mathcal{B}}{t^2}\|u_0\|^2_{D^{\alpha,2}}\Big)\\
&\le t^3\Big(\int_{\{u_0>\beta\}}\frac{f(u_0)}{t^2}u_0\mathrm{d}x - \int_{\{tu_0>\beta\}}\frac{f(tu_0)}{t^3}u_0\mathrm{d}x + b\|u_0\|^4_{D^{\alpha,2}} - \frac{b\mathcal{B}}{t^2}\|u_0\|^2_{D^{\alpha,2}}\Big).
\end{aligned}\tag{8.27}$$

Similarly, we can also obtain

$$h'(t) \ge t^3\Big(\int_{\{u_0>\beta\}}\frac{f(u_0)}{t^2}u_0\mathrm{d}x - \int_{\{tu_0\ge\beta\}}\frac{f(tu_0)}{t^3}u_0\mathrm{d}x + b\|u_0\|^4_{D^{\alpha,2}} - \frac{b\mathcal{B}}{t^2}\|u_0\|^2_{D^{\alpha,2}}\Big).$$

Define $\mathcal{H} : (0, +\infty) \to \mathbb{R}$ by

$$\mathcal{H}(t) := \int_{\{u_0>\beta\}}\frac{f(u_0)}{t^2}u_0\mathrm{d}x - \int_{\{tu_0>\beta\}}\frac{f(tu_0)}{t^3}u_0\mathrm{d}x + b\|u_0\|^4_{D^{\alpha,2}} - \frac{b\mathcal{B}}{t^2}\|u_0\|^2_{D^{\alpha,2}}.$$

Obviously, $\mathcal{H}(1) < 0$ due to $\|u_0\|^2_{D^{\alpha,2}} < \mathcal{B}$.

It follows from (f_1)–(f_3) and (8.27) that there exists $\tilde{t} \in (0,1)$ such that $\mathcal{H}(\tilde{t}) = 0$ and $h'(t) < 0$ for all $t \in (\tilde{t}, +\infty) \cap I^c$. Based on (8.25) and the above facts, h has a global maximum in $t = t^* \in (0, \tilde{t}]$. Thus, assertion (a) is obtained and then the claim is true. Recalling the definition of c_β, we obtain

$$c_\beta \le I_\beta(t^* u_0). \tag{8.28}$$

Observe by $|\Lambda| = 0$ and the definition of H that

$$\int_{\{u_0\le\beta\}} F_H(u_0)\mathrm{d}x = 0. \tag{8.29}$$

Since t^* is a global maximum point of $h(t)$, there exist a sequence $\{t_n\} \subset (t^*, +\infty) \cap I^c$ with $t_n \to t^*$ and $\rho_n \in \partial B(t_n u_0)$ such that

$$h'(t_n) = t\Big(\int_{\mathbb{R}^3} a|(-\Delta)^{\frac{\alpha}{2}}u_0|^2\mathrm{d}x + \int_{\mathbb{R}^3}V(x)u_0^2\mathrm{d}x\Big) + t^3 b\|u_0\|^4_{D^{\alpha,2}} - \int_{\mathbb{R}^3}\rho_n u_0\mathrm{d}x \le 0,$$

which implies by (8.28), (8.29), the definition of I_β and $t_n \in (0,1)$ that

$$\begin{aligned}
c_\beta \le& I_\beta(t^* u_0) \le I_\beta(t_n u_0) - \frac{1}{4}h'(t_n)t_n + o(1)\\
\le& \frac{t_n^2}{4}\|u_0\|^2 - \int_{\mathbb{R}^3} F_H(t_n u_0)\mathrm{d}x + \frac{1}{4}\int_{\mathbb{R}^3}\rho_n t_n u_0\mathrm{d}x + o(1).
\end{aligned}$$

It follows that

$$
\begin{aligned}
c_\beta \leq & \frac{t_n^2}{4}\|u_0\|^2 + \int_{\{t_n u_0 \geq \beta\}} \frac{1}{4} f(t_n u_0) t_n u_0 \mathrm{d}x + o(1) \\
& - \int_{\{t_n u_0 > \beta\}} [F(t_n u_0) - F(\beta)] \mathrm{d}x - \int_{\{t_n u_0 \leq \beta\}} F_H(t_n u_0) \mathrm{d}x \\
= & \frac{t_n^2}{4}\|u_0\|^2 + \int_{\{t_n u_0 \geq \beta\}} \left[\frac{1}{4} f(t_n u_0) t_n u_0 - F(t_n u_0)\right] \mathrm{d}x + \int_{\{t_n u_0 \geq \beta\}} F(\beta) \mathrm{d}x + o(1) \\
< & \frac{1}{4}\|u_0\|^2 + \int_{\{u_0 > \beta\}} \left[\frac{1}{4} f(u_0) u_0 - F(u_0)\right] \mathrm{d}x + \int_{\{u_0 > \beta\}} F(\beta) \mathrm{d}x \\
\leq & \frac{1}{4}\|u_0\|^2 + \int_{\mathbb{R}^3} \left[\frac{1}{4} \underline{f}_H(u_0) u_0 - F_H(u_0)\right] \mathrm{d}x + \int_{\{u_0 \leq \beta\}} F_H(u_0) \mathrm{d}x \\
\leq & \liminf_{n \to \infty} \left\{\frac{1}{4}\|u_n\|^2 + \int_{\mathbb{R}^3} \left[\frac{1}{4} \underline{f}_H(u_n) u_n - F_H(u_n)\right] \mathrm{d}x\right\} \\
\leq & \liminf_{n \to \infty} \left[I_\beta(u_n) - \frac{1}{4}\langle \omega_n, u_n \rangle\right] < c_\beta,
\end{aligned}
\tag{8.30}
$$

which is a contradiction. Here, ω_n has been defined in Lemma 8.5. We conclude that Case 2 does not occur and the proof is complete. □

8.4 Problem (K_ϵ): qualitative and asymptotic analysis

Consider the Hilbert space

$$
E_\epsilon := \left\{u \in H^\alpha(\mathbb{R}^3) : \int_{\mathbb{R}^3} V(\epsilon x) u^2 \mathrm{d}x < \infty\right\}
$$

endowed with the norm

$$
\|u\|_\epsilon^2 = \int_{\mathbb{R}^3} [a|(-\Delta)^{\frac{\alpha}{2}} u|^2 + V(\epsilon x) u^2] \mathrm{d}x.
$$

The energy functional associated with problem (K_ϵ) is given by

$$
I_{\epsilon,\beta}(u) = \frac{1}{2}\|u\|_\epsilon^2 + \frac{b}{4}\left(\int_{\mathbb{R}^3} |(-\Delta)^{\frac{\alpha}{2}} u|^2 \mathrm{d}x\right)^2 - \int_{\mathbb{R}^3} F_H(u) \mathrm{d}x, \quad u \in E_\epsilon.
$$

Using the same argument as Lemma 8.4, we can prove that $I_{\epsilon,\beta}$ has the mountain pass geometry. The mountain pass level of $I_{\epsilon,\beta}$ is denoted by $c_{\epsilon,\beta}$, namely

$$
c_{\epsilon,\beta} := \inf_{\gamma \in \Gamma_\epsilon} \max_{s \in [0,1]} I_{\epsilon,\beta}(\gamma(s)) > 0,
$$

where

$$
\Gamma_\epsilon := \{\gamma \in C([0,1], E_\epsilon) \ : \ \gamma(0) = 0, \, I_{\epsilon,\beta}(\gamma(1)) < 0\}.
$$

8.4.1 Existence

Define $I_{V_0} : H^\alpha(\mathbb{R}^3) :\to \mathbb{R}$ by

$$
I_{V_0}(u) := \frac{1}{2}\int_{\mathbb{R}^3} (a|(-\Delta)^{\frac{\alpha}{2}} u|^2 + V_0 u^2) \mathrm{d}x + \frac{b}{4}\left(\int_{\mathbb{R}^3} |(-\Delta)^{\frac{\alpha}{2}} u|^2 \mathrm{d}x\right)^2 - \int_{\mathbb{R}^3} F(u) \mathrm{d}x,
$$

We know that $\{u_n\}$ is bounded in $H^\alpha(\mathbb{R}^3)$. Take $v_n := u_n(x+\tilde{y}_n)$ such that $v_n \rightharpoonup v \neq 0$ in $H^\alpha(\mathbb{R}^3)$ and $v_n(x) \to v(x)$ a.e., in $\mathbb{R}^3$. Then v_n solves the following equation

$$\begin{cases} \left(a + b\int_{\mathbb{R}^3} |(-\Delta)^{\frac{\alpha}{2}} v_n|^2 \mathrm{d}x\right)(-\Delta)^\alpha v_n + V_n(x)v_n = \tilde{\rho}_n \quad \text{a. e. in } \mathbb{R}^3, \\ v_n \in H^\alpha(\mathbb{R}^3), \quad v_n > 0 \quad \text{in } \mathbb{R}^3 \end{cases} \tag{8.37}$$

with $V_n(x) = V(\epsilon_n x + \epsilon_n \tilde{y}_n)$ and $\tilde{\rho}_n(x) \in [\underline{f}_H(v_n(x)), \overline{f}_H(v_n(x))]$ a.e. in $\mathbb{R}^3$.

Lemma 8.12 *The sequence $\{v_n\}$ has a convergent subsequence in $H^\alpha(\mathbb{R}^3)$. Moreover, up to a subsequence, $y_n := \epsilon_n \tilde{y}_{\epsilon_n} \to y^* \in \Theta$, where $\Theta := \{x \in \mathbb{R}^3 : V(x) = V_0\}$.*

Proof. For each v_n, choosing $t_n > 0$ such that $\bar{v}_n := t_n v_n \in \mathcal{N}_{V_0}$, we deduce from $0 \in \partial I_{\epsilon_n,\eta_n}(u_n)$ and Lemma 8.9 that

$$\begin{aligned} &I_{V_0}(t_n v_n) \\ &\leq \frac{t_n^2}{2}\int_{\mathbb{R}^3} a|(-\Delta)^{\alpha/2} v_n|^2 + V_n(x)v_n^2)\mathrm{d}x + \frac{t_n^4}{4} b\left(\int_{\mathbb{R}^3} |(-\Delta)^{\frac{\alpha}{2}} v_n|^2 \mathrm{d}x\right)^2 - \int_{\mathbb{R}^3} F_H(t_n v_n)\mathrm{d}x \\ &= I_{\epsilon_n,\eta_n}(t_n u_n) \leq I_{\epsilon_n,\eta_n}(u_n) = c_{V_0} + o(1). \end{aligned}$$

Since $I_{V_0}(\bar{v}_n) \geq c_{V_0}$, we obtain that $\lim_{n\to\infty} I_{V_0}(\bar{v}_n) = c_{V_0}$.

We now show that $\{t_n\}$ is bounded. We argue by contradiction. In this case, since $\|u_n\|_{\epsilon_n}^2$ is bounded uniformly for n, we obtain that $I_{V_0}(\bar{v}_n) \to -\infty$ as $t_n \to +\infty$. This is not possible since $I_{V_0}(\bar{v}_n) \geq c_{V_0}$ for all $n \in \mathbb{N}$. Hence, $\{t_n\}$ is bounded. Up to a subsequence, we assume that $t_n \to t \geq 0$. If $t = 0$, then $\bar{v}_n \to 0$ in $H^\alpha(\mathbb{R}^3)$, because $\{v_n\}$ is bounded in $H^\alpha(\mathbb{R}^3)$. Hence $I_{V_0}(\bar{v}_n) \to 0$ as $n \to \infty$, which contradicts $c_{V_0} > 0$. So $\bar{v}_n := t_n v_n \rightharpoonup \bar{v}$ in $H^\alpha(\mathbb{R}^3) \setminus \{0\}$. By uniqueness, we deduce that $\bar{v} = tv$. Using Ekeland's variational principle, we obtain that $\{\bar{v}_n\} \subset \mathcal{N}_{V_0}$ is a Palais-Smale sequence of I_{V_0}, that is,

$$I_{V_0}(\bar{v}_n) \to c_{V_0} \quad I'_{V_0}(\bar{v}_n) \to 0 \quad \text{in } H^\alpha(\mathbb{R}^3). \tag{8.38}$$

Although I_{V_0} is of C^1 class, we can still use similar arguments as in Lemma 8.9 to obtain $\bar{v}_n \to \bar{v}$ in $H^\alpha(\mathbb{R}^3)$ and $\bar{v} \in \mathcal{N}_{V_0}$, and so $v_n \to v$ in $H^\alpha(\mathbb{R}^3)$.

Let us show that $y_n := \epsilon_n \tilde{y}_n$ is bounded. If not, then $|y_n| \to \infty$. It follows from $\bar{v}_n \in \mathcal{N}_{V_0}$, the Fatou lemma and Lemma 8.9 that

$$\begin{aligned} c_{V_0} \leq &I_{V_0}(\bar{v}) < I_{V_\infty}(tv) - \frac{1}{4} I'_{V_0}(tv)tv \\ = &\int_{\mathbb{R}^3} \left[\frac{a}{4}|(-\Delta)^{\alpha/2} tv|^2 + \left(\frac{1}{2}V_\infty - \frac{1}{4}V_0\right)t^2 v^2\right]\mathrm{d}x + \int_{\mathbb{R}^3}\left(f(tv)tv - \frac{1}{4}F(tv)\right)\mathrm{d}x \\ \leq &\liminf_{n\to\infty}\left(\frac{1}{4}\int_{\mathbb{R}^3} |(-\Delta)^{\alpha/2} t_n v_n|^2 \mathrm{d}x + \int_{\mathbb{R}^3}\left(\frac{1}{2}V(\epsilon_n x + y_n) - \frac{1}{4}V_0\right)t_n^2 v_n^2 \mathrm{d}x\right. \\ &\left. + \int_{\mathbb{R}^3}\left(f(t_n v_n)t_n v_n - \frac{1}{4}F_H(t_n v_n)\right)\mathrm{d}x\right) = \liminf_{n\to\infty} I_{\epsilon_n,\beta_n}(t_n u_n) \\ \leq &\liminf_{n\to\infty} I_{\epsilon_n,\beta_n}(u_n) = c_{V_0}, \end{aligned}$$

which is a contradiction. Thus, $\{y_n\}$ is bounded. Up to a subsequence, we can assume that $y_n \to y^*$. Moreover, since $f(t) = 0$ for all $t \leq 0$, we have as $\beta \to 0$

$$\underline{f}_H(t), \overline{f}_H(t) \to \begin{cases} 0, & \text{if } t < 0, \\ f(t), & \text{if } t \geq 0. \end{cases}$$

Hence, based on the facts that $\tilde{\rho}_n(x) \in [\underline{f}_H(v_n(x)), \overline{f}_H(v_n(x))]$ a.e. in $\mathbb{R}^3$ and $v_n \to v$ in $H^\alpha(\mathbb{R}^3)$, using the Lebesgue dominated convergence theorem, we deduce that for any $\eta \in C_0^\infty(\mathbb{R}^3)$

$$\int_{\mathbb{R}^3} \tilde{\rho}_n \eta \mathrm{d}x \to \int_{\mathbb{R}^3} f(v)\eta \mathrm{d}x, \quad \text{as } n \to \infty.$$

For each $\eta \in C_0^\infty(\mathbb{R}^3)$, we then deduce from (8.37) and $v_n \to v$ in $H^\alpha(\mathbb{R}^3)$ that

$$\begin{aligned}&\lim_{n\to\infty}\left((a+b\|v_n\|^2_{D^{\alpha,2}})\int_{\mathbb{R}^3}(-\Delta)^{\alpha/2}v_n(-\Delta)^{\alpha/2}\eta \mathrm{d}x + \int_{\mathbb{R}^3} V_n(x)v_n\eta \mathrm{d}x - \int_{\mathbb{R}^3}\tilde{\rho}_n\eta \mathrm{d}x\right)\\ =&(a+b\|v\|^2_{D^{\alpha,2}})\int_{\mathbb{R}^3}(-\Delta)^{\alpha/2}v(-\Delta)^{\alpha/2}\eta \mathrm{d}x + \int_{\mathbb{R}^3} V(y^*)v\eta \mathrm{d}x - \int_{\mathbb{R}^3} f(v)\eta \mathrm{d}x.\end{aligned}$$

Therefore, the limit v of the sequence $\{v_n\}$ solves the equation

$$(a+b\|v\|^2_{D^{\alpha,2}})(-\Delta)^\alpha v + V(y^*)v = f(v) \quad \text{in } \mathbb{R}^3. \tag{8.39}$$

Define the functional

$$I_{y^*}(u) := \frac{1}{2}\int_{\mathbb{R}^3}(a|(-\Delta)^{\alpha/2}u|^2 + V(y^*)u^2)\mathrm{d}x + \frac{b}{4}\|v\|^4_{D^{\alpha,2}} - \int_{\mathbb{R}^3} F(u)\mathrm{d}x.$$

If $V(y^*) > V_0$, then we can get a contradiction by similar arguments as above. So $V(y^*) = V_0$ and $I_{y^*}(v) = c_{V_0}$. It follows that v is a positive ground state solutions of problem (8.39). The proof is complete. □

We can use the Moser iteration method [120] in the order to prove the following L^∞-estimate for problem (8.37).

Lemma 8.13 *Let $u_n \in E_{\epsilon_n}$ be a solution of problem* (K_{ϵ_n}) *with $\epsilon_n \to 0^+$ and $\beta_n \to 0$. Then $v_n = u_n(\cdot + \tilde{y}_n) \in L^\infty(\mathbb{R}^3)$.*

Proof. Since $|\Lambda^*| = 0$ with $\Lambda^* := \{x \in \mathbb{R}^3 : v_n(x) = \beta\}$, our argument is similar to that in Lemma 3.2 of Ambrosio [3]. So we omit the details of the proof. □

Now, we claim that there exists $c > 0$ such that $\|v_n\|_\infty \geq c > 0$. Otherwise, $\|v_n\|_\infty \to 0$ as $n \to \infty$. By (f_1)–(f_3) and the definition of f_H, for n large enough, we have

$$\int_{\mathbb{R}^3}[a|(-\Delta)^{\alpha/2}v_n|^2 + V_0 v_n^2]\mathrm{d}x \leq \int_{\mathbb{R}^3}\frac{V_0}{2}v_n^2 \mathrm{d}x,$$

which is a contradiction. Hence, from Lemma 8.13 there exist $c, C > 0$ independent of n such that

$$c \leq \|v_n\|_\infty \leq C. \tag{8.40}$$

Observe from (8.37) that v_n solves the problem

$$\begin{cases}(-\Delta)^\alpha v_n + v_n = \chi_n & \text{a.e. in } \mathbb{R}^3,\\ v_n \in H^\alpha(\mathbb{R}^3), \quad v_n > 0 & \text{in } \mathbb{R}^3,\end{cases} \tag{8.41}$$

where

$$\chi_n(x) := \left(a + b\int_{\mathbb{R}^3}|(-\Delta)^{\frac{\alpha}{2}}v_n|^2\mathrm{d}x\right)^{-1}[\tilde{\rho}_n - V_n(x)v_n] + v_n$$

and $V_n(x) = V(\epsilon_n x + \epsilon_n \tilde{y}_n)$ and $\tilde{\rho}_n(x) \in [\underline{f}_H(v_n(x)), \overline{f}_H(v_n(x))]$ a.e. in $\mathbb{R}^3$.

From (8.40) and the definitions of $\underline{f}_H, \overline{f}_H$, we deduce that $\{\tilde{\rho}_n\}$ is bounded in $L^\infty(\mathbb{R}^3)$. Using the fact that $v_n \to v$ in $H^\alpha(\mathbb{R}^3)$, we see that there exists $\tilde{\rho}^*(x) \in [\underline{f}_H(v(x)), \overline{f}_H(v(x))]$

(ii) *Nonsmooth dynamics*: by introducing the Heaviside function, the reaction term becomes discontinuous. This moves the problem from the classical smooth variational framework into the realm of nonsmooth critical point theory. The case studied in this chapter involves the Heaviside function, which creates a single jump at $u = \beta$. Problems with multiple discontinuities or more general discontinuous nonlinearities are completely open and relevant for multi-threshold physical models.

(iii) *Concentration phenomena*: analysis establishes of how solutions behave as $\epsilon \to 0$ and $\beta \to 0$ is central. The analysis shows that positive solutions concentrate near minima of the potential $V(x)$, connecting the singularly perturbed problem to a limit autonomous equation. A finer problem is to determine the asymptotic shape of the solutions as $\epsilon \to 0$ and the convergence rate of the maximum points $x_{\epsilon,\beta}$ to the minimum of the potential V.

The content of this chapter is based on the results established in [108], while concentration properties for Kirchhoff equations with exponential growth were subsequently obtained by Sun, Fu and Liang [143].

Some open problems inspired by the analysis developed in this chapter are described in what follows.

Open Problem 1. The analysis in this chapter assumes a specific nonlocal term $M(t) = a + bt$. A broad open area is to consider general Kirchhoff functions $M(t)$ that may be degenerate (that is, $M(0) = 0$) or singular, which would alter the problem's geometry significantly.

The following problem is concerned with the competition of the small parameters.

Open Problem 2. The analysis treats ϵ and β as small positive parameters, but their relative scaling could lead to different concentration behaviors and limit equations, which is unexplored. For instance, consider the case $\beta = \epsilon^\gamma$ for some $\gamma > 0$.

What about the singular limit problem? This is what about the next problem is concerned with.

Open Problem 3. What happens when the fractional order α converges to 1^-? One can establish that the family of corresponding fractional solutions converges to a solution of a classical (local) Kirchhoff equation with discontinuity?

The next research direction deals with the investigation of sign-changing solutions. We assume that this analysis requires adapting methods like the Nehari manifold or fountain theorems to the nonsmooth fractional Kirchhoff setting.

Open Problem 4. In this chapter it is established the existence of positive solutions. A challenging next step is investigating nodal solutions or the existence of infinitely many solutions.

Deepening the regularity of solutions is a very interesting open problem, which we formulate in what follows.

Open Problem 5. This chapter establishes existence results, but the smoothness of these solutions is unknown. Does the combination of a nonlocal operator and a discontinuous nonlinearity allow for classical solutions, or what is the optimal Hölder regularity?

Another unexplored direction is about a possible optimal notion of solution.

Open Problem 6. The analysis developed in this chapter is based on the concept of solution in the Filippov's inclusion sense, which seems to be relevant for physical applications (phase transitions or plasticity). Develop a similar analysis with other concepts of solutions, like viscosity solutions.

Finally, we suggest to develop a thorough numerical analysis of solutions to Kirchhoff-type equations with discontinuous reaction.

Open Problem 7. Develop robust numerical schemes to approximate solutions of fractional Kirchhoff problems with discontinuous nonlinearities.

8.7 Glossary

Heaviside Function: It is also called the unit step function and is a discontinuous function widely used in engineering, physics, and mathematics to model sudden changes, switches, or signals that turn on at a certain point.

Discontinuous Reaction in Kirchhoff Equations: This usually arises in models where the reaction term changes abruptly, often modeled with Heaviside or sign functions.

Differential Inclusion: It is a basic framework for dealing with discontinuous ODEs/PDEs, for instance, fractional Kirchhoff equations with discontinuous reaction.

Nonsmooth Critical Point Theory: It is the main analytical method used to solve variational problems with discontinuous nonlinearities.

Clarke Subdifferential: It is the natural extension of the derivative for dealing with functions that are not differentiable in the classical sense. For a locally Lipschitz function f, the Clarke subdifferential at a point is defined in two steps: definition of the generalized directional derivative and definition of the subdifferential.

Mountain Pass Theorem for Locally Lipschitz Functions: It extends the classical Ambrosetti–Rabinowitz theorem to a nonsmooth setting via Clarke's critical point theory.

9

Mass critical fractional Kirchhoff equations

Why mass critical fractional Kirchhoff equations are important? Because this class of Kirchhoff problems represents a delicate and significant frontier in pure and applied nonlinear analysis, which is important for modeling quantum phenomena and other patterns in the real world.

This chapter focuses on the constraint minimization problem associated with the fractional Kirchhoff equation

$$\begin{cases} \left(a + b \displaystyle\int_{\mathbb{R}^N} |(-\Delta)^{\frac{s}{2}} u|^2 dx\right) (-\Delta)^s u + |x|^2 u = \mu u + \beta u^{\frac{8s}{N}+1} & \text{in } \mathbb{R}^N, \\ \displaystyle\int_{\mathbb{R}^N} |u|^2 dx = 1, \end{cases}$$

where $s \in (N/4, 1), N = 2, 3$, $a \geq 0, b > 0$ are constants, $\mu \in \mathbb{R}$ is the corresponding Lagrange multiplier and $(-\Delta)^s$ is the fractional Laplacian operator, $8s/N + 1$ is the corresponding mass critical exponent.

The purpose in this chapter is threefold:

(i) To establish the existence and non-existence of the L^2-constraint minimizers to the degenerate fractional Kirchhoff problem: the existence is proved under certain conditions involving the mass parameter and the Kirchhoff coefficient, while no minimizers exists when the mass exceeds a certain critical value.

(ii) To prove some concentration behaviors of constraint minimizers: as the mass approaches a critical threshold, the minimizers concentrate around a global minimum point of an auxiliary potential related to the Kirchhoff term.

(iii) To reveal the local uniqueness of constraint minimizers under double nonlocal effect: for sufficiently small mass, the minimizer is shown to be unique (up to translation and phase) and radially symmetric.

In particular, we establish some energy estimates, decay properties and uniform regularity to find that the maximal point of constraint minimizer concentrates on the bottom point of the homogeneous potential. Furthermore, we introduce several new techniques based on the combination of the localization method of $(-\Delta)^s$ and by establishing the nonlocal Pohozaev identity, which allows to get over some new challenges due to the nonlocal property of $(-\Delta)^s$ and the fact that $(\int_{\mathbb{R}^N} |(-\Delta)^{\frac{s}{2}} u|^2 dx)(-\Delta)^s u$ does not vanish as $a \searrow 0$.

In the analysis of this problem, three fundamental features interact:

(a) *Fractional Kirchhoff term*: the equation is nonlocal in both the fractional Laplacian operator and the Kirchhoff coefficient.

(b) *Mass critical exponent*: the nonlinearity power corresponds to the critical case in the sense of the Gagliardo–Nirenberg inequality, which leads to delicate compactness and blow-up behavior.

DOI: 10.1201/9781003797159-9

(c) *Constraint minimization*: solutions are obtained by minimizing the energy functional subject to a fixed L^2-norm.

9.1 Statement of the problem

This chapter is devoted to the study of the following class of mass critical fractional Kirchhoff equations with mass constraint

$$\begin{cases} \left(a+b\int_{\mathbb{R}^N}|(-\Delta)^{\frac{s}{2}}u|^2dx\right)(-\Delta)^s u+|x|^2u=\mu u+\beta u^{\frac{8s}{N}+1} \quad \text{in } \mathbb{R}^N, \\ \int_{\mathbb{R}^N}|u|^2dx=1, \end{cases} \tag{9.1}$$

where $s \in (N/4, 1), N = 2, 3$, $a \geq 0, b > 0$ are constants, $\mu \in \mathbb{R}$ is the corresponding Lagrange multiplier, $(-\Delta)^s$ is the fractional Laplacian operator defined by

$$(-\Delta)^s u(x) := C_s P.V. \int_{\mathbb{R}^N} \frac{u(x)-u(y)}{|x-y|^{N+2s}}dy = C_s \lim_{\varepsilon\to 0}\int_{\mathbb{R}^N\setminus B_\varepsilon(x)} \frac{u(x)-u(y)}{|x-y|^{N+2s}}dy,$$

for $x \in \mathbb{R}^N$, where *P.V.* is the principal value, and C_s is a normalization constant.

A natural question in the analysis developed in this chapter is whether we can establish the concentration of solutions and the local uniqueness of constraint minimizers of the energy associated to (9.1) as $a \searrow 0$? This corresponds to the *degenerate case* in problem (9.1). To the best of our knowledge, there are no such results in this field. Two intuitive difficulties arise in this analysis. On the one hand, some standard regularity theories for classical elliptic problems are invalid due to the nonlocal properties of fractional Laplace operators. On the other hand, the nonlocal Kirchhoff term will not vanish as $a \searrow 0$ and it will cause great difficulty concerning the uniqueness since it could influence the rate of decay of solutions sequence.

The energy functional associated to the weighted Kirchhoff equation (9.1) is defined by

$$E_{a,\beta}(u) := \int_{\mathbb{R}^N}(a|(-\Delta)^{\frac{s}{2}}u|^2+|x|^2u^2)dx+\frac{b}{2}\left(\int_{\mathbb{R}^N}|(-\Delta)^{\frac{s}{2}}u|^2dx\right)^2-\frac{N\beta}{N+4s}\int_{\mathbb{R}^N}|u|^{\frac{8s+2N}{N}}dx,$$

for all $u \in H$, where the weighted fractional space H is defined by

$$H := \left\{u \in H^s(\mathbb{R}^N) : \int_{\mathbb{R}^N}|x|^2u^2 < \infty\right\}.$$

One can associate the normalized solution of equation (9.1) with the following constrained minimization problem

$$e(a,\beta) := \inf_{\{u\in H, \|u\|_2^2=1\}} E_{a,\beta}(u).$$

9.1.1 Exact threshold, blow-up and uniqueness for small masses

In order to consider the asymptotic properties of constraint minimizers, we first need to figure out their existence under the degenerate condition, that is, if $a = 0$. Similar to the classical case, fractional Gagliardo–Nirenberg–Sobolev inequality plays a crucial role in verifying the existence of a constraint minimizer. Du, Tian, Wang and Zhang [47] established the following fractional Gagliardo–Nirenberg–Sobolev inequality

$$\int_{\mathbb{R}^N} |u(x)|^{\frac{8s+2N}{N}} dx \le \frac{N+4s}{N\|\varphi\|_2^{\frac{8s}{N}}} \left(\int_{\mathbb{R}^N} |(-\Delta)^{\frac{s}{2}} u(x)|^2 dx\right)^2 \left(\int_{\mathbb{R}^N} |u(x)|^2 dx\right)^{\frac{4s-N}{N}}, \tag{9.2}$$

where $\varphi(x)$ is the unique radial positive ground state solution of the following equation

$$2(-\Delta)^s u + \frac{4s-N}{N} u - u^{\frac{8s}{N}+1} = 0 \quad \text{in } \mathbb{R}^N, \tag{9.3}$$

with

$$\int_{\mathbb{R}^N} |(-\Delta)^{\frac{s}{2}} \varphi(x)|^2 dx = \int_{\mathbb{R}^N} |\varphi(x)|^2 dx = \frac{N}{N+4s} \int_{\mathbb{R}^N} |\varphi(x)|^{\frac{8s+2N}{N}} dx \tag{9.4}$$

and

$$\frac{C_1}{1+|x|^{N+2s}} \le \varphi(x) \le \frac{C_2}{1+|x|^{N+2s}} \quad \text{for } x \in \mathbb{R}^N. \tag{9.5}$$

The first result in this chapter gives a sharp existence/non-existence property based on the mass constraint. More precisely, the next theorem establishes the exact threshold for the existence of minimizers, proving that there exists at least one minimizer at the threshold and there is no minimizer at the threshold and above it.

Theorem 9.1 *Let $\varphi(x) > 0$ be the unique positive solution of problem (9.3). Then we have the following properties.*

(i) If $0 < \beta < \beta^* := \frac{b}{2}\|\varphi\|_2^{\frac{8s}{N}}$, then there exists at least one minimizer to $e(0,\beta)$.

(ii) If $\beta \ge \beta^*$, then $e(0,\beta)$ has no minimizer.

Moreover, $e(0,\beta^*) = 0$ and $e(0,\beta) = -\infty$ for all $\beta > \beta^*$.

Next, we consider the concentration phenomenon of constraint minimizer as $a \searrow 0$. In such a way, we characterize the blow-up profile and concentration at the origin as the mass approaches the critical value.

Theorem 9.2 *Let u_k be a nonnegative minimizer of $e(a_k, \beta^*)$ with $a_k \to 0$ as $k \to \infty$. Then there exists a subsequence of $\{u_k\}$, still denoted by $\{u_k\}$ such that each u_k has a unique global maximum point $\bar{z}_k$ with*

$$\lim_{k\to\infty} \bar{z}_k = 0 \tag{9.6}$$

and

$$\lim_{k\to\infty} \bar{\varepsilon}_k^{\frac{N}{2}} u_k(\bar{\varepsilon}_k x + \bar{z}_k) = \left(\frac{b}{2\beta^*}\right)^{\frac{N}{8s}} \varphi(x) \quad \text{in } L^\infty(\mathbb{R}^N), \tag{9.7}$$

where $\bar{\varepsilon}_k$ is defined by

$$\bar{\varepsilon}_k := \left[\frac{s a_k \|\varphi\|_2^2}{\int_{\mathbb{R}^N} |x|^2 \varphi^2(x) dx}\right]^{\frac{1}{2+2s}}. \tag{9.8}$$

Moreover, we have

$$\lim_{k\to\infty} \frac{\bar{z}_k}{\bar{\varepsilon}_k} = 0. \tag{9.9}$$

We notice that in the process of establishing the concentration phenomenon of constraint minimizers, some classical local elliptic theories such as De Giorgi-Nash-Moser theory, regularity theory and comparison principle are no longer applicable. Therefore, some uniform regularity and decay estimates for the solutions to the fractional Kirchhoff equation are established by taking full advantage of the Bessel kernel and employing the Morse iteration. In addition, for the sake of subsequent uniqueness results, as in Theorem 9.2, we need to verify the convergence of minimizers in $L^\infty(\mathbb{R}^N)$ instead of $H^s(\mathbb{R}^N)$, where the most convergence results of minimizers to fractional nonlinear equations were established in the latter space.

Based on the above concentration phenomenon, the following result establishes the uniqueness of the constraint minimizer for small masses.

Theorem 9.3 *Let u_k be a nonnegative minimizer of $e(a_k, \beta^*)$ with $a_k \to 0$ as $k \to \infty$, then the minimizer u_k is unique as $k \to \infty$.*

9.1.2 Comments on Theorems 9.2 and 9.3

The basic idea to prove Theorem 9.3 is to get a contradiction by establishing nonlocal Pohozaev identities. First, select two suitable $\hat{u}_1$ and $\hat{u}_2$ which are scaling functions of two different constraint minimizers u_1 and u_2. By studying the properties of fractional Laplacian, we can decompose $\bar{\eta} := (\hat{u}_1 - \hat{u}_2)/\|\hat{u}_1 - \hat{u}_2\|_\infty$ as the linear combination of some functions consisting of the unique radial positive ground state solution to (9.3). Next, we consider the localization method introduced by Caffarelli and Silvestre [16]. More precisely, for a function $u \in H^s(\mathbb{R}^N)$, set

$$\tilde{u}(x,t) = \mathcal{P}_s[u] = \int_{\mathbb{R}^N} \mathcal{P}_s(x-z,t)u(z)dz, \quad (x,t) \in \mathbb{R}^{N+1}_+ := \{(x,t) : x \in \mathbb{R}^N, t > 0\}, \tag{9.10}$$

where

$$\mathcal{P}_s(x,t) = \beta_s \frac{t^{2s}}{(|x|^2 + t^2)^{\frac{N+2s}{2}}} \tag{9.11}$$

with a constant β_s such that $\int_{\mathbb{R}^N} \mathcal{P}_s(x,1)dx = 1$. Then $\tilde{u}$ satisfies

$$\begin{cases} \operatorname{div}(t^{1-2s}\nabla \tilde{u}) = 0 & x \in \mathbb{R}^{N+1}_+, \\ -\lim\limits_{t\to 0} t^{1-2s}\partial_t \tilde{u}(x,t) = \omega_s(-\Delta)^s u(x) & x \in \mathbb{R}^N, \end{cases} \tag{9.12}$$

in the distribution sense, where $\omega_s = 2^{1-s}\Gamma(1-s)/\Gamma(s)$. Without loss of generality, we may assume $\omega_s = 1$. Taking advantage of decay estimates and blow up analysis, we obtain a contradiction which indicates the local uniqueness as a closes to zero sufficiently by establishing two nonlocal Pohozaev identities.

Compared to the classical case, constraint minimizers in our problem only satisfy polynomial decay (it can be exponential decay in the classical case) which requires more precision in our estimates. In the process of our research, we find that the scaling functions used by Hu and Tang [73] cannot meet our requirement of estimates due to the polynomial decay and the non-vanishing of the nonlocal fractional Kirchhoff term. The testing function defined in this chapter, which is defined by

$$\hat{u}_{ik}(x) := \left(\frac{2\beta^*}{b}\right)^{\frac{N}{8s}} \bar{\varepsilon}_k^{\frac{N}{2}} u_{ik}(\bar{\varepsilon}_k^{\tau} x) \quad i = 1, 2,$$

achieves the best decay speed, which improves the accuracy of our estimates. As a new attempt, we believe it will be useful in the relevant research of fractional nonlinear equations.

Some technique difficulties appear, for instance, the appearance of $(-\Delta)^s$ makes the classical elliptic regularity invalid, we have to set the decay estimate of $|\nabla u|$ (where u stands for the constraint minimizer). Moreover, the property of the linearized operator $\Gamma := (-\Delta)^s + \frac{4s-N}{2N} - \frac{8s+N}{2N}\varphi^{\frac{8s}{N}}$ is also an important topic of the research in this chapter.

Although the corresponding harmonic extension problem can help us to establish Pohozaev identities, we have to deal with several new integral terms that never appeared in the classical local elliptic problems. As a result, the nonlocal Pohozaev identity we establish may be the first result of the fractional Kirchhoff equation and we believe it will be useful in the relevant research of fractional nonlinear equations.

9.2 Existence and non-existence of minimizers

In this section, we discuss the existence and non-existence of the constraint minimizer to $e(0,\beta)$ and we shall distinguish three cases. We first recall the following compact embedding, see the monograph by Molica Bisci, Rădulescu and Servadei [119] for details.

Lemma 9.1 *The embedding $H \hookrightarrow L^q(\mathbb{R}^N)$ is compact for all $q \in [2, 2_s^*)$.*

Next, we give the proof of the existence/non-existence property.

Proof of Theorem 9.1. Fix $\beta \in (0,\beta^*)$. By the Gagliardo–Nirenberg–Sobolev given in (9.2), we have for all $u \in H$ with $\|u\|_2 = 1$

$$\begin{aligned} E_{0,\beta}(u) &= \int_{\mathbb{R}^N} |x|^2u^2dx + \frac{b}{2}\left(\int_{\mathbb{R}^N} |(-\Delta)^{\frac{s}{2}}u|^2dx\right)^2 - \frac{N\beta}{N+4s}\int_{\mathbb{R}^N} |u|^{\frac{8s+2N}{N}}dx \\ &\geq \left(\frac{b}{2} - \frac{\beta}{\|\varphi\|_2^{\frac{8s}{N}}}\right)\left(\int_{\mathbb{R}^N} |(-\Delta)^{\frac{s}{2}}u|^2dx\right)^2, \end{aligned} \tag{9.13}$$

which implies that $E_{0,\beta}(u)$ is bounded from below in this case. Let $\{u_n\} \in H$ be a minimizing sequence of $e(0,\beta)$ for some $\beta \in (0,\beta^*)$. By relation (9.13), sequences $\{\int_{\mathbb{R}^N} |(-\Delta)^{\frac{s}{2}}u_n|^2dx\}$ and $\{\int_{\mathbb{R}^N} |x|^2u_n^2dx\}$ are bounded uniformly in n. By the weak lower semi-continuity of the norm in H and Lemma 9.1, we conclude that $e(0,\beta)$ has at least one minimizer.

For $\beta > \beta^*$, we take advantage of the cut-off function to prove the non-existence of minimizer of $e(0,\beta)$ in this case. Let $\omega \in C_0^\infty(\mathbb{R}^2)$ be a cut-off function with $0 \leq \omega \leq 1$, $\omega(x) = 1$ when $|x| \leq 1$, $\omega(x) = 0$ when $|x| > 2$ and $|\nabla\omega| \leq 2$, and define the following trail function

$$U_\tau(x) := \frac{A_\tau \tau^{\frac{N}{2}}}{\|\varphi\|_2}\omega\left(\frac{x-x_0}{\tau^2}\right)\varphi(\tau|x-x_0|), \tag{9.14}$$

where $A_\tau > 0$ is chosen such that $\|U_\tau(x)\|_2^2 = 1$. From the polynomial decay of φ, we have

$$\frac{1}{A_\tau^2} = 1 + \frac{1}{\|\varphi\|_2^2}\int_{\mathbb{R}^N\setminus B_{\tau^3}(0)}\left[\omega^2\left(\frac{x}{\tau^3}\right) - 1\right]\varphi^2dx = 1 + O(\tau^{-3N+2s}) \quad \text{as } \tau \to \infty. \tag{9.15}$$

Following (9.4), [47, Lemma 3.2] and (9.15), we obtain

$$\begin{aligned}\int_{\mathbb{R}^N} |(-\Delta)^{\frac{s}{2}} U_\tau(x)|^2 dx &= \frac{A_\tau^2 \tau^{2s}}{\|\varphi\|_2^2} \int_{\mathbb{R}^N}\int_{\mathbb{R}^N} \frac{\left|\omega\left(\frac{x}{\tau^3}\right)\varphi(x) - \omega\left(\frac{y}{\tau^3}\right)\varphi(y)\right|^2}{|x-y|^{N+2s}} dxdy \\ &\le \frac{A_\tau^2 \tau^{2s}}{\|\varphi\|_2^2}\left(\int_{\mathbb{R}^N} |(-\Delta)^{\frac{s}{2}}\varphi(x)|^2 dx + O(\tau^{-12s})\right) \\ &= \tau^{2s} + O(\tau^{-10s}) \quad \text{as } \tau \to \infty,\end{aligned} \tag{9.16}$$

$$\begin{aligned}\int_{\mathbb{R}^N} |U_\tau(x)|^{\frac{8s+2N}{N}} dx &= \frac{A_\tau^{\frac{8s+2N}{N}} \tau^{4s}}{\|\varphi\|_2^{\frac{8s+2N}{N}}} \int_{\mathbb{R}^N} \omega^{\frac{8s+2N}{N}}\left(\frac{x}{\tau^3}\right)\varphi^{\frac{8s+2N}{N}} dx \\ &= \frac{(N+4s)\tau^{4s}}{N\|\varphi\|_2^{\frac{8s}{N}}} + O\left(\tau^{-3N-36s-\frac{48s^2}{N}}\right) \quad \text{as } \tau \to \infty,\end{aligned} \tag{9.17}$$

and

$$\int_{\mathbb{R}^N} |x|^2 |U_\tau(x)|^2 dx = \frac{A_\tau^2}{\|\varphi\|_2^2}\int_{B_{2\tau^3}(0)} \left|\frac{x}{\tau} + x_0\right|^2 \omega^2\left(\frac{x}{\tau^3}\right)\varphi^2 dx \to |x_0|^2 \quad \text{as } \tau \to \infty. \tag{9.18}$$

Therefore, we derive from (9.16)–(9.18) that for $\beta > \beta^*$,

$$e(0,\beta) \le E_{0,\beta}(u) \le \left(\frac{b}{2} - \frac{\beta}{\|\varphi\|_2^{\frac{8s}{N}}}\right)\tau^{4s} + |x_0|^2 + O(\tau^{-8s}) + o(1) \to -\infty \quad \text{as } \tau \to \infty, \tag{9.19}$$

which implies that $e(0,\beta)$ has no minimizer for $\beta > \beta^*$.

For $\beta = \beta^*$, we may assume that there exists a positive constraint minimizer v, then taking $x_0 = 0$ in (9.14), we can deduce from (9.19) that $e(0,\beta) \le 0$. Moreover, (9.13) shows that

$$E_{0,\beta^*}(v) \ge \int_{\mathbb{R}^N} |x|^2 v^2 dx \ge 0.$$

Then we obtain $e(0,\beta^*) = 0$ and $v(x) = 0$ in $\mathbb{R}^N \setminus \{0\}$, which is a contradiction to the fact that $\int_{\mathbb{R}^N} |v|^2 dx = 1$. □

9.3 Concentration phenomenon

In this section, we fix $\beta = \beta^*$ and focus on some concentration phenomenon of the minimizer u_a to $e(a,\beta^*)$. Similar to the proof of Theorem 9.1, we can prove that for any $a > 0$, there exists a nonnegative minimizer u_a of $e(a,\beta^*)$. On the other hand, there is no minimizer to $e(0,\beta^*)$ as $a = 0$. Therefore, a nature question is what happens in the limiting case, that is, if $a \searrow 0$. We shall establish Theorem 9.2 dealing with the concentration phenomenon of u_a as $a \searrow 0$. As a bedding, the following property describes the limiting behavior of the energy as $a \searrow 0$.

Lemma 9.2 *Let u_a be a nonnegative minimizer of $e(a,\beta^*)$, then*

$$\begin{cases} e(a,\beta^*) \to e(0,\beta^*) = 0 \quad \text{as } a \searrow 0, \\ \int_{\mathbb{R}^N} |(-\Delta)^{\frac{s}{2}} u_a|^2 dx \to +\infty \quad \text{as } a \searrow 0, \\ a\int_{\mathbb{R}^N} |(-\Delta)^{\frac{s}{2}} u_a|^2 dx \to 0 \quad \text{as } a \searrow 0, \\ \int_{\mathbb{R}^N} |x|^2 |u_a|^2 dx \to 0 \quad \text{as } a \searrow 0. \end{cases}$$

Proof. In view of Theorem 9.1 and (9.2), we have

$$e(a,\beta^*) = E_{a,\beta^*}(u_a) \geq \int_{\mathbb{R}^N} (a|(-\Delta)^{\frac{s}{2}}u_a|^2 + |x|^2 u_a^2)dx > e(0,\beta^*) = 0. \tag{9.20}$$

Defining $U_\tau(x)$ as in relation (9.14), it follows from (9.15)–(9.18) that

$$0 < e(a,\beta^*) \leq E_{a,\beta^*}(U_\tau) \leq a\tau^{2s} + |x_0|^2 + Ca\tau^{-10s} + C\tau^{-8s} + o(1) \quad \text{as } \tau \to \infty. \tag{9.21}$$

Taking $x_0 = (0,0,\ldots 0)$ and letting $\tau = a^{-\frac{1}{4s}}$, then (9.21) shows that

$$0 < e(a,\beta^*) \to 0 \quad \text{as } a \searrow 0. \tag{9.22}$$

This relation and (9.20) yield

$$a\int_{\mathbb{R}^N} |(-\Delta)^{\frac{s}{2}}u_a|^2 dx \to 0, \quad \int_{\mathbb{R}^N} |x|^2|u_a|^2 dx \to 0 \quad \text{as } a \searrow 0. \tag{9.23}$$

We now prove that

$$\int_{\mathbb{R}^N} |(-\Delta)^{\frac{s}{2}}u_a|^2 dx \to +\infty \quad \text{as } a \searrow 0. \tag{9.24}$$

On the contrary, we assume that (9.24) is false, then relation (9.23) shows that there exists a sequence $\{a_k\}$ with $a_k \to 0$ as $k \to \infty$ such that $\{u_k\}$ is bounded in H, where $u_k := u_{a_k}$. Thus, Lemma 9.1 implies that there exists $u_0 \in H$ such that $u_k \rightharpoonup u_0$ weakly in H and $u_k \to u_0$ strongly in $L^p(\mathbb{R}^N)$ as $k \to \infty$ for $p \in [2, 2_s^*)$. It follows that

$$e(0,\beta^*) \leq \liminf_{k\to\infty} E_{a_k,\beta^*}(u_k) = \lim_{k\to\infty} e(a_k,\beta^*) = 0 = e(0,\beta^*).$$

This shows that u_0 is a minimizer of $e(0,\beta^*)$, which is a contradiction. □

Next, we investigate the asymptotic behavior of the minimizers (up to scaling) and we give some properties of the trail function.

Lemma 9.3 *Let u_k be a nonnegative minimizer of $e(a_k,\beta^*)$ with $a_k \to 0$ as $k \to \infty$, then there exist a sequence $\{y_k\}$, $K_0 > 0$ and $\delta > 0$ such that the normalized function*

$$w_k(x) = \varepsilon_k^{\frac{N}{2}} u_k(\varepsilon_k x + \varepsilon_k y_k) \quad \textit{where} \quad \varepsilon_k := \left(\int_{\mathbb{R}^N} |(-\Delta)^{\frac{s}{2}}u_k|^2 dx\right)^{-\frac{1}{2s}} \tag{9.25}$$

satisfies

$$\liminf_{k\to\infty} \int_{B_{K_0}(0)} |w_k|^2 dx \geq \delta > 0. \tag{9.26}$$

Moreover, for any sequence $\{a_k\}$ with $a_k \to 0$ as $k \to \infty$, there exists a subsequence, still denoted by $\{a_k\}$, such that $z_k := \varepsilon_k y_k \to 0$ as $k \to \infty$. In addition, for any $\rho > 0$ small enough

$$u_k(x) = \varepsilon_k^{-\frac{N}{2}} w_k\left(\frac{x - z_k}{\varepsilon_k}\right) \to 0 \quad \textit{as } k \to \infty \textit{ for } x \in B_\rho^c(0), \tag{9.27}$$

and, for some $y_0 \in \mathbb{R}^N$,

$$w_k(x) \to w_0(x) := \left(\frac{b}{2\beta^*}\right)^{\frac{N}{8s}} \varphi(|x - y_0|) \quad \textit{as } k \to \infty \textit{ in } H^s(\mathbb{R}^N). \tag{9.28}$$

Proof. We divide this proof into six steps.

Step 1. Defining $\tilde{w}_k(x) := \varepsilon_k^{\frac{N}{2}} u_k(\varepsilon_k x)$, we derive from Lemma 9.2 that

$$\begin{aligned}\int_{\mathbb{R}^N} |\tilde{w}_k|^{\frac{8s+2N}{N}} dx =& \varepsilon_k^{4s} \int_{\mathbb{R}^N} |u_k|^{\frac{8s+2N}{N}} dx \\ =& \frac{\varepsilon_k^{4s}(N+4s)}{N\beta^*} \Big[\int_{\mathbb{R}^N} (a_k|(-\Delta)^{\frac{s}{2}} u_k|^2 + |x|^2 u_k^2) dx \\ &+ \frac{b}{2} \left(\int_{\mathbb{R}^N} |(-\Delta)^{\frac{s}{2}} u_k|^2 dx \right)^2 - e(a_k, \beta^*) \Big] \\ \to& \frac{b(N+4s)}{2N\beta^*} \quad \text{as } k \to \infty.\end{aligned} \tag{9.29}$$

We now claim that there exist a sequence $\{y_k\} \subset \mathbb{R}^N$ and $K_0, \delta > 0$ such that

$$\liminf_{\varepsilon_k \to 0} \int_{B_{K_0}(y_k)} |\tilde{w}_k|^2 dx \geq \delta > 0. \tag{9.30}$$

Suppose by contradiction that for any $K > 0$, there exists a subsequence of $\{\tilde{w}_k\}$, still denoted by $\{\tilde{w}_k\}$, satisfying

$$\lim_{k \to \infty} \sup_{y \in \mathbb{R}^N} \int_{B_K(y)} |\tilde{w}_k|^2 dx = 0.$$

By the fact that $s \in (N/4, 1)$, we deduce from the vanishing lemma that $\tilde{w}_k \to 0$ in $L^{\frac{8s+2N}{N}}(\mathbb{R}^N)$ as $k \to \infty$, which contradicts (9.29). Then (9.30) shows that relation (9.26) holds.

Step 2. Applying Lemma 9.2 and (9.25), we have

$$\int_{\mathbb{R}^N} |x|^2 u_k^2 dx = \int_{\mathbb{R}^N} |\varepsilon_k x + \varepsilon_k y_k|^2 w_k^2 dx \to 0 \quad \text{as } k \to \infty,$$

which implies that

$$0 = \liminf_{k \to \infty} \int_{\mathbb{R}^N} |\varepsilon_k x + \varepsilon_k y_k|^2 w_k^2 dx \geq \liminf_{k \to \infty} \int_{B_{K_0(0)}} |\varepsilon_k x + \varepsilon_k y_k|^2 w_k^2 dx.$$

Since $|x|^2 \to \infty$ as $|x| \to \infty$, then (9.26) shows that $\{\varepsilon_k y_k\}$ is bounded in $\mathbb{R}^N$. Up to a subsequence if necessary, there exists a $x_0 \in \mathbb{R}^N$ such that $z_k := \varepsilon_k y_k \to x_0$ as $k \to \infty$. By Fatou's lemma, it then follows from (9.26) that

$$\liminf_{k \to \infty} \int_{\mathbb{R}^N} |\varepsilon_k x + \varepsilon_k y_k|^2 w_k^2 dx \geq |x_0|^2 \int_{B_{K_0(0)}} \lim_{k \to \infty} w_k^2 dx \geq \frac{|x_0|^2 \delta}{2},$$

which shows that $|x_0| = 0$.

Step 3. Since u_k is a nonnegative minimizer of $e(a_k, \beta^*)$, then it satisfies the following fractional Kirchhoff equation

$$\left(a_k + b \int_{\mathbb{R}^N} |(-\Delta)^{\frac{s}{2}} u_k|^2 dx \right) (-\Delta)^s u_k + |x|^2 u_k = \mu_k u_k + \beta^* u_k^{\frac{8s}{N}+1} \quad \text{in } \mathbb{R}^N, \tag{9.31}$$

where $\mu_k \in \mathbb{R}$ is the Lagrange multiplier. It is easy to check that

$$\mu_k = \int_{\mathbb{R}^N} (a_k|(-\Delta)^{\frac{s}{2}} u_k|^2 + |x|^2 u_k^2) dx + b \left(\int_{\mathbb{R}^N} |(-\Delta)^{\frac{s}{2}} u_k|^2 dx \right)^2 - \beta^* \int_{\mathbb{R}^N} u_k^{\frac{8s+2N}{N}} dx \tag{9.32}$$

and

$$e(a_k,\beta^*)=\int_{\mathbb{R}^N}(a_k|(-\Delta)^{\frac{s}{2}}u_k|^2+|x|^2u_k^2)dx+\frac{b}{2}\left(\int_{\mathbb{R}^N}|(-\Delta)^{\frac{s}{2}}u_k|^2dx\right)^2-\frac{N\beta^*}{N+4s}\int_{\mathbb{R}^N}u_k^{\frac{8s+2N}{N}}dx. \tag{9.33}$$

Combining (9.25) and (9.33), it yields from Lemma 9.2 that

$$\begin{aligned}&\int_{\mathbb{R}^N}|(-\Delta)^{\frac{s}{2}}w_k|^2dx=\int_{\mathbb{R}^N}|w_k|^2=1,\\ &\int_{\mathbb{R}^N}w_k^{\frac{8s+2N}{N}}dx=\varepsilon_k^{4s}\int_{\mathbb{R}^N}u_k^{\frac{8s+2N}{N}}dx\to\frac{b(N+4s)}{2N\beta^*}\quad\text{as } k\to\infty.\end{aligned} \tag{9.34}$$

Following (9.32) and (9.34), we have

$$\mu_k\varepsilon_k^{4s}\to\frac{(N-4s)b}{2N}\quad\text{as } k\to\infty. \tag{9.35}$$

By straightforward computation, we obtain that $w_k(x)$ satisfies the following equation

$$(a_k\varepsilon_k^{2s}+b)(-\Delta)^s w_k+\varepsilon_k^{4s}|\varepsilon_k x+\varepsilon_k y_k|^2 w_k=\mu_k\varepsilon_k^{4s}w_k+\beta^* w_k^{\frac{8s}{N}+1}\quad\text{in } \mathbb{R}^N. \tag{9.36}$$

Applying (9.34), we conclude that $\{w_k\}$ is bounded in $H^s(\mathbb{R}^N)$. Passing to a subsequence, there exists $w_0\in H^s(\mathbb{R}^N)$ such that

$$w_k\rightharpoonup w_0\geq 0\quad\text{in } H^s(\mathbb{R}^N)\text{ as } k\to\infty. \tag{9.37}$$

Since w_k satisfies (9.36), using (9.35) and passing to weak limit, we obtain

$$b(-\Delta)^s w_0=\frac{(N-4s)b}{2N}w_0+\beta^* w_0^{\frac{8s}{N}+1}\quad\text{in } \mathbb{R}^N \tag{9.38}$$

in weak sense. Furthermore, (9.26) implies that $w_0\not\equiv 0$. By nonlinearity regularity, we have that $w_0\in C^{2,\alpha}(\mathbb{R}^N)$ for some $\alpha\in(0,1)$. Then, following [46, Lemma 3.2], we have

$$(-\Delta)^s w_0(x)=-\frac{C_s}{2}\int_{\mathbb{R}^N}\frac{w_0(x+y)+w_0(x-y)-2w_0(x)}{|y|^{N+2s}}dy\quad\text{in } \mathbb{R}^N.$$

Assume that there exists $x_0\in\mathbb{R}^N$ such that $w_0(x_0)=0$, it then follows from $w_0(x)\geq 0$ and $w_0\not\equiv 0$ that

$$(-\Delta)^s w_0(x_0)=-\frac{C_s}{2}\int_{\mathbb{R}^N}\frac{w_0(x_0+y)+w_0(x_0-y)}{|y|^{N+2s}}dy<0.$$

However, from (9.38) we know that $(-\Delta)^s w_0(x_0)=0$, which is a contradiction. Hence, $w(x)>0$ for any $x\in\mathbb{R}^N$. Comparing (9.3) and (9.38), the uniqueness of positive radial solution of (9.3) implies that

$$w_0(x)=\left(\frac{b}{2\beta^*}\right)^{\frac{N}{8s}}\varphi(|x-y_0|),\quad\text{for some } y_0\in\mathbb{R}^N. \tag{9.39}$$

Moreover, we derive from (9.2) and (9.39) that

$$\int_{\mathbb{R}^N}|(-\Delta)^{\frac{s}{2}}w_0|^2dx=\int_{\mathbb{R}^N}|w_0|^2\,dx=1. \tag{9.40}$$

In view of (9.34), (9.37) and (9.40), one has

$$w_k \to w_0 := \left(\frac{b}{2\beta^*}\right)^{\frac{N}{8s}} \varphi(|x-y_0|) \quad \text{for some } y_0 \in \mathbb{R}^N.$$

Step 4. Now we show that $\|w_k\|_\infty < \infty$ uniformly for large k. In fact, from (9.35), (9.36) and (9.40), we have

$$(-\Delta)^s w_k \leq \frac{2\beta^*}{b} w_k^{\frac{8s}{N}+1}. \tag{9.41}$$

Define

$$h(t) :\equiv h_{T,\gamma}(t) = \begin{cases} 0 & t \leq 0, \\ t^\gamma & 0 < t \leq T, \\ \gamma T^{\gamma-1}(t-T) + T^\gamma & t > T, \end{cases}$$

where $\gamma > 1$ and $T > 0$. It is easy to check that $h(t)$ is convex and Lipschitz continuous, so

$$(-\Delta)^s h(w_k) \leq h'(w_k)(-\Delta)^s w_k$$

in the weak sense. Clearly, $h \in C^1(\mathbb{R}) \cap W^{1,\infty}(\mathbb{R})$ and it is positive, increasing and convex.

Set

$$\tilde{h}(t) := \int_0^t (h'(r))^2 dr,$$

which is positive, increasing, convex and belongs to $C^1(\mathbb{R}) \cap W^{1,\infty}(\mathbb{R})$, then we have $\tilde{h}'(t) = (h'(t))^2$ and $\tilde{h}(t) - \tilde{h}(r) = \tilde{h}'(t)(t-r)$ for $t, r \in \mathbb{R}$. Also, by the Jensen inequality, the fact $0 \leq h(t) \leq |t|^\gamma$ and $0 \leq th'(t) \leq \gamma h(t)$ for $r, t \in \mathbb{R}$, we deduce $0 \leq \tilde{h}(t) \leq \|h'\|_\infty |t|^\gamma$ and $|h(t) - h(r)|^2 \leq (\tilde{h}(t) - \tilde{h}(r))(t-r)$. Thus, we conclude $\tilde{h}(w_k) \in H^s(\mathbb{R}^N)$ since $\tilde{h}$ is Lipschitz continuous and $\tilde{h}(0) = 0$.

Assume that $1 < \gamma \leq 2_s^*/2$, it has $\tilde{h}(w_k) \leq \|h'\|_\infty |w_k|^\gamma \in L^2(\mathbb{R}^N)$. By using the Sobolev inequality, relation (9.41) and the properties of $\tilde{h}(t)$, we can see that

$$\begin{aligned}
\|h(w_p)\|_{2_s^*}^2 &\leq C\|(-\Delta)^{\frac{s}{2}} h(w_k)\|_2^2 \\
&\leq C \int_{\mathbb{R}^N} \int_{\mathbb{R}^N} \frac{(\tilde{h}(w_k(x)) - \tilde{h}(w_k(y)))(w_k(x) - w_k(y))}{|x-y|^{N+2s}} dxdy \\
&= C \int_{\mathbb{R}^N} (-\Delta)^{\frac{s}{2}} w_k (-\Delta)^{\frac{s}{2}} \tilde{h}(w_k) dx \leq C \int_{\mathbb{R}^N} (w_k + w_k^{2_s^*-1}) \tilde{h}(w_k) dx \\
&\leq C \int_{\mathbb{R}^N} (w_k + w_k^{2_s^*-1}) w_k \tilde{h}'(w_k) dx \leq C \int_{\mathbb{R}^N} (w_k + w_k^{2_s^*-1}) w_k (h'(w_k))^2 dx \\
&\leq \gamma^2 C \int_{\mathbb{R}^N} (1 + w_k^{2_s^*-2})(h(w_k))^2 dx \\
&= \gamma^2 C \int_{\mathbb{R}^N} (h(w_k))^2 dx + \gamma^2 C \int_{\mathbb{R}^N} w_k^{2_s^*-2} (h(w_k))^2 dx.
\end{aligned} \tag{9.42}$$

Now let $R > 0$ be fixed, by the Hölder's inequality, we have

$$\begin{aligned}
\int_{\mathbb{R}^N} (h(w_k))^2 w_k^{2_s^*-2} dx &= \int_{\{w_k \leq R\}} (h(w_k))^2 w_k^{2_s^*-2} dx + \int_{\{w_k > R\}} (h(w_k))^2 w_k^{2_s^*-2} dx \\
\leq R^{2_s^*-2} \|h(w_k)\|_2^2 &+ \left(\int_{\{w_k > R\}} w_k^{2_s^*} dx \right)^{\frac{2_s^*-2}{2_s^*}} \|h(w_k)\|_{2_s^*}^2.
\end{aligned} \tag{9.43}$$

Since $\{w_k\}$ converges strongly in $H^s(\mathbb{R}^N)$, then $\{w_k\}$ converges strongly in $L^{2_s^*}(\mathbb{R}^N)$, so we can choose $R > 0$ sufficiently large such that

$$\left(\int_{\{w_k>R\}} w_k^{2_s^*}dx\right)^{\frac{2_s^*-2}{2_s^*}} \leq \frac{1}{2\gamma^2 C}.$$

It follows from (9.42), (9.43) and the fact $h(t) \leq |t|^\gamma$ that

$$\|h(w_k)\|_{2_s^*}^2 \leq 2\gamma^2 C(1 + R^{2_s^*-2})\|h(w_k)\|_2^2 \leq 2\gamma^2 C(1 + R^{2_s^*-2})\|w_k\|_{2\gamma}^{2\gamma}.$$

From the Fatou's lemma and $\lim_{T\to+\infty} h_{T,\gamma}(t) = t^\gamma$, we deduce that

$$\begin{aligned}\|w_k\|_{2_s^*\gamma}^{2\gamma} &= \left(\int_{\mathbb{R}^N} \liminf_{T\to+\infty} h_{T,\gamma}^{2_s^*}(w_k)dx\right)^{\frac{2}{2_s^*}} \leq \liminf_{T\to+\infty}\left(\int_{\mathbb{R}^N} h_{T,\gamma}^{2_s^*}(w_k)dx\right)^{\frac{2}{2_s^*}}\\ &\leq 2\gamma^2 C(1 + R^{2_s^*-2})\|w_k\|_{2\gamma}^{2\gamma},\end{aligned}$$

which implies that $w_k \in L^{2_s^*\gamma}(\mathbb{R}^N)$ since $1 < \gamma \leq \frac{2_s^*}{2}$. Iterating this argument with $\gamma_0 := \frac{2_s^*}{2}$, $\gamma_i := \frac{2_s^*}{2}\gamma_{i-1}$ and let $\gamma_i \to +\infty$, we can obtain that $w_k \in L^r(\mathbb{R}^N)$ where $r \in [2, +\infty)$. By Fatou's lemma, (9.42) and the fact $h(t) \leq |t|^\gamma$, we have

$$\|w_k\|_{2_s^*\gamma}^{2\gamma} \leq \gamma^2 C\int_{\mathbb{R}^N}(w_k^{2\gamma} + w_k^{2_s^*-2+2\gamma})dx. \tag{9.44}$$

By Hölder's inequality and Young's inequality, we get that

$$\int_{\mathbb{R}^N} w_k^{2_s^*-2+2\gamma}dx \leq \|w_k^{2_s^*-2}\|_{\frac{N}{s}}\|w_k^\gamma\|_2\|w_k^\gamma\|_{2_s^*} \leq \|w_k^{2_s^*-2}\|_{\frac{N}{s}}\left(\frac{1}{2\epsilon}\|w_k\|_{2\gamma}^{2\gamma} + \frac{\epsilon}{2}\|w_k\|_{2_s^*\gamma}^{2\gamma}\right). \tag{9.45}$$

Set

$$L := \|w_k^{2_s^*-2}\|_{\frac{2}{N}},$$

then the fact $w_k \in L^r(\mathbb{R}^N)$ shows that $L < +\infty$, where $r \in [2, +\infty)$. Let $\epsilon = 1/(L\gamma^2 C)$. It follows from (9.44) and (9.45) that

$$\|w_k\|_{2_s^*\gamma}^{2\gamma} \leq 2\gamma^2 C\left(1 + \frac{1}{2L^2\gamma^2 C}\right)\|w_k\|_{2\gamma}^{2\gamma} \leq C\gamma^4\|w_k\|_{2\gamma}^{2\gamma}. \tag{9.46}$$

Letting $2\gamma_i = 2_s^*\gamma_{i-1}$ in (9.46), we have

$$\|w_k\|_{2_s^*\gamma_i} \leq (C\gamma_i^4)^{\frac{1}{2\gamma_i}}\|w_k\|_{2_s^*\gamma_{i-1}},$$

which implies that

$$\|w_k\|_{2_s^*\gamma_i} \leq \prod_{j=0}^{i}(C\gamma_j^4)^{\frac{1}{2\gamma_j}}\|w_k\|_{2_s^*\gamma_0}. \tag{9.47}$$

Letting $i \to +\infty$ in (9.47), we conclude that

$$\|w_k\|_\infty \leq e^{\sum_{j=0}^{\infty}\frac{\log\left(C\left(\frac{2_s^*}{2}\right)^{4j}\gamma_0^4\right)}{2\left(\frac{2_s^*}{2}\right)^{j}\gamma_0}}\|w_k\|_{2_s^*\gamma_0} \leq C.$$

Step 5. Next, we prove that $w_k \to 0$ as $|x| \to \infty$ uniformly for large k. We rewrite problem (9.36) as follows

$$(-\Delta)^s w_k + w_k = h_k(x) \quad \text{in } \mathbb{R}^N,$$

where

$$h_k := w_k + (\varepsilon_k^{2s} a_k + b)^{-1} \left(\mu_k \varepsilon_k^{4s} w_k + \beta^* w_k^{\frac{8s}{N}+1} - \varepsilon_k^{4s} |\varepsilon_k x + \varepsilon_k y_k|^2 w_k \right).$$

Then $h_k \in L^\infty(\mathbb{R}^N)$ for large k, thus by the interpolation on the L^p-spaces and the convergence of $\{w_k\}$ in $H^s(\mathbb{R}^N)$, there exists $h \in L^r(\mathbb{R}^N)$ such that $h_k \to h$ in $L^r(\mathbb{R}^N)$ as $k \to \infty$ for $r \geq 2$. We have

$$w_k = \int_{\mathbb{R}^N} \mathcal{K}(x-y) h_k(x) dy,$$

where $\mathcal{K}$ is a Bessel potential and it satisfies

($\mathcal{K}_1$) $\mathcal{K}$ is positive, radially symmetric and smooth in $\mathbb{R}^N \backslash \{0\}$.

($\mathcal{K}_2$) There exists $C > 0$ such that $\mathcal{K}(x) \leq \frac{C}{|x|^{N+2s}}$ for $x \in \mathbb{R}^N \backslash \{0\}$.

($\mathcal{K}_3$) $\mathcal{K} \in L^r(\mathbb{R}^N)$ for $r \in [1, N/(N-2s))$.

Now argue as in the proof of Lemma 6.4 in Teng and Cheng [147], we conclude that

$$\lim_{|x| \to \infty} w_k(x) \to 0 \quad \text{as } k \to \infty. \tag{9.48}$$

Step 6. We now show that (9.27) holds. In fact, Lemma 9.2 shows that there exists $M > 0$ such that

$$a_k \varepsilon_k^{2s} + b \leq M \quad \text{uniformly for large } k. \tag{9.49}$$

Following the proof of Lemma 4.3 in Felmer, Quaas and Tan [51], there exists a function $\mathcal{X}$ such that

$$0 < \mathcal{X}(x) < \frac{1}{1+|x|^{N+2s}}$$

and

$$(-\Delta)^s \mathcal{X} + \frac{(4s-N)b}{4NM} \mathcal{X} = 0 \quad \text{in } \mathbb{R}^N \backslash B_{R_1}(0) \text{ for some } R_1 > 0.$$

Following (9.35), (9.36), (9.48) and (9.49), we obtain for large k,

$$\begin{aligned}
(-\Delta)^s w_k + \frac{(4s-N)b}{4NM} w_k &\leq (-\Delta)^s w_k + \frac{(4s-N)b w_k}{4N(a_k \varepsilon_k^{2s} + b)} \\
&= \frac{-\varepsilon_k^{4s} |\varepsilon_k x + \varepsilon_k y_k|^2 w_k + \mu_k \varepsilon_k^{4s} w_k + \beta^* w_k^{\frac{8s}{N}+1} + \frac{(4s-N)b}{4N} w_k}{a_k \varepsilon_k^{2s} + b} \\
&\leq 0
\end{aligned}$$

for $|x| \geq R_2$ with R_2 large enough. By the comparison principle, we conclude that

$$w_k(x) \leq \frac{C}{1+|x|^{N+2s}} \quad \text{uniformly for large } k. \tag{9.50}$$

For any $x \in B_\rho^c(0)$, we have

$$\frac{|x - z_k|}{\varepsilon_k} \geq \frac{|x|}{2\varepsilon_k} \geq \frac{\rho}{2\varepsilon_k} \to +\infty \quad \text{as } k \to \infty. \tag{9.51}$$

We now prove that relation (9.53) holds. It follows from (9.5) and (9.50) that for any $\epsilon > 0$, there exists a constant $R_\epsilon > 0$ independent k, such that

$$|\bar{w}_k(x)|, \quad \left| \left(\frac{b}{2\beta^*}\right)^{\frac{N}{8s}} \varphi(x) \right| < \frac{\epsilon}{4} \quad \text{for any } |x| > R_\epsilon \text{ as } k \to \infty,$$

which yields that

$$\sup_{|x|>R_\epsilon} |\bar{w}_k(x) - \bar{w}_0(x)| \le \sup_{|x|>R_\epsilon} \left(|\bar{w}_k(x)| + \left| \left(\frac{b}{2\beta^*}\right)^{\frac{N}{8s}} \varphi(x) \right| \right) < \frac{\epsilon}{2} \quad \text{as } k \to \infty,$$

from which and (9.62) we can obtain (9.53). □

Following the proof of Proposition 9.1, we now address Theorem 9.2 on the local properties of concentration points. Before proving Theorem 9.2, we first establish several auxiliary properties.

Claim 1. Define $\lambda_0 := \int_{\mathbb{R}^N} |x|^2 \varphi^2(x)dx$, then we have

$$\limsup_{a\to 0} \frac{e(a,\beta^*)}{a^{\frac{1}{1+s}}} \le (1+s)\left(\frac{\lambda_0}{s\|\varphi\|_2^2}\right)^{\frac{s}{1+s}}.$$

Indeed, let $U_\tau(x)$ be the trail function defined by (9.14). Taking $x_0 = 0$, we deduce from (9.16) and (9.17) that

$$\int_{\mathbb{R}^N} |x|^2 |U_\tau(x)|^2 dx = \frac{\lambda_0}{\tau^2\|\varphi\|_2^2}(1+o(1)) \quad \text{as } \tau \to \infty, \tag{9.63}$$

and

$$e(a,\beta^*) = E_{a,\beta^*}(U_\tau) \le a\tau^{2s} + aC\tau^{-10s} + \frac{\lambda_0}{\tau^2\|\varphi\|_2^2}(1+o(1)) + C\tau^{-8s} \quad \text{as } \tau \to \infty. \tag{9.64}$$

Let $\tau = \left(\frac{\lambda_0}{as\|\varphi\|_2^2}\right)^{\frac{1}{2+2s}}$, then $\tau \to \infty$ as $a \searrow 0$. We thus obtain from (9.63) and (9.64) that

$$e(a,\beta^*) \le a\left(\frac{\lambda_0}{as\|\varphi\|_2^2}\right)^{\frac{s}{1+s}} + \frac{\lambda_0}{\|\varphi\|_2^2}\left(\frac{\lambda_0}{as\|\varphi\|_2^2}\right)^{\frac{-1}{1+s}}(1+o(1)) + C\tau^{-8s} + o(1) \quad \text{as } a \searrow 0,$$

which shows that

$$\limsup_{a\to 0} \frac{e(a,\beta^*)}{a^{\frac{1}{1+s}}} \le (1+s)\left(\frac{\lambda_0}{s\|\varphi\|_2^2}\right)^{\frac{s}{1+s}}.$$

Claim 2. Let u_k be a nonnegative minimizer of $e(a_k,\beta^*)$ as in Proposition 9.1 with $a_k \to 0$ as $k \to \infty$, then the unique global maximum point $\bar{z}_k$ of u_k satisfies $\bar{z}_k \to 0$ as $k \to \infty$ and $\left\{\frac{\bar{z}_k}{\varepsilon_k}\right\}$ is bounded.

In fact, if $\left|\frac{\bar{z}_k}{\varepsilon_k}\right| \to \infty$ as $k \to \infty$, then for any $M > 0$ large enough, we have

$$\liminf_{k\to\infty} \frac{1}{\varepsilon_k^2}\int_{\mathbb{R}^N} |x|^2 u_k^2 dx = \liminf_{k\to\infty} \int_{\mathbb{R}^N} \left|x + \frac{\bar{z}_k}{\varepsilon_k}\right|^2 \bar{w}_k^2 dx \ge M. \tag{9.65}$$

Following (9.2), we deduce by calculation that there exists a constant $C > 0$ such that

$$e(a_k,\beta^*) = E_{a_k,\beta^*}(u_k) \ge a_k\varepsilon_k^{-2s} + M\varepsilon_k^2 \ge CM^{\frac{s}{1+s}} a_k^{\frac{1}{1+s}},$$

which is a contradiction in view of Claim 1 by choosing $M > 0$ large enough.

Proof of Theorem 9.2: Actually, Claim 2 shows that there exists $z_0 \in \mathbb{R}^N$ such that

$$\frac{\bar{z}_k}{\varepsilon_k} \to z_0 \quad \text{as } k \to \infty. \tag{9.66}$$

Since $\varphi(x)$ is a radial decreasing function and decays polynomially as $|x| \to \infty$, we obtain

$$\begin{aligned} \liminf_{k\to\infty} \frac{1}{\varepsilon_k^2} \int_{\mathbb{R}^N} |x|^2 u_k^2 dx &= \frac{1}{\|\varphi\|_2^2} \int_{\mathbb{R}^N} |x + z_0|^2 \varphi^2(x) dx \\ &\geq \frac{1}{\|\varphi\|_2^2} \int_{\mathbb{R}^N} |x|^2 \varphi^2(x) dx = \frac{\lambda_0}{\|\varphi\|_2^2}. \end{aligned} \tag{9.67}$$

Together with the Young's inequality, we derive from (9.64) that

$$\begin{aligned} e(a_k, \beta^*) = E_{a_k,\beta^*}(u_k) &\geq a_k \varepsilon_k^{-2s} + \frac{\lambda_0 \varepsilon_k^2}{\|\varphi\|_2^2}(1 + o(1)) \\ &\geq (1+s) a_k^{\frac{1}{1+s}} (1 + o(1)) \left(\frac{s\|\varphi\|_2^2}{\lambda_0} \right)^{\frac{-s}{1+s}}. \end{aligned}$$

Hence

$$\liminf_{k\to\infty} \frac{e(a_k, \beta^*)}{a_k^{\frac{1}{1+s}}} \geq (1+s) \left(\frac{\lambda_0}{s\|\varphi\|_2^2} \right)^{\frac{s}{1+s}}, \tag{9.68}$$

where the equality holds if and only if

$$\lim_{k\to\infty} \frac{\varepsilon_k}{\bar{\varepsilon}_k} = 1 \quad \text{where } \bar{\varepsilon}_k := \left(\frac{s a_k \|\varphi\|_2^2}{\lambda_0} \right)^{\frac{1}{2+2s}}. \tag{9.69}$$

In view of (9.68) and Claim 1, it follows that

$$\liminf_{k\to\infty} \frac{e(a_k, \beta^*)}{a_k^{\frac{1}{1+s}}} = (1+s) \left(\frac{\lambda_0}{s\|\varphi\|_2^2} \right)^{\frac{s}{1+s}}.$$

Therefore, (9.53) and (9.69) show that (9.7) holds.

Next, we prove (9.9). Indeed, from the definition of $e(a_k, \beta^*)$, we know that

$$\int_{\mathbb{R}^N} |x|^2 u_k^2(x) dx \leq \int_{\mathbb{R}^N} |x|^2 u_k^2(x + \bar{z}_k) dx. \tag{9.70}$$

By the decay estimate of $\bar{w}_k$ and φ, we deduce from (9.69) that

$$\int_{\mathbb{R}^N} |x|^2 u_k^2(x + \bar{z}_k) dx = \frac{\varepsilon_k^2}{\|\varphi\|_2^2}(1 + o(1)) \int_{\mathbb{R}^N} |x|^2 \varphi^2(x) dx \quad \text{as } k \to \infty. \tag{9.71}$$

Note that

$$\int_{\mathbb{R}^N} |x|^2 u_k^2(x) dx \geq \int_{B_{\frac{1}{\sqrt{\varepsilon_k}}}(0)} |\varepsilon_k x + \bar{z}_k|^2 \bar{w}_k^2(x) dx = \frac{(1 + o(1))\varepsilon_k^2}{\|\varphi\|_2^2} \int_{B_{\frac{1}{\sqrt{\varepsilon_k}}}(0)} \left| x + \frac{\bar{z}_k}{\varepsilon_k} \right|^2 \varphi^2(x) dx. \tag{9.72}$$

On the contrary, we assume that there exists a constant $\rho > 0$ and a subsequence of $\left\{ \frac{\bar{z}_k}{\varepsilon_k} \right\}$ with $\frac{|\bar{z}_k|}{\varepsilon_k} \geq \rho > 0$ as $k \to \infty$. Then relations (9.70)–(9.72) show that

$$\frac{\varepsilon_k^2}{\|\varphi\|_2^2} \int_{B_{\frac{1}{\sqrt{\varepsilon_k}}}(0)} \left| x + \frac{\bar{z}_k}{\varepsilon_k} \right|^2 \varphi^2(x) dx \leq \frac{(1 + o(1))\varepsilon_k^2}{\|\varphi\|_2^2} \int_{\mathbb{R}^N} |x|^2 \varphi^2(x) dx \quad \text{as } k \to \infty.$$

Applying the Fatou's lemma, we have

$$\int_{\mathbb{R}^N} |x|^2\varphi^2(x)dx < \int_{\mathbb{R}^N} \liminf_{k\to\infty} \left| x + \frac{\bar{z}_k}{\varepsilon_k} \right|^2 \varphi^2(x)dx \le \int_{\mathbb{R}^N} |x|^2\varphi^2(x)dx,$$

which is a contradiction. Thus, $\frac{\bar{z}_k}{\varepsilon_k} \to 0$ as $k \to \infty$, from which and relation (9.69) we can obtain (9.9). This completes the proof. □

9.4 Local uniqueness

This section is devoted to studying the local uniqueness of nonnegative minimizers when a is small enough. For this purpose, let $\{a_k\}$ be a sequence satisfies $a_k \to 0$ as $k \to \infty$. By contradiction, let u_{1k} and u_{2k} be two different normalized minimizers to $e(a_k, \beta^*)$ with $\|u_{ik}\|_2^2 = 1$ for $i = 1, 2$. Let $\bar{z}_{1k}$ and $\bar{z}_{2k}$ be the unique global maximal point of u_{1k} and u_{2k} as $k \to \infty$, respectively. We rewrite (9.31) as

$$\left(a_k + b\int_{\mathbb{R}^N} |(-\Delta)^{\frac{s}{2}}u_{ik}|^2dx\right)(-\Delta)^s u_{ik} + |x|^2u_{ik} = \mu_{ik}u_{ik} + \beta^* u_{ik}^{\frac{8s}{N}+1} \quad \text{in } \mathbb{R}^N, i = 1, 2, \tag{9.73}$$

where $\mu_{ik} \in \mathbb{R}$ is the corresponding Lagrange multiplier. Define

$$\bar{u}_{ik}(x) := \left(\frac{2\beta^*}{b}\right)^{\frac{N}{8s}} \bar{\varepsilon}_k^{\frac{N}{2}} u_{ik}(\bar{\varepsilon}_k x + \bar{z}_{1k}) \quad i = 1, 2, \tag{9.74}$$

and

$$\bar{\eta}_k(x) := \frac{\bar{u}_{1k}(x) - \bar{u}_{2k}(x)}{\|\bar{u}_{1k} - \bar{u}_{2k}\|_\infty}. \tag{9.75}$$

One can check that $\bar{u}_{ik}$ and $\bar{\eta}_k$ satisfy

$$\begin{aligned} &\left[a_k\bar{\varepsilon}_k^{2s} + b\left(\frac{b}{2\beta^*}\right)^{\frac{N}{4s}} \int_{\mathbb{R}^N} |(-\Delta)^{\frac{s}{2}}\bar{u}_{ik}|^2dx\right](-\Delta)^s\bar{u}_{ik} + \bar{\varepsilon}_k^{4s}|\bar{\varepsilon}_k x + \bar{z}_{1k}|^2\bar{u}_{ik} \\ =&\mu_{ik}\bar{\varepsilon}_k^{4s}\bar{u}_{ik} + \frac{b}{2}\bar{u}_{ik}^{\frac{8s}{N}+1} \quad \text{in } \mathbb{R}^N, i = 1, 2, \end{aligned} \tag{9.76}$$

and

$$\begin{aligned} &\left[a_k\bar{\varepsilon}_k^{2s} + b\left(\frac{b}{2\beta^*}\right)^{\frac{N}{4s}} \int_{\mathbb{R}^N} |(-\Delta)^{\frac{s}{2}}\bar{u}_{2k}|^2dx\right](-\Delta)^s\bar{\eta}_k + \bar{\varepsilon}_k^{4s}|\bar{\varepsilon}_k x + \bar{z}_{1k}|^2\bar{\eta}_k \\ &+ b\left(\frac{b}{2\beta^*}\right)^{\frac{N}{4s}} \int_{\mathbb{R}^N} (-\Delta)^{\frac{s}{2}}(\bar{u}_{1k} + \bar{u}_{2k})(-\Delta)^{\frac{s}{2}}\bar{\eta}_k dx(-\Delta)^{\frac{s}{2}}\bar{u}_{1k} \\ =&\mu_{1k}\bar{\varepsilon}_k^{4s}\bar{\eta}_k + \frac{\bar{\varepsilon}_k^{4s}\bar{u}_{2k}(\mu_{1k} - \mu_{2k})}{\|\bar{u}_{1k} - \bar{u}_{2k}\|_\infty} + \frac{(8s + N)b}{2N}[A\bar{u}_{1k} + (1 - A)\bar{u}_{2k}]^{\frac{8s}{N}}\bar{\eta}_k, \end{aligned} \tag{9.77}$$

where $A \in (0, 1)$.

The following result establishes the polynomial decay of $\bar{u}_{ik}$ and $|\nabla\bar{u}_{ik}|$.

Lemma 9.5 *Assume that $u_{ik}(i=1,2)$ are two minimizers of $e(a_k,\beta^*)$ with $a_k \to 0$ as $k\to\infty$ and $\bar{u}_{ik}$ is defined by (9.74), then there exist $R>0$ large enough and $2s>\rho>0$ small enough such that*

$$\bar{u}_{ik}(x) \le \frac{C}{1+|x|^{N+2s}} \quad \textit{as } k\to\infty, \tag{9.78}$$

and

$$|\nabla \bar{u}_{ik}(x)| \le \frac{C}{1+|x|^{N+2s}} \quad \textit{for } |x|>R \textit{ as } k\to\infty. \tag{9.79}$$

Proof. Relation (9.78) follows from the decay estimate of u_{ik} in Section 9.3. We now show that (9.79) is true. In fact, $\bar{u}_{ik}\in C^{1,\alpha}(\mathbb{R}^N)$ for some $\alpha\in(0,1)$ as $k\to\infty$. In particular, $|D\bar{u}_{ik}|\in L^\infty(\mathbb{R}^N)$ as $k\to\infty$. Differentiating (9.76), we get

$$\begin{aligned}&\left[a_k\bar{\varepsilon}_k^{2s}+b\left(\frac{b}{2\beta^*}\right)^{\frac{N}{4s}}\int_{\mathbb{R}^N}|(-\Delta)^{\frac{s}{2}}\bar{u}_{ik}|^2dx\right](-\Delta)^s\frac{\partial \bar{u}_{ik}}{\partial x_j}+\bar{\varepsilon}_k^{4s}|\bar{\varepsilon}_k x+\bar{z}_{1k}|^2\frac{\partial \bar{u}_{ik}}{\partial x_j}\\ =&\mu_{ik}\bar{\varepsilon}_k^{4s}\frac{\partial \bar{u}_{ik}}{\partial x_j}+\frac{b}{2}\frac{8s+N}{N}\bar{u}_{ik}^{\frac{8s}{N}}\frac{\partial \bar{u}_{ik}}{\partial x_j}-2\bar{\varepsilon}_k^{4s+1}(\bar{\varepsilon}_k x_j+\bar{z}_{1k}^j)\bar{u}_{ik},\end{aligned} \tag{9.80}$$

where $i=1,2$ and $j=1,2,\ldots,N$. By (9.59) and (9.69), we have

$$\mu_{ik}\bar{\varepsilon}_k^{4s}\to\frac{(N-4s)b}{2N} \quad \text{as } k\to\infty \text{ and } i=1,2. \tag{9.81}$$

Using the fact that $\bar{z}_{1k}\to 0$ as $k\to\infty$, we derive from (9.78) and (9.81) that there exists $\vartheta>0$ large such that

$$\inf_{|x|\ge\vartheta}\left[\frac{-\mu_{ik}\bar{\varepsilon}_k^{4s}-\frac{b}{2}\frac{8s+N}{N}\bar{u}_{ik}^{\frac{8s}{N}}+\bar{\varepsilon}_k^{4s}|\bar{\varepsilon}_k x+\bar{z}_{1k}|^2}{a_k\bar{\varepsilon}_k^{2s}+b\left(\frac{b}{2\beta^*}\right)^{\frac{N}{4s}}\int_{\mathbb{R}^N}|(-\Delta)^{\frac{s}{2}}\bar{u}_{ik}|^2dx}\right]=:\rho_0>0 \quad \text{as } k\to\infty \text{ and } i=1,2. \tag{9.82}$$

Let $v\in H^s(\mathbb{R}^N)$ be a positive radial solution of

$$(-\Delta)^s v+\rho_0 v=v^{\frac{8s}{N}+1} \quad \text{in } \mathbb{R}^N. \tag{9.83}$$

From Frank, Lenzmann and Silvestre [59], we know that $v\in C(\mathbb{R}^N)$ and there exist constants $C_2>C_1>0$ such that

$$\frac{C_1}{1+|x|^{N+2s}}\le v(x)\le\frac{C_2}{1+|x|^{N+2s}} \quad x\in\mathbb{R}^N. \tag{9.84}$$

For large k and $\vartheta_0>\vartheta$, set

$$\bar{v}:=\left(1+\frac{\|D\check{u}_{ik}\|_\infty}{\inf_{|x|\le\vartheta_0}v}\right)v.$$

In view of (9.80) and (9.83), it follows that for all $i=1,2$ and $j=1,2,\ldots,N$,

$$\begin{aligned}&(-\Delta)^s\left(\frac{\partial \bar{u}_{ik}}{\partial x_j}-\bar{v}\right)+\left[\frac{-\mu_{ik}\bar{\varepsilon}_k^{4s}-\frac{b}{2}\frac{8s+N}{N}\bar{u}_{ik}^{\frac{8s}{N}}+\bar{\varepsilon}_k^{4s}|\bar{\varepsilon}_k x+\bar{z}_{1k}|^2}{a_k\bar{\varepsilon}_k^{2s}+b\left(\frac{b}{2\beta^*}\right)^{\frac{N}{4s}}\int_{\mathbb{R}^N}|(-\Delta)^{\frac{s}{2}}\bar{u}_{ik}|^2dx}\right]\frac{\partial \bar{u}_{ik}}{\partial x_j}-\rho_0\bar{v}\\ =&\frac{-2\bar{\varepsilon}_k^{4s+1}(\bar{\varepsilon}_k x_j+\bar{z}_{1k}^j)\bar{u}_{ik}}{a_k\bar{\varepsilon}_k^{2s}+b\left(\frac{b}{2\beta^*}\right)^{\frac{N}{4s}}\int_{\mathbb{R}^N}|(-\Delta)^{\frac{s}{2}}\bar{u}_{ik}|^2dx}-\left(1+\frac{\|D\check{u}_{ik}\|_\infty}{\inf_{|x|\le\vartheta_0}v}\right)v^{\frac{8s}{N}+1}.\end{aligned} \tag{9.85}$$

Testing (9.85) with $\left(\frac{\partial \bar{u}_{ik}}{\partial x_j}-\bar{v}\right)^+$, we get

$$\begin{aligned}
&\int_{\mathbb{R}^N}(-\Delta)^{\frac{s}{2}}\left(\frac{\partial \bar{u}_{ik}}{\partial x_j}-\bar{v}\right)(-\Delta)^{\frac{s}{2}}\left(\frac{\partial \bar{u}_{ik}}{\partial x_j}-\bar{v}\right)^+dx\\
&+\int_{\mathbb{R}^N}\left\{\left[\frac{-\mu_{ik}\bar{\varepsilon}_k^{4s}-\frac{b}{2}\frac{8s+N}{N}\bar{u}_{ik}^{\frac{8s}{N}}+\bar{\varepsilon}_k^{4s}|\bar{\varepsilon}_k x+\bar{z}_{1k}|^2}{a_k\bar{\varepsilon}_k^{2s}+b\left(\frac{b}{2\beta^*}\right)^{\frac{N}{4s}}\int_{\mathbb{R}^N}|(-\Delta)^{\frac{s}{2}}\bar{u}_{ik}|^2dx}\right]\frac{\partial \bar{u}_{ik}}{\partial x_j}-\rho_0\bar{v}\right\}\left(\frac{\partial \bar{u}_{ik}}{\partial x_j}-\bar{v}\right)^+dx\\
=&\int_{\mathbb{R}^N}\left[\frac{-2\bar{\varepsilon}_k^{4s+1}(\bar{\varepsilon}_k x_j+\bar{z}_{1k}^j)\bar{u}_{ik}}{a_k\bar{\varepsilon}_k^{2s}+b\left(\frac{b}{2\beta^*}\right)^{\frac{N}{4s}}\int_{\mathbb{R}^N}|(-\Delta)^{\frac{s}{2}}\bar{u}_{ik}|^2dx}-\left(1+\frac{\|D\check{u}_{ik}\|_\infty}{\inf_{|x|\le\vartheta_0}v}\right)v^{\frac{8s}{N}+1}\right]\left(\frac{\partial \bar{u}_{ik}}{\partial x_j}-\bar{v}\right)^+dx,
\end{aligned}\tag{9.86}$$

where $i=1,2$ and $j=1,2,\ldots,N$. According to the definition of $\bar{v}$, we can see that $\left(\frac{\partial \bar{u}_{ik}}{\partial x_j}-\bar{v}\right)^+=0$ in $B_{\vartheta_0}(0)$ as $k\to\infty$. Moreover, we have that as $k\to\infty$

$$\int_{\mathbb{R}^N}\left[\frac{-2\bar{\varepsilon}_k^{4s+1}(\bar{\varepsilon}_k x_j+\bar{z}_{1k}^j)\bar{u}_{ik}}{a_k\bar{\varepsilon}_k^{2s}+b\left(\frac{b}{2\beta^*}\right)^{\frac{N}{4s}}\int_{\mathbb{R}^N}|(-\Delta)^{\frac{s}{2}}\bar{u}_{ik}|^2dx}-\left(1+\frac{\|D\check{u}_{ik}\|_\infty}{\inf_{|x|\le\vartheta_0}v}\right)v^{\frac{8s}{N}+1}\right]\left(\frac{\partial \bar{u}_{ik}}{\partial x_j}-\bar{v}\right)^+dx\le 0,\tag{9.87}$$

where $i=1,2$ and $j=1,2,\ldots,N$. By the fact that $\vartheta_0>\vartheta$, we obtain from (9.82) that

$$\begin{aligned}
&\int_{\mathbb{R}^N}\left\{\left[\frac{-\mu_{ik}\bar{\varepsilon}_k^{4s}-\frac{b}{2}\frac{8s+N}{N}\bar{u}_{ik}^{\frac{8s}{N}}+\bar{\varepsilon}_k^{4s}|\bar{\varepsilon}_k x+\bar{z}_{1k}|^2}{a_k\bar{\varepsilon}_k^{2s}+b\left(\frac{b}{2\beta^*}\right)^{\frac{N}{4s}}\int_{\mathbb{R}^N}|(-\Delta)^{\frac{s}{2}}\bar{u}_{ik}|^2dx}\right]\frac{\partial \bar{u}_{ik}}{\partial x_j}-\rho_0\bar{v}\right\}\left(\frac{\partial \bar{u}_{ik}}{\partial x_j}-\bar{v}\right)^+dx\\
=&\int_{B^c_{\vartheta_0}(0)}\left\{\left[\frac{-\mu_{ik}\bar{\varepsilon}_k^{4s}-\frac{b}{2}\frac{8s+N}{N}\bar{u}_{ik}^{\frac{8s}{N}}+\bar{\varepsilon}_k^{4s}|\bar{\varepsilon}_k x+\bar{z}_{1k}|^2}{a_k\bar{\varepsilon}_k^{2s}+b\left(\frac{b}{2\beta^*}\right)^{\frac{N}{4s}}\int_{\mathbb{R}^N}|(-\Delta)^{\frac{s}{2}}\bar{u}_{ik}|^2dx}\right]\frac{\partial \bar{u}_{ik}}{\partial x_j}-\rho_0\bar{v}\right\}\left(\frac{\partial \bar{u}_{ik}}{\partial x_j}-\bar{v}\right)^+dx\\
\ge&\int_{B^c_{\vartheta_0}(0)}\left(\rho_0\frac{\partial \bar{u}_{ik}}{\partial x_j}-\rho_0\bar{v}\right)\left(\frac{\partial \bar{u}_{ik}}{\partial x_j}-\bar{v}\right)^+dx\\
=&\rho_0\int_{\mathbb{R}^N}\left|\left(\frac{\partial \bar{u}_{ik}}{\partial x_j}-\bar{v}\right)^+\right|^2dx\quad\text{as } k\to\infty,\ i=1,2\text{ and }j=1,2,\ldots,N.
\end{aligned}\tag{9.88}$$

Let $\omega_i:=\frac{\partial \bar{u}_{ik}}{\partial x_j}-\bar{v}$, where $i=1,2$ and $j=1,2,\ldots,N$. By straightforward computation we have

$$\begin{aligned}
&\int_{\mathbb{R}^N}\int_{\mathbb{R}^N}\frac{(\omega_i(x)-\omega_i(y))(\omega_i^+(x)-\omega_i^+(y))}{|x-y|^{N+2s}}dxdy\\
=&\int_{\{\omega_i(x)\ge0\}}\int_{\{\omega_i(y)<0\}}\frac{(\omega_i(x)-\omega_i(y))\omega_i(x)}{|x-y|^{N+2s}}dxdy\\
&+\int_{\{\omega_i(x)<0\}}\int_{\{\omega_i(y)\ge0\}}\frac{(\omega_i(y)-\omega_i(x))\omega_i(y)}{|x-y|^{N+2s}}dxdy\\
&+\int_{\{\omega_i(x)\ge0\}}\int_{\{\omega_i(y)\ge0\}}\frac{(\omega_i(x)-\omega_i(y))^2}{|x-y|^{N+2s}}dxdy
\end{aligned}\tag{9.89}$$

and

$$\int_{\mathbb{R}^N}\int_{\mathbb{R}^N}\frac{|\omega_i^+(x)-\omega_i^+(y)|^2}{|x-y|^{N+2s}}dxdy=\int_{\{\omega_i(x)\geq 0\}}\int_{\{\omega_i(y)<0\}}\frac{\omega_i^2(x)}{|x-y|^{N+2s}}dxdy$$
$$+\int_{\{\omega_i(x)<0\}}\int_{\{\omega_i(y)\geq 0\}}\frac{\omega_i^2(y)}{|x-y|^{N+2s}}dxdy \tag{9.90}$$
$$+\int_{\{\omega_i(x)\geq 0\}}\int_{\{\omega_i(y)\geq 0\}}\frac{(\omega_i(x)-\omega_i(y))^2}{|x-y|^{N+2s}}dxdy.$$

Following (9.89) and (9.90), it follows that

$$\int_{\mathbb{R}^N}(-\Delta)^{\frac{s}{2}}\left(\frac{\partial\bar{u}_{ik}}{\partial x_j}-\bar{v}\right)(-\Delta)^{\frac{s}{2}}\left(\frac{\partial\bar{u}_{ik}}{\partial x_j}-\bar{v}\right)^+dx\geq\int_{\mathbb{R}^N}\left|(-\Delta)^{\frac{s}{2}}\left(\frac{\partial\bar{u}_{ik}}{\partial x_j}-\bar{v}\right)^+\right|^2dx. \tag{9.91}$$

It then follows from (9.86)–(9.88) and (9.91) that for $i=1,2$ and $j=1,2,\ldots,N$

$$\int_{\mathbb{R}^N}\left|(-\Delta)^{\frac{s}{2}}\left(\frac{\partial\bar{u}_{ik}}{\partial x_j}-\bar{v}\right)^+\right|^2dx+\rho_0\int_{\mathbb{R}^N}\left|\left(\frac{\partial\bar{u}_{ik}}{\partial x_j}-\bar{v}\right)^+\right|^2dx\leq 0\quad\text{as } k\to\infty,$$

which implies that $\left(\frac{\partial\bar{u}_{ik}}{\partial x_j}-\bar{v}\right)^+=0$ in $\mathbb{R}^3$. Thus, we have

$$\frac{\partial\bar{u}_{ik}}{\partial x_j}\leq\bar{v}\leq\frac{C}{1+|x|^{N+2s}}\quad\text{as } k\to\infty,\ i=1,2\text{ and } j=1,2,\ldots,N.$$

By the same arguments as above, we can also obtain that $-\frac{\partial\bar{u}_{ik}}{\partial x_j}\leq\frac{C}{1+|x|^{N+2s}}$ as $k\to\infty$, $i=1,2$ and $j=1,2,\ldots,N$. This completes the proof of Lemma 9.5. □

To achieve the optimal decay frequency, denote

$$\hat{u}_{ik}(x):=\left(\frac{2\beta^*}{b}\right)^{\frac{N}{8s}}\bar{\varepsilon}_k^{\frac{N}{2}}u_{ik}(\bar{\varepsilon}_k^{\tau}x)\quad\text{for } i=1,2, \tag{9.92}$$

and

$$\hat{\eta}_k(x):=\frac{\hat{u}_{1k}(x)-\hat{u}_{2k}(x)}{\|\hat{u}_{1k}-\hat{u}_{2k}\|_\infty}, \tag{9.93}$$

where $\tau<0$ satisfy

$$\begin{cases}\tau<\min\left\{-2s,\frac{-4s-6}{4s}\right\} & N=2,\\ -4s-4<\tau<\min\left\{-2s,\frac{-4s-5}{4s+2}\right\} & N=3.\end{cases} \tag{9.94}$$

In view of (9.74), (9.76), (9.92) and (9.93), it follows that

$$\left[a_k\bar{\varepsilon}_k^{2s(2-\tau)}+b\left(\frac{b}{2\beta^*}\right)^{\frac{N}{4s}}\bar{\varepsilon}_k^{(\tau-1)(N-4s)}\int_{\mathbb{R}^N}|(-\Delta)^{\frac{s}{2}}\hat{u}_{ik}|^2dx\right](-\Delta)^s\hat{u}_{ik}$$
$$+\bar{\varepsilon}_k^{4s}|\bar{\varepsilon}_k^{\tau}x|^2\hat{u}_{ik}=\mu_{ik}\bar{\varepsilon}_k^{4s}\hat{u}_{ik}+\frac{b}{2}\hat{u}_{ik}^{\frac{8s}{N}+1}\quad\text{in }\mathbb{R}^N,\ i=1,2 \tag{9.95}$$

and

$$
\begin{aligned}
&2a_k\bar{\varepsilon}_k^{2s(2-\tau)}(-\Delta)^s\hat{\eta}_k + b\left(\frac{b}{2\beta^*}\right)^{\frac{N}{4s}}\bar{\varepsilon}_k^{(\tau-1)(N-4s)}\int_{\mathbb{R}^N}\left(|(-\Delta)^{\frac{s}{2}}\hat{u}_{1k}|^2 + |(-\Delta)^{\frac{s}{2}}\hat{u}_{2k}|^2\right)dx(-\Delta)^s\hat{\eta}_k \\
&+ b\left(\frac{b}{2\beta^*}\right)^{\frac{N}{4s}}\bar{\varepsilon}_k^{(\tau-1)(N-4s)}\int_{\mathbb{R}^N}(-\Delta)^{\frac{s}{2}}(\hat{u}_{1k}+\hat{u}_{2k})(-\Delta)^{\frac{s}{2}}\hat{\eta}_k dx(-\Delta)^s(\hat{u}_{1k}+\hat{u}_{2k}) + 2\bar{\varepsilon}_k^{4s}|\bar{\varepsilon}_k^{\tau}x|^2\hat{\eta}_k \\
=&\bar{\varepsilon}_k^{4s}(\mu_{1k}+\mu_{2k})\hat{\eta}_k + \frac{\bar{\varepsilon}_k^{4s}(\mu_{1k}-\mu_{2k})}{\|\hat{u}_{1k}-\hat{u}_{2k}\|_\infty}(\hat{u}_{1k}+\hat{u}_{2k}) + \frac{(8s+N)b}{N}[A\hat{u}_{1k}+(1-A)\hat{u}_{1k}]^{\frac{8s}{N}}\hat{\eta}_k.
\end{aligned}
\tag{9.96}
$$

Following (9.74)–(9.76), we have

$$
\begin{aligned}
&\frac{\bar{\varepsilon}_k^{4s}(\mu_{1k}-\mu_{2k})}{\|\hat{u}_{1k}-\hat{u}_{2k}\|_\infty}(\hat{u}_{1k}+\hat{u}_{2k}) \\
=&\frac{b}{2}\left(\frac{2\beta^*}{b}\right)^{-\frac{N}{2s}}\int_{\mathbb{R}^N}\left(|(-\Delta)^{\frac{s}{2}}\bar{u}_{1k}|^2 + |(-\Delta)^{\frac{s}{2}}\bar{u}_{2k}|^2\right)dx \\
&\times\int_{\mathbb{R}^N}(-\Delta)^{\frac{s}{2}}(\bar{u}_{1k}+\bar{u}_{2k})(-\Delta)^{\frac{s}{2}}\bar{\eta}_k dx(\hat{u}_{1k}+\hat{u}_{2k}) \\
&-\frac{4s\beta^*}{N}\left(\frac{2\beta^*}{b}\right)^{-1-\frac{N}{4s}}\int_{\mathbb{R}^N}\left(\bar{u}_{1k}^{\frac{4s+N}{N}} + \bar{u}_{2k}^{\frac{4s+N}{N}}\right)[A\bar{u}_{1k}+(1-A)\bar{u}_{2k}]^{\frac{4s}{N}}\bar{\eta}_k dx(\hat{u}_{1k}+\hat{u}_{2k}).
\end{aligned}
\tag{9.97}
$$

Next, we claim that for any $x_0 \in \mathbb{R}^N$, there exists a small constant $\delta > 0$ such that

$$
\int_{\partial B_\delta(x_0)}\left[\bar{\varepsilon}_k^{2s(1-\tau)}|(-\Delta)^{\frac{s}{2}}\hat{\eta}_k|^2 + \bar{\varepsilon}_k^{4s}|\bar{\varepsilon}_k^{\tau}x|^2\hat{\eta}_k^2 + \hat{\eta}_k^2\right]dS \le C\bar{\varepsilon}_k^{N(1-\tau)} \quad \text{as } k \to \infty. \tag{9.98}
$$

In fact, multiplying (9.96) by $\hat{\eta}_k$ and integrating over $\mathbb{R}^N$, we have

$$
\begin{aligned}
&\left[2a_k\bar{\varepsilon}_k^{2s(2-\tau)} + b\left(\frac{b}{2\beta^*}\right)^{\frac{N}{4s}}\bar{\varepsilon}_k^{(\tau-1)(N-4s)}\int_{\mathbb{R}^N}(|(-\Delta)^{\frac{s}{2}}\hat{u}_{1k}|^2 + |(-\Delta)^{\frac{s}{2}}\hat{u}_{2k}|^2)dx\right]\int_{\mathbb{R}^N}|(-\Delta)^{\frac{s}{2}}\hat{\eta}_k|^2 dx \\
&+ 2\bar{\varepsilon}_k^{4s}\int_{\mathbb{R}^N}|\bar{\varepsilon}_k^{\tau}x|^2\hat{\eta}_k^2 dx - \bar{\varepsilon}_k^{4s}(\mu_{1k}+\mu_{2k})\int_{\mathbb{R}^N}\hat{\eta}_k^2 dx =: A_1 + A_2 + A_3 + A_4,
\end{aligned}
\tag{9.99}
$$

where

$$
A_1 := -b\left(\frac{b}{2\beta^*}\right)^{\frac{N}{4s}}\bar{\varepsilon}_k^{(\tau-1)(N-4s)}\left(\int_{\mathbb{R}^N}(-\Delta)^{\frac{s}{2}}(\hat{u}_{1k}+\hat{u}_{2k})(-\Delta)^{\frac{s}{2}}\hat{\eta}_k dx\right)^2,
$$

$$
\begin{aligned}
A_2 :=& \frac{b}{2}\left(\frac{2\beta^*}{b}\right)^{-\frac{N}{2s}}\bar{\varepsilon}_k^{(1-\tau)N}\int_{\mathbb{R}^N}(|(-\Delta)^{\frac{s}{2}}\bar{u}_{1k}|^2 + |(-\Delta)^{\frac{s}{2}}\bar{u}_{2k}|^2)dx \\
&\times\int_{\mathbb{R}^N}(-\Delta)^{\frac{s}{2}}(\bar{u}_{1k}+\bar{u}_{2k})(-\Delta)^{\frac{s}{2}}\bar{\eta}_k dx\int_{\mathbb{R}^N}(\bar{u}_{1k}+\bar{u}_{2k})\bar{\eta}_k dx,
\end{aligned}
$$

$$
\begin{aligned}
A_3 :=& -\frac{4s\beta^*}{N}\left(\frac{2\beta^*}{b}\right)^{-1-\frac{N}{4s}}\bar{\varepsilon}_k^{(1-\tau)N}\int_{\mathbb{R}^N}\left(\bar{u}_{1k}^{\frac{4s+N}{N}} + \bar{u}_{2k}^{\frac{4s+N}{N}}\right) \\
&\times[A\bar{u}_{1k}+(1-A)\bar{u}_{2k}]^{\frac{4s}{N}}\bar{\eta}_k dx\int_{\mathbb{R}^N}(\bar{u}_{1k}+\bar{u}_{2k})\bar{\eta}_k dx,
\end{aligned}
$$

$$
A_4 := \frac{(8s+N)b}{N}\bar{\varepsilon}_k^{(1-\tau)N}\int_{\mathbb{R}^N}[A\bar{u}_{1k}+(1-A)\bar{u}_{2k}]^{\frac{8s}{N}}\bar{\eta}_k^2 dx.
$$

By the fact that $\|\bar{u}_{ik}\|_{H^s(\mathbb{R}^N)} \leq C$ as $k \to \infty$, we can derive from the Hölder's inequality that

$$\begin{aligned}
|A_1| &\leq C\bar{\varepsilon}_k^{(\tau-1)(N-4s)} \int_{\mathbb{R}^N} |(-\Delta)^{\frac{s}{2}}(\hat{u}_{1k}+\hat{u}_{2k})|^2 dx \int_{\mathbb{R}^N} |(-\Delta)^{\frac{s}{2}}\hat{\eta}_k|^2 dx \\
&\leq C\bar{\varepsilon}_k^{(\tau-1)(N-4s)+(1-\tau)(N-2s)} \int_{\mathbb{R}^N} |(-\Delta)^{\frac{s}{2}}(\bar{u}_{1k}+\bar{u}_{2k})|^2 dx \int_{\mathbb{R}^N} |(-\Delta)^{\frac{s}{2}}\hat{\eta}_k|^2 dx \\
&\leq C\bar{\varepsilon}_k^{2s(1-\tau)} \|\bar{u}_{1k}+\bar{u}_{2k}\|^2_{H^s(\mathbb{R}^N)} \int_{\mathbb{R}^N} |(-\Delta)^{\frac{s}{2}}\hat{\eta}_k|^2 dx \\
&\leq C\bar{\varepsilon}_k^{2s(1-\tau)} \int_{\mathbb{R}^N} |(-\Delta)^{\frac{s}{2}}\hat{\eta}_k|^2 dx \quad \text{as } k \to \infty.
\end{aligned} \tag{9.100}$$

Using the same argument of (9.100), we can obtain that

$$\begin{aligned}
|A_2| &\leq C\bar{\varepsilon}_k^{(1-\tau)N} \|\bar{\eta}_k\|^2_{H^s(\mathbb{R}^N)} \\
&\leq C\bar{\varepsilon}_k^{(1-\tau)N} \left[\bar{\varepsilon}_k^{(N-2s)(\tau-1)} \int_{\mathbb{R}^N} |(-\Delta)^{\frac{s}{2}}\hat{\eta}_k|^2 dx + \bar{\varepsilon}_k^{N(\tau-1)} \int_{\mathbb{R}^N} \hat{\eta}_k^2 dx\right] \\
&\leq C\bar{\varepsilon}_k^{2s(1-\tau)} \int_{\mathbb{R}^N} |(-\Delta)^{\frac{s}{2}}\hat{\eta}_k|^2 dx + C\int_{\mathbb{R}^N} \hat{\eta}_k^2 dx \quad \text{as } k \to \infty.
\end{aligned} \tag{9.101}$$

Noticing that $|\bar{\eta}_k| \leq 1$, we conclude from Lemma 9.5 that

$$|A_3 + A_4| \leq C\bar{\varepsilon}_k^{(1-\tau)N} \quad \text{as } k \to \infty. \tag{9.102}$$

Thus, from (9.8), (9.35) and (9.99)–(9.102), we obtain that

$$\begin{aligned}
&\bar{\varepsilon}_k^{2s(1-\tau)} \int_{\mathbb{R}^N} |(-\Delta)^{\frac{s}{2}}\hat{\eta}_k|^2 dx + \bar{\varepsilon}_k^{4s} \int_{\mathbb{R}^N} |\bar{\varepsilon}_k^{\tau} x|^2 \hat{\eta}_k^2 dx - (\mu_{1k}+\mu_{2k})\bar{\varepsilon}_k^{4s} \int_{\mathbb{R}^N} \hat{\eta}_k^2 dx \\
\leq & C\left[\bar{\varepsilon}_k^{2s(1-\tau)} \int_{\mathbb{R}^N} |(-\Delta)^{\frac{s}{2}}\hat{\eta}_k|^2 dx + \int_{\mathbb{R}^N} \hat{\eta}_k^2 dx + \bar{\varepsilon}_k^{2(1-\tau)}\right] \quad \text{as } k \to \infty.
\end{aligned}$$

Combining this result with Lemma 4.5 of Cao, Li and Luo [20], we obtain that (9.98) holds. Thus, we obtain immediately the following estimate

$$\|\bar{\eta}_k\|_{H^s(\mathbb{R}^N)} \leq C \quad \text{as } k \to \infty. \tag{9.103}$$

In order to deduce the limiting behaviors of $\bar{\eta}_k$, let us first study some basic properties of the ground state φ, which will be used later.

Lemma 9.6 *Let $\varphi > 0$ be the unique radial positive solution of (9.3) and let Γ denote the corresponding linearized operator given by*

$$\Gamma := (-\Delta)^s + \frac{4s-N}{2N} - \frac{8s+N}{2N}\varphi^{\frac{8s}{N}}.$$

Then we have the following properties:

(i) φ is nondegenerate in $H^s(\mathbb{R}^N)$ in the sense that there holds

$$Ker\Gamma = span\left\{\frac{\partial\varphi}{\partial x_1}, \frac{\partial\varphi}{\partial x_2}, \dots, \frac{\partial\varphi}{\partial x_N}\right\};$$

(ii) $\Gamma\left(\frac{N}{4}\varphi + x\cdot\nabla\varphi\right) = -\frac{(4s-N)s}{N}\varphi$ *and* $\Gamma(x\cdot\nabla\varphi) = 2s(-\Delta)^s\varphi$.

Proof. According to [59, Theorem 3], we know that (i) holds. Let $w(x) > 0$ be the radial positive solution of

$$(-\Delta)^s w + w - w^{\frac{8s}{N}+1} = 0. \tag{9.104}$$

Define

$$\mathcal{L} := (-\Delta)^s + 1 - \frac{8s+N}{N} w^{\frac{8s}{N}}.$$

Following [59, (7.2) and (7.3)], we have that $\mathcal{L}\left(\frac{N}{4}w + x \cdot \nabla w\right) = -2sw$. Thus, (9.104) shows that

$$(-\Delta)^s(x \cdot \nabla w) + x \cdot \nabla w - \frac{8s+N}{N} w^{\frac{8s}{N}}(x \cdot \nabla w) = 2sw^{\frac{8s}{N}+1} - 2sw. \tag{9.105}$$

In view of (9.3) and (9.104), it follows that

$$w(x) = \left(\frac{N}{4s-N}\right)^{\frac{N}{8s}} \varphi\left[\left(\frac{2N}{4s-N}\right)^{\frac{1}{2s}} x\right], \tag{9.106}$$

then

$$(-\Delta)^s(x \cdot \nabla\varphi) + \frac{4s-N}{2N}(x \cdot \nabla\varphi) - \frac{8s+N}{2N}\varphi^{\frac{8s}{N}}(x \cdot \nabla\varphi) = s\varphi^{\frac{8s+N}{N}} - \frac{(4s-N)s}{N}\varphi. \tag{9.107}$$

Moreover, since $\varphi(x) > 0$ is the unique solution to (9.3), a straightforward calculation shows that

$$\begin{aligned}\Gamma\left(\frac{N}{4}\varphi + x \cdot \nabla\varphi\right) =& (-\Delta)^s(x \cdot \nabla\varphi) + \frac{4s-N}{2N}(x \cdot \nabla\varphi) - s\varphi^{\frac{8s+N}{N}} \\ & - \frac{8s+N}{2N}\varphi^{\frac{8s}{N}}(x \cdot \nabla\varphi) = -\frac{(4s-N)s}{N}\varphi,\end{aligned}$$

in view of (9.107). This completes the proof of Lemma 9.6. □

Lemma 9.7 *Under the assumptions of Theorem 9.3, there exists $\bar{\eta}_0 \in C^1(\mathbb{R}^N)$ such that*

$$\bar{\eta}_k \to \bar{\eta}_0 \quad in\ C^1(\mathbb{R}^N)\ as\ k \to \infty, \tag{9.108}$$

and

$$\bar{\eta}_0(x) = d_0\varphi + \bar{d}_0(x \cdot \nabla\varphi) + \sum_{i=1}^N d_i \frac{\partial\varphi}{\partial x_i}, \tag{9.109}$$

where $d_0, \bar{d}_0, d_1, \ldots, d_N$ are all constants.

Proof. By the fact that $\|\bar{u}_{ik}\|_{H^s(\mathbb{R}^N)} \le C$ as $k \to \infty$ for $i = 1, 2$, we deduce from (9.103) that

$$\begin{aligned}&\int_{\mathbb{R}^N} (-\Delta)^{\frac{s}{2}}(\bar{u}_{1k} + \bar{u}_{2k})(-\Delta)^{\frac{s}{2}}\bar{\eta}_k dx \\ \le& \left(\int_{\mathbb{R}^N} |(-\Delta)^{\frac{s}{2}}(\bar{u}_{1k} + \bar{u}_{2k})|^2 dx\right)^{\frac{1}{2}} \left(\int_{\mathbb{R}^N} |(-\Delta)^{\frac{s}{2}}\bar{\eta}_k|^2 dx\right)^{\frac{1}{2}} \\ \le& \|\bar{u}_{1k} + \bar{u}_{2k}\|_{H^s(\mathbb{R}^N)}\|\bar{\eta}_k\|_{H^s(\mathbb{R}^N)} \le C \quad \text{as } k \to \infty.\end{aligned} \tag{9.110}$$

Since $\bar{z}_{1k} \to 0$ and $\|\bar{u}_{1k}\|_\infty \le C$ as $k \to \infty$, we derive from (9.59) and (9.76) that

$$\begin{aligned}|(-\Delta)^s\bar{u}_{1k}| =& \Bigg| \left(a_k\bar{\varepsilon}_k^{2s} + b\left(\frac{b}{2\beta^*}\right)^{\frac{N}{4s}} \int_{\mathbb{R}^N} |(-\Delta)^{\frac{s}{2}}\bar{u}_{1k}|^2 dx\right)^{-1} \\ & \times [-\bar{\varepsilon}_k^{4s}|\bar{\varepsilon}_k x + \bar{z}_{1k}|^2\bar{u}_{1k} + \mu_{2k}\bar{\varepsilon}_k^{4s}\bar{u}_{1k} + \frac{b}{2}\bar{u}_{1k}^{\frac{8s}{N}+1}]\Bigg| \le C \quad \text{as } k \to \infty.\end{aligned} \tag{9.111}$$

Combining (9.73) with (9.74) yields that

$$\begin{aligned}\mu_{ik}\bar{\varepsilon}_k^{4s} =& e(a_k,\beta^*)\bar{\varepsilon}_k^{4s} + \frac{b}{2}\left(\frac{2\beta^*}{b}\right)^{-\frac{N}{2s}}\left(\int_{\mathbb{R}^N}|(-\Delta)^{\frac{s}{2}}\bar{u}_{ik}|^2dx\right)^2 \\ &- \frac{4s\beta^*}{N+4s}\left(\frac{2\beta^*}{b}\right)^{-1-\frac{N}{4s}}\int_{\mathbb{R}^N}\bar{u}_{ik}^{\frac{8s+2N}{N}}dx \quad i=1,2.\end{aligned} \tag{9.112}$$

This relation together with (9.75) implies that

$$\begin{aligned}&\bar{\varepsilon}_k^{4s}\bar{u}_{2k}\frac{\mu_{1k}-\mu_{2k}}{\|\bar{u}_{1k}-\bar{u}_{2k}\|_\infty} = \frac{b}{2}\left(\frac{2\beta^*}{b}\right)^{-\frac{N}{2s}}\int_{\mathbb{R}^N}(|(-\Delta)^{\frac{s}{2}}\bar{u}_{1k}|^2+|(-\Delta)^{\frac{s}{2}}\bar{u}_{2k}|^2)dx \\ &\times\int_{\mathbb{R}^N}(-\Delta)^{\frac{s}{2}}(\bar{u}_{1k}+\bar{u}_{2k})(-\Delta)^{\frac{s}{2}}\bar{\eta}_k dx\bar{u}_{2k} \\ &-\frac{4s\beta^*}{N}\left(\frac{2\beta^*}{b}\right)^{-1-\frac{N}{4s}}\int_{\mathbb{R}^N}\left(\bar{u}_{1k}^{\frac{4s+N}{N}}+\bar{u}_{2k}^{\frac{4s+N}{N}}\right)[A\bar{u}_{1k}+(1-A)\bar{u}_{2k}]^{\frac{4s}{N}}\bar{\eta}_k dx\bar{u}_{2k}.\end{aligned} \tag{9.113}$$

By the fact that $|\bar{\eta}_k|\le 1$, $\|\bar{u}_{2k}\|_\infty\le C$ and $\|\bar{u}_{ik}\|_{H^s(\mathbb{R}^N)}\le C$ as $k\to\infty$ for $i=1,2$, we conclude from (9.78) and (9.113) that

$$\left|\bar{\varepsilon}_k^{4s}\bar{u}_{2k}\frac{\mu_{1k}-\mu_{2k}}{\|\bar{u}_{1k}-\bar{u}_{2k}\|_\infty}\right|\le C \quad \text{as } k\to\infty. \tag{9.114}$$

We now rewrite problem (9.77) as

$$(-\Delta)^s\bar{\eta}_k = \bar{F}_k(x) \quad \text{in } \mathbb{R}^N,$$

where

$$\begin{aligned}\bar{F}_k(x) =& \left(a_k\bar{\varepsilon}_k^{2s}+b\left(\frac{b}{2\beta^*}\right)^{\frac{N}{4s}}\int_{\mathbb{R}^N}|(-\Delta)^{\frac{s}{2}}\bar{u}_{2k}|^2dx\right)^{-1} \\ &\times\Bigg[-b\Big(\frac{b}{2\beta^*}\Big)^{\frac{N}{4s}}\int_{\mathbb{R}^N}(-\Delta)^{\frac{s}{2}}(\bar{u}_{1k}+\bar{u}_{2k})(-\Delta)^{\frac{s}{2}}\bar{\eta}_k dx(-\Delta)^s\bar{u}_{1k} \\ &-\bar{\varepsilon}_k^{4s}|\bar{\varepsilon}_k x+\bar{z}_{1k}|^2\bar{\eta}_k+\mu_{1k}\bar{\varepsilon}_k^{4s}\bar{\eta}_k+\frac{\bar{\varepsilon}_k^{4s}(\mu_{1k}-\mu_{2k})\bar{u}_{2k}}{\|\bar{u}_{1k}-\bar{u}_{2k}\|_\infty} \\ &+\frac{(8s+N)b}{2N}[A\bar{u}_{1k}+(1-A)\bar{u}_{2k}]^{\frac{8s}{N}}\bar{\eta}_k\Bigg].\end{aligned}$$

Since $|\bar{\eta}_k|\le 1$, $\|\bar{u}_{ik}\|_\infty\le C$ and $\|\bar{u}_{ik}\|_{H^s(\mathbb{R}^N)}\le C$ as $k\to\infty$ for $i=1,2$, we deduce from (9.110), (9.111) and (9.114) that $\bar{F}_k(x)\in L^\infty(\mathbb{R}^N)$ as $k\to\infty$. Thus, using Proposition 2.9 of Silvestre [138], we know that $\|\bar{\eta}_k\|_{C^{1,\alpha}(\mathbb{R}^N)}\le C$ for some $\alpha\in(0,1)$ as $k\to\infty$. Passing to a subsequence, there exists some function $\bar{\eta}_0\in C^{1,\alpha}(\mathbb{R}^N)$ such that relation (9.108) holds.

Moreover, (9.53), (9.69) and (9.74) show that

$$\int_{\mathbb{R}^N}|(-\Delta)^{\frac{s}{2}}\bar{u}_{ik}|^2dx\to\left(\frac{2\beta^*}{b}\right)^{\frac{N}{4s}} \quad \text{as } k\to\infty \text{ for } i=1,2. \tag{9.115}$$

Taking $k \to \infty$ in (9.77), we then derive from (9.113) and (9.115) that

$$(-\Delta)^s\bar{\eta}_0 + \frac{4s-N}{2N}\bar{\eta}_0 - \frac{8s+N}{2N}\varphi^{\frac{8s}{N}}\bar{\eta}_0 = -2\left(\frac{b}{2\beta^*}\right)^{\frac{N}{4s}}\int_{\mathbb{R}^N}(-\Delta)^{\frac{s}{2}}\varphi(-\Delta)^{\frac{s}{2}}\bar{\eta}_0dx(-\Delta)^s\varphi$$
$$+2\left(\frac{2\beta^*}{b}\right)^{-\frac{N}{4s}}\int_{\mathbb{R}^N}(-\Delta)^{\frac{s}{2}}\varphi(-\Delta)^{\frac{s}{2}}\bar{\eta}_0dx\varphi - \frac{8s\beta^*}{Nb}\left(\frac{2\beta^*}{b}\right)^{-1-\frac{N}{4s}}\int_{\mathbb{R}^N}\varphi^{\frac{8s+N}{N}}\bar{\eta}_0dx\varphi, \tag{9.116}$$

from which and Lemma 9.6 we deduce that (9.109) holds. □

Until now, we have decomposed $\bar{\eta}_0$ and we need to verify that $d_0, \bar{d}_0, d_1, \ldots, d_N = 0$. Hereafter, we consider the relationships between the parameters.

Lemma 9.8 *Assume that $d_0, \bar{d}_0, d_1, \ldots, d_N$ are defined in (9.109), then*

$$\bar{d}_0 = \frac{2}{N}d_0.$$

Proof. We first claim that

$$d_1\int_{\mathbb{R}^N}(-\Delta)^{\frac{s}{2}}\frac{\partial\varphi}{\partial x_1}(-\Delta)^{\frac{s}{2}}\varphi dx + \cdots + d_N\int_{\mathbb{R}^N}(-\Delta)^{\frac{s}{2}}\frac{\partial\varphi}{\partial x_N}(-\Delta)^{\frac{s}{2}}\varphi dx = 0. \tag{9.117}$$

Indeed, from (9.3) we have

$$\begin{aligned} &d_i\int_{\mathbb{R}^N}(-\Delta)^{\frac{s}{2}}\frac{\partial\varphi}{\partial x_i}(-\Delta)^{\frac{s}{2}}\varphi dx + d_i\frac{4s-N}{2N}\int_{\mathbb{R}^N}\varphi\frac{\partial\varphi}{\partial x_i}dx \\ &-\frac{d_i}{2}\int_{\mathbb{R}^N}\varphi^{\frac{8s+N}{N}}\frac{\partial\varphi}{\partial x_i}dx = 0 \quad i = 1, 2, \ldots, N. \end{aligned} \tag{9.118}$$

Applying Lemma 9.6, we conclude that

$$\begin{aligned} &d_1\int_{\mathbb{R}^N}(-\Delta)^{\frac{s}{2}}\frac{\partial\varphi}{\partial x_1}(-\Delta)^{\frac{s}{2}}\varphi dx + \cdots + d_N\int_{\mathbb{R}^N}(-\Delta)^{\frac{s}{2}}\frac{\partial\varphi}{\partial x_N}(-\Delta)^{\frac{s}{2}}\varphi dx \\ &+ d_1\frac{4s-N}{2N}\int_{\mathbb{R}^N}\varphi\frac{\partial\varphi}{\partial x_1}dx + \cdots + d_N\frac{4s-N}{2N}\int_{\mathbb{R}^N}\varphi\frac{\partial\varphi}{\partial x_N}dx \\ &-\frac{d_1(8s+N)}{2N}\int_{\mathbb{R}^N}\varphi^{\frac{8s+N}{N}}\frac{\partial\varphi}{\partial x_1}dx - \cdots - \frac{d_N(8s+N)}{2N}\int_{\mathbb{R}^N}\varphi^{\frac{8s+N}{N}}\frac{\partial\varphi}{\partial x_N}dx = 0. \end{aligned} \tag{9.119}$$

In view of (9.118) and (9.119), it follows that

$$d_1\int_{\mathbb{R}^N}\varphi^{\frac{8s+N}{N}}\frac{\partial\varphi}{\partial x_1}dx + \cdots + d_N\int_{\mathbb{R}^N}\varphi^{\frac{8s+N}{N}}\frac{\partial\varphi}{\partial x_N}dx = 0. \tag{9.120}$$

Moreover, we know that $\int_{\mathbb{R}^N}\varphi\frac{\partial\varphi}{\partial x_i}dx = 0$ for the season that φ is even in x_i while $\frac{\partial\varphi}{\partial x_i}$ is odd. Thus, from (9.119) and (9.120), we get that (9.117) holds. Following (9.109) and (9.117), we have

$$\begin{aligned} &\int_{\mathbb{R}^N}(-\Delta)^{\frac{s}{2}}\varphi(-\Delta)^{\frac{s}{2}}\bar{\eta}_0dx \\ =&\int_{\mathbb{R}^N}(-\Delta)^{\frac{s}{2}}\varphi(-\Delta)^{\frac{s}{2}}\left(d_0\varphi + \bar{d}_0(x\cdot\nabla\varphi) + \sum_{i=1}^N d_i\frac{\partial\varphi}{\partial x_i}\right)dx \\ =&d_0\int_{\mathbb{R}^N}|(-\Delta)^{\frac{s}{2}}\varphi|^2dx + \bar{d}_0\int_{\mathbb{R}^N}(-\Delta)^{\frac{s}{2}}\varphi(-\Delta)^{\frac{s}{2}}(x\cdot\nabla\varphi)dx. \end{aligned} \tag{9.121}$$

Taking advantage of Lemma 9.6, we deduce from (9.4) that

$$\begin{aligned}
&\int_{\mathbb{R}^N}(-\Delta)^{\frac{s}{2}}\varphi(-\Delta)^{\frac{s}{2}}(x\cdot\nabla\varphi)dx = 2s\int_{\mathbb{R}^N}|(-\Delta)^{\frac{s}{2}}\varphi|^2dx \\
&+\frac{N-4s}{2N}\int_{\mathbb{R}^N}\varphi(x\cdot\nabla\varphi)dx+\frac{8s+N}{2N}\int_{\mathbb{R}^N}\varphi^{\frac{8s+N}{N}}(x\cdot\nabla\varphi)dx \\
&=2s\int_{\mathbb{R}^N}|(-\Delta)^{\frac{s}{2}}\varphi|^2dx+\frac{N-4s}{4N}\int_{\mathbb{R}^N}x\cdot\nabla\varphi^2dx+\frac{8s+N}{16s+4N}\int_{\mathbb{R}^N}x\cdot\nabla\varphi^{\frac{8s+2N}{N}}dx \\
&=2s\int_{\mathbb{R}^N}|(-\Delta)^{\frac{s}{2}}\varphi|^2dx-\frac{N-4s}{4}\int_{\mathbb{R}^N}\varphi^2dx+\frac{(8s+N)(-N)}{16s+4N}\int_{\mathbb{R}^N}\varphi^{\frac{8s+2N}{N}}dx \\
&=2s\left(\frac{2\beta^*}{b}\right)^{\frac{N}{4s}}+\frac{4s-N}{4}\left(\frac{2\beta^*}{b}\right)^{\frac{N}{4s}}-\frac{(8s+N)(4s+N)}{16s+4N}\left(\frac{2\beta^*}{b}\right)^{\frac{N}{4s}} \\
&=\left(s-\frac{N}{2}\right)\left(\frac{2\beta^*}{b}\right)^{\frac{N}{4s}}.
\end{aligned} \tag{9.122}$$

Also, we get from (9.4) and (9.120) that

$$\begin{aligned}
\int_{\mathbb{R}^N}\varphi^{\frac{8s+N}{N}}\bar{\eta}_0dx &= \int_{\mathbb{R}^N}\varphi^{\frac{8s+N}{N}}\left(d_0\varphi+\bar{d}_0(x\cdot\nabla\varphi)+\sum_{i=1}^N d_i\frac{\partial\varphi}{\partial x_i}\right)dx \\
&=\frac{N+4s}{N}d_0\left(\frac{2\beta^*}{b}\right)^{\frac{N}{4s}}-\frac{N^2\bar{d}_0}{8s+2N}\int_{\mathbb{R}^N}\varphi^{\frac{8s+2N}{N}}dx \\
&=\frac{N+4s}{N}d_0\left(\frac{2\beta^*}{b}\right)^{\frac{N}{4s}}-\frac{N\bar{d}_0}{2}\left(\frac{2\beta^*}{b}\right)^{\frac{N}{4s}}.
\end{aligned} \tag{9.123}$$

Hence, we derive from Lemma 9.6, (9.116), (9.121), (9.122) and (9.123) that

$$\begin{aligned}
\Gamma(\bar{\eta}_0) =& -2\left(\frac{b}{2\beta^*}\right)^{\frac{N}{4s}}\left[d_0\left(\frac{2\beta^*}{b}\right)^{\frac{N}{4s}}+\left(s-\frac{N}{2}\right)\bar{d}_0\left(\frac{2\beta^*}{b}\right)^{\frac{N}{4s}}\right](-\Delta)^s\varphi \\
&+2\left(\frac{2\beta^*}{b}\right)^{-\frac{N}{4s}}\left[d_0\left(\frac{2\beta^*}{b}\right)^{\frac{N}{4s}}+\left(s-\frac{N}{2}\right)\bar{d}_0\left(\frac{2\beta^*}{b}\right)^{\frac{N}{4s}}\right]\varphi \\
&-\frac{8s\beta^*}{Nb}\left(\frac{2\beta^*}{b}\right)^{-1-\frac{N}{4s}}\left[\frac{N+4s}{N}\left(\frac{2\beta^*}{b}\right)^{\frac{N}{4s}}d_0-\frac{N}{2}\left(\frac{2\beta^*}{b}\right)^{\frac{N}{4s}}\bar{d}_0\right]\varphi \\
=&\,\Gamma\left(d_0\varphi+\bar{d}_0(x\cdot\nabla\varphi)+\sum_{i=1}^N d_i\frac{\partial\varphi}{\partial x_i}\right)=d_0\Gamma(\varphi)+\bar{d}_0\Gamma(x\cdot\nabla\varphi) \\
=&\,\frac{4s(N-4s)}{N^2}d_0\varphi-\frac{8s}{N}d_0(-\Delta)^s\varphi+2s\bar{d}_0(-\Delta)^s\varphi,
\end{aligned}$$

which implies that

$$\bar{d}_0=\frac{2}{N}d_0.$$

This completes the proof of Lemma 9.8. □

From [68], we have the following elementary estimate.

Lemma 9.9 *Let $\rho>\theta>0$ be two constants. Suppose $(y-x)^2+t^2\geq\rho^2$, $t>0$ and $\alpha>N$. Then, when $\beta>N$, it holds that*

$$\int_{\mathbb{R}^N\setminus B_\theta(y-x)}\frac{dz}{(t+|z|)^\alpha|y-z-x|^\beta}\leq C\left[\frac{1}{(1+|y-x|)^\beta}\frac{1}{t^{\alpha-N}}+\frac{1}{(1+|y-x|)^\beta}\frac{1}{\theta^{\beta-N}}\right],$$

where $C > 0$ is a constant independent of θ. Moreover, as $\epsilon \to 0$, we have

$$\int_{\mathbb{R}^N} \frac{dz}{(t+|z|)^\alpha(1+\frac{|y-z-x|}{\epsilon})^\beta} \le C\epsilon^N \left[\frac{1}{(1+|y-x|)^\beta}\frac{1}{t^{\alpha-N}} + \frac{1}{(1+|y-x|)^\alpha}\right].$$

From the above results, we obtain the following property about the parameter d_i, $i = 1, 2, \dots, N$.

Lemma 9.10 *Assume that d_i are defined in (9.109), where $i = 1, 2, \dots, N$, it holds*

$$d_1 = d_2 = \dots = d_N = 0.$$

Proof. We first write the extension of $\hat{u}_{ik}$, then

$$\begin{cases} \qquad \mathrm{div}(\mathrm{t}^{1-2s}\nabla\widetilde{\hat{\mathrm{u}}_{\mathrm{ik}}}) = 0 \quad \text{in } \mathbb{R}^{N+1}_+, \\ - \quad \left[a_k\bar{\varepsilon}_k^{2s(2-\tau)} + b\left(\frac{b}{2\beta^*}\right)^{\frac{N}{4s}}\bar{\varepsilon}_k^{(\tau-1)(N-4s)}\int_{\mathbb{R}^N}|(-\Delta)^{\frac{s}{2}}\hat{u}_{ik}|^2dx\right]\lim_{t\to 0}\partial_t\widetilde{\hat{u}_{ik}}(x,t) \\ = \quad -\bar{\varepsilon}_k^{4s}|\bar{\varepsilon}_k^{\tau}x|^2\hat{u}_{ik} + \mu_{ik}\bar{\varepsilon}_k^{4s}\hat{u}_{ik} + \frac{b}{2}\hat{u}_{ik}^{\frac{8s}{N}+1} \quad \text{in } \mathbb{R}^N, \end{cases} \tag{9.124}$$

where $i = 1, 2$. For writing convenience, we give the following definitions

$$\begin{aligned} &B_\delta(\bar{\varepsilon}_k^{-\tau}\bar{z}_{1k}) := \{x \in \mathbb{R}^N : |x - \bar{\varepsilon}_k^{-\tau}\bar{z}_{1k}| \le \delta\} \subseteq \mathbb{R}^N, \\ &\mathbf{B}_\delta^+(\bar{\varepsilon}_k^{-\tau}\bar{z}_{1k}) := \{X = (x,t) : |X - (\bar{\varepsilon}_k^{-\tau}\bar{z}_{1k}, 0)| \le \delta, t > 0\} \subseteq \mathbb{R}^{N+1}_+, \\ &\partial'\mathbf{B}_\delta^+(\bar{\varepsilon}_k^{-\tau}\bar{z}_{1k}) := \{X = (x,t) : |X - (\bar{\varepsilon}_k^{-\tau}\bar{z}_{1k}, 0)| \le \delta, t = 0\} \subseteq \mathbb{R}^N, \\ &\partial''\mathbf{B}_\delta^+(\bar{\varepsilon}_k^{-\tau}\bar{z}_{1k}) := \{X = (x,t) : |X - (\bar{\varepsilon}_k^{-\tau}\bar{z}_{1k}, 0)| = \delta, t > 0\} \subseteq \mathbb{R}^{N+1}_+, \\ &\partial\mathbf{B}_\delta^+(\bar{\varepsilon}_k^{-\tau}\bar{z}_{1k}) := \partial'\mathbf{B}_\delta^+(\bar{\varepsilon}_k^{-\tau}\bar{z}_{1k}) \cup \partial''\mathbf{B}_\delta^+(\bar{\varepsilon}_k^{-\tau}\bar{z}_{1k}), \end{aligned}$$

where $\delta > 0$ small is given by (9.98) and $\tau < 0$ is given by (9.94). Multiplying (9.124) by $\frac{\partial\widetilde{\hat{u}_{ik}}}{\partial x_j}$, where $i = 1, 2$ and $j = 1, 2, \dots, N$, and integrating over $B_\delta(\bar{\varepsilon}_k^{-\tau}\bar{z}_{1k})$, we see that

$$\begin{aligned} &\left[a_k\bar{\varepsilon}_k^{2s(2-\tau)} + b\left(\frac{b}{2\beta^*}\right)^{\frac{N}{4s}}\bar{\varepsilon}_k^{(\tau-1)(N-4s)}\int_{\mathbb{R}^N}|(-\Delta)^{\frac{s}{2}}\hat{u}_{ik}|^2dx\right] \\ &\times\left[-\int_{\partial''\mathbf{B}_\delta^+(\bar{\varepsilon}_k^{-\tau}\bar{z}_{1k})} t^{1-2s}\frac{\partial\widetilde{\hat{u}_{ik}}}{\partial\nu}\frac{\partial\widetilde{\hat{u}_{ik}}}{\partial x_j} + \frac{1}{2}\int_{\partial''\mathbf{B}_\delta^+(\bar{\varepsilon}_k^{-\tau}\bar{z}_{1k})} t^{1-2s}|\nabla\widetilde{\hat{u}_{ik}}|^2\nu_j\right] \\ &+\bar{\varepsilon}_k^{4s}\int_{B_\delta(\bar{\varepsilon}_k^{-\tau}\bar{z}_{1k})}|\bar{\varepsilon}_k^{\tau}x|^2\hat{u}_{ik}\frac{\partial\hat{u}_{ik}}{\partial x_j}dx \\ =&\frac{\mu_{ik}\bar{\varepsilon}_k^{4s}}{2}\int_{\partial B_\delta(\bar{\varepsilon}_k^{-\tau}\bar{z}_{1k})}\hat{u}_{ik}^2\nu_j dS + \frac{Nb}{2(8s+2N)}\int_{\partial B_\delta(\bar{\varepsilon}_k^{-\tau}\bar{z}_{1k})}\hat{u}_{ik}^{\frac{8s+2N}{N}}\nu_j dS. \end{aligned} \tag{9.125}$$

Notice that

$$\begin{aligned} \int_{B_\delta(\bar{\varepsilon}_k^{-\tau}\bar{z}_{1k})}|\bar{\varepsilon}_k^{\tau}x|^2\hat{u}_{ik}\frac{\partial\hat{u}_{ik}}{\partial x_j}dx &= \frac{1}{2}\int_{\partial B_\delta(\bar{\varepsilon}_k^{-\tau}\bar{z}_{1k})}|\bar{\varepsilon}_k^{\tau}x|^2\hat{u}_{ik}^2\nu_j dS \\ &-\frac{1}{2}\int_{B_\delta(\bar{\varepsilon}_k^{-\tau}\bar{z}_{1k})}\frac{\partial|\bar{\varepsilon}_k^{\tau}x|^2}{\partial x_j}\hat{u}_{ik}^2dx. \end{aligned} \tag{9.126}$$

Using this relation together with (9.125), we deduce that

$$
\begin{aligned}
&\frac{1}{2}\bar{\varepsilon}_k^{4s}\int_{B_\delta(\bar{\varepsilon}_k^{-\tau}\bar{z}_{1k})}\frac{\partial|\bar{\varepsilon}_k^{\tau}x|^2}{\partial x_j}\hat{u}_{ik}^2dx\\
=&\left[a_k\bar{\varepsilon}_k^{2s(2-\tau)}+b\left(\frac{b}{2\beta^*}\right)^{\frac{N}{4s}}\bar{\varepsilon}_k^{(\tau-1)(N-4s)}\int_{\mathbb{R}^N}|(-\Delta)^{\frac{s}{2}}\hat{u}_{ik}|^2dx\right]\\
&\times\left[-\int_{\partial''\mathbf{B}_\delta^+(\bar{\varepsilon}_k^{-\tau}\bar{z}_{1k})}t^{1-2s}\frac{\partial\widetilde{\hat{u}_{ik}}}{\partial\nu}\frac{\partial\widetilde{\hat{u}_{ik}}}{\partial x_j}+\frac{1}{2}\int_{\partial''\mathbf{B}_\delta^+(\bar{\varepsilon}_k^{-\tau}\bar{z}_{1k})}t^{1-2s}|\nabla\widetilde{\hat{u}_{ik}}|^2\nu_j\right]\\
&+\frac{\bar{\varepsilon}_k^{4s}}{2}\int_{\partial B_\delta(\bar{\varepsilon}_k^{-\tau}\bar{z}_{1k})}|\bar{\varepsilon}_k^{\tau}x|^2\hat{u}_{ik}^2\nu_jdS-\frac{\mu_{ik}\bar{\varepsilon}_k^{4s}}{2}\int_{\partial B_\delta(\bar{\varepsilon}_k^{-\tau}\bar{z}_{1k})}\hat{u}_{ik}^2\nu_jdS\\
&-\frac{Nb}{2(8s+2N)}\int_{\partial B_\delta(\bar{\varepsilon}_k^{-\tau}\bar{z}_{1k})}\hat{u}_{ik}^{\frac{8s+2N}{N}}\nu_jdS.
\end{aligned}
\tag{9.127}
$$

Applying (9.74), (9.92) and Lemma 9.5, we can obtain that

$$
\begin{aligned}
|\hat{u}_{ik}(x)|+|\nabla\hat{u}_{ik}(x)|&=|\bar{u}_{ik}(\bar{\varepsilon}_k^{\tau-1}x-\bar{\varepsilon}_k^{-1}\bar{z}_{1k})|+\bar{\varepsilon}_k^{\tau-1}|\nabla\bar{u}_{ik}(\bar{\varepsilon}^{\tau-1}x-\bar{\varepsilon}_k^{-1}\bar{z}_{1k})|\\
&\le\frac{C}{1+|\bar{\varepsilon}_k^{\tau-1}x-\bar{\varepsilon}_k^{-1}\bar{z}_{1k}|^{N+2s}}+\frac{C\bar{\varepsilon}_k^{\tau-1}}{1+|\bar{\varepsilon}_k^{\tau-1}x-\bar{\varepsilon}_k^{-1}\bar{z}_{1k}|^{N+2s}}\\
&\le\frac{C\bar{\varepsilon}_k^{\tau-1}}{\left(1+|\frac{x-\bar{\varepsilon}_k^{-\tau}\bar{z}_{1k}}{\bar{\varepsilon}_k^{1-\tau}}|^2\right)^{\frac{N+2s}{2}}}\quad\text{as } k\to\infty \text{ and } i=1,2.
\end{aligned}
\tag{9.128}
$$

Following (9.10), (9.11), (9.128) and Lemma 9.9, we derive that for $|x-\bar{\varepsilon}_k^{-\tau}\bar{z}_{1k}|+t^2=\delta^2$ and $t>0$, it has

$$
\begin{aligned}
|\widetilde{\hat{u}_{ik}}(x,t)|&\le C\bar{\varepsilon}_k^{\tau-1}\int_{\mathbb{R}^N}\frac{t^{2s}}{(|x-\xi|+t)^{N+2s}}\frac{1}{\left(1+|\frac{\xi-\bar{\varepsilon}_k^{-\tau}\bar{z}_{1k}}{\bar{\varepsilon}_k^{1-\tau}}|\right)^{N+2s}}d\xi\\
&\le\frac{C\bar{\varepsilon}_k^{(1-\tau)(N-1)}}{(1+|x-\bar{\varepsilon}_k^{-\tau}\bar{z}_{1k}|)^{N+2s}}\quad\text{as } k\to\infty \text{ and } i=1,2.
\end{aligned}
\tag{9.129}
$$

Moreover, from (9.10), (9.11), (9.128), Lemma 9.5 and Lemma 9.9, we also obtain for $|x-\bar{\varepsilon}_k^{-\tau}\bar{z}_{1k}|+t^2=\delta^2$ and $t>0$

$$
\begin{aligned}
\left|\frac{\partial}{\partial x_j}\widetilde{\hat{u}_{ik}}(x,t)\right|=&\left|\int_{\mathbb{R}^N}\frac{1}{(1+|z|^2)^{\frac{N+2s}{2}}}\frac{\partial}{\partial x_j}\hat{u}_{ik}(x-tz)dz\right|\\
\le& C\bar{\varepsilon}_k^{\tau-1}\int_{\mathbb{R}^N}\frac{1}{(1+|z|^2)^{\frac{N+2s}{2}}}\frac{1}{\left(1+|\frac{x-tz-\bar{\varepsilon}_k^{-\tau}\bar{z}_{1k}}{\bar{\varepsilon}_k^{1-\tau}}|^2\right)^{\frac{N+2s}{2}}}dz\\
\le& C\bar{\varepsilon}_k^{\tau-1}\int_{\mathbb{R}^N}\frac{t^{2s}}{(t+|x-\xi|)^{N+2s}}\frac{1}{\left(1+|\frac{\xi-\bar{\varepsilon}_k^{-\tau}\bar{z}_{1k}}{\bar{\varepsilon}_k^{1-\tau}}|\right)^{N+2s}}d\xi\\
\le&\frac{C\bar{\varepsilon}_k^{(1-\tau)(N-1)}}{(1+|x-\bar{\varepsilon}_k^{-\tau}\bar{z}_{1k}|)^{N+2s}}\quad\text{as } k\to\infty,\ i=1,2 \text{ and } j=1,2,\ldots,N.
\end{aligned}
\tag{9.130}
$$

Under the same assumptions, we can also get

$$
\left|\frac{\partial}{\partial\nu}\widetilde{\hat{u}_{ik}}(x,t)\right|\le\frac{C\bar{\varepsilon}_k^{(1-\tau)(N-1)}}{(1+|x-\bar{\varepsilon}_k^{-\tau}\bar{z}_{1k}|)^{N+2s}}\quad\text{as } k\to\infty \text{ and } i=1,2.
\tag{9.131}
$$

Clearly we have

$$
\begin{aligned}
|\hat{\xi}_k(x)| + |\nabla\hat{\xi}_k(x)| = & \left| \frac{\hat{u}_{1k}(x) - \hat{u}_{2k}(x)}{\left(\frac{2\beta^*}{b}\right)^{\frac{N}{8s}} \bar{\varepsilon}_k^{\frac{N}{2}} \|u_{1k}(\bar{\varepsilon}_k^{\tau}x) - u_{2k}(\bar{\varepsilon}_k^{\tau}x)\|_\infty} \right| \\
& + \left| \frac{\nabla\hat{u}_{1k}(x) - \nabla\hat{u}_{2k}(x)}{\left(\frac{2\beta^*}{b}\right)^{\frac{N}{8s}} \bar{\varepsilon}_k^{\frac{N}{2}} \|u_{1k}(\bar{\varepsilon}_k^{\tau}x) - u_{2k}(\bar{\varepsilon}_k^{\tau}x)\|_\infty} \right| \\
& \leq \frac{C\bar{\varepsilon}_k^{\tau-1-\frac{N}{2}}}{\left(1 + |\frac{x - \bar{\varepsilon}_k^{-\tau}\bar{z}_{1k}}{\bar{\varepsilon}_k^{1-\tau}}|^2\right)^{\frac{N+2s}{2}}} \quad \text{as } k \to \infty.
\end{aligned}
$$

We then deduce from (9.129)–(9.131) that for $|x - \bar{\varepsilon}_k^{-\tau}\bar{z}_{1k}| + t^2 = \delta^2$ and $t > 0$

$$
|\widetilde{\hat{\eta}_k}(x,t)|, \left|\frac{\partial}{\partial x_j}\widetilde{\hat{\eta}_k}(x,t)\right|, \left|\frac{\partial}{\partial \nu}\widetilde{\hat{\eta}_k}(x,t)\right| \leq \frac{C\bar{\varepsilon}_k^{(N-1)(1-\tau)-\frac{N}{2}}}{(1 + |x - \bar{\varepsilon}_k^{-\tau}\bar{z}_{1k}|)^{N+2s}} \text{ as } k \to \infty \text{ and } j = 1, 2, \ldots, N. \tag{9.132}
$$

Following (9.127), we have

$$
\frac{\bar{\varepsilon}_k^{4s}}{2} \int_{B_\delta(\bar{\varepsilon}_k^{-\tau}\bar{z}_{1k})} \frac{\partial|\bar{\varepsilon}_k^{\tau}x|^2}{\partial x_j}(\hat{u}_{1k} + \hat{u}_{2k})\hat{\eta}_k dx =: B_1 + B_2 + B_3 + B_4 + B_5 + B_6, \tag{9.133}
$$

where

$$
\begin{aligned}
B_1 :=& \frac{\bar{\varepsilon}_k^{4s}}{2} \int_{\partial B_\delta(\bar{\varepsilon}_k^{-\tau}\bar{z}_{1k})} |\bar{\varepsilon}_k^{\tau}x|^2(\hat{u}_{1k} + \hat{u}_{2k})\hat{\eta}_k\nu_j dS, \\
B_2 :=& -\frac{\mu_{1k}\bar{\varepsilon}_k^{4s}}{2} \int_{\partial B_\delta(\bar{\varepsilon}_k^{-\tau}\bar{z}_{1k})} (\hat{u}_{1k} + \hat{u}_{2k})\hat{\eta}_k\nu_j dS - \frac{\bar{\varepsilon}_k^{4s}(\mu_{1k} - \mu_{2k})}{2\|\hat{u}_{1k} - \hat{u}_{2k}\|_\infty} \int_{\partial B_\delta(\bar{\varepsilon}_k^{-\tau}\bar{z}_{1k})} \hat{u}_{2k}^2\nu_j dS, \\
B_3 :=& -\frac{b}{4} \int_{\partial B_\delta(\bar{\varepsilon}_k^{-\tau}\bar{z}_{1k})} \left(\hat{u}_{1k}^{\frac{4s+N}{N}} + \hat{u}_{2k}^{\frac{4s+N}{N}}\right) [A\hat{u}_{1k} + (1-A)\hat{u}_{2k}]^{\frac{4s}{N}} \hat{\eta}_k\nu_j dS, \\
B_4 :=& -\left[a_k\bar{\varepsilon}_k^{2s(2-\tau)} + b\left(\frac{b}{2\beta^*}\right)^{\frac{N}{4s}} \bar{\varepsilon}_k^{(\tau-1)(N-4s)} \int_{\mathbb{R}^N} |(-\Delta)^{\frac{s}{2}}\hat{u}_{1k}|^2 dx\right] \\
& \times \int_{\partial''\mathbf{B}_\delta^+(\bar{\varepsilon}_k^{-\tau}\bar{z}_{1k})} t^{1-2s}\left(\frac{\partial\widetilde{\hat{\eta}_k}}{\partial x_j}\frac{\partial\widetilde{\hat{u}_{1k}}}{\partial \nu} + \frac{\partial\widetilde{\hat{\eta}_k}}{\partial \nu}\frac{\partial\widetilde{\hat{u}_{2k}}}{\partial x_j}\right), \\
B_5 :=& \frac{1}{2}\left[a_k\bar{\varepsilon}_k^{2s(2-\tau)} + b\left(\frac{b}{2\beta^*}\right)^{\frac{N}{4s}} \bar{\varepsilon}_k^{(\tau-1)(N-4s)} \int_{\mathbb{R}^N} |(-\Delta)^{\frac{s}{2}}\hat{u}_{1k}|^2 dx\right] \\
& \times \int_{\partial''\mathbf{B}_\delta^+(\bar{\varepsilon}_k^{-\tau}\bar{z}_{1k})} t^{1-2s}\nabla(\widetilde{\hat{u}_{1k}} + \widetilde{\hat{u}_{2k}})\nabla\widetilde{\hat{\eta}_k}\nu_j, \\
B_6 :=& b\left(\frac{b}{2\beta^*}\right)^{\frac{N}{4s}} \bar{\varepsilon}_k^{(\tau-1)(N-4s)} \int_{\mathbb{R}^N} (-\Delta)^{\frac{s}{2}}(\hat{u}_{1k} + \hat{u}_{2k})(-\Delta)^{\frac{s}{2}}\hat{\eta}_k dx \\
& \times \left[\frac{1}{2}\int_{\partial''\mathbf{B}_\delta^+(\bar{\varepsilon}_k^{-\tau}\bar{z}_{1k})} t^{1-2s}|\nabla\widetilde{\hat{u}_{2k}}|^2\nu_j - \int_{\partial''\mathbf{B}_\delta^+(\bar{\varepsilon}_k^{-\tau}\bar{z}_{1k})} t^{1-2s}\frac{\partial\widetilde{\hat{u}_{2k}}}{\partial \nu}\frac{\partial\widetilde{\hat{u}_{2k}}}{\partial x_j}\right].
\end{aligned}
$$

Let us estimate each of term the above one by one. By the fact $\bar{z}_{1k} \to 0$ as $k \to \infty$, it follows from Lemma 9.5 and (9.98) that

$$\begin{aligned}|B_1| \le& C\bar{\varepsilon}_k^{4s}\left(\int_{\partial B_\delta(\bar{\varepsilon}_k^{-\tau}\bar{z}_{1k})}|\bar{\varepsilon}_k^{\tau}x|^2\hat{\eta}_k dS\right)^{\frac{1}{2}}\left(\int_{\partial B_\delta(\bar{\varepsilon}_k^{-\tau}\bar{z}_{1k})}|\bar{\varepsilon}_k^{\tau}x|^2|\hat{u}_{1k}+\hat{u}_{2k}|^2 dS\right)^{\frac{1}{2}}\\ \le& C\bar{\varepsilon}_k^{4s+\frac{(1-\tau)N-4s}{2}}\left(\int_{\partial B_\delta(0)}|\bar{\varepsilon}_k^{\tau}x+\bar{z}_{1k}|^2|\hat{u}_{1k}(x+\bar{\varepsilon}_k^{-\tau}\bar{z}_{1k})+\hat{u}_{2k}(x+\bar{\varepsilon}_k^{-\tau}\bar{z}_{1k})|^2 dS\right)^{\frac{1}{2}}\\ \le& C\bar{\varepsilon}_k^{4s+\frac{(1-\tau)N-4s}{2}}\left(\int_{\partial B_\delta(0)}|\bar{\varepsilon}_k^{\tau}x+\bar{z}_{1k}|^2|\bar{u}_{1k}(\bar{\varepsilon}_k^{\tau-1}x)+\bar{u}_{2k}(\bar{\varepsilon}_k^{\tau-1}x)|^2 dS\right)^{\frac{1}{2}}\\ \le& C\bar{\varepsilon}_k^{4s+\frac{(1-\tau)N-4s}{2}+\tau+(1-\tau)(N+2s)} \quad \text{as } k\to\infty.\end{aligned} \tag{9.134}$$

Using the fact $\mu_{ik}\bar{\varepsilon}_k^{-4s}\to\frac{(N-4s)b}{2N}$ as $k\to\infty$, we deduce from Lemma 9.5 and (9.98) that

$$\begin{aligned}|B_2| \le& C\left(\int_{\partial B_\delta(\bar{\varepsilon}_k^{-\tau}\bar{z}_{1k})}|\hat{u}_{1k}+\hat{u}_{2k}|^2 dS\right)^{\frac{1}{2}}\left(\int_{\partial B_\delta(\bar{\varepsilon}_k^{-\tau}\bar{z}_{1k})}\hat{\eta}_k^2 dS\right)^{\frac{1}{2}}+C\int_{\partial B_\delta(\bar{\varepsilon}_k^{-\tau}\bar{z}_{1k})}|\hat{u}_{2k}|^2 dS\\ \le& C\left(\int_{\partial B_\delta(0)}|\hat{u}_{1k}(x+\bar{\varepsilon}_k^{-\tau}\bar{z}_{1k})+\hat{u}_{2k}(x+\bar{\varepsilon}_k^{-\tau}\bar{z}_{1k})|^2 dS\right)^{\frac{1}{2}}\left(\int_{\partial B_\delta(\bar{\varepsilon}_k^{-\tau}\bar{z}_{1k})}\hat{\eta}_k^2 dS\right)^{\frac{1}{2}}\\ &+C\int_{\partial B_\delta(0)}|\hat{u}_{2k}(x+\bar{\varepsilon}_k^{-\tau}\bar{z}_{1k})|^2 dS\\ \le& C\bar{\varepsilon}_k^{\frac{(1-\tau)N}{2}}\left(\int_{\partial B_\delta(0)}|\bar{u}_{1k}(\bar{\varepsilon}_k^{\tau-1}x)+\bar{u}_{2k}(\bar{\varepsilon}_k^{\tau-1}x)|^2 dS\right)^{\frac{1}{2}}+C\int_{\partial B_\delta(0)}|\bar{u}_{2k}(\bar{\varepsilon}_k^{\tau-1}x)|^2 dS\\ \le& C\bar{\varepsilon}_k^{\frac{(1-\tau)N}{2}+(1-\tau)(N+2s)} \quad \text{as } k\to\infty.\end{aligned} \tag{9.135}$$

Similarly, we also have

$$\begin{aligned}|B_3| \le& C\int_{\partial B_\delta(0)}\left|\bar{u}_{1k}^{\frac{4s+N}{N}}(\bar{\varepsilon}_k^{\tau-1}x)+\bar{u}_{2k}^{\frac{4s+N}{N}}(\bar{\varepsilon}_k^{\tau-1}x)\right||A\bar{u}_{1k}(\bar{\varepsilon}_k^{\tau-1}x)+(1-A)\bar{u}_{2k}(\bar{\varepsilon}_k^{\tau-1}x)|^{\frac{4s}{N}} dS\\ \le& C\bar{\varepsilon}_k^{(1-\tau)(N+2s)\frac{8s+N}{N}} \quad \text{as } k\to\infty.\end{aligned} \tag{9.136}$$

Noticing that $\|\bar{u}_{ik}\|_{H^s(\mathbb{R}^N)}\le C$ as $k\to\infty$ for $i=1,2$, it then follows from (9.130)–(9.132) that

$$\begin{aligned}|B_4| \le& C\bar{\varepsilon}_k^{(1-\tau)2s}\int_{\partial''\mathbf{B}_\delta^+(\bar{\varepsilon}_k^{-\tau}\bar{z}_{1k})}t^{1-2s}\left(\frac{\partial\widetilde{\hat{\eta}_k}}{\partial x_j}\frac{\partial\widetilde{\hat{u}_{1k}}}{\partial\nu}+\frac{\partial\widetilde{\hat{\eta}_k}}{\partial\nu}\frac{\partial\widetilde{\hat{u}_{2k}}}{\partial x_j}\right)\\ \le& C\bar{\varepsilon}_k^{(1-\tau)(2s+2N-2)-\frac{N}{2}}\int_{\partial''\mathbf{B}_\delta^+(\bar{\varepsilon}_k^{-\tau}\bar{z}_{1k})}t^{1-2s}\left(\frac{1}{(1+|x-\bar{\varepsilon}_k^{-\tau}\bar{z}_{1k}|)^{N+2s}}\right)^2\\ \le& C\bar{\varepsilon}_k^{(1-\tau)(2s+2N-2)-\frac{N}{2}} \quad \text{as } k\to\infty.\end{aligned} \tag{9.137}$$

Similarly

$$|B_5| \le C\bar{\varepsilon}_k^{(1-\tau)(2s+2N-2)-\frac{N}{2}} \quad \text{as } k\to\infty. \tag{9.138}$$

Applying the Hölder's inequality, we derive from (9.130)–(9.132) that

$$\begin{aligned}|B_6| \leq & C\bar{\varepsilon}_k^{(1-\tau)(2s+2N-2)} \left(\int_{\mathbb{R}^N} |(-\Delta)^{\frac{s}{2}}(\bar{u}_{1k}+\bar{u}_{2k})|^2 dx\right)^{\frac{1}{2}} \left(\int_{\mathbb{R}^N} |(-\Delta)^{\frac{s}{2}}\bar{\eta}_k|^2 dx\right)^{\frac{1}{2}} \\ & \times \int_{\partial''\mathbf{B}_\delta^+(\bar{\varepsilon}_k^{-\tau}\bar{z}_{1k})} t^{1-2s}\left(\frac{1}{(1+|x-\bar{\varepsilon}_k^{-\tau}\bar{z}_{1k}|)^{N+2s}}\right)^2 \\ \leq & C\|\bar{u}_{1k}+\bar{u}_{2k}\|_{H^s(\mathbb{R}^N)}\|\bar{\eta}_k\|_{H^s(\mathbb{R}^N)}\bar{\varepsilon}_k^{(1-\tau)(2s+2N-2)} \leq C\bar{\varepsilon}_k^{(1-\tau)(2s+2N-2)},\end{aligned} \tag{9.139}$$

as $k \to \infty$. For simplicity, we define

$$\begin{cases} q_1 := (1-\tau)\left(\frac{3N}{2}+2s\right)+\tau+2s, \\ q_2 := (1-\tau)\left(\frac{3N}{2}+2s\right), \\ q_3 := (1-\tau)(N+2s)\frac{8s+N}{N}, \\ q_4 := (1-\tau)(2s+2N-2)-\frac{N}{2}, \\ q_5 := (1-\tau)(2s+2N-2). \end{cases}$$

By (9.94), we can see that $q_1, q_2, q_3, q_4, q_5 > 0$, $0 < q_1 < q_2 < q_3$, $0 < q_4 < q_5$ and $0 < q_4 < q_1$. Then by the fact that $\frac{\bar{z}_{1k}}{\bar{\varepsilon}_k} \to 0$ and $\bar{\eta}_k \to \bar{\eta}_0$ in $C^1(\mathbb{R}^N)$ as $k \to \infty$, we can get from Lemma 9.5, (9.5) and (9.133)–(9.139) that

$$\begin{aligned} & O\left(\bar{\varepsilon}_k^{(1-\tau)(2s+2N-2)-\frac{N}{2}}\right) = \frac{\bar{\varepsilon}_k^{4s}}{2}\int_{B_\delta(\bar{\varepsilon}_k^{-\tau}\bar{z}_{1k})} \frac{\partial|\bar{\varepsilon}_k^{\tau}x|^2}{\partial x_j}(\hat{u}_{1k}+\hat{u}_{2k})\hat{\eta}_k dx = \\ & \frac{\bar{\varepsilon}_k^{4s}}{2}\int_{B_\delta(\bar{\varepsilon}_k^{-\tau}\bar{z}_{1k})} \frac{\partial|\bar{\varepsilon}_k^{\tau}x|^2}{\partial x_j}[\bar{u}_{1k}(\bar{\varepsilon}_k^{\tau-1}x-\bar{\varepsilon}_k^{-1}\bar{z}_{1k})+ \\ & \bar{u}_{2k}(\bar{\varepsilon}_k^{\tau-1}x-\bar{\varepsilon}_k^{-1}\bar{z}_{1k})]\bar{\eta}_k(\bar{\varepsilon}_k^{\tau-1}x-\bar{\varepsilon}_k^{-1}\bar{z}_{1k})dx = \\ & \frac{\bar{\varepsilon}_k^{4s+(1-\tau)N+\tau-1}}{2}\int_{B_{\bar{\varepsilon}_k^{\tau-1}\delta}(0)} \frac{\partial|\varepsilon_k x+z_{1k}|^2}{\partial x_j}(\bar{u}_{1k}+\bar{u}_{2k})\bar{\eta}_k dx = \\ & \bar{\varepsilon}_k^{4s+(1-\tau)N+\tau+1}\int_{B_{\bar{\varepsilon}_k^{\tau-1}\delta}(0)} \left|x_j+\frac{\bar{z}_{1k}^j}{\bar{\varepsilon}_k}\right|(\bar{u}_{1k}+\bar{u}_{2k})\bar{\eta}_k dx = \\ & O\left(\bar{\varepsilon}_k^{4s+(1-\tau)N+\tau+1}\right)(1+o(1))\int_{\mathbb{R}^N} 2x_j\varphi\bar{\eta}_0 dx, \end{aligned} \tag{9.140}$$

where $j = 1, 2, \ldots, N$ and $\bar{z}_{1k}^j = (\bar{z}_{1k}^1, \bar{z}_{1k}^2, \ldots, \bar{z}_{1k}^N) \in \mathbb{R}^N$. Applying (9.94) and the fact that $\frac{N}{4} < s < 1$, we can get that $(1-\tau)(2s+2N-2)-\frac{N}{2} > (1-\tau)N+4s+2 > 4s+(1-\tau)N+\tau+1 > 0$. We thus obtain from (9.140), Lemma 9.7 and Lemma 9.8 that

$$0 = 2\int_{\mathbb{R}^N} x_j\varphi\left[d_0\varphi+\bar{d}_0(x\cdot\nabla\varphi)+\sum_{i=1}^N d_i\frac{\partial\varphi}{\partial x_i}\right]dx = -d_j\int_{\mathbb{R}^N}\varphi^2 dx \quad j = 1, 2, \ldots, N,$$

which shows that $d_j = 0$ for $j = 1, 2, \ldots, N$. □

Lemma 9.11 *Assume that d_0 and $\bar{d}_0$ are defined in (9.109), then $d_0 = \bar{d}_0 = 0$.*

Proof. Multiplying (9.124) by $[X - (\bar{\varepsilon}_k^{-\tau}\bar{z}_{1k}, 0)] \cdot \nabla\widetilde{\hat{u}_{ik}}$, where $i = 1, 2$, and integrating over

$B_\delta(\bar{\varepsilon}_k^{-\tau}\bar{z}_{1k})$, other assumptions are same as before, then we have

$$\begin{aligned}
&-\left[a_k\bar{\varepsilon}_k^{2s(2-\tau)}+b\left(\frac{b}{2\beta^*}\right)^{\frac{N}{4s}}\bar{\varepsilon}_k^{(\tau-1)(N-4s)}\int_{\mathbb{R}^N}|(-\Delta)^{\frac{s}{2}}\hat{u}_{ik}|^2dx\right]\\
&\times\int_{\partial''\mathbf{B}_\delta^+(\bar{\varepsilon}_k^{-\tau}\bar{z}_{1k})}t^{1-2s}\frac{\partial\widetilde{\hat{u}_{ik}}}{\partial\nu}[X-(\bar{\varepsilon}_k^{-\tau}\bar{z}_{1k},0)]\cdot\nabla\widetilde{\hat{u}_{ik}}\\
&+\frac{1}{2}\left[a_k\bar{\varepsilon}_k^{2s(2-\tau)}+b\left(\frac{b}{2\beta^*}\right)^{\frac{N}{4s}}\bar{\varepsilon}_k^{(\tau-1)(N-4s)}\int_{\mathbb{R}^N}|(-\Delta)^{\frac{s}{2}}\hat{u}_{ik}|^2dx\right]\\
&\times\int_{\partial''\mathbf{B}_\delta^+(\bar{\varepsilon}_k^{-\tau}\bar{z}_{1k})}t^{1-2s}|\nabla\widetilde{\hat{u}_{ik}}|^2[X-(\bar{\varepsilon}_k^{-\tau}\bar{z}_{1k},0)]\cdot\nu\\
&+\frac{2s-N}{2}\left[a_k\bar{\varepsilon}_k^{2s(2-\tau)}+b\left(\frac{b}{2\beta^*}\right)^{\frac{N}{4s}}\bar{\varepsilon}_k^{(\tau-1)(N-4s)}\int_{\mathbb{R}^N}|(-\Delta)^{\frac{s}{2}}\hat{u}_{ik}|^2dx\right]\\
&\times\int_{\partial\mathbf{B}_\delta^+(\bar{\varepsilon}_k^{-\tau}\bar{z}_{1k})}t^{1-2s}\frac{\partial\widetilde{\hat{u}_{ik}}}{\partial\nu}\widetilde{\hat{u}_{ik}}\\
=&\int_{B_\delta(\bar{\varepsilon}_k^{-\tau}\bar{z}_{1k})}\left[-\bar{\varepsilon}_k^{4s}|\bar{\varepsilon}_k^{\tau}x|^2\hat{u}_{ik}+\mu_{ik}\bar{\varepsilon}_k^{4s}\hat{u}_{ik}+\frac{b}{2}\hat{u}_{ik}^{\frac{8s}{N}+1}\right](x-\bar{\varepsilon}_k^{-\tau}\bar{z}_{1k})\cdot\nabla\hat{u}_{ik}dx. \qquad (9.141)
\end{aligned}$$

By direct computation, we have

$$\begin{aligned}
&\frac{2s-N}{2}\left[a_k\bar{\varepsilon}_k^{2s(2-\tau)}+b\left(\frac{b}{2\beta^*}\right)^{\frac{N}{4s}}\bar{\varepsilon}_k^{(\tau-1)(N-4s)}\int_{\mathbb{R}^N}|(-\Delta)^{\frac{s}{2}}\hat{u}_{ik}|^2dx\right]\\
&\times\int_{\partial\mathbf{B}_\delta^+(\bar{\varepsilon}_k^{-\tau}\bar{z}_{1k})}t^{1-2s}\frac{\partial\widetilde{\hat{u}_{ik}}}{\partial\nu}\widetilde{\hat{u}_{ik}}\\
=&\frac{2s-N}{2}\left[a_k\bar{\varepsilon}_k^{2s(2-\tau)}+b\left(\frac{b}{2\beta^*}\right)^{\frac{N}{4s}}\bar{\varepsilon}_k^{(\tau-1)(N-4s)}\int_{\mathbb{R}^N}|(-\Delta)^{\frac{s}{2}}\hat{u}_{ik}|^2dx\right] \qquad (9.142)\\
&\times\int_{\partial''\mathbf{B}_\delta^+(\bar{\varepsilon}_k^{-\tau}\bar{z}_{1k})}t^{1-2s}\frac{\partial\widetilde{\hat{u}_{ik}}}{\partial\nu}\widetilde{\hat{u}_{ik}}\\
&+\frac{2s-N}{2}\int_{B_\delta(\bar{\varepsilon}_k^{-\tau}\bar{z}_{1k})}\left[-\bar{\varepsilon}_k^{4s}|\bar{\varepsilon}_k^{\tau}x|^2\hat{u}_{ik}+\mu_{ik}\bar{\varepsilon}_k^{4s}\hat{u}_{ik}+\frac{b}{2}\hat{u}_{ik}^{\frac{8s}{N}+1}\right]\hat{u}_{ik}dx.
\end{aligned}$$

Integrating by parts, we see that

$$\begin{aligned}
&\int_{B_\delta(\bar{\varepsilon}_k^{-\tau}\bar{z}_{1k})}\left[-\bar{\varepsilon}_k^{4s}|\bar{\varepsilon}_k^{\tau}x|^2\hat{u}_{ik}+\mu_{ik}\bar{\varepsilon}_k^{4s}\hat{u}_{ik}+\frac{b}{2}\hat{u}_{ik}^{\frac{8s}{N}+1}\right](x-\bar{\varepsilon}_k^{-\tau}\bar{z}_{1k})\cdot\nabla\hat{u}_{ik}dx\\
=&-\frac{\bar{\varepsilon}_k^{4s}}{2}\int_{B_\delta(\bar{\varepsilon}_k^{-\tau}\bar{z}_{1k})}|\bar{\varepsilon}_k^{\tau}x|^2(x-\bar{\varepsilon}_k^{-\tau}\bar{z}_{1k})\cdot\nabla\hat{u}_{ik}^2dx\\
&+\frac{\mu_{ik}\bar{\varepsilon}_k^{4s}}{2}\int_{B_\delta(\bar{\varepsilon}_k^{-\tau}\bar{z}_{1k})}(x-\bar{\varepsilon}_k^{-\tau}\bar{z}_{1k})\cdot\nabla\hat{u}_{ik}^2dx \qquad (9.143)\\
&+\frac{bN}{2(8s+2N)}\int_{B_\delta(\bar{\varepsilon}_k^{-\tau}\bar{z}_{1k})}(x-\bar{\varepsilon}_k^{-\tau}\bar{z}_{1k})\cdot\nabla\hat{u}_{ik}^{\frac{8s+2N}{N}}dx,
\end{aligned}$$

$$\begin{aligned}
&\int_{B_\delta(\bar\varepsilon_k^{-\tau}\bar z_{1k})} |\bar\varepsilon_k^\tau x|^2 (x-\bar\varepsilon_k^{-\tau}\bar z_{1k})\cdot\nabla\hat u_{ik}^2 dx \\
&= \int_{\partial B_\delta(\bar\varepsilon_k^{-\tau}\bar z_{1k})} |\bar\varepsilon_k^\tau x|^2 (x-\bar\varepsilon_k^{-\tau}\bar z_{1k})\cdot\nu\hat u_{ik}^2 dS \\
&\quad - \int_{B_\delta(\bar\varepsilon_k^{-\tau}\bar z_{1k})} \left[\nabla|\bar\varepsilon_k^\tau x|^2\cdot(x-\bar\varepsilon_k^{-\tau}\bar z_{1k}) + N|\bar\varepsilon_k^\tau x|^2\right]\hat u_{ik}^2 dx,
\end{aligned} \tag{9.144}$$

$$\begin{aligned}
&\int_{B_\delta(\bar\varepsilon_k^{-\tau}\bar z_{1k})} (x-\bar\varepsilon_k^{-\tau}\bar z_{1k})\cdot\nabla\hat u_{ik}^2 dx \\
&= \int_{\partial B_\delta(\bar\varepsilon_k^{-\tau}\bar z_{1k})} (x-\bar\varepsilon_k^{-\tau}\bar z_{1k})\cdot\nu\hat u_{ik}^2 dS - N\int_{B_\delta(\bar\varepsilon_k^{-\tau}\bar z_{1k})} \hat u_{ik}^2 dx,
\end{aligned} \tag{9.145}$$

$$\begin{aligned}
&\int_{B_\delta(\bar\varepsilon_k^{-\tau}\bar z_{1k})} (x-\bar\varepsilon_k^{-\tau}\bar z_{1k})\cdot\nabla\hat u_{ik}^{\frac{8s+2N}{N}} dx \\
&= \int_{\partial B_\delta(\bar\varepsilon_k^{-\tau}\bar z_{1k})} (x-\bar\varepsilon_k^{-\tau}\bar z_{1k})\cdot\nu\hat u_{ik}^{\frac{8s+2N}{N}} dS - N\int_{B_\delta(\bar\varepsilon_k^{-\tau}\bar z_{1k})} \hat u_{ik}^{\frac{8s+2N}{N}} dx.
\end{aligned} \tag{9.146}$$

Following (9.141)–(9.146), we derive that

$$\begin{aligned}
&-\left[a_k\bar\varepsilon_k^{2s(2-\tau)} + b\left(\frac{b}{2\beta^*}\right)^{\frac{N}{4s}}\bar\varepsilon_k^{(\tau-1)(N-4s)}\int_{\mathbb{R}^N}|(-\Delta)^{\frac{s}{2}}\hat u_{ik}|^2 dx\right] \\
&\times\int_{\partial''\mathbf{B}_\delta^+(\bar\varepsilon_k^{-\tau}\bar z_{1k})} t^{1-2s}\frac{\partial\widetilde{\hat u_{ik}}}{\partial\nu}[X-(\bar\varepsilon_k^{-\tau}\bar z_{1k},0)]\cdot\nabla\widetilde{\hat u_{ik}} \\
&+\frac{1}{2}\left[a_k\bar\varepsilon_k^{2s(2-\tau)} + b\left(\frac{b}{2\beta^*}\right)^{\frac{N}{4s}}\bar\varepsilon_k^{(\tau-1)(N-4s)}\int_{\mathbb{R}^N}|(-\Delta)^{\frac{s}{2}}\hat u_{ik}|^2 dx\right] \\
&\times\int_{\partial''\mathbf{B}_\delta^+(\bar\varepsilon_k^{-\tau}\bar z_{1k})} t^{1-2s}|\nabla\widetilde{\hat u_{ik}}|^2[X-(\bar\varepsilon_k^{-\tau}\bar z_{1k},0)]\cdot\nu \\
&+\frac{2s-N}{2}\left[a_k\bar\varepsilon_k^{2s(2-\tau)} + b\left(\frac{b}{2\beta^*}\right)^{\frac{N}{4s}}\bar\varepsilon_k^{(\tau-1)(N-4s)}\int_{\mathbb{R}^N}|(-\Delta)^{\frac{s}{2}}\hat u_{ik}|^2 dx\right] \\
&\times\int_{\partial''\mathbf{B}_\delta^+(\bar\varepsilon_k^{-\tau}\bar z_{1k})} t^{1-2s}\frac{\partial\widetilde{\hat u_{ik}}}{\partial\nu}\widetilde{\hat u_{ik}} \\
=&-\frac{\bar\varepsilon_k^{4s}}{2}\int_{\partial B_\delta(\bar\varepsilon_k^{-\tau}\bar z_{1k})}|\bar\varepsilon_k^\tau x|^2(x-\bar\varepsilon_k^{-\tau}\bar z_{1k})\cdot\nu\hat u_{ik}^2 dS \\
&+\frac{\bar\varepsilon_k^{4s}}{2}\int_{B_\delta(\bar\varepsilon_k^{-\tau}\bar z_{1k})}\nabla|\bar\varepsilon_k^\tau x|^2\cdot(x-\bar\varepsilon_k^{-\tau}\bar z_{1k})\hat u_{ik}^2 dx - \frac{b(8s^2-2sN)}{16s+4N}\int_{B_\delta(\bar\varepsilon_k^{-\tau}\bar z_{1k})}\hat u_{ik}^{\frac{8s+2N}{N}} dx \\
&+s\bar\varepsilon_k^{4s}\int_{B_\delta(\bar\varepsilon_k^{-\tau}\bar z_{1k})}|\bar\varepsilon_k^\tau x|^2\hat u_{ik}^2 dx + \frac{\mu_{ik}\bar\varepsilon_k^{4s}}{2}\int_{\partial B_\delta(\bar\varepsilon_k^{-\tau}\bar z_{1k})}(x-\bar\varepsilon_k^{-\tau}\bar z_{1k})\cdot\nu\hat u_{ik}^2 dS \\
&-s\mu_{ik}\bar\varepsilon_k^{4s}\int_{B_\delta(\bar\varepsilon_k^{-\tau}\bar z_{1k})}\hat u_{ik}^2 dx + \frac{bN}{2(8s+2N)}\int_{\partial B_\delta(\bar\varepsilon_k^{-\tau}\bar z_{1k})}(x-\bar\varepsilon_k^{-\tau}\bar z_{1k})\cdot\nu\hat u_{ik}^{\frac{8s+2N}{N}} dS.
\end{aligned} \tag{9.147}$$

Moreover, we deduce from (9.112) that

$$\begin{aligned}\mu_{ik}\bar{\varepsilon}_k^{4s}\int_{\mathbb{R}^N}\hat{u}_{ik}^2dx=&\left(\frac{2\beta^*}{b}\right)^{\frac{N}{4s}}\bar{\varepsilon}_k^{(1-\tau)N+4s}e(a_k,\beta^*)\\&+\frac{b}{2}\left(\frac{2\beta^*}{b}\right)^{-\frac{N}{4s}}\bar{\varepsilon}_k^{(\tau-1)(N-4s)}\left(\int_{\mathbb{R}^N}|(-\Delta)^{\frac{s}{2}}\hat{u}_{ik}|^2dx\right)^2-\frac{2sb}{N+4s}\int_{\mathbb{R}^N}\hat{u}_{ik}^{\frac{8s+2N}{N}}dx,\end{aligned}\tag{9.148}$$

from which and (9.147), we conclude that

$$\begin{aligned}&-\left[a_k\bar{\varepsilon}_k^{2s(2-\tau)}+b\left(\frac{b}{2\beta^*}\right)^{\frac{N}{4s}}\bar{\varepsilon}_k^{(\tau-1)(N-4s)}\int_{\mathbb{R}^N}|(-\Delta)^{\frac{s}{2}}\hat{u}_{ik}|^2dx\right]\\&\times\int_{\partial''\mathbf{B}_\delta^+(\bar{\varepsilon}_k^{-\tau}\bar{z}_{1k})}t^{1-2s}\frac{\partial\widetilde{\hat{u}_{ik}}}{\partial\nu}[X-(\bar{\varepsilon}_k^{-\tau}\bar{z}_{1k},0)]\cdot\nabla\widetilde{\hat{u}_{ik}}\\&+\frac{1}{2}\left[a_k\bar{\varepsilon}_k^{2s(2-\tau)}+b\left(\frac{b}{2\beta^*}\right)^{\frac{N}{4s}}\bar{\varepsilon}_k^{(\tau-1)(N-4s)}\int_{\mathbb{R}^N}|(-\Delta)^{\frac{s}{2}}\hat{u}_{ik}|^2dx\right]]\\&\times\int_{\partial''\mathbf{B}_\delta^+(\bar{\varepsilon}_k^{-\tau}\bar{z}_{1k})}t^{1-2s}|\nabla\widetilde{\hat{u}_{ik}}|^2[X-(\bar{\varepsilon}_k^{-\tau}\bar{z}_{1k},0)]\cdot\nu\\&+\frac{2s-N}{2}\left[a_k\bar{\varepsilon}_k^{2s(2-\tau)}+b\left(\frac{b}{2\beta^*}\right)^{\frac{N}{4s}}\bar{\varepsilon}_k^{(\tau-1)(N-4s)}\int_{\mathbb{R}^N}|(-\Delta)^{\frac{s}{2}}\hat{u}_{ik}|^2dx\right]\\&\times\int_{\partial''\mathbf{B}_\delta^+(\bar{\varepsilon}_k^{-\tau}\bar{z}_{1k})}t^{1-2s}\frac{\partial\widetilde{\hat{u}_{ik}}}{\partial\nu}\widetilde{\hat{u}_{ik}}+\frac{\bar{\varepsilon}_k^{4s}}{2}\int_{\partial B_\delta(\bar{\varepsilon}_k^{-\tau}\bar{z}_{1k})}|\bar{\varepsilon}_k^{\tau}x|^2(x-\bar{\varepsilon}_k^{-\tau}\bar{z}_{1k})\cdot\nu\hat{u}_{ik}^2dS\\&-\frac{\bar{\varepsilon}_k^{4s}}{2}\int_{B_\delta(\bar{\varepsilon}_k^{-\tau}\bar{z}_{1k})}\nabla|\bar{\varepsilon}_k^{\tau}x|^2\cdot(x-\bar{\varepsilon}_k^{-\tau}\bar{z}_{1k})\hat{u}_{ik}^2dx-s\bar{\varepsilon}_k^{4s}\int_{B_\delta(\bar{\varepsilon}_k^{-\tau}\bar{z}_{1k})}|\bar{\varepsilon}_k^{\tau}x|^2\hat{u}_{ik}^2dx\\=&\frac{\mu_{ik}\bar{\varepsilon}_k^{4s}}{2}\int_{\partial B_\delta(\bar{\varepsilon}_k^{-\tau}\bar{z}_{1k})}(x-\bar{\varepsilon}_k^{-\tau}\bar{z}_{1k})\cdot\nu\hat{u}_{ik}^2dS+s\mu_{ik}\bar{\varepsilon}_k^{4s}\int_{\mathbb{R}^N\setminus B_\delta(\bar{\varepsilon}_k^{-\tau}\bar{z}_{1k})}\hat{u}_{ik}^2dx\\&+\frac{bN}{2(8s+2N)}\int_{\partial B_\delta(\bar{\varepsilon}_k^{-\tau}\bar{z}_{1k})}(x-\bar{\varepsilon}_k^{-\tau}\bar{z}_{1k})\cdot\nu\hat{u}_{ik}^{\frac{8s+2N}{N}}dS\\&+\frac{b(8s^2-2sN)}{16s+4N}\int_{\mathbb{R}^N\setminus B_\delta(\bar{\varepsilon}_k^{-\tau}\bar{z}_{1k})}\hat{u}_{ik}^{\frac{8s+2N}{N}}dx-s\left(\frac{2\beta^*}{b}\right)^{\frac{N}{4s}}\bar{\varepsilon}_k^{(\tau-1)(N-4s)}e(a_k,\beta^*)\\&-\frac{sb}{2}\left(\frac{2\beta^*}{b}\right)^{-\frac{N}{4s}}\bar{\varepsilon}_k^{(\tau-1)(N-4s)}\left(\int_{\mathbb{R}^N}|(-\Delta)^{\frac{s}{2}}\hat{u}_{ik}|^2dx\right)^2+\frac{Nsb}{8s+2N}\int_{\mathbb{R}^N}\hat{u}_{ik}^{\frac{8s+2N}{N}}dx.\end{aligned}$$

This shows there exists $A\in(0,1)$ such that

$$\begin{aligned}&\frac{sb}{2}\left(\frac{2\beta^*}{b}\right)^{-\frac{N}{4s}}\bar{\varepsilon}_k^{(\tau-1)(N-4s)}\int_{\mathbb{R}^N}(|(-\Delta)^{\frac{s}{2}}\hat{u}_{1k}|^2+|(-\Delta)^{\frac{s}{2}}\hat{u}_{2k}|^2)dx\\&\times\int_{\mathbb{R}^N}|(-\Delta)^{\frac{s}{2}}(\hat{u}_{1k}+\hat{u}_{2k})(-\Delta)^{\frac{s}{2}}\hat{\eta}_kdx\\&-\frac{sb}{2}\int_{\mathbb{R}^N}\left(\hat{u}_{1k}^{\frac{4s+N}{N}}+\hat{u}_{2k}^{\frac{4s+N}{N}}\right)[A\hat{u}_{1k}+(1-A)\hat{u}_{2k}]^{\frac{4s}{N}}\hat{\eta}_kdx\\=:&D_1+D_2+D_3+D_4,\end{aligned}\tag{9.149}$$

where

$$D_1 := \frac{\left[a_k \bar{\varepsilon}_k^{2s(2-\tau)} + b\left(\frac{b}{2\beta^*}\right)^{\frac{N}{4s}} \bar{\varepsilon}_k^{(\tau-1)(N-4s)} \int_{\mathbb{R}^N} |(-\Delta)^{\frac{s}{2}} \hat{u}_{2k}|^2 dx\right] F_2}{\|\hat{u}_{1k} - \hat{u}_{2k}\|_\infty}$$
$$- \frac{\left[a_k \bar{\varepsilon}_k^{2s(2-\tau)} + b\left(\frac{b}{2\beta^*}\right)^{\frac{N}{4s}} \bar{\varepsilon}_k^{(\tau-1)(N-4s)} \int_{\mathbb{R}^N} |(-\Delta)^{\frac{s}{2}} \hat{u}_{1k}|^2 dx\right] F_1}{\|\hat{u}_{1k} - \hat{u}_{2k}\|_\infty}$$
$$- \frac{\bar{\varepsilon}_k^{4s}}{2} \int_{\partial B_\delta(\bar{\varepsilon}_k^{-\tau} \bar{z}_{1k})} |\bar{\varepsilon}_k^\tau x|^2 (x - \bar{\varepsilon}_k^{-\tau} \bar{z}_{1k}) \cdot \nu (\hat{u}_{2k} + \hat{u}_{1k}) \hat{\eta}_k dS$$
$$+ \frac{b(8s^2 - 2sN)}{4N} \int_{\mathbb{R}^N \setminus B_\delta(\bar{\varepsilon}_k^{-\tau} \bar{z}_{1k})} \left(\hat{u}_{1k}^{\frac{4s+N}{N}} + \hat{u}_{2k}^{\frac{4s+N}{N}}\right) [A\hat{u}_{1k} + (1-A)\hat{u}_{2k}]^{\frac{4s}{N}} \hat{\eta}_k dx$$
$$+ \frac{\mu_{1k} \bar{\varepsilon}_k^{4s}}{2} \int_{\partial B_\delta(\bar{\varepsilon}_k^{-\tau} \bar{z}_{1k})} (x - \bar{\varepsilon}_k^{-\tau} \bar{z}_{1k}) \cdot \nu (\hat{u}_{2k} + \hat{u}_{1k}) \hat{\eta}_k dS$$
$$+ \frac{\bar{\varepsilon}_k^{4s} (\mu_{1k} - \mu_{2k})}{2\|\hat{u}_{1k} - \hat{u}_{2k}\|_\infty} \int_{\partial B_\delta(\bar{\varepsilon}_k^{-\tau} \bar{z}_{1k})} (x - \bar{\varepsilon}_k^{-\tau} \bar{z}_{1k}) \cdot \nu \hat{u}_{2k}^2 dS$$
$$+ s\mu_{1k} \bar{\varepsilon}_k^{4s} \int_{\mathbb{R}^N \setminus B_\delta(\bar{\varepsilon}_k^{-\tau} \bar{z}_{1k})} (\hat{u}_{2k} + \hat{u}_{1k}) \hat{\eta}_k dx + \frac{s\bar{\varepsilon}_k^{4s} (\mu_{1k} - \mu_{2k})}{\|\hat{u}_{1k} - \hat{u}_{2k}\|_\infty} \int_{\mathbb{R}^N \setminus B_\delta(\bar{\varepsilon}_k^{-\tau} \bar{z}_{1k})} \hat{u}_{2k}^2 dx$$
$$+ \frac{b}{4} \int_{\partial B_\delta(\bar{\varepsilon}_k^{-\tau} \bar{z}_{1k})} (x - \bar{\varepsilon}_k^{-\tau} \bar{z}_{1k}) \cdot \nu \left(\hat{u}_{1k}^{\frac{4s+N}{N}} + \hat{u}_{2k}^{\frac{4s+N}{N}}\right) [A\hat{u}_{1k} + (1-A)\hat{u}_{2k}]^{\frac{4s}{N}} \hat{\eta}_k dS,$$

$$F_i := - \int_{\partial'' \mathbf{B}_\delta^+(\bar{\varepsilon}_k^{-\tau} \bar{z}_{1k})} t^{1-2s} \frac{\partial \widetilde{\hat{u}_{ik}}}{\partial \nu} [X - (\bar{\varepsilon}_k^{-\tau} \bar{z}_{1k}, 0)] \cdot \nabla \widetilde{\hat{u}_{ik}}$$
$$+ \frac{1}{2} \int_{\partial'' \mathbf{B}_\delta^+(\bar{\varepsilon}_k^{-\tau} \bar{z}_{1k})} t^{1-2s} |\nabla \widetilde{\hat{u}_{ik}}|^2 [X - (\varepsilon_k{}^{\tau} \bar{z}_{1k}, 0)] \cdot \nu + \frac{2s - N}{2} \int_{\partial'' \mathbf{B}_\delta^+(\bar{\varepsilon}_k^{-\tau} \bar{z}_{1k})} t^{1-2s} \frac{\partial \widetilde{\hat{u}_{ik}}}{\partial \nu} \bar{\hat{u}}_{ik},$$

$$D_2 := \frac{\bar{\varepsilon}_k^{4s}}{2} \int_{B_\delta(\bar{\varepsilon}_k^{-\tau} \bar{z}_{1k})} [\nabla |\bar{\varepsilon}_k^\tau x|^2 \cdot x] (\hat{u}_{2k} + \hat{u}_{1k}) \hat{\eta}_k dx,$$

$$D_3 := -\frac{\bar{\varepsilon}_k^{4s}}{2} \int_{B_\delta(\bar{\varepsilon}_k^{-\tau} \bar{z}_{1k})} [\nabla |\bar{\varepsilon}_k^\tau x|^2 \cdot (\bar{\varepsilon}_k^{-\tau} \bar{z}_{1k})] (\hat{u}_{2k} + \hat{u}_{1k}) \hat{\eta}_k dx,$$

$$D_4 := s\bar{\varepsilon}_k^{4s} \int_{B_\delta(\bar{\varepsilon}_k^{-\tau} \bar{z}_{1k})} |\bar{\varepsilon}_k^\tau x|^2 (\hat{u}_{2k} + \hat{u}_{1k}) \hat{\eta}_k dx.$$

The same arguments of $|B_1| - |B_6|$ give that

$$D_1 = O\left(\bar{\varepsilon}_k^{(1-\tau)(2s+2N-2) - \frac{N}{2}}\right) \quad \text{as } k \to \infty. \tag{9.150}$$

By the fact $\bar{\varepsilon}_k^{-\tau} \bar{z}_{1k} \to 0$ as $k \to \infty$, we conclude from (9.140) that

$$\frac{\bar{\varepsilon}_k^{4s}}{2} \int_{B_\delta(\bar{\varepsilon}_k^{-\tau} \bar{z}_{1k})} [\nabla |\bar{\varepsilon}_k^\tau x|^2 \cdot (\bar{\varepsilon}_k^{-\tau} \bar{z}_{1k})] (\hat{u}_{2k} + \hat{u}_{1k}) \hat{\eta}_k dx$$
$$= \frac{\bar{\varepsilon}_k^{4s}}{2} \sum_{i=1}^N \bar{\varepsilon}_k^{-\tau} \bar{z}_{1k}^{(i)} \int_{B_\delta(\bar{\varepsilon}_k^{-\tau} \bar{z}_{1k})} \frac{\partial |\bar{\varepsilon}_k^\tau x|^2}{\partial x_i} (\hat{u}_{2k} + \hat{u}_{1k}) \hat{\eta}_k dx = O\left(\bar{\varepsilon}_k^{(1-\tau)(2s+2N-2) - \frac{N}{2}}\right) \quad \text{as } k \to \infty,$$

which shows that

$$D_3 = O\left(\bar{\varepsilon}_k^{(1-\tau)(2s+2N-2) - \frac{N}{2}}\right)) \quad \text{as } k \to \infty. \tag{9.151}$$

By the fact that $\nabla|\bar{\varepsilon}_k^\tau x|^2\cdot x=2|\bar{\varepsilon}_k^\tau x|^2$ and $\frac{\bar{z}_{1k}}{\bar{\varepsilon}_k}\to 0$ as $k\to\infty$, we deduce from Lemma 9.5 and Lemma 9.7 that

$$\begin{aligned}
D_2=&\bar{\varepsilon}_k^{4s+2\tau}\int_{B_\delta(\bar{\varepsilon}_k^{-\tau}\bar{z}_{1k})}|x|^2(\hat{u}_{2k}+\hat{u}_{1k})\hat{\eta}_k dx\\
=&\bar{\varepsilon}_k^{4s+2\tau}\int_{B_\delta(\bar{\varepsilon}_k^{-\tau}\bar{z}_{1k})}|x|^2\left(\bar{u}_{1k}(\bar{\varepsilon}_k^{\tau-1}x-\bar{\varepsilon}_k^{-1}\bar{z}_{1k})+\bar{u}_{2k}(\bar{\varepsilon}_k^{\tau-1}x-\bar{\varepsilon}_k^{-1}\bar{z}_{1k})\right)\bar{\eta}_k(\bar{\varepsilon}_k^{\tau-1}x-\bar{\varepsilon}_k^{-1}\bar{z}_{1k})dx\\
=&\bar{\varepsilon}_k^{4s+2+(1-\tau)N}\int_{B_{\bar{\varepsilon}_k^{\tau-1}\delta}(0)}\left|x+\frac{\bar{z}_{1k}}{\bar{\varepsilon}_k}\right|^2(\bar{u}_{1k}+\bar{u}_{2k})\bar{\eta}_k dx\\
=&2(1+o(1))\bar{\varepsilon}_k^{4s+2+(1-\tau)N}\int_{\mathbb{R}^N}|x|^2\varphi\bar{\eta}_0 dx
\end{aligned}\tag{9.152}$$

and

$$\begin{aligned}
D_4=&s\bar{\varepsilon}_k^{4s}\int_{B_\delta(\bar{\varepsilon}_k^{-\tau}\bar{z}_{1k})}|\bar{\varepsilon}_k^\tau x|^2\left(\bar{u}_{1k}(\bar{\varepsilon}_k^{\tau-1}x-\bar{\varepsilon}_k^{-1}\bar{z}_{1k})+\bar{u}_{2k}(\bar{\varepsilon}_k^{\tau-1}x-\bar{\varepsilon}_k^{-1}\bar{z}_{1k})\right)\bar{\eta}_k(\bar{\varepsilon}_k^{\tau-1}x-\bar{\varepsilon}_k^{-1}\bar{z}_{1k})dx\\
=&s\bar{\varepsilon}_k^{4s+2+(1-\tau)N}\int_{B_{\bar{\varepsilon}_k^{\tau-1}\delta}(0)}\left|x+\frac{\bar{z}_{1k}}{\bar{\varepsilon}_k}\right|^2(\bar{u}_{1k}+\bar{u}_{2k})\bar{\eta}_k dx\\
=&2s(1+o(1))\bar{\varepsilon}_k^{4s+2+(1-\tau)N}\int_{\mathbb{R}^N}|x|^2\varphi\bar{\eta}_0 dx.
\end{aligned}\tag{9.153}$$

Using (9.4) and Lemma 9.7, we conclude that

$$\begin{aligned}
&\frac{bs}{2}\left(\frac{2\beta^*}{b}\right)^{-\frac{N}{4s}}\bar{\varepsilon}_k^{(N-4s)(\tau-1)}\int_{\mathbb{R}^N}(|(-\Delta)^{\frac{s}{2}}\hat{u}_{1k}|^2+|(-\Delta)^{\frac{s}{2}}\hat{u}_{2k}|^2)dx\int_{\mathbb{R}^N}(-\Delta)^{\frac{s}{2}}(\hat{u}_{1k}+\hat{u}_{2k})(-\Delta)^{\frac{s}{2}}\hat{\eta}_k dx\\
=&\frac{bs}{2}\left(\frac{2\beta^*}{b}\right)^{-\frac{N}{4s}}\bar{\varepsilon}_k^{(N-4s)(\tau-1)+2(1-\tau)(N-2s)}\int_{\mathbb{R}^N}(|(-\Delta)^{\frac{s}{2}}\bar{u}_{1k}|^2+|(-\Delta)^{\frac{s}{2}}\bar{u}_{2k}|^2)dx\\
&\times\int_{\mathbb{R}^N}(-\Delta)^{\frac{s}{2}}(\bar{u}_{1k}+\bar{u}_{2k})(-\Delta)^{\frac{s}{2}}\bar{\eta}_k dx\\
=&2bs(1+o(1))\bar{\varepsilon}_k^{(1-\tau)N}\int_{\mathbb{R}^N}(-\Delta)^{\frac{s}{2}}\varphi(-\Delta)^{\frac{s}{2}}\bar{\eta}_0 dx
\end{aligned}\tag{9.154}$$

and

$$\begin{aligned}
&\frac{bs}{2}\int_{\mathbb{R}^N}\left(\hat{u}_{1k}^{\frac{4s+N}{N}}+\hat{u}_{2k}^{\frac{4s+N}{N}}\right)[A\hat{u}_{1k}+(1-A)\hat{u}_{2k}]^{\frac{4s}{N}}\hat{\eta}_k dx\\
=&\frac{bs}{2}\bar{\varepsilon}_k^{(1-\tau)N}\int_{\mathbb{R}^N}\left(\bar{u}_{1k}^{\frac{4s+N}{N}}+\bar{u}_{2k}^{\frac{4s+N}{N}}\right)[A\bar{u}_{1k}+(1-A)\bar{u}_{2k}]^{\frac{4s}{N}}\bar{\eta}_k dx\\
=&bs(1+o(1))\bar{\varepsilon}_k^{(1-\tau)N}\int_{\mathbb{R}^N}\varphi^{\frac{8s+N}{N}}\bar{\eta}_0 dx.
\end{aligned}\tag{9.155}$$

Following (9.121)–(9.123) and (9.149)–(9.155), we have

$$\begin{aligned}
&2bs(1+o(1))\bar{\varepsilon}_k^{(1-\tau)N}\left[\left(\frac{2\beta^*}{b}\right)^{\frac{4s}{N}}d_0+(s-\frac{N}{2})\left(\frac{2\beta^*}{b}\right)^{\frac{N}{4s}}\bar{d}_0\right]\\
&-bs(1+o(1))\bar{\varepsilon}_k^{(1-\tau)N}\left[\frac{N+4s}{N}\left(\frac{2\beta^*}{b}\right)^{\frac{N}{4s}}d_0-\frac{N}{2}\left(\frac{2\beta^*}{b}\right)^{\frac{N}{4s}}\bar{d}_0\right]\\
=&O\left(\bar{\varepsilon}_k^{(1-\tau)(2s+2N-2)-\frac{N}{2}}\right)+2(1+s)(1+o(1))\bar{\varepsilon}_k^{(1-\tau)N+4s+2}\int_{\mathbb{R}^N}|x|^2\varphi\bar{\eta}_0 dx.
\end{aligned}\tag{9.156}$$

By (9.94), we know that $(1-\tau)(2s+2N-2)-\frac{N}{2}>(1-\tau)N+4s+2>0$. We thus derive from Lemma 9.8 and (9.156) that

$$\int_{\mathbb{R}^N}|x|^2\varphi\bar{\eta}_0dx=0,$$

from which and Lemma 9.7 and Lemma 9.10, we obtain that

$$\begin{aligned}0&=\int_{\mathbb{R}^N}|x|^2\varphi\left[\frac{N}{2}\bar{d}_0\varphi+\bar{d}_0(x\cdot\varphi)\right]dx=\frac{\bar{d}_0}{2}\int_{\mathbb{R}^N}[N|x|^2\varphi^2+|x|^2(x\cdot\nabla\varphi^2)]dx\\&=-\bar{d}_0\int_{\mathbb{R}^N}|x|^2\varphi^2dx,\end{aligned}$$

which shows that $\bar{d}_0=0$, thus, $d_0=0$. □

Proof of Theorem 9.3: Let q_k be a point satisfying $|\bar{\eta}_k(q_k)|=\|\bar{\eta}_k\|_\infty=1$, we know that $|q_k|\le C$ uniformly in k. Then Lemma 9.7 implies that $\bar{\eta}_0\not\equiv0$ on $\mathbb{R}^N$. From Lemma 9.7, Lemma 9.10 and Lemma 9.11, we know that $\bar{\eta}_0\equiv0$. Thus, our assumption that $u_{1k}\not\equiv u_{2k}$ is false. This completes the proof of Theorem 9.3. □

9.5 Historical comments

The first feature of the problem studied in this chapter is the appearance of the fractional Laplace operator. Such type of operators, which has special properties and the connection with the Fourier transform, arise in a quite natural way in many different applications, such as optimization, finance, phase transitions, stratified materials, anomalous diffusion, crystal dislocation, soft thin films, semipermeable membranes, flame propagation, conservation laws and water waves.

When $s=1$, the classical Kirchhoff type equation is analogous to the stationary case of equations that arise in the study of string or membrane vibrations, namely,

$$u_{tt}-M\left(\int_{\mathbb{R}^N}|\nabla u|^2dx\right)\Delta u=h(x,u)\quad\text{in }\mathbb{R}^N,$$

where u denotes the displacement, and $h(x,u)$ the external force, such equations were proposed by Kirchhoff [88] in 1883 to describe the transversal oscillations of a stretched string, furthermore, taking into account the subsequent change in string length caused by oscillations. The most typical feature of such an equation is the appearance of the term $\left(\int_{\mathbb{R}^N}|\nabla u|^2dx\right)\Delta u$, which makes Kirchhoff's model a nonlocal one and more challenging to deal with.

The solvability of the Kirchhoff equation (9.1) has been well-studied in general dimensions by various authors and its main feature is the normalization condition

$$\int_{\mathbb{R}^N}|u(x)|^2dx=\|u\|_2^2=1,$$

the solution of such an equation under which is commonly called normalized solution accordingly. From a physical point of view, the normalization condition can also be regarded as a mass conservation, which makes such problems more practical and physically significant, for example, it can represent the total number of atoms or the power source in nonlinear optics. Therefore, the research of the correlation properties about the normalized solution has become an important topic in the research of nonlinear equations.

Among the investigations of the above problems, due to the appearance of normalization condition and Lagrange multiplier μ, the common idea is to use the constraint minimization method on a constrained set, more specially on some submanifolds. For the classical case, the energy functional of the following form contains a lot of classical models

$$\begin{aligned}\bar{E}_{a,b,p,q}(u) &:= \int_{\mathbb{R}^N} (a|\nabla u|^2 + V(x)|u|^2)dx \\ &+ \frac{b}{2}\left(\int_{\mathbb{R}^N} |\nabla u|^2 dx\right)^2 - \frac{2p}{q+2}\int_{\mathbb{R}^N} |u|^{q+2}dx \end{aligned} \tag{9.157}$$

and its constraint minimization problem

$$\bar{e}(a,b,p,q,c) := \inf_{\{\|u\|_2^2 = c^2\}} \bar{E}_{a,b,p,q}(u) \tag{9.158}$$

has been studied in many different contexts. For instance, as $N = 2$, $a = 1$, $b = 0$ and $q = 2$, the functional (9.157) turns to the Gross-Pitaevskii energy, which describes Bose-Einstein condensates of a dilute gas with attractive interactions, we refer to Guo, Zeng and Zhou [67]. For the classical Laplacian, the L^2-critical exponent $2 + 4/N$ plays an important role in the Schrödinger equation, when $b = 0$ and $2 < q < 2+4/N$, the energy functional $\bar{E}_{a,b,p,q}$ is bounded below on the constrained set.

Back to the classical Kirchhoff equation, that is a, and b are arbitrary but fixed, and its corresponding L^2-critical exponent becomes $2 + 8/N$. Ye [157] first demonstrated the existence and non-existence of normalized solutions to a kind of Kirchhoff equation when $s = 1$, then a series of subsequent study has been done on the existence of normalized solutions to the nonlinear Kirchhoff equation, for L^2-critical problem with $s = 1$. Luo and Wang [112] obtained the multiplicity existence of normalized solutions to a kind of Kirchoff equation when $N = 3$. Zeng, Zhang, Zhang and Zhong [160] used a global branch approach for handling the nonlinearities in a unified way, and obtained the existence of normalized solutions to the Kirchhoff equation with general nonlinearities.

If $N = 2$, $a = 1$, $b = 0$ and $q = 2$ in the energy functional defined in (9.157), Guo, Zeng and Zhou [67] explored the limiting behavior of minimizers for the Gross-Pitaevskii energy functional (9.158) by some techniques consisting of blow-up estimate and elliptic regularity theory, then they also proved the local uniqueness of constraint minimizer by establishing Pohozaev identities and by contradiction. When $b \neq 0$ and $N = 2$, the energy functional (9.157) becomes a classical Kirchhoff type energy functional. Guo and Zhou [66] considered the limiting behavior of constraint minimizers as $b \to 0^+$, in particular, they first illustrated the existence and nonexistence of constraint minimizers for different parameters p, and then they considered concentration behaviors of the minimizers of (9.158) as $b \to 0^+$ and also the uniqueness of the minimizers of problem (9.158) for b sufficiently close to 0 were shown. For the concentration phenomenon and local uniqueness of constraint minimizers of (9.158) as $a \searrow 0$, recently, Hu and Tang [73] overcame the difficulties in estimates and blow-up analysis caused by the nonlocal term in the process of establishing decay estimate and setting Pohozaev identities, they improved the concentration phenomenon in the sense of that they provided the accurate estimating exponent instead of $o(1)$, and then they finally established the corresponding results. Motivated by [73], Guo, Liu and Zhao [64] studied the normalized solutions for a class of Kirchhoff equations on a suitable weighted Sobolev space, and they investigated the limit behaviors of the normalized solutions for this equation as $(a,b) \to (0^+, 0^+)$, furthermore, they also discussed the uniqueness of the normalized solution when a, b close to 0.

As for nonlinear equations involving fractional Laplacian, Cingolani, Gallo and Tanaka [36] studied the following fractional problem

$$\begin{cases} (-\Delta)^s u + \mu u = g(u) \quad \text{in } \mathbb{R}^N, \\ \displaystyle\int_{\mathbb{R}^N} |u|^2 dx = m, \\ u \in H_r^s(\mathbb{R}^N), \end{cases}$$

where $N \geq 2$, $s \in (0,1)$, $m > 0$, μ is an unknown Lagrange multiplier and $g \in C(\mathbb{R}, \mathbb{R})$ satisfies Berestycki-Lions type conditions. Using a Lagrangian formulation of the above problem, they proved the existence of a weak solution with prescribed mass when g has L^2-subcritical growth. Luo and Zhang [111] proved some existence and nonexistence results about the normalized solutions of the fractional nonlinear Schrödinger equations with combined nonlinearities

$$\begin{cases} (-\Delta)^s u = \lambda u + \mu|u|^{q-2}u + |u|^{p-2}u \quad \text{in } \mathbb{R}^N, \\ \displaystyle\int_{\mathbb{R}^N} |u|^2 dx = m, \end{cases}$$

where $0 < s < 1$, $N \geq 2$, $\mu \in \mathbb{R}$ and $2 < q < p < 2_s^* = \frac{2N}{N-2s}$. Next, Zhen and Zhang [163] considered the case of $p = 2_s^*$. We also refer to He and Zou [71] for important contributions to the analysis of concentration properties of solutions to the fractional critical Schrödinger equation.

Back to our concerned fractional Kirchhoff equation, there are a few literature studies on this issue. Huang and Zhang [74] provided a thorough classification for the existence of solutions to the fractional Kirchhoff functional on the L^2-normalized manifold when f is pure power nonlinearity, taking into account p. Chen and Huang [35] considered the existence and nonexistence of solutions to the fractional Kirchhoff equation with an external potential V and doubly critical exponents: critical Sobolev exponent and the fractional Gagliardo–Nirenberg–Sobolev critical exponent. By decomposing the Pohozaev set and constructing a fiber map, Liu, Chen and Yang [102] established the existence and properties of normalized ground states to fractional Kirchhoff equation with combined nonlinearities. Kong and Chen [89] studied the existence of normalized ground states for nonlinear fractional Kirchhoff equations with Sobolev critical exponent and mixed nonlinearities in $\mathbb{R}^3$, and analyzed the asymptotic behavior of the obtained normalized solutions.

9.6 Final remarks, perspectives, and open problems

This chapter investigates normalized solutions for a class of fractional Kirchhoff equations with a mass-critical exponent. The novel perspective in this analysis lies in tackling the *double nonlocal effect* that combines the fractional Laplace operator $(-\Delta)^s u$ with the nonlocal Kirchhoff term $\int_{\mathbb{R}^N} |(-\Delta)^{\frac{s}{2}} u|^2 dx$. The analysis carried out in this chapter has a threefold purpose, as described below.

(i) *Existence and non-existence*: the main results establish conditions for the existence of constraint minimizers, particularly in the degenerate case $a = 0$.

(ii) *Concentration behavior*: it is obtained that the maximizer of these solutions concentrates at the bottom point of the homogeneous potential $|x|^2$.

(iii) *Local uniqueness*: the analysis reveals the local uniqueness of these minimizers under the double nonlocal effect.

The content of this chapter is based on the results established in [104], while concentration properties for Kirchhoff equations with exponential growth were subsequently obtained by Sun, Fu and Liang [143]. We also refer to Jin, Tang and Hu [86] for the case of nonautonomous critical Kirchhoff equations with convolution nonlinearity.

Some open problems and future research directions inspired by the analysis developed in this chapter are described in what follows.

Open Problem 1. Theorem 9.2 establishes concentration at the potential's bottom point. An open problem is to determine the exact blow-up rate and the limit profile of the concentrated solutions, which is expected to be related to the ground state of a limiting equation.

The analysis in this chapter focuses on the homogeneous potential $|x|^2$. A major extension would be to consider more general external potentials. We propose the following interesting open problem.

Open Problem 2. Study problem (9.1) for more general potentials affecting the linear term and establish how their geometry (for instance, number and shape of minima) affects the concentration and multiplicity of minimizers.

This chapter deals with stationary solutions. The following research direction is concerned with the associated evolution problem.

Open Problem 3. Study the dynamical stability of the normalized solutions under the time-dependent flow of the fractional Kirchhoff equation.

The next open problem proposes to extend this analysis to higher dimensions and other exponents. This is a challenging research direction.

Open Problem 4. The analysis carried out in this chapter is set for dimensions $N = 2, 3$. Extend Theorems 9.1, 9.2 and 9.3 on concentration and uniqueness to higher dimensions or to problems with different combinations of critical exponents.

9.7 Glossary

Constraint Minimization: It is a fundamental method in the calculus of variations used to find points (called minimizers) where a given functional attains its smallest value, but only among functions that satisfy an additional condition, or constraint.

Localization Method: This is a powerful mathematical tool used to transform a problem involving a nonlocal operator (like the fractional Laplacian) into a local one in a higher-dimensional space, making it much easier to analyze.

A

Fractional Sobolev spaces

Let Ω be an open set in $\mathbb{R}^N$ and $p \in [1, +\infty)$. For any $s > 0$ we define the fractional Sobolev space $W^{s,p}(\Omega)$.

If $s \geq 1$ is a positive integer we denote by $W^{s,p}(\Omega)$ the classical Sobolev space equipped with the standard norm

$$\|u\|_{W^{s,p}(\Omega)} := \sum_{0 \leq |\alpha| \leq s} \|D^\alpha u\|_{L^p(\Omega)},$$

for every $u \in W^{s,p}(\Omega)$, where $\|\cdot\|_{L^p(\Omega)}$ denotes the usual norm in the Lebesgue space $L^p(\Omega)$ and D^α stands for the α-distributional derivative. This section is devoted to the definition of fractional Sobolev spaces, that is, we are interested in the case where s is not a positive integer.

For a fixed $s \in (0, 1)$, we recall that the Sobolev space $W^{s,p}(\Omega)$ is defined as follows

$$W^{s,p}(\Omega) := \left\{ u \in L^p(\Omega) : \frac{|u(x) - u(y)|}{|x - y|^{\frac{N}{p}+s}} \in L^p(\Omega \times \Omega) \right\}.$$

This space is endowed with the natural norm

$$\|u\|_{W^{s,p}(\Omega)} := \left(\int_\Omega |u(x)|^p \, dx + \int_\Omega \int_\Omega \frac{|u(x) - u(y)|^p}{|x - y|^{N+sp}} \, dx \, dy \right)^{1/p},$$

where

$$[u]_{W^{s,p}(\Omega)} := \left(\int_\Omega \int_\Omega \frac{|u(x) - u(y)|^p}{|x - y|^{N+sp}} \, dx \, dy \right)^{1/p}$$

is the *Gagliardo seminorm* of u.

When $s > 1$ and $s \notin \mathbb{N}$ we can write $s = m + \sigma$, where $m \in \mathbb{N}$ and $\sigma \in (0, 1)$. We can define $W^{s,p}(\Omega)$ as follows

$$W^{s,p}(\Omega) := \left\{ u \in W^{m,p}(\Omega) : D^\alpha u \in W^{\sigma,p}(\Omega) \text{ for any } \alpha \text{ s.t. } |\alpha| = m \right\}.$$

In this case, $W^{s,p}(\Omega)$ is endowed with the norm

$$\|u\|_{W^{s,p}(\Omega)} := \left(\|u\|^p_{W^{m,p}(\Omega)} + \sum_{|\alpha|=m} \|D^\alpha u\|^p_{W^{\sigma,p}(\Omega)} \right)^{1/p},$$

for every $u \in W^{s,p}(\Omega)$.

The fractional Sobolev space $W^{s,p}(\Omega)$ is well defined and is a Banach space for every $s > 0$.

DOI: 10.1201/9781003797159-A

As in the classical case (if s is a positive integer), any function in the fractional Sobolev space $W^{s,p}(\mathbb{R}^N)$ can be approximated by a sequence of smooth functions with compact support. Indeed, for any $s > 0$

$$\overline{C_0^\infty(\mathbb{R}^N)}^{\|\cdot\|_{W^{s,p}(\mathbb{R}^N)}} = W^{s,p}(\mathbb{R}^N),$$

that is, the function space $C_0^\infty(\mathbb{R}^N)$ is dense in $W^{s,p}(\mathbb{R}^N)$.

In general, if $\Omega \subset \mathbb{R}^N$, the space $C_0^\infty(\Omega)$ is not dense in $W^{s,p}(\Omega)$. Hence, we denote by $W_0^{s,p}(\Omega)$ the closure of $C_0^\infty(\Omega)$ with respect to the norm $\|\cdot\|_{W^{s,p}(\Omega)}$, that is,

$$W_0^{s,p}(\Omega) := \overline{C_0^\infty(\Omega)}^{\|\cdot\|_{W^{s,p}(\Omega)}}.$$

With this definition, we can construct $W^{s,p}(\Omega)$ also when $s < 0$. Indeed, for $s < 0$ and $p \in (1, +\infty)$, we can define

$$W^{s,p}(\Omega) := \left(W_0^{-s,q}(\Omega)\right)',$$

hence $W^{s,p}(\Omega)$ is the dual space of $W_0^{-s,q}(\Omega)$, where $1/p + 1/q = 1$.

An important case corresponds to $p = 2$. Denote

$$H^s(\Omega) := W^{s,2}(\Omega)$$

for any $s \in (0,1)$.

In this case, the corresponding fractional Sobolev space turns out to be a Hilbert space. Indeed, the inner product on $H^s(\Omega)$ defined by

$$\langle u, v\rangle_{H^s(\Omega)} := \int_\Omega u(x)v(x)\,dx + \int_\Omega\int_\Omega \frac{(u(x)-u(y))(v(x)-v(y))}{|x-y|^{N+2s}}\,dx\,dy,$$

for any $u, v \in H^s(\Omega)$.

We remark that for every $s \in (0,1)$ we have

$$H^s(\mathbb{R}^N) := W^{s,2}(\mathbb{R}^N) = \left\{u \in L^2(\mathbb{R}^N) : [u]_{W^{s,2}(\mathbb{R}^N)} < +\infty\right\}.$$

The space $H^s(\mathbb{R}^n)$ can be defined in an alternative way via the Fourier transform, namely

$$\widehat{H}^s(\mathbb{R}^N) := \left\{u \in L^2(\mathbb{R}^N) : \int_{\mathbb{R}^N} (1 + |\xi|^{2s})|\mathcal{F}u(\xi)|^2\,dx < +\infty\right\},$$

for any $s > 0$ and

$$\widehat{H}^s(\mathbb{R}^N) := \left\{u \in \mathcal{S}' : \int_{\mathbb{R}^N} (1 + |\xi|^2)^s|\mathcal{F}u(\xi)|^2\,dx < +\infty\right\},$$

for every $s < 0$.

Sobolev Embeddings

We now recall some basic embedding properties of fractional Sobolev spaces into Lebesgue spaces.

Theorem A.1 *Let $p \in [1, +\infty)$ and let Ω be an open set in $\mathbb{R}^N$. Then the following properties hold true:*

(i) *if* $0 < s \leq s' < 1$, *then the embedding*

$$W^{s',p}(\Omega) \hookrightarrow W^{s,p}(\Omega)$$

is continuous. Hence, there exists a constant $C_1(N, s, p) \geq 1$ *such that*

$$\|u\|_{W^{s,p}(\Omega)} \leq C_1(N, s, p)\|u\|_{W^{s',p}(\Omega)}$$

for any $u \in W^{s',p}(\Omega)$;

(ii) *if* $0 < s < 1$ *and* Ω *is of class* $C^{0,1}$ *and with bounded boundary* $\partial\Omega$, *then the embedding*

$$W^{1,p}(\Omega) \hookrightarrow W^{s,p}(\Omega)$$

is continuous. Hence, there exists a constant $C_2(N, s, p) \geq 1$ *such that*

$$\|u\|_{W^{s,p}(\Omega)} \leq C_2(N, s, p)\|u\|_{W^{1,p}(\Omega)},$$

for any $u \in W^{1,p}(\Omega)$;

(iii) *if* $s' \geq s > 1$ *and* Ω *is of class* $C^{0,1}$, *then the embedding*

$$W^{s',p}(\Omega) \hookrightarrow W^{s,p}(\Omega)$$

is continuous.

Theorem A.2 *Let* $s \in (0, 1)$ *and* $p \in [1, +\infty)$ *such that* $sp < N$.
Then, there exists a positive constant $C := C(N, p, s)$ *such that for any* $u \in W^{s,p}(\mathbb{R}^N)$

$$\|u\|^p_{L^{p^*_s}(\mathbb{R}^n)} \leq C \int_{\mathbb{R}^N} \int_{\mathbb{R}^N} \frac{|u(x) - u(y)|^p}{|x - y|^{N+ps}} \, dx \, dy\,,$$

where the constant

$$p^*_s := \frac{pN}{N - sp}$$

is the fractional critical exponent. Consequently, the space $W^{s,p}(\mathbb{R}^N)$ *is continuously embedded in* $L^q(\mathbb{R}^N)$ *for any* $q \in [p, p^*_s]$. *Moreover, the embedding* $W^{s,p}(\mathbb{R}^N) \hookrightarrow L^q_{\text{loc}}(\mathbb{R}^N)$ *is compact for every* $q \in [p, p^*_s)$.

Theorem A.3 *Let* $s \in (0, 1)$ *and* $p \in [1, +\infty)$ *such that* $sp = N$.
Then, there exists a positive constant $C := C(N, p, s)$ *such that for any* $u \in W^{s,p}(\mathbb{R}^N)$

$$\|u\|_{L^q(\mathbb{R}^N)} \leq C\|u\|_{W^{s,p}(\mathbb{R}^N)},$$

for any $q \in [p, +\infty)$, *that is, the space* $W^{s,p}(\mathbb{R}^N)$ *is continuously embedded in* $L^q(\mathbb{R}^N)$ *for any* $q \in [p, +\infty)$.

Theorem A.4 *Let* $s \in (0, 1)$, $p \in [1, +\infty)$ *such that* $sp > N$ *and* Ω *be a* $C^{0,1}$ *domain of* $\mathbb{R}^N$. *Then there exists a positive constant* $C := C(N, p, s, \Omega)$ *such that for any* $u \in W^{s,p}(\Omega)$

$$\|u\|_{C^{0,\alpha}(\Omega)} \leq C\|u\|_{W^{s,p}(\Omega)},$$

with $\alpha := (sp - N)/p$, *that is, the space* $W^{s,p}(\Omega)$ *is continuously embedded in* $C^{0,\alpha}(\Omega)$.

Corollary A.1 *Let* $s \in (0, 1)$, $p \in [1, +\infty)$ *such that* $sp > N$ *and* Ω *be a* $C^{0,1}$ *bounded domain of* $\mathbb{R}^N$. *Then the embedding*

$$W^{s,p}(\Omega) \hookrightarrow C^{0,\beta}(\Omega)$$

is compact for every $\beta < \alpha$, *with* $\alpha := (sp - N)/p$.

The Fractional Laplace Operator

Let $s \in (0,1)$ and define the operator $(-\Delta)^s : \mathcal{S} \to L^2(\mathbb{R}^N)$ by

$$(-\Delta)^s u(x) := C(N,s) \lim_{\varepsilon \to 0^+} \int_{\mathbb{R}^N \setminus B(x,\varepsilon)} \frac{u(x)-u(y)}{|x-y|^{N+2s}}\, dy, \quad x \in \mathbb{R}^N, \tag{A.1}$$

where $\mathcal{S}$ is the Schwartz space of rapidly decaying $C^\infty(\mathbb{R}^N)$ functions and $C(N,s)$ is the following normalization constant

$$C(N,s) := \left(\int_{\mathbb{R}^N} \frac{1-\cos(\zeta_1)}{|\zeta|^{N+2s}}\, d\zeta \right)^{-1}, \tag{A.2}$$

with $\zeta = (\zeta_1, \zeta')$, $\zeta' \in \mathbb{R}^{N-1}$.

The operator defined in (A.1) is called the *fractional Laplacian*. We can also write

$$(-\Delta)^s u(x) := C(N,s)\, \text{P.V.} \int_{\mathbb{R}^N} \frac{u(x)-u(y)}{|x-y|^{N+2s}}\, dy, \quad x \in \mathbb{R}^N, \tag{A.3}$$

where

$$\text{P.V.} \int_{\mathbb{R}^N} \frac{u(x)-u(y)}{|x-y|^{n+2s}}\, dy := \lim_{\varepsilon \to 0^+} \int_{\mathbb{R}^N \setminus B(x,\varepsilon)} \frac{u(x)-u(y)}{|x-y|^{N+2s}}\, dy.$$

The singular integral given in (A.3) can be written as a weighted second order differential quotient as follows:

Proposition A.1 *Let $s \in (0,1)$. Then, for any $u \in \mathcal{S}$*

$$(-\Delta)^s u(x) = -\frac{1}{2} C(N,s) \int_{\mathbb{R}^N} \frac{u(x+y)+u(x-y)-2u(x)}{|y|^{N+2s}}\, dy, \quad x \in \mathbb{R}^N.$$

The constant $C(N,s)$ defined in (A.2) can be written as

$$C(N,s) = \frac{s(1-s)}{A(N,s)B(s)},$$

where

$$A(N,s) := \begin{cases} 1 & \text{if } N = 1, \\ \displaystyle\int_{\mathbb{R}^{N-1}} \frac{1}{(1+|\eta'|^2)^{\frac{N+2s}{2}}}\, d\eta' & \text{if } N \geq 2, \end{cases}$$

and

$$B(s) := s(1-s) \int_{\mathbb{R}} \frac{1-\cos t}{|t|^{1+2s}}\, dt.$$

B

Basic inequalities and theorems

In Chapter 6, there are used the following elementary inequalities, see Simon [139]: there exist positive constants c_p and C_p depending only on p such that for all $\xi, \eta \in \mathbb{R}^N$,

$$\begin{aligned} &|\xi - \eta|^p \leq c_p(|\xi|^{p-2}\xi - |\eta|^{p-2}\eta) \cdot (\xi - \eta) \quad \text{for } p \geq 2, \\ &|\xi - \eta|^p \leq C_p[(|\xi|^{p-2}\xi - |\eta|^{p-2}\eta) \cdot (\xi - \eta)]^{\frac{p}{2}}(|\xi|^p + |\eta|^p)^{\frac{2-p}{2}} \quad \text{for } 1 < p < 2. \end{aligned} \tag{B.1}$$

The Gagliardo–Nirenberg inequality is a fundamental estimate in the theory of Sobolev spaces that interpolates between different norms of a function and its derivatives. For more details, we refer to the pioneering papers Gagliardo [60] and Nirenberg [124], as well as to Weinstein [151]. The Gagliardo–Nirenberg inequality has important applications to various fields, such as:

(i) *regularity theory*: interpolate between norms in bootstrap arguments;

(ii) *PDE existence proofs*: control nonlinear terms in energy estimates;

(iii) *decay estimates*: for instance, in diffusive PDEs via Nash inequality;

(iv) *fluid mechanics*: Ladyzhenskaya's inequality bounds in Navier–Stokes equations.

In Chapter 3, it is used the following version of the Gagliardo–Nirenberg inequality.

Lemma B.1 *Let $p \in (2, 6]$. Then there exists a sharp constant $C_p = \left(\frac{p}{2\|U_p\|_2^{p-2}}\right)^{\frac{1}{p}}$ such that*

$$\|u\|_p \leq C_p \|\nabla u\|_2^{\gamma_p} \|u\|_2^{1-\gamma_p}, \quad \forall u \in H^1(\mathbb{R}^3),$$

where $\gamma_p := 3\left(\frac{1}{2} - \frac{1}{p}\right)$ and U_p is the unique positive solution of

$$-\Delta U + \left(\frac{1}{\gamma_p} - 1\right) U = \frac{2}{p\gamma_p} |U|^{p-2} U, \quad \in \mathbb{R}^3.$$

The following basic inequality established by Trudinger [148] and Moser [120] plays a key role in the treatment of partial differential equations with critical exponential growth in the two-dimensional case. The following result is known as the *Trudinger-Moser inequality*.

Lemma B.2 *The following inequalities hold true.*

(i) If $\alpha > 0$ and $u \in H^1(\mathbb{R}^2)$, then

$$\int_{\mathbb{R}^2} \left(e^{\alpha u^2} - 1\right) dx < \infty.$$

(ii) If $u \in H^1(\mathbb{R}^2)$, $\|\nabla u\|_2^2 \leq 1$, $\|u\|_2 \leq M < \infty$, and $\alpha < 4\pi$, then there exists a constant $\mathcal{C}(M, \alpha)$, which depends only on M and α, such that

$$\int_{\mathbb{R}^2} \left(e^{\alpha u^2} - 1\right) dx \leq \mathcal{C}(M, \alpha).$$

DOI: 10.1201/9781003797159-B

Fatou's lemma is a fundamental result used several times in this volume. It describes the relationship between the integral of a limit of functions and the limit of the integrals. In simple terms, it provides a condition under which we can say that *the integral of the limit is less than or equal to the limit of the integrals.* The rigorous statement of the Fatou lemma is given in what follows.

Lemma B.3 *Let $\Omega \subset \mathbb{R}^N$ be an open set and let $\{f_n\}_{n\geq 1}$ be a sequence of functions in $L^1(\Omega)$ that satisfy*

(a) for all n, $f_n \geq 0$ a.e.;

(b) $\sup_n \int_\Omega f_n dx < \infty$.

For almost all $x \in \Omega$, we set $f(x) = \liminf_{n\to\infty} f_n(x) \leq +\infty$. Then $f \in L^1(\Omega)$ and

$$\int_\Omega f(x)dx \leq \liminf_{n\to\infty} \int_\Omega f_n(x)dx.$$

The Fatou lemma is usually one step in proving stronger results that allow to swap the limit and the integral under certain conditions. One of these basic results is the dominated convergence theorem, which we state in what follows.

Theorem B.1 *Let $\Omega \subset \mathbb{R}^N$ be an open set and let $\{f_n\}_{n\geq 1}$ be a sequence of functions in $L^1(\Omega)$ that satisfy*

(a) $f_n(x) \to f(x)$ a.e. in Ω;

(b) there is a function $g \in L^1(\Omega)$ such that for all n, $|f_n(x)| \leq g(x)$ a.e. in Ω.

Then $f \in L^1(\Omega)$ and $\|f_n - f\|_{L^1} \to 0$.

A more general result than the Fatou lemma is the following Brezis–Lieb lemma, which establishes a more precise evaluation of the gap between the integral of a sequence of functions and the integral of its pointwise limit, see Brezis and Lieb [14] for details.

Lemma B.4 *Let $\Omega \subset \mathbb{R}^N$ be an open set and let $1 \leq p < \infty$. Assume that $\{f_n\}_{n\geq 1}$ is a bounded sequence of functions in $L^p(\Omega)$, convergent to f almost everywhere. Then*

$$\int_\Omega |f_n(x)|^p dx - \int_\Omega |f_n(x) - f(x)|^p dx - \int_\Omega |f(x)|^p dx \to 0 \text{ as } n \to \infty.$$

Ekeland's variational principle [48] was established in 1974 and is the nonlinear version of the Bishop–Phelps theorem, with its main feature of how to use the norm completeness and a partial ordering to obtain a point where a linear functional achieves its supremum on a closed bounded convex set. A major consequence of Ekeland's variational principle is that even if it is not always possible to minimize a nonnegative C^1 functional Φ on a Banach space; however, there is always a minimizing sequence $\{u_n\}_{n\geq 1}$ such that $\Phi'(u_n) \to 0$ as $n \to \infty$. We now state the original version of Ekeland's variational principle.

Theorem B.2 *Let (M, d) be a complete metric space and assume that $\Phi : M \to (-\infty, \infty]$, $\Phi \not\equiv \infty$, is a lower semicontinuous functional that is bounded from below.*

Then, for every $\varepsilon > 0$ and for any $z_0 \in M$, there exists $z \in M$ such that

(i) $\Phi(z) \leq \Phi(z_0) - \varepsilon\, d(z, z_0)$;

(ii) $\Phi(x) \geq \Phi(z) - \varepsilon\, d(x, z)$, for any $x \in M$.

The following consequence of Ekeland's variational principle establishes the existence of almost critical points for bounded from below C^1-functionals defined on Banach spaces. In order words, Ekeland's variational principle can be viewed as a generalization of the Fermat theorem, which asserts that interior extrema points of a smooth functional are, necessarily, critical points of this functional.

Corollary B.1 *Let E be a Banach space and let $\Phi : E \to \mathbb{R}$ be a C^1 functional that is bounded from below. Then, for any $\varepsilon > 0$, there exists $z \in E$ such that*

$$\Phi(z) \le \inf_E \Phi + \varepsilon \qquad \text{and} \qquad \|\Phi'(z)\|_{E^\star} \le \varepsilon.$$

The mountain pass theorem was established by Ambrosetti and Rabinowitz in [1]. Their original proof relies on some deep deformation techniques developed by Palais and Smale [125], [126], who put the main ideas of the Morse theory into the framework of differential topology on infinite dimensional manifolds. In this way, Palais and Smale replaced the finite dimensionality assumption with an appropriate compactness hypothesis, which is stated in the following.

Let E be a real Banach space. A functional $J : E \to \mathbb{R}$ of class C^1 satisfies the Palais–Smale condition if any sequence $\{u_n\}_{n\ge 1}$ in E is relatively compact, provided that

$$\sup_{n\ge 1} |J(u_n)| < \infty \quad \text{and} \quad \lim_{n\to\infty} \|J'(u_n)\|_{E^\star} = 0\,.$$

The mountain pass theorem is a powerful tool for proving the existence of critical points of energy functionals, hence of weak solutions to wide classes of nonlinear problems. The precise statement of this result is the following.

Theorem B.3 *Let E be a real Banach space and assume that $J : E \to \mathbb{R}$ is a C^1 functional that satisfies the following conditions: There exist positive constants α and R such that*

(i) $J(0) = 0$ and $J(v) \ge \alpha$ for all $v \in E$ with $\|v\| = R$;

(ii) $J(v_0) \le 0$, for some $v_0 \in E$ with $\|v_0\| > R$.

Set

$$\Gamma := \{p \in C([0,1]; E) : p(0) = 0 \text{ and } p(1) = v_0\}$$

and

$$c := \inf_{p\in\Gamma} \max_{t\in[0,1]} J(p(t))\,.$$

Then there exists a sequence $\{u_n\}_{n\ge 1}$ in E such that $J(u_n) \to c$ and $J'(u_n) \to 0$ as $n \to \infty$. Moreover, if J satisfies the Palais–Smale condition, then c is a nontrivial critical value of J, that is, there exists $u \in E$ such that $J(u) = c$ and $J'(u) = 0$.

The elementary meaning of the mountain pass theorem is the following. The set Γ denotes the family of all continuous paths that join the "villages" 0 and v_0. Hypotheses (i) and (ii) imply $c \ge \alpha > \max\{J(0), J(v_0)\}$. In fact, in the proof of the mountain pass theorem, it is used only the basic assumption $c > \max\{J(0), J(v_0)\}$. We also observe that the geometric conditions of the mountain pass theorem imply the existence of two valleys around these villages, but also of a mountain range represented by the sphere centered at the origin and of radius R. Under these hypotheses, Theorem B.3 states that there exists a lowest mountain pass at the level c.

The following version of the mountain pass theorem, in the abstract setting of locally Lipschitz functionals, was established by Rădulescu [134].

Theorem B.4 *Let E be a real Banach space and assume that $I : E \to \mathbb{R}$ is a locally Lipschitz functional with $I(0) = 0$ and satisfying:*

(i) there exist $r > 0$ and $\rho > 0$ such that $I(u) \geq \rho$ for $\|u\| = r$, $u \in E$;

(ii) there exists $e \in E \setminus B_r(0)$ with $I(e) < 0$.

Set

$$c := \inf_{\gamma \in \Gamma} \max_{s \in [0,1]} I(\gamma(s)) > 0,$$

where

$$\Gamma := \{\gamma \in C([0,1], E) \ : \ \gamma(0) = 0,\ I(\gamma(1)) < 0\}.$$

Then $c \geq \rho$ and there is a sequence $\{u_n\} \subset E$ verifying $I(u_n) \to c$ and $I'(u_n) \to 0$ in E'.

Bibliography

[1] Antonio Ambrosetti and Paul H. Rabinowitz. Dual variational methods in critical point theory and applications. *J. Functional Analysis*, 14:349–381, 1973.

[2] Vincenzo Ambrosio. An existence result for a fractional Kirchhoff-Schrödinger-Poisson system. *Z. Angew. Math. Phys.*, 69(2):Paper No. 30, 13, 2018.

[3] Vincenzo Ambrosio. Concentrating solutions for a fractional Kirchhoff equation with critical growth. *Asymptot. Anal.*, 116(3-4):249–278, 2020.

[4] Vincenzo Ambrosio and Teresa Isernia. Multiplicity and concentration results for some nonlinear Schrödinger equations with the fractional p-Laplacian. *Discrete Contin. Dyn. Syst.*, 38(11):5835–5881, 2018.

[5] Vincenzo Ambrosio and Teresa Isernia. A multiplicity result for a fractional Kirchhoff equation in $\mathbb{R}^N$ with a general nonlinearity. *Commun. Contemp. Math.*, 20(5):1750054, 17, 2018.

[6] Vincenzo Ambrosio, Teresa Isernia, and Vicenţiu D. Rădulescu. Concentration of positive solutions for a class of fractional p-Kirchhoff type equations. *Proc. Roy. Soc. Edinburgh Sect. A*, 151(2):601–651, 2021.

[7] A. Arosio and S. Panizzi. On the well-posedness of the Kirchhoff string. *Trans. Amer. Math. Soc.*, 348(1):305–330, 1996.

[8] Giuseppina Autuori, Francesca Colasuonno, and Patrizia Pucci. Lifespan estimates for solutions of polyharmonic Kirchhoff systems. *Math. Models Methods Appl. Sci.*, 22(2):1150009, 36, 2012.

[9] Vieri Benci and Giovanna Cerami. Multiple positive solutions of some elliptic problems via the Morse theory and the domain topology. *Calc. Var. Partial Differential Equations*, 2(1):29–48, 1994.

[10] S. Bernstein. Sur une classe d'équations fonctionnelles aux dérivées partielles. *Bull. Acad. Sci. URSS. Sér. Math. [Izvestia Akad. Nauk SSSR]*, 4:17–26, 1940.

[11] Bartosz Bieganowski and Jarosław Mederski. Normalized ground states of the nonlinear Schrödinger equation with at least mass critical growth. *J. Funct. Anal.*, 280(11):Paper No. 108989, 26, 2021.

[12] Bartosz Bieganowski, Jarosław Mederski, and Jacopo Schino. Normalized solutions to at least mass critical problems: singular polyharmonic equations and related curl-curl problems. *J. Geom. Anal.*, 34(10):Paper No. 322, 32, 2024.

[13] C. Brändle, E. Colorado, A. de Pablo, and U. Sánchez. A concave-convex elliptic problem involving the fractional Laplacian. *Proc. Roy. Soc. Edinburgh Sect. A*, 143(1):39–71, 2013.

[14] Haïm Brezis and Elliott Lieb. A relation between pointwise convergence of functions and convergence of functionals. *Proc. Amer. Math. Soc.*, 88(3):486–490, 1983.

[15] Haïm Brezis and Elliott H. Lieb. Minimum action solutions of some vector field equations. *Comm. Math. Phys.*, 96(1):97–113, 1984.

[16] Luis Caffarelli and Luis Silvestre. An extension problem related to the fractional Laplacian. *Comm. Partial Differential Equations*, 32(7-9):1245–1260, 2007.

[17] Li Cai and Vicenţiu D. Rădulescu. Normalized solutions for (p,q)-Laplacian equations with mass supercritical growth. *J. Differential Equations*, 391:57–104, 2024.

[18] Li Cai and Vicenţiu D. Rădulescu. Normalized solutions for quasilinear (p,q)-equations. *Rev. R. Acad. Cienc. Exactas Fís. Nat. Ser. A Mat. RACSAM*, 119(3):Paper No. 74, 28, 2025.

[19] Li Cai and Vicenţiu D. Rădulescu. Solutions with prescribed mass to Kirchhoff equations: generic double-behavior nonlinearities. *Nonlinearity*, 38(10):Paper No. 105004, 32, 2025.

[20] Daomin Cao, Shuanglong Li, and Peng Luo. Uniqueness of positive bound states with multi-bump for nonlinear Schrödinger equations. *Calc. Var. Partial Differential Equations*, 54(4):4037–4063, 2015.

[21] G. F. Carrier. On the non-linear vibration problem of the elastic string. *Quart. Appl. Math.*, 3:157–165, 1945.

[22] M. M. Cavalcanti, V. N. Domingos Cavalcanti, and J. A. Soriano. Global existence and uniform decay rates for the Kirchhoff-Carrier equation with nonlinear dissipation. *Adv. Differential Equations*, 6(6):701–730, 2001.

[23] Kung Ching Chang. Variational methods for nondifferentiable functionals and their applications to partial differential equations. *J. Math. Anal. Appl.*, 80(1):102–129, 1981.

[24] Xiaojun Chang and Zhi-Qiang Wang. Nodal and multiple solutions of nonlinear problems involving the fractional Laplacian. *J. Differential Equations*, 256(8):2965–2992, 2014.

[25] Sitong Chen, Peng Jin, Vicenţiu D. Rădulescu, and Xin'ao Zhou. Normalized solutions for the upper critical Choquard equation with nonlocal perturbations. *Math. Ann.*, 394(1), 2026.

[26] Sitong Chen, Dongdong Qin, Vicenţiu D. Rădulescu, and Xianhua Tang. Ground states for quasilinear equations of N-Laplacian type with critical exponential growth and lack of compactness. *Sci. China Math.*, 68(6):1323–1354, 2025.

[27] Sitong Chen, Vicenţiu D. Rădulescu, and Xianhua Tang. Normalized solutions of nonautonomous Kirchhoff equations: sub- and super-critical cases. *Appl. Math. Optim.*, 84(1):773–806, 2021.

[28] Sitong Chen, Vicenţiu D. Rădulescu, and Xianhua Tang. Multiple normalized solutions for the planar Schrödinger-Poisson system with critical exponential growth. *Math. Z.*, 306(3):Paper No. 50, 32, 2024.

[29] Sitong Chen, Vicenţiu D. Rădulescu, Xianhua Tang, and Lixi Wen. Planar Kirchhoff equations with critical exponential growth and trapping potential. *Math. Z.*, 302(2):1061–1089, 2022.

[30] Sitong Chen, Vicenţiu D. Rădulescu, Xianhua Tang, and Lixi Wen. Planar Schrödinger equations with critical exponential growth. *Calc. Var. Partial Differential Equations*, 63(9):Paper No. 243, 46, 2024.

[31] Sitong Chen, Vicenţiu D. Rădulescu, Xianhua Tang, and Shuai Yuan. Normalized solutions for Schrödinger equations with critical exponential growth in $\mathbb{R}^2$. *SIAM J. Math. Anal.*, 55(6):7704–7740, 2023.

[32] Sitong Chen and Xianhua Tang. Axially symmetric solutions for the planar Schrödinger-Poisson system with critical exponential growth. *J. Differential Equations*, 269(11):9144–9174, 2020.

[33] Sitong Chen and Xianhua Tang. Normalized solutions for Kirchhoff equations with Sobolev critical exponent and mixed nonlinearities. *Math. Ann.*, 391(2):2783–2836, 2025.

[34] Sitong Chen, Xianhua Tang, and Jiuyang Wei. Improved results on planar Kirchhoff-type elliptic problems with critical exponential growth. *Z. Angew. Math. Phys.*, 72(1):Paper No. 38, 18, 2021.

[35] Wenjing Chen and Xiaomeng Huang. The existence of normalized solutions for a fractional Kirchhoff-type equation with doubly critical exponents. *Z. Angew. Math. Phys.*, 73(6):Paper No. 226, 18, 2022.

[36] S. Cingolani, M. Gallo, and K. Tanaka. Normalized solutions for fractional nonlinear scalar field equations via Lagrangian formulation. *Nonlinearity*, 34(6):4017–4056, 2021.

[37] F. H. Clarke. *Optimization and nonsmooth analysis*, volume 5 of *Classics in Applied Mathematics*. Society for Industrial and Applied Mathematics (SIAM), Philadelphia, PA, second edition, 1990.

[38] P. D'Ancona and S. Spagnolo. Global solvability for the degenerate Kirchhoff equation with real analytic data. *Invent. Math.*, 108(2):247–262, 1992.

[39] Piero D'Ancona and Yoshihiro Shibata. On global solvability of nonlinear viscoelastic equations in the analytic category. *Math. Methods Appl. Sci.*, 17(6):477–486, 1994.

[40] Pietro d'Avenia, Marco Squassina, and Marianna Zenari. Fractional logarithmic Schrödinger equations. *Math. Methods Appl. Sci.*, 38(18):5207–5216, 2015.

[41] D. G. de Figueiredo, O. H. Miyagaki, and B. Ruf. Elliptic equations in $\mathbf{R}^2$ with nonlinearities in the critical growth range. *Calc. Var. Partial Differential Equations*, 3(2):139–153, 1995.

[42] Leandro M. Del Pezzo and Alexander Quaas. Spectrum of the fractional p-Laplacian in $\mathbb{R}^N$ and decay estimate for positive solutions of a Schrödinger equation. *Nonlinear Anal.*, 193:111479, 24, 2020.

[43] Manuel del Pino and Patricio L. Felmer. Local mountain passes for semilinear elliptic problems in unbounded domains. *Calc. Var. Partial Differential Equations*, 4(2):121–137, 1996.

[44] Shengbing Deng and Wenshan Luo. Multiplicity and concentration of solutions for a fractional magnetic Kirchhoff equation with competing potentials. *Ann. Henri Poincaré*, 25(7):3499–3528, 2024.

[45] Yinbin Deng and Wei Shuai. Sign-changing multi-bump solutions for Kirchhoff-type equations in $\mathbb{R}^3$. *Discrete Contin. Dyn. Syst.*, 38(6):3139–3168, 2018.

[46] Eleonora Di Nezza, Giampiero Palatucci, and Enrico Valdinoci. Hitchhiker's guide to the fractional Sobolev spaces. *Bull. Sci. Math.*, 136(5):521–573, 2012.

[47] Miao Du, Lixin Tian, Jun Wang, and Fubao Zhang. Existence of normalized solutions for nonlinear fractional Schrödinger equations with trapping potentials. *Proc. Roy. Soc. Edinburgh Sect. A*, 149(3):617–653, 2019.

[48] Ivar Ekeland. On the variational principle. *J. Math. Anal. Appl.*, 47:324–353, 1974.

[49] Maria J. Esteban and Pierre-Louis Lions. Stationary solutions of nonlinear Schrödinger equations with an external magnetic field. In *Partial differential equations and the calculus of variations, Vol. I*, volume 1 of *Progr. Nonlinear Differential Equations Appl.*, pages 401–449. Birkhäuser Boston, Boston, MA, 1989.

[50] Lawrence C. Evans. *Partial differential equations*, volume 19 of *Graduate Studies in Mathematics*. American Mathematical Society, Providence, RI, second edition, 2010.

[51] Patricio Felmer, Alexander Quaas, and Jinggang Tan. Positive solutions of the nonlinear Schrödinger equation with the fractional Laplacian. *Proc. Roy. Soc. Edinburgh Sect. A*, 142(6):1237–1262, 2012.

[52] G. M. Figueiredo and J. R. Santos Júnior. Existence of a least energy nodal solution for a Schrödinger-Kirchhoff equation with potential vanishing at infinity. *J. Math. Phys.*, 56(5):051506, 18, 2015.

[53] Giovany M. Figueiredo, Norihisa Ikoma, and João R. Santos Júnior. Existence and concentration result for the Kirchhoff type equations with general nonlinearities. *Arch. Ration. Mech. Anal.*, 213(3):931–979, 2014.

[54] Giovany M. Figueiredo and João R. Santos Júnior. Multiplicity and concentration behavior of positive solutions for a Schrödinger-Kirchhoff type problem *via* penalization method. *ESAIM Control Optim. Calc. Var.*, 20(2):389–415, 2014.

[55] Giovany M. Figueiredo and Uberlandio B. Severo. Ground state solution for a Kirchhoff problem with exponential critical growth. *Milan J. Math.*, 84(1):23–39, 2016.

[56] Alessio Fiscella and Patrizia Pucci. Kirchhoff-Hardy fractional problems with lack of compactness. *Adv. Nonlinear Stud.*, 17(3):429–456, 2017.

[57] Alessio Fiscella and Enrico Valdinoci. A critical Kirchhoff type problem involving a nonlocal operator. *Nonlinear Anal.*, 94:156–170, 2014.

[58] Andreas Floer and Alan Weinstein. Nonspreading wave packets for the cubic Schrödinger equation with a bounded potential. *J. Funct. Anal.*, 69(3):397–408, 1986.

[59] Rupert L. Frank, Enno Lenzmann, and Luis Silvestre. Uniqueness of radial solutions for the fractional Laplacian. *Comm. Pure Appl. Math.*, 69(9):1671–1726, 2016.

[60] Emilio Gagliardo. Ulteriori proprietà di alcune classi di funzioni in più variabili. *Ricerche Mat.*, 8:24–51, 1959.

[61] Liu Gao and Zhong Tan. Existence results for fractional Kirchhoff problems with magnetic field and supercritical growth. *J. Math. Phys.*, 64(3):Paper No. 031503, 14, 2023.

[62] Marius Ghergu and Vicenţiu D. Rădulescu. *Nonlinear PDEs. Mathematical models in biology, chemistry and population genetics.* Springer Monographs in Mathematics. Springer, Heidelberg, 2012.

[63] David Gilbarg and Neil S. Trudinger. *Elliptic partial differential equations of second order.* Classics in Mathematics. Springer-Verlag, Berlin, 2001. Reprint of the 1998 edition.

[64] Helin Guo, Haolin Liu, and Lingling Zhao. Concentration behavior and local uniqueness of normalized solutions for Kirchhoff type equation. *Z. Angew. Math. Phys.*, 75(3):Paper No. 89, 19, 2024.

[65] Helin Guo, Yimin Zhang, and Huansong Zhou. Blow-up solutions for a Kirchhoff type elliptic equation with trapping potential. *Commun. Pure Appl. Anal.*, 17(5):1875–1897, 2018.

[66] Helin Guo and Huan-Song Zhou. Properties of the minimizers for a constrained minimization problem arising in Kirchhoff equation. *Discrete Contin. Dyn. Syst.*, 41(3):1023–1050, 2021.

[67] Yujin Guo, Xiaoyu Zeng, and Huan-Song Zhou. Energy estimates and symmetry breaking in attractive Bose-Einstein condensates with ring-shaped potentials. *Ann. Inst. H. Poincaré C Anal. Non Linéaire*, 33(3):809–828, 2016.

[68] Yuxia Guo, Jianjun Nie, Miaomiao Niu, and Zhongwei Tang. Local uniqueness and periodicity for the prescribed scalar curvature problem of fractional operator in $\mathbb{R}^N$. *Calc. Var. Partial Differential Equations*, 56(4):Paper No. 118, 41, 2017.

[69] Xiaoming He, Vicenţiu D. Rădulescu, and Wenming Zou. Normalized ground states for the critical fractional Choquard equation with a local perturbation. *J. Geom. Anal.*, 32(10):Paper No. 252, 51, 2022.

[70] Xiaoming He and Wenming Zou. Existence and concentration behavior of positive solutions for a Kirchhoff equation in $\mathbb{R}^3$. *J. Differential Equations*, 252(2):1813–1834, 2012.

[71] Xiaoming He and Wenming Zou. Existence and concentration result for the fractional Schrödinger equations with critical nonlinearities. *Calc. Var. Partial Differential Equations*, 55(4):Art. 91, 39, 2016.

[72] Yi He, Gongbao Li, and Shuangjie Peng. Concentrating bound states for Kirchhoff type problems in $\mathbb{R}^3$ involving critical Sobolev exponents. *Adv. Nonlinear Stud.*, 14(2):483–510, 2014.

[73] Tingxi Hu and Chun-Lei Tang. Limiting behavior and local uniqueness of normalized solutions for mass critical Kirchhoff equations. *Calc. Var. Partial Differential Equations*, 60(6):Paper No. 210, 26, 2021.

[74] Xiaomeng Huang and Yimin Zhang. Existence and uniqueness of minimizers for L^2-constrained problems related to fractional Kirchhoff equation. *Math. Methods Appl. Sci.*, 43(15):8763–8775, 2020.

[75] Antonio Iannizzotto, Sunra Mosconi, and Marco Squassina. Global Hölder regularity for the fractional p-Laplacian. *Rev. Mat. Iberoam.*, 32(4):1353–1392, 2016.

[76] Slim Ibrahim, Nader Masmoudi, and Kenji Nakanishi. Trudinger-Moser inequality on the whole plane with the exact growth condition. *J. Eur. Math. Soc. (JEMS)*, 17(4):819–835, 2015.

[77] Leonelo Iturriaga and Eugenio Massa. Sobolev versus Hölder local minimizers in degenerate Kirchhoff type problems. *J. Differential Equations*, 269(5):4381–4405, 2020.

[78] Louis Jeanjean. Existence of solutions with prescribed norm for semilinear elliptic equations. *Nonlinear Anal.*, 28(10):1633–1659, 1997.

[79] Louis Jeanjean and Sheng-Sen Lu. On global minimizers for a mass constrained problem. *Calc. Var. Partial Differential Equations*, 61(6):Paper No. 214, 18, 2022.

[80] Louis Jeanjean and Kazunaga Tanaka. A remark on least energy solutions in $\mathbf{R}^N$. *Proc. Amer. Math. Soc.*, 131(8):2399–2408, 2003.

[81] Louis Jeanjean, Jianjun Zhang, and Xuexiu Zhong. A global branch approach to normalized solutions for the Schrödinger equation. *J. Math. Pures Appl. (9)*, 183:44–75, 2024.

[82] Louis Jeanjean, Jianjun Zhang, and Xuexiu Zhong. Normalized ground states for a coupled Schrödinger system: mass super-critical case. *NoDEA Nonlinear Differential Equations Appl.*, 31(5):Paper No. 85, 26, 2024.

[83] Chao Ji and Vicenţiu D. Rădulescu. Multiplicity and concentration of solutions to the nonlinear magnetic Schrödinger equation. *Calc. Var. Partial Differential Equations*, 59(4):Paper No. 115, 28, 2020.

[84] Chao Ji and Vicenţiu D. Rădulescu. Concentration phenomena for magnetic Kirchhoff equations with critical growth. *Discrete Contin. Dyn. Syst.*, 41(12):5551–5577, 2021.

[85] Chao Ji and Vicenţiu D. Rădulescu. Multiplicity and concentration of solutions for Kirchhoff equations with magnetic field. *Adv. Nonlinear Stud.*, 21(3):501–521, 2021.

[86] Peng Jin, Xianhua Tang, and Die Hu. Solutions with prescribed mass for nonautonomous critical Kirchhoff equations with convolution nonlinearity. *Discrete Contin. Dyn. Syst. Ser. B*, 31:277–303, 2026.

[87] Peng Jing, Vicenţiu D. Rădulescu, Xianhua Tang, and Lixi Wen. Least energy solutions for Choquard equations involving vanishing potentials and exponential growth. *Rev. Real Acad. Cienc. Exactas Fis. Nat. Ser. A-Mat.*, 120(2):Paper No. 35, 2026.

[88] George Kirchhoff. *Mechanik*. Teubner. Teubner, Leipzig, 1883.

[89] Lingzheng Kong and Haibo Chen. Normalized ground states for fractional Kirchhoff equations with Sobolev critical exponent and mixed nonlinearities. *J. Math. Phys.*, 64(6):Paper No. 061501, 13, 2023.

[90] Nikolai Laskin. Fractional quantum mechanics and Lévy path integrals. *Phys. Lett. A*, 268(4-6):298–305, 2000.

[91] Gongbao Li, Peng Luo, Shuangjie Peng, Chunhua Wang, and Chang-Lin Xiang. A singularly perturbed Kirchhoff problem revisited. *J. Differential Equations*, 268(2):541–589, 2020.

[92] Gongbao Li, Xiao Luo, and Tao Yang. Normalized solutions to a class of Kirchhoff equations with Sobolev critical exponent. *Ann. Fenn. Math.*, 47(2):895–925, 2022.

[93] Quanqing Li, Vicenţiu D. Rădulescu, Jian Zhang, and Xin Zhao. Normalized solutions of the autonomous Kirchhoff equation with Sobolev critical exponent: sub- and supercritical cases. *Proc. Amer. Math. Soc.*, 151(2):663–678, 2023.

[94] Quanqing Li, Vicenţiu D. Rădulescu, and Wen Zhang. Normalized ground states for the Sobolev critical Schrödinger equation with at least mass critical growth. *Nonlinearity*, 37(2):Paper No. 025018, 28, 2024.

[95] Yimei Li and Jiguang Bao. Fractional Hardy-Hénon equations on exterior domains. *J. Differential Equations*, 266(2-3):1153–1175, 2019.

[96] Sihua Liang and Vicenţiu D. Rădulescu. Least-energy nodal solutions of critical Kirchhoff problems with logarithmic nonlinearity. *Anal. Math. Phys.*, 10(4):Paper No. 45, 31, 2020.

[97] Sihua Liang, Xueqi Sun, Shuaishuai Liang, and Vicenţiu D. Rădulescu. Normalized solutions of the critical Schrödinger-Bopp-Podolsky system with logarithmic nonlinearity. *Trans. London Math. Soc.*, 12(1):Paper No. e70021, 41, 2025.

[98] Elliott H. Lieb and Michael Loss. *Analysis*, volume 14 of *Graduate Studies in Mathematics*. American Mathematical Society, Providence, RI, second edition, 2001.

[99] J.-L. Lions. On some questions in boundary value problems of mathematical physics. In *Contemporary developments in continuum mechanics and partial differential equations (Proc. Internat. Sympos., Inst. Mat., Univ. Fed. Rio de Janeiro, Rio de Janeiro, 1977)*, volume 30 of *North-Holland Math. Stud.*, pages 284–346. North-Holland, Amsterdam-New York, 1978.

[100] P.-L. Lions. The concentration-compactness principle in the calculus of variations. The locally compact case. I. *Ann. Inst. H. Poincaré Anal. Non Linéaire*, 1(2):109–145, 1984.

[101] Haidong Liu and Vicenţiu D. Rădulescu. Normalized solutions of quasilinear problems with nonlocal reaction. *Bulletin of Mathematical Sciences*, 16, 2026.

[102] Lintao Liu, Haibo Chen, and Jie Yang. Normalized solutions to the fractional Kirchhoff equations with a perturbation. *Appl. Anal.*, 102(4):1229–1249, 2023.

[103] Lintao Liu, Vicenţiu D. Rădulescu, and Shuai Yuan. Constraint minimizers of mass critical fractional Kirchhoff equations: concentration and uniqueness. *Nonlinearity*, 38(4):Paper No. 045008, 46, 2025.

[104] Lintao Liu, Vicenţiu D. Rădulescu, and Shuai Yuan. Constraint minimizers of mass critical fractional Kirchhoff equations: concentration and uniqueness. *Nonlinearity*, 38(4):Paper No. 045008, 46, 2025.

[105] Tiantian Liu and Aixia Qian. Concentration phenomena for a fractional Choquard-Kirchhoff equation with magnetic field. *J. Math. Phys.*, 66(6):Paper No. 061501, 23, 2025.

[106] Zhisu Liu, Yijun Lou, and Jianjun Zhang. A perturbation approach to studying sign-changing solutions of Kirchhoff equations with a general nonlinearity. *Ann. Mat. Pura Appl. (4)*, 201(3):1229–1255, 2022.

[107] Zhisu Liu, Zigen Ouyang, and Jianjun Zhang. Existence and multiplicity of sign-changing standing waves for a gauged nonlinear Schrödinger equation in $\mathbb{R}^2$. *Nonlinearity*, 32(8):3082–3111, 2019.

[108] Zhisu Liu, Vicenţiu D. Rădulescu, and Ziqing Yuan. Concentration of solutions for fractional Kirchhoff equations with discontinuous reaction. *Z. Angew. Math. Phys.*, 73(5):Paper No. 211, 23, 2022.

[109] Zhisu Liu, Vicenţiu D. Rădulescu, and Jianjun Zhang. A planar Schrödinger-Newton system with Trudinger-Moser critical growth. *Calc. Var. Partial Differential Equations*, 62(4):Paper No. 122, 31, 2023.

[110] Zhisu Liu, Vicenţiu D. Rădulescu, and Jianjun Zhang. Groundstates of the planar Schrödinger-Poisson system with potential well and lack of symmetry. *Proc. Roy. Soc. Edinburgh Sect. A*, 154(4):993–1023, 2024.

[111] Haijun Luo and Zhitao Zhang. Normalized solutions to the fractional Schrödinger equations with combined nonlinearities. *Calc. Var. Partial Differential Equations*, 59(4):Paper No. 143, 35, 2020.

[112] Xiao Luo and Qingfang Wang. Existence and asymptotic behavior of high energy normalized solutions for the Kirchhoff type equations in $\mathbb{R}^3$. *Nonlinear Anal. Real World Appl.*, 33:19–32, 2017.

[113] Jarosław Mederski. Nonradial solutions of nonlinear scalar field equations. *Nonlinearity*, 33(12):6349–6380, 2020.

[114] Jarosław Mederski and Jacopo Schino. Least energy solutions to a cooperative system of Schrödinger equations with prescribed L^2-bounds: at least L^2-critical growth. *Calc. Var. Partial Differential Equations*, 61(1):Paper No. 10, 31, 2022.

[115] Xiang Mingqi, Vicenţiu D. Rădulescu, and Binlin Zhang. Combined effects for fractional Schrödinger-Kirchhoff systems with critical nonlinearities. *ESAIM Control Optim. Calc. Var.*, 24(3):1249–1273, 2018.

[116] Xiang Mingqi, Vicenţiu D. Rădulescu, and Binlin Zhang. Nonlocal Kirchhoff diffusion problems: local existence and blow-up of solutions. *Nonlinearity*, 31(7):3228–3250, 2018.

[117] Xiang Mingqi, Vicenţiu D. Rădulescu, and Binlin Zhang. Fractional Kirchhoff problems with critical Trudinger-Moser nonlinearity. *Calc. Var. Partial Differential Equations*, 58(2):Paper No. 57, 27, 2019.

[118] Carlo Miranda. Un'osservazione su un teorema di Brouwer. *Boll. Un. Mat. Ital. (2)*, 3:5–7, 1940.

[119] Giovanni Molica Bisci, Vicenţiu D. Rădulescu, and Raffaella Servadei. *Variational methods for nonlocal fractional problems*, volume 162 of *Encyclopedia of Mathematics and its Applications*. Cambridge University Press, Cambridge, 2016.

[120] J. Moser. A sharp form of an inequality by N. Trudinger. *Indiana Univ. Math. J.*, 20:1077–1092, 1970/71.

[121] Dumitru Motreanu and Vicenţiu D. Rădulescu. *Variational and non-variational methods in nonlinear analysis and boundary value problems*, volume 67 of *Nonconvex Optimization and its Applications*. Kluwer Academic Publishers, Dordrecht, 2003.

[122] D. Naimen and C. Tarsi. Multiple solutions of a Kirchhoff type elliptic problem with the Trudinger-Moser growth. *Adv. Differential Equations*, 22(11-12):983–1012, 2017.

[123] Wei-Ming Ni and Izumi Takagi. On the shape of least-energy solutions to a semilinear Neumann problem. *Comm. Pure Appl. Math.*, 44(7):819–851, 1991.

[124] L. Nirenberg. On elliptic partial differential equations. *Ann. Scuola Norm. Sup. Pisa Cl. Sci. (3)*, 13:115–162, 1959.

[125] R. S. Palais and S. Smale. A generalized Morse theory. *Bull. Amer. Math. Soc.*, 70:165–172, 1964.

[126] Richard S. Palais. Lusternik-Schnirelman theory on Banach manifolds. *Topology*, 5:115–132, 1966.

[127] Kanishka Perera and Zhitao Zhang. Nontrivial solutions of Kirchhoff-type problems via the Yang index. *J. Differential Equations*, 221(1):246–255, 2006.

[128] S. I. Pohožaev. A certain class of quasilinear hyperbolic equations. *Mat. Sb. (N.S.)*, 96(138):152–166, 168, 1975.

[129] Patrizia Pucci and Vicenţiu D. Rădulescu. Progress in nonlinear Kirchhoff problems [Editorial]. *Nonlinear Anal.*, 186:1–5, 2019.

[130] Patrizia Pucci and Sara Saldi. Critical stationary Kirchhoff equations in $\mathbb{R}^N$ involving nonlocal operators. *Rev. Mat. Iberoam.*, 32(1):1–22, 2016.

[131] Patrizia Pucci, Mingqi Xiang, and Binlin Zhang. Multiple solutions for nonhomogeneous Schrödinger-Kirchhoff type equations involving the fractional p-Laplacian in $\mathbb{R}^N$. *Calc. Var. Partial Differential Equations*, 54(3):2785–2806, 2015.

[132] Shijie Qi and Wenming Zou. Exact number of positive solutions for the Kirchhoff equation. *SIAM J. Math. Anal.*, 54(5):5424–5446, 2022.

[133] Paul H. Rabinowitz. On a class of nonlinear Schrödinger equations. *Z. Angew. Math. Phys.*, 43(2):270–291, 1992.

[134] Vicenţiu D. Rădulescu. Mountain pass theorems for nondifferentiable functions and applications. *Proc. Japan Acad. Ser. A Math. Sci.*, 69(6):193–198, 1993.

[135] Simone Secchi. Ground state solutions for nonlinear fractional Schrödinger equations in $\mathbb{R}^N$. *J. Math. Phys.*, 54(3):031501, 17, 2013.

[136] Wei Shuai. Sign-changing solutions for a class of Kirchhoff-type problem in bounded domains. *J. Differential Equations*, 259(4):1256–1274, 2015.

[137] Luis Silvestre. Hölder estimates for solutions of integro-differential equations like the fractional Laplace. *Indiana Univ. Math. J.*, 55(3):1155–1174, 2006.

[138] Luis Silvestre. Regularity of the obstacle problem for a fractional power of the Laplace operator. *Comm. Pure Appl. Math.*, 60(1):67–112, 2007.

[139] Jacques Simon. Régularité de la solution d'une équation non linéaire dans $\mathbb{R}^N$. In *Journées d'Analyse Non Linéaire (Proc. Conf., Besançon, 1977)*, volume 665 of *Lecture Notes in Math.*, pages 205–227. Springer, Berlin, 1978.

[140] Nicola Soave. Normalized ground states for the NLS equation with combined nonlinearities: the Sobolev critical case. *J. Funct. Anal.*, 279(6):108610, 43, 2020.

[141] Michael Struwe. *Variational methods.* Springer-Verlag, Berlin, 1990. Applications to nonlinear partial differential equations and Hamiltonian systems.

[142] Jijiang Sun, Lin Li, Matija Cencelj, and Boštjan Gabrovšek. Infinitely many sign-changing solutions for Kirchhoff type problems in $\mathbb{R}^3$. *Nonlinear Anal.*, 186:33–54, 2019.

[143] Xueqi Sun, Yongqiang Fu, and Sihua Liang. Multiplicity and concentration of solutions for Kirchhoff equations with exponential growth. *Bull. Math. Sci.*, 14(3):Paper No. 2450004, 42, 2024.

[144] Andrzej Szulkin and Tobias Weth. The method of Nehari manifold. In *Handbook of nonconvex analysis and applications*, pages 597–632. Int. Press, Somerville, MA, 2010.

[145] Xianhua Tang and Sitong Chen. Singularly perturbed Choquard equations with nonlinearity satisfying Berestycki-Lions assumptions. *Adv. Nonlinear Anal.*, 9(1):413–437, 2020.

[146] Xianhua Tang and Bitao Cheng. Ground state sign-changing solutions for Kirchhoff type problems in bounded domains. *J. Differential Equations*, 261(4):2384–2402, 2016.

[147] Kaimin Teng and Yiqun Cheng. Multiplicity and concentration of nontrivial solutions for fractional Schrödinger-Poisson system involving critical growth. *Nonlinear Anal.*, 202:Paper No. 112144, 32, 2021.

[148] Neil S. Trudinger. On imbeddings into Orlicz spaces and some applications. *J. Math. Mech.*, 17:473–483, 1967.

[149] Le Xuan Truong. The Nehari manifold for fractional p-Laplacian equation with logarithmic nonlinearity on whole space. *Comput. Math. Appl.*, 78(12):3931–3940, 2019.

[150] Da-Bin Wang. Least energy sign-changing solutions of Kirchhoff-type equation with critical growth. *J. Math. Phys.*, 61(1):011501, 19, 2020.

[151] Michael I. Weinstein. Nonlinear Schrödinger equations and sharp interpolation estimates. *Comm. Math. Phys.*, 87(4):567–576, 1982/83.

[152] Tobias Weth. Energy bounds for entire nodal solutions of autonomous superlinear equations. *Calc. Var. Partial Differential Equations*, 27(4):421–437, 2006.

[153] Michel Willem. *Minimax theorems*, volume 24 of *Progress in Nonlinear Differential Equations and their Applications.* Birkhäuser Boston, Inc., Boston, MA, 1996.

[154] Mingqi Xiang, Die Hu, and Di Yang. Least energy solutions for fractional Kirchhoff problems with logarithmic nonlinearity. *Nonlinear Anal.*, 198:111899, 20, 2020.

[155] Weihong Xie and Haibo Chen. Existence and multiplicity of normalized solutions for the nonlinear Kirchhoff type problems. *Comput. Math. Appl.*, 76(3):579–591, 2018.

[156] Shuai Yao, Haibo Chen, Vicenţiu D. Rădulescu, and Juntao Sun. Normalized solutions for lower critical Choquard equations with critical Sobolev perturbation. *SIAM J. Math. Anal.*, 54(3):3696–3723, 2022.

[157] Hongyu Ye. The existence of normalized solutions for L^2-critical constrained problems related to Kirchhoff equations. *Z. Angew. Math. Phys.*, 66(4):1483–1497, 2015.

[158] Hongyu Ye. The sharp existence of constrained minimizers for a class of nonlinear Kirchhoff equations. *Math. Methods Appl. Sci.*, 38(13):2663–2679, 2015.

[159] Hongyu Ye. The mass concentration phenomenon for L^2-critical constrained problems related to Kirchhoff equations. *Z. Angew. Math. Phys.*, 67(2):Art. 29, 16, 2016.

[160] Xiaoyu Zeng, Jianjun Zhang, Yimin Zhang, and Xuexiu Zhong. On the Kirchhoff equation with prescribed mass and general nonlinearities. *Discrete Contin. Dyn. Syst. Ser. S*, 16(11):3394–3409, 2023.

[161] Jian Zhang and Wenming Zou. Multiplicity and concentration behavior of solutions to the critical Kirchhoff-type problem. *Z. Angew. Math. Phys.*, 68(3):Paper No. 57, 27, 2017.

[162] Penghui Zhang and Zhiqing Han. Normalized ground states for Kirchhoff equations in $\mathbb{R}^3$ with a critical nonlinearity. *J. Math. Phys.*, 63(2):Paper No. 021505, 15, 2022.

[163] Maoding Zhen and Binlin Zhang. Normalized ground states for the critical fractional NLS equation with a perturbation. *Rev. Mat. Complut.*, 35(1):89–132, 2022.

Index

For Product Safety Concerns and Information please contact our EU representative GPSR@taylorandfrancis.com Taylor & Francis Verlag GmbH, Kaufingerstraße 24, 80331 München, Germany